Optimization, Dynamics, and Economic Analysis

Essays in Honor of Gustav Feichtinger

Professor Gustav Feichtinger

Engelbert J. Dockner · Richard F. Hartl
Mikulas Luptačik · Gerhard Sorger
Editors

Optimization, Dynamics, and Economic Analysis

Essays in Honor of Gustav Feichtinger

With 75 Figures
and 18 Tables

Springer-Verlag Berlin Heidelberg GmbH

Prof. Dr. Engelbert J. Dockner
Department of Business, University of Vienna
Brünnerstraße 72, A-1210 Vienna, Austria
Email: engelbert.dockner@univie.ac.at

Prof. Dr. Richard F. Hartl
Department of Business, University of Vienna
Brünnerstraße 72, A-1210 Vienna, Austria
Email: richard.hartl@univie.ac.at

Prof. Dr. Mikulas Luptačik
Vienna University of Economics and Business Administration
Augasse 2–6, A-1090 Vienna, Austria
Email: mikulas.luptacik@wu-wien.ac.at

Prof. Dr. Gerhard Sorger
Department of Economics, Queen Mary & Westfield College
Mile End Road, GB-London E1 4NS, United Kingdom
Email: g.sorger@qmw.ac.uk

ISBN 978-3-642-63327-0

Cataloging-in-Publication Data applied for
Die Deutsche Bibliothek – CIP-Einheitsaufnahme
Optimization, Dynamics, and Economic Analysis: Essays in Honor of Gustav Feichtinger: with 18
tables / Engelbert J. Dockner; Richard F. Hartl; Mikulas Luptačik; Gerhard Sorger (eds.). – Heidel-
berg; New York: Physica-Verl., 2000
 ISBN 978-3-642-63327-0   ISBN 978-3-642-57684-3 (eBook)
 DOI 10.1007/978-3-642-57684-3

© Springer-Verlag Berlin Heidelberg 2000
Originally published by Physica-Verlag Heidelberg in 2000
Softcover reprint of the hardcover 1st edition 2000

The use of general descriptive names, registered names, trademarks, etc. in this publication does
not imply, even in the absence of a specific statement, that such names are exempt from the
relevant protective laws and regulations and therefore free for general use.

Cover design: Erich Kirchner, Heidelberg

SPIN 10764266   88/2202-5 4 3 2 1 0 – Printed on acid-free paper

Preface by the Editors

The scope of a book with the title "Optimization, Dynamics and Economic Analysis" seems to be too broad and most likely gives the impression of lacking a focus. There is a lot to this point of view. Things are very different, however, with the present volume. This book is dedicated to Professor Gustav Feichtinger, our academic teacher, friend and coauthor, in which we want to honor his academic contributions. Those of the readers who know Professor Feichtinger will agree that a book with the title "Optimization, Dynamics and Economic Analysis" best describes his many contributions to the literature.

One of the most admirable things in the academic career of Professor Feichtinger is his broad interest in static and dynamic optimization, dynamic game theory, economic dynamics, theory of the firm and population dynamics. He was and still is working in all these fields of operations research and economic analysis and much of his published work was influential to most of the authors represented in this Festschrift. The different papers covered here should shed some light on those topics where we think his contributions are most important. The Festschrift is divided into the following five parts, each of which covers a certain area of academic research by Professor Feichtinger over the last thirty years:

- Optimization and dynamic games,
- Economic dynamics,
- Theory of the firm,
- Population dynamics, and
- Miscellaneous applications of economic theory to fields including the theory of addiction and crime.

We have grouped the papers in this Festschrift according to these topics.

It is fair to say that Professor Feichtinger is best known for his contributions in dynamic optimization and optimal control theory including differential games. We would like to mention some of his publications in this field that best characterize his work there. Since the end of the seventies he was convinced that a sound understanding of economic decision making requires a dynamic multiperiod framework. This has stimulated his interest in both optimal control and differential game analysis. His aim, however, was not just to make use of Pontryagin's maximum principle in specific areas of applications but to understand the general mechanisms that drive the optimal intertemporal decisions in control problems.

This interest has led him to apply phase diagram analysis to optimal control models through which the derivation of optimal decisions is possible even when the state dynamics and the objective function are characterized by general nonlinear functions. Very quickly the Vienna School of Optimal Control headed by Professor Feichtinger emerged and gained international reputation in saddle point analysis of optimal control models. Professor Feichtinger's book coauthored with Richard F. Hartl [B9]* nicely summarizes his many contributions in that area.

---

* Numbers here refer to the list of publications presented at the end of this volume.

Other publications include edited volumes [B6 – B8, B10 – B16] and journal articles like [99, 111].

His motivation of describing interactions between economic agents initiated his work on differential games. Here he found a framework by which he not only was able to study intertemporal decision making but also strategic interactions of dominant players. The structural analysis of optimal control models triggered his ideas to apply phase diagram analysis to differential games and led to the introduction of so called qualitatively solvable differential games (see Dockner et al. [100]). These games are characterized by a separability between state and control variables and have the property that the open-loop Nash equilibria also qualify as Markov perfect equilibria. Many interesting game applications are characterized by the property of state separability and hence allowed through phase diagram analysis interesting insights into the game interactions; see e.g. [88, 97, 102, 126]

The interest of Professor Feichtinger in phase diagram analysis of control models made it necessary to analyze the stability properties of the canonical system associated with Pontryagin's maximum principle. In the mid eighties this has triggered the question by him of whether or not saddle-point stability is the only possibility for optimal solutions in control models. Inspired by the work of Benhabib and Nishimura he directed his interest towards nonlinear dynamics and chaos theory. A series of very interesting papers emerged [see 107, 115, 118, 121, 124, 125, 128, 129, 160] in which he found that under a variety of model structures limit cycles can be the outcome of an optimal control problem. In addition to limit cycles he was also interested in other irregular dynamics including deterministic chaos [147, 154, 156, 159]. The papers here are characterized by the continuous effort to identify economic mechanisms that lead decision makers to choose optimally a policy that results in complicated dynamic behavior.

It is necessary to point out that Professor Feichtinger also made important contributions to the dynamic theory of the firm. Many of his optimal control and differential game models deal with firm specific issues like dynamic price and advertising policies, optimal R & D investment, production and inventory controls and dynamic investment behavior. Professor Feichtinger published too many papers in this field to mention them separately, but many of them provided interesting insights into dynamic allocation problems, because his goal always was to solve a model in its most general form to reach conclusions that do not depend on specific forms of demand, production or utility functions [73, 75, 76, 79, 83, 98, 153].

Early work in population dynamics [B2] includes research on the representation of demographic quantities as stochastic processes and more generally the description and analysis of demographic processes using stochastic models (in particular Markov models). Applications range from individual based micro processes as fertility and nuptiality, competing risk models, stochastic decrement tables to macro models with special reference to the stable population model.

Influenced by the demographic trends of the seventies his focus of research during the seventies and early eighties shifted towards demographic and economic impacts of stationary and declining populations [B3 - B5]. At the same time several contributions to formal demography have been published. Specific topics include research on quasi-stable populations, statistical measurement of the family life cycle, the relation between period and cohort measures of demographic processes and optimal growth of stable populations in neoclassical models.

During the late eighties and the nineties most of his research in population dynamics has been devoted to demo-economics with special focus on non-linear relations between demographic and economic/environmental dynamics. These include for instance models on self generated fertility waves as discussed by Easterlin which relate fertility behavior to opportunities on the labor market and the level of consumption aspiration formed when young. Further topics include models on the escape from the Malthusian trap, and the interrelationships between population and resource dynamics in primitive societies.

His strong interest in optimal control theory also led to the analysis of intertemporal decision models that can be classified as non-standard applications. In the meantime the whole group around him at the Institute for Econometrics, OR, and Systems Theory at the Vienna University of Technology developed a special reputation for this non-standard optimal control models [103, 104]. He worked on problems of rational addiction [145], a game between a thief and the police [86] and more recently on models of the economics of crime in general [208]. This work documents a nice aspect of Professor Feichtinger's contribution to OR and economics. He constantly was and still is looking for areas of applications in which a sound dynamic analysis helps us to better understand the trade-offs that are involved and to give corresponding policy advice.

Professor Feichtinger has not only had a significant impact on the scientific community through his research and his publications, he has also paved the way into the world of scientific research for many scholars, who hold now senior research and teaching positions in various universities or research institutes. For some of them he was an advisor in the early stages of their academic careers when they finished their doctoral studies. Others had their first academic positions at the Institute for Econometrics, OR, and Systems Theory before they went on to other universities. And yet for others he acted as a catalyst in forming their own research agenda, even if he was never formally involved in their career as a teacher, advisor, or colleague. This remarkable influence is rooted in Professor Feichtinger's ability to encourage others for interesting research topics. Not only is his enthusiasm for research infectious, he is also more than willing to discuss possible solutions to open problems with anybody who is interested in those problems. We believe it is safe to say that the editors of this volume along with many other former students, coworkers, or colleagues owe part of their success in academia to the early encouragement by and the constant interaction with Professor Feichtinger. On the occasion of his 60th birthday we would like to thank him for this positive influence through the present Festschrift.

We as the team of editors would also like to thank all the contributors in this volume, the referees, Alexia Fürnkranz-Prskawetz for help in writing this preface, Thomas Fent for technical assistance, and Werner A. Müller from Springer Verlag for including this volume into Physica Verlag's programme.

Engelbert J. Dockner, University of Vienna
Richard F. Hartl, University of Vienna
Mikulas Luptacik, Vienna University of Economics
Gerhard Sorger, University of London

Contents

Optimization and Dynamic Games

Economic Dynamics

Theory of the Firm

Population Dynamics

Miscellaneous Applications

Copositivity Aspects of Standard Quadratic Optimization Problems

Immanuel M. Bomze

University of Vienna, Universitätsstraße 5, A-1010 Wien, Austria,
immanuel.bomze@univie.ac.at

Summary. The aim of this paper is to make transparent the close relationship between copositivity (a property of symmetric matrices which is an extension of positive-semidefiniteness), and standard quadratic problems which form a central class in quadratic optimization. Apart from theoretical interest, the results presented may have immediate consequences in implementing efficient global optimization procedures, e.g., along the lines of primal-dual interior-point methods which have become increasingly popular recently. Accounts on algorithms and important real-life applications are also provided.

1. Introduction

A *standard quadratic optimization problem (StQP)* consists of finding global maximizers of a (possibly indefinite) quadratic form over the *standard simplex* Δ in n-dimensional Euclidean space $\mathbb{R}^n$,

$$\Delta = \left\{ \mathbf{x} \in \mathbb{R}^n : x_i \geq 0 \text{ for all } i \in N, \ \mathbf{e}^\top \mathbf{x} = 1 \right\},$$

where $N = \{1, \ldots, n\}$; $\mathbf{e} = [1, \ldots, 1]^\top \in \mathbb{R}^n$; and the symbol $^\top$ denotes transposition. Hence a StQP can be written as a (global) quadratic optimization problem of the form

$$\max \left\{ \mathbf{x}^\top R \mathbf{x} : \mathbf{x} \in \Delta \right\}, \tag{1.1}$$

where R is an arbitrary symmetric $n \times n$ matrix. Since the maximizers of (1.1) remain the same if R is replaced with $R + \gamma \mathbf{e}\mathbf{e}^\top$ where γ is an arbitrary constant, one may assume without loss of generality that all entries of R are positive. This assumption will become important in an algorithm treated in Section 4. Furthermore, the question of finding maximizers of a general quadratic function $\mathbf{x}^\top Q \mathbf{x} + 2\mathbf{c}^\top \mathbf{x}$ over Δ can be homogenized in a similar way by considering the rank-two update $R = Q + \mathbf{e}\mathbf{c}^\top + \mathbf{c}\mathbf{e}^\top$ in (1.1) which has the same objective values on Δ.

StQPs have diverse direct real-life applications (see Section 5), and also arise as auxiliary problems in many important optimization procedures (cf. Section 3). One of them is as follows: suppose that *branch-and-bound* schemes with *simplicial partitions* [11] are applied to a *general quadratic optimization problem (QP)* of the form

$$\max \left\{ f(\mathbf{x}) = \tfrac{1}{2} \mathbf{x}^\top Q \mathbf{x} + \mathbf{c}^\top \mathbf{x} : \mathbf{x} \in M \right\}, \tag{1.2}$$

where $M = \{\mathbf{x} \in \mathbb{R}^n : A\mathbf{x} \leq \mathbf{b}\}$ with A an $m \times n$ matrix and Q a symmetric $n \times n$ matrix. A subproblem then is of the form $\max\{f(\mathbf{x}) : \mathbf{x} \in P\}$ with $P \cap M \neq \emptyset$ and $P = \text{conv}\{\mathbf{v}_0, \ldots, \mathbf{v}_n\}$ for some points $\mathbf{v}_i \in \mathbb{R}^n$ (if all vertices of M are easy to determine, one could even take them rather than the $\mathbf{v}_i$). With the $n \times (n+1)$ matrix $U = [\mathbf{v}_0, \ldots, \mathbf{v}_n]$, the subproblem reduces to the StQP

$$\max\{\mathbf{y}^\top R\mathbf{y} : \mathbf{y} \in \Delta\}$$

where $R = U^\top QU + \mathbf{ec}^\top U + U^\top \mathbf{ce}^\top$ is a symmetric $(n+1) \times (n+1)$ matrix and $\Delta \subseteq \mathbb{R}^{n+1}$. Efficient bounds can thus be obtained with one of the algorithms for obtaining (local) solutions to a StQP; see Section 4 for more details.

Recall that a symmetric $n \times n$ matrix $\overline{Q}$ is said to be *copositive* if $\overline{Q}$ satisfies $\mathbf{v}^\top \overline{Q}\mathbf{v} \geq 0$ if $\mathbf{v} \geq \mathbf{o}$. As usual, the relation $\mathbf{v} \geq \mathbf{o}$ means $v_i \geq 0$ for all i. For reasons which will be obvious soon, we denote the set of all copositive $n \times n$ matrices with $\mathcal{K}^*$. Evidently, this set is a convex, but non-polyhedral cone in the space of all symmetric $n \times n$ matrices which contains the cone of all positive-semidefinite matrices of that kind.

The paper is organized as follows: in Section 2, we establish equivalence of StQPs with *copositive programming*, a variant of the familiar *semidefinite programming (SDP)* problem, where definiteness is replaced with copositivity. Section 3 is devoted to global optimality conditions for general quadratic problems, where copositivity and StQPs arise in a quite natural way. As already indicated above, Sections 4 and 5 treat several algorithms and applications, respectively.

2. Copositive programming and StQPs

In this section, we want to establish the close connection between StQPs and copositivity. To make this evident also from a practical perspective, we extend the usual SDP approach to recast a StQP into a linear optimization problem on a cone $\mathcal{K}$ which is the (pre-) dual of the cone $\mathcal{K}^*$ of all copositive symmetric $n \times n$ matrices, with respect to the duality

$$\langle R, S \rangle = \text{trace}\,(RS) = \sum_{i,j} r_{ij}s_{ij}$$

operating on pairs (R, S) of symmetric $n \times n$ matrices. This formulation allows to employ *interior-point algorithms*, similar to the methods used in SDP. Both cones $\mathcal{K}^*$ and $\mathcal{K}$ have non-empty interiors, and the latter can be described as follows [10, 13]:

$$\mathcal{K} = \text{conv}\,\{\mathbf{x}\mathbf{x}^\top : \mathbf{x} \geq \mathbf{o}\},$$

the convex hull of all symmetric rank-one matrices, i.e. dyadic products, generated by non-negative vectors. Note that dropping the non-negativity requirement, we would arrive at the positive-semidefinite case. The next result is essential in establishing the equivalence between the indicated class of cone optimization problems and StQPs. Now let $E = \mathbf{ee}^\top$ be the $n \times n$ matrix of all ones, and denote by

$$\mathcal{L} = \{X \in \mathcal{K} : \langle E, X \rangle = 1\}.$$

Lemma 2.1. *The extremal points of the set $\mathcal{L}$ are exactly the rank-one matrices $X = \mathbf{xx}^\top$ with $\mathbf{x} \in \Delta$.*

Proof. All $X = \mathbf{xx}^\top$ with $\mathbf{x} \in \Delta$ belong to $\mathcal{L}$ because of $\langle E, X \rangle = (\mathbf{e}^\top\mathbf{x})^2 = 1$. Now suppose that for such X we have $X = (1-\lambda)Y + \lambda Z$ for some $Z, Y \in \mathcal{L}$ and some λ with $0 < \lambda < 1$. Choose an orthogonal basis $\{\mathbf{x}_1, \mathbf{x}_2, \ldots, \mathbf{x}_n\}$ of $\mathbb{R}^n$ with $\mathbf{x} = \mathbf{x}_n$. Then since Z and Y also are positive-semidefinite, we get from

$$(1 - \lambda)\mathbf{x}_i{}^\top Y\mathbf{x}_i + \lambda\mathbf{x}_i{}^\top Z\mathbf{x}_i = \mathbf{x}_i{}^\top X\mathbf{x}_i = (\mathbf{x}_i^\top\mathbf{x})^2 = 0$$

that $\mathbf{x}_i{}^\top Z\mathbf{x}_i = \mathbf{x}_i{}^\top Y\mathbf{x}_i = 0$ for all $i < n$ and therefore both Z and Y have rank one. As both belong to $\mathcal{K}$, we thus obtain $Z = \mathbf{zz}^\top$ and $Y = \mathbf{yy}^\top$ for some $\mathbf{z}, \mathbf{y} \in \mathbb{R}^n$ with $\mathbf{z} \geq \mathbf{o}$ and $\mathbf{y} \geq \mathbf{o}$. But then we obtain $\mathbf{x}_i^\top\mathbf{z} = \mathbf{x}_i^\top\mathbf{y} = 0$ for all $i < n$, so that Z and Y must be positive multiples of $\mathbf{xx}^\top$. The requirement $\langle E, Z \rangle = \langle E, Y \rangle = 1$ shows that $Z = Y = \mathbf{xx}^\top = X$. Hence X is an extremal point of $\mathcal{L}$. To prove the converse, suppose that X is an extremal point of $\mathcal{L} \subset \mathcal{K}$. Then $X = \sum_{i=1}^m \lambda_i\mathbf{x}_i\mathbf{x}_i^\top$ with $\mathbf{x}_i \in \mathbb{R}^n \setminus \{\mathbf{o}\}$, $\mathbf{x}_i \geq \mathbf{o}$, and $\lambda_i \geq 0$ for all i as well as $\sum_{i=1}^m \lambda_i = 1$. Since $X \in \mathcal{L}$, we get

$$1 = \langle E, X \rangle = \sum_{i=1}^m \lambda_i(\mathbf{e}^\top\mathbf{x}_i)^2 \tag{2.1}$$

where $\mathbf{e}^\top\mathbf{x}_i > 0$ for all i. Now put $\mathbf{y}_i = (\mathbf{e}^\top\mathbf{x}_i)^{-1}\mathbf{x}_i \in \Delta$, so that $Y_i = \mathbf{y}_i\mathbf{y}_i^\top \in \mathcal{L}$ for all i. Hence

$$X = \sum_{i=1}^m \lambda_i(\mathbf{e}^\top\mathbf{x}_i)^2 Y_i$$

is, due to (2.1), a convex combination of matrices Y_i in $\mathcal{L}$, whence by the extremality assumption $X = Y_1$ is of the form stated. $\qquad\square$

Theorem 2.1. *The StQP (1.1) is equivalent to the copositive program*

$$\max\{\langle R, X \rangle : X \in \mathcal{L}\} = \max\{\langle R, X \rangle : X \in \mathcal{K}, \langle E, X \rangle = 1\}. \tag{2.2}$$

Proof. By convexity of feasible set and maximand in (2.2), one solution X^* has to be an extremal point of the set $\mathcal{L}$. Due to Lemma 2.1, we know that $X^* = \mathbf{x}^*\mathbf{x}^{*\top}$, the dyadic product of a vector $\mathbf{x}^* \in \Delta$ with itself. Thus

$$\max_{\mathbf{x}\in\Delta} \mathbf{x}^\top R\mathbf{x} \leq \max_{X\in\mathcal{L}}\langle R, X\rangle = \langle R, X^*\rangle = \mathbf{x}^{*\top} R\mathbf{x}^*.$$

Hence the StQP (1.1) is equivalent to the problem (2.2). □

In general, SDPs are mostly used as a relaxation only for solving non-linear optimization problems. The reason for this is the larger feasible set of the SDP, which a priori does not rule out the possibility that also the (global) SDP solution cannot be transferred directly to the original problem, so that the global maximum of the latter is always overestimated by the SDP solution. Although also here the feasible set $\mathcal{L}$ of (2.2) "is larger than" the feasible set Δ of the StQP (1.1), observe that in this case, (2.2) is no relaxation but indeed an exact reformulation of the StQP (1.1), due to the extremality result in Lemma 2.1.

It is easy to see that the dual formulation [16] of (2.2) is

$$\min\{y \in \mathbb{R} : yE - R \in \mathcal{K}^*\}, \tag{2.3}$$

which is the task to find the smallest y such that $yE - R$ is copositive. Thus the dual problem is related to the question of eigenvalue bounds (replace E with the identity matrix and "copositive" with "positive-semidefinite"), and this can be exploited in primal-dual SDP-style algorithms for solving (2.2), along with shortcuts for copositivity detection (see, e.g. [3]). Given a powerful classical SDP technology is available, another shortcut for detecting copositivity of a matrix Q is as follows: search for a positive semi-definite matrix S such that $P = Q - S$ has no negative entries. This makes use of the well-known result that the sum $P + S$ of a nonnegative, and a positive-semidefinite matrix is necessarily copositive. While the converse is not true in general, this nevertheless could serve as a promising heuristic for reducing the duality gap, if problem (2.3) is replaced with

$$\min\{y \in \mathbb{R} : yE - R - S \text{ is nonnegative for some p.s.d. } S\}.$$

3. Global optimality in QPs and copositivity

Copositivity arises quite naturally both in global optimality conditions and also in procedures which enable an escape from inefficient local solutions of general QPs, e.g. in the form (1.2).

We now formulate a characterization of global optimality of a *Karush-Kuhn-Tucker* point $\overline{\mathbf{x}}$ for (1.2): first add a trivial non-binding constraint, i.e. the most elementary strict inequality

$$0 < 1,$$

to obtain slacks $\overline{\mathbf{u}}$ as follows: denote by $\mathbf{a}_i^\top$ the i-th row of A and put $\mathbf{a}_0 = \mathbf{o}$. Similarly put $b_0 = 1$ and enrich $\overline{A} = [\mathbf{a}_0|A^\top]^\top = [\mathbf{o}, \mathbf{a}_1, \ldots, \mathbf{a}_m]^\top$ as well as $\overline{\mathbf{b}} = [b_0|\mathbf{b}^\top]^\top = [1, b_1, \ldots, b_m]^\top$. Finally, define $\overline{\mathbf{u}} = \overline{\mathbf{b}} - \overline{A}\overline{\mathbf{x}} \geq \mathbf{o}$. Then perform, for any $i \in \{0, \ldots, m\}$, a rank-one update of $\overline{A}$ and a rank-two update of Q, using the current gradient $\mathbf{g} = \nabla f(\overline{\mathbf{x}}) = Q\overline{\mathbf{x}} + \mathbf{c}$ of the objective:

$$D_i = \overline{\mathbf{u}}\,\mathbf{a}_i^\top - \overline{u}_i\overline{A} \quad \text{and} \quad Q_i = -\mathbf{a}_i\,\mathbf{g}^\top - \mathbf{g}\,\mathbf{a}_i^\top - \overline{u}_iQ.$$

This gives a symmetric $n \times n$ matrix Q_i and a matrix D_i which is effectively $m \times n$ since its ith row is zero. Denoting by $J(\overline{\mathbf{x}})$ the set of all non-binding constraints, the following result is proved in [2]:

A Karush-Kuhn-Tucker point $\overline{\mathbf{x}}$ of (1.2) is a global solution if and only if for all $i \in J(\overline{\mathbf{x}}) = \{i \in N \cup \{0\} : \overline{u}_i > 0\}$,

$$\mathbf{v}^\top Q_i\mathbf{v} \geq 0 \quad \text{if} \quad D_i\mathbf{v} \geq \mathbf{o}. \tag{3.1}$$

If $\mathbf{v}^\top Q_i\mathbf{v} < 0$ for some $\mathbf{v}$ with $D_i\mathbf{v} \geq \mathbf{o}$, then

$$\tilde{\mathbf{x}} = \overline{\mathbf{x}} + \lambda\mathbf{v} \tag{3.2}$$

is an improving feasible point for $\lambda = \overline{u}_i/(\mathbf{a}_i^\top\mathbf{v})$ (if $\mathbf{a}_i^\top\mathbf{v} = 0$, i.e. $\lambda = +\infty$, this means that (1.2) is unbounded and has therefore no finite optimal solution).

Determining whether or not (3.1) is satisfied, amounts to the question whether or not

$$\max\{\mathbf{v}^\top(-Q_i)\mathbf{v} : D_i\mathbf{v} \geq \mathbf{o}\} \leq 0.$$

This is a typical (general) copositivity question which is decomposable [7, 8] into problems of the form

$$\max\{\mathbf{x}^\top R\mathbf{x} : \mathbf{x} \geq \mathbf{o}\}, \tag{3.3}$$

where the constraint $\mathbf{e}^\top\mathbf{x} = \sum_i x_i = 1$ can be added without loss of generality, rendering a StQP. In fact, in order to determine an improving feasible direction (3.2), by homogeneity of (3.3) it is not necessary to solve this problem to optimality, but rather sufficient to determine a feasible point $\mathbf{x} \in \Delta$ with $\mathbf{x}^\top R\mathbf{x} > 0$.

If the original problem (1.2) is itself already a StQP, then all checks of (3.1) can be reduced to a single one: if $\overline{\mathbf{x}} \in \Delta$ is any feasible point, then $\overline{\mathbf{x}}$ is a global maximizer of $\mathbf{x}^\top Q\mathbf{x}$ over Δ if and only if the matrix $\overline{Q} = (\overline{\mathbf{x}}^\top Q\overline{\mathbf{x}})\mathbf{e}\mathbf{e}^\top - Q$ is copositive.

4. Algorithms for StQPs

Quadratic optimization problems like (1.1) are *NP*-hard [11], even regarding the detection of local solutions. Nevertheless, there are several procedures which try to exploit favourable data constellations in a systematic way, and to avoid the worst-case behaviour whenever possible. Examples for this type of algorithms are specified below.

First we concentrate on the evolutionary approach to finding local solutions of StQP (1.1). To this end, assume that R is a symmetric $n \times n$ matrix with positive entries only and consider the following dynamical system operating on Δ:

$$\dot{x}_i(\tau) = x_i(\tau)[(R\mathbf{x}(\tau))_i - \mathbf{x}(\tau)^\top R\mathbf{x}(\tau)], \ i \in N, \tag{4.1}$$

where a dot signifies derivative w.r.t. time τ, and a discrete time version

$$x_i(\tau + 1) = x_i(\tau) \frac{[R\mathbf{x}(\tau)]_i}{\mathbf{x}(\tau)^\top R\mathbf{x}(\tau)}, \quad i \in N. \tag{4.2}$$

The *stationary points* under (4.1) and (4.2) coincide, and all local solutions of (1.1) are among these (see below). A stationary point $\bar{\mathbf{x}}$ is said to be *asymptotically stable*, if every solution to (4.1) or (4.2) which starts close enough to $\bar{\mathbf{x}}$, will converge to $\bar{\mathbf{x}}$ as $\tau \nearrow \infty$. Now the following results hold (for proofs and further characterization results linking optimization theory, evolutionary game theory, and qualitative theory of dynamical systems, see [6] and the references therein):

- the objective $f(\mathbf{x}(\tau))$ increases strictly along non-constant trajectories of (4.1) and (4.2);
- all trajectories converge to a stationary point;
- all *Karush-Kuhn-Tucker* points and hence all local solutions of (1.1) are stationary points under (4.1) and (4.2);
- if no principal minor of $R = R^\top$ vanishes, then with probability one (regarding the choice of $\mathbf{x}(0)$, the starting point), any trajectory of (4.1) converges to one of the *strict* local solutions $\bar{\mathbf{x}}$ of (1.1), which coincide with the asymptotically stable points under (4.1) and (4.2);
- further, $\mathbf{y}^\top R\mathbf{y} < \bar{\mathbf{x}}^\top R\bar{\mathbf{x}}$ for all $\mathbf{y} \in \Delta$ with $\mathbf{y} \neq \bar{\mathbf{x}}$ but $y_i = 0$ if $\bar{x}_i = 0$.

Although strictly increasing objective values are guaranteed as trajectories under (4.1) or (4.2) are followed, one could get stuck in an inefficient local solution of (1.1). One possibility to escape is the *Genetic Engineering via Negative Fitness (GENF)* approach [6] described in the sequel. From the properties above, a strict local solution $\bar{\mathbf{x}}$ of (1.1) must be a global one if all $\bar{x}_i > 0$. Consequently, at an inefficient local solution necessarily $\bar{x}_i = 0$ for some i. In the usual genetic interpretation of the dynamics (4.1) and (4.2), this means that some alleles die out during the natural selection process, and these are therefore unfit in the environment currently prevailing. The escape step now artificially re-introduces some alleles which would have gone extinct during the natural selection process, and restarts with a smaller subproblem which will yield an improvement if $\bar{\mathbf{x}}$ is inefficient, as follows:

GENF procedure to escape from inefficient local solutions in StQPs

1. Given a local solution $\overline{x}$ to (1.1), remove all alleles which are not unfit, i.e. all $i \in S = \{i \in N : \overline{x}_i > 0\}$;
2. determine a (local) fitness *minimizer* y in the reduced problem, i.e. consider problem (1.1) with R replaced by

$$\overline{R} = [\gamma_S - r_{ij}]_{i,j \in N \setminus S}$$

   where $\gamma_S = \max_{i,j \in N \setminus S} r_{ij}$ (thus $\overline{R}$ has no negative entries either);
3. with a local solution y of this auxiliary problem, put

$$J = \{j \in N \setminus S : y_j > 0\}$$

   and denote by m the cardinality of J;
4. for all $s \in S$ and $t \in J$, consider the reduced problem $\mathcal{P}_{t \to s}$, i.e. problem (1.1) in $n - m$ variables for the $(n - m) \times (n - m)$ matrix $R_{t \to s}$ obtained from R as follows: replace r_{si} with r_{ti} and remove all other $j \in J$;
5. $\overline{x}$ is a global solution to the master problem (1.1) if and only if for all $(s, t) \in S \times J$, the maximum of $\mathcal{P}_{t \to s}$ does not exceed the current best value $\overline{x}^\top R \overline{x}$;
6. in the negative, i.e. if $u^\top R_{t \to s} u > \overline{x}^\top R \overline{x}$ for some $u \in \Delta \subset \mathbb{R}^{n-m}$, and if $j \in J$ is chosen such that for all $q \in J$,

$$\sum_{p \notin J \cup \{s\}} (r_{jp} - r_{qp}) u_p \geq \tfrac{1}{2}(r_{qq} - r_{jj}) u_s,$$

then a strictly improving feasible point $\tilde{x}$ is obtained as follows:

$$\tilde{x}_q = \begin{cases} u_t & \text{if } q = j, \\ 0 & \text{if } q \in J \cup \{s\} \setminus \{j\}, \\ u_q & \text{if } q \in N \setminus J. \end{cases}$$

In view of the possible combinatorial explosion in effort with increasing number of variables, this dimension reducing strategy seems to be promising: if k is the size of S, the above result yields a series of km StQPs in $n - m$ variables rather than in n. We are now ready to describe the algorithm which stops after finitely many repetitions, since it yields strict local solutions with strictly increasing objective values [6].

Replicator dynamics algorithm for StQPs

1. Start with $x(0) = [\frac{1}{n}, \ldots, \frac{1}{n}]^\top$ or nearby, iterate (4.2) until convergence;
2. the limit $\overline{x} = \lim_{\tau \to \infty} x(\tau)$ is a strict local solution with probability one; call the GENF procedure to improve the objective, if possible; denote the improving point by $\tilde{x}$;
3. repeat 1., starting with $x(0) = \tilde{x}$.

A different approach towards global solutions of StQPs uses familiar branch-and-bound schemes. For ease of exposition, now consider the minimization StQP

$$\min\{\mathbf{x}^\top Q \mathbf{x} : \mathbf{x} \in \Delta\}, \tag{4.3}$$

and assume again without loss of generality that Q has only positive entries. If one applies a usual simplicial partition to Δ, all subproblems are again StQPs. To obtain lower bounds for these problems, convex minorants for the objective $\mathbf{x}^\top Q \mathbf{x}$ on Δ may be used, e.g. quadratic forms $\mathbf{x}^\top F \mathbf{x}$ with F positive-semidefinite (or some related matrix $\hat{F}$ which ensures that the minorant is convex necessarily over Δ only), where F is chosen such that the gap between the objectives is small. This can be accomplished by requiring diag $F = $ diag Q and that $\sum_{i,j}[q_{ij} - f_{ij}]$ is small, while the minorant condition is guaranteed by the requirement $f_{ij} \leq q_{ij}$ for all $i, j \in N$. Therefore one arrives at a classical SDP which can be solved by the usual methods. The resulting matrix F then gives a convex problem, so that the desired lower bound for (4.3),

$$\min\{\mathbf{x}^\top F \mathbf{x} : \mathbf{x} \in \Delta\},$$

can be obtained efficiently, e.g. via *local search* techniques or *linear complementarity* approaches. For details and results see [17].

5. Applications

An important application for StQPs is the search for a maximum weight clique arising in computer vision, pattern recognition and robotics (see [5] for a more detailed account): consider an undirected graph $\mathcal{G} = (N, \mathcal{E})$ with n nodes, and a weight vector $\mathbf{w} = [w_1, \ldots, w_n]^\top$ of positive weights w_i associated to the nodes $i \in N$. A *clique* S is a subset of N which corresponds to a complete subgraph of $\mathcal{G}$ (i.e. any pair of different nodes in S is an edge in $\mathcal{E}$). A clique S is said to be *maximal* if there is no larger clique containing S. Every clique S in $\mathcal{G}$ has a weight $W(S) = \sum_{i \in S} w_i$. The *maximum weight clique problem (MWCP)* consists of finding a clique in the graph which has largest total weight. The classical (unweighted) maximum clique problem is a special case with $\mathbf{w} = \mathbf{e}$. To reformulate the MWCP as a StQP, consider the following class of symmetric $n \times n$ matrices: let

$$\mathcal{C}(\mathcal{E}) = \{(c_{ij})_{i,j \in N} : c_{ij} = 0 \text{ if } (i,j) \in \mathcal{E}\},$$

as well as $\mathcal{C}_+(\mathcal{G}) = \{C \in \mathcal{C}(\mathcal{E}) : C^\top = C \text{ and } c_{ij} \geq c_{ii} + c_{jj} \text{ if } (i,j) \notin \mathcal{E}\}$, and form the class

$$\mathcal{C}(\mathcal{G}, \mathbf{w}) = \left\{C \in \mathcal{C}_+(\mathcal{G}) : c_{ii} = \frac{1}{2w_i} \text{ for all } i\right\}.$$

Now consider the (minimization) StQP

$$\min\{\mathbf{x}^\top C\mathbf{x} : \mathbf{x} \in \Delta\} \qquad (5.1)$$

for some matrix $C \in \mathcal{C}(\mathcal{G}, \mathbf{w})$. Given a subset $S \subseteq N$, define the S-face of Δ as

$$\Delta_S = \{\mathbf{x} \in \Delta : x_i = 0 \text{ if } i \notin S\}$$

and its *weighted barycenter* as $\mathbf{x}^S = \sum_{i \in S} \frac{w_i}{W(S)} \mathbf{e}_i \in \Delta_S$ with $\{\mathbf{e}_i : i \in N\}$ the vertices of Δ, the standard basis vectors. Then the following assertions hold [4]:

- A point $\mathbf{x} \in \Delta$ is a local solution to problem (5.1) if and only if $\mathbf{x} = \mathbf{x}^S$, where S is a maximal clique.
- A vector $\mathbf{x} \in \Delta$ is a global solution to problem (5.1) if and only if $\mathbf{x} = \mathbf{x}^S$, where S is a maximum weight clique.
- Moreover, all local (and hence global) solutions to (5.1) are strict.

Note that a different class used in [9] does not share these properties. The class $\mathcal{C}(\mathcal{G}, \mathbf{w})$ is isomorphic to the positive orthant in $\binom{n}{2} - e$ dimensions where e is the cardinality of $\mathcal{E}$. This class is a polyhedral pointed cone with its apex given by the matrix $C^{\mathcal{G}, \mathbf{w}}$ with entries

$$c_{ij}^{\mathcal{G}, \mathbf{w}} = \begin{cases} \frac{1}{2w_i} & \text{if } i = j\,, \\ \frac{1}{2w_i} + \frac{1}{2w_j} & \text{if } i \neq j \text{ and } (i,j) \notin \mathcal{E}\,, \\ 0 & \text{otherwise.} \end{cases}$$

In the unweighted case $\mathbf{w} = \mathbf{e}$, we get $C^{\mathcal{G}, \mathbf{e}} = \frac{1}{2}I + A_{\overline{\mathcal{G}}}$ where $A_{\overline{\mathcal{G}}}$ is the adjacency matrix of the complement graph $\overline{\mathcal{G}}$, and I the $n \times n$ identity matrix. Therefore (5.1) can be seen as a regularized generalization of the original approach of T.S. Motzkin and E.G. Straus [15].

Another application of StQP is concerned with the mean/variance *portfolio selection* problem (see, e.g. [14]) which can be formalized as follows: suppose there are n securities to invest in, at an amount expressed in relative shares $x_i \geq 0$ of an investor's budget. Thus, the budget constraint reads $\mathbf{e}^\top \mathbf{x} = 1$, and the set of all feasible portfolios (investment plans) is given by Δ. Now, given the expected return m_i of security i during the forthcoming period, and an $n \times n$ covariance matrix V across all securities, the investor faces the multiobjective problem to maximize the expected return $\mathbf{m}^\top \mathbf{x}$ and simultaneously minimize the risk $\mathbf{x}^\top V\mathbf{x}$ associated with her decision $\mathbf{x}$.

One of the most popular approaches to such type of problems in general applications is that the user prespecifies a parameter β which in her eyes balances the benefits of high return and low risk, i.e. consider the parametric QP

$$\max\{f_\beta(\mathbf{x}) = \mathbf{m}^\top \mathbf{x} - \beta \mathbf{x}^\top V\mathbf{x} : \mathbf{x} \in \Delta\}\,.$$

For fixed β, this is a StQP. Anyhow, the question remains how to choose β. In finance applications, the notion of market portfolio is used to determine a reasonable value for this parameter. This emerges more or less from an exogenous artefact, namely by introducing a completely risk-free asset which is used to scale return versus risk [12]. An alternative, purely endogenous derivation of market portfolio could use a result of M.J. Best and B. Ding [1] who consider the problem

$$\max_{\beta>0} \max_{\mathbf{x}\in\Delta} \frac{1}{\beta} f_\beta(\mathbf{x}), \tag{5.2}$$

and show how optimal solutions $(\beta^*, \mathbf{x}^*)$ for (5.2) emerge from a single StQP (1.1) with, e.g. $R = 2\mathbf{m}\mathbf{m}^\top - V$.

References

1. Best, M.J. and Ding, B. (1997): Global and local quadratic minimization. J. Global Optimization **10**, 77–90.
2. Bomze, I.M. (1992): Copositivity conditions for global optimality in indefinite quadratic programming problems. Czechoslovak J. Oper. Res., **1**, 7–19.
3. Bomze, I.M. (1996): Block pivoting and shortcut strategies for detecting copositivity. Linear Algebra and its Applications **248**, 161–184.
4. Bomze, I.M. (1998): On standard quadratic optimization problems. J. Global Optimization **13**, 369–387.
5. Bomze, I.M., Budinich, M., Pardalos, P.M. and Pelillo, M. (1999): The maximum clique problem, in: D.-Z. Du and P. M. Pardalos (eds), Handbook of Combinatorial Optimization (Suppl. Vol. A), Kluwer Academic Publishers, Boston, MA (to appear).
6. Bomze, I.M. and Stix, V. (1999): Genetical engineering via negative fitness: evolutionary dynamics for global optimization. Annals of Operations Research **89**, 297–318.
7. Cohen, J. and Hickey, T. (1979): Two algorithms for determining volumes of convex polyhedra. J. ACM **26**, 401–414.
8. Danninger, G. (1990): A recursive algorithm to detect (strict) copositivity of a symmetric matrix, in: U. Rieder, P. Gessner, A. Peyerimhoff and F.J. Radermacher (eds), Methods of Operations Research **62**, 45–52. Hain, Meisenheim.
9. Gibbons, L.E., Hearn, D.W., Pardalos, P.M. and Ramana, M.V. (1997): Continuous characterizations of the maximum clique problem. Math. Oper. Res. **22**, 754–768.
10. Hall, M., jr. and Newman, M. (1963): Copositive and completely positive quadratic forms. Proc. Cambridge Philos. Soc. **59**, 329–339.
11. Horst, R., Pardalos, P.M. and Thoai, N.V. (1995): Introduction to Global Optimization, Kluwer Academic, Dordrecht.
12. Huang, C.-F. and Litzenberger, R.H. (1988): Foundations for financial economics, 6th printing. North Holland, Amsterdam.
13. Markham, T.L. (1971): Factorization of completely positive matrices. Proc. Cambridge Philos. Soc. **69**, 53–58.
14. Markowitz, H.M. (1995): The general mean-variance portfolio selection problem, in: S.D. Howison, F.P. Kelly and P. Wilmott (eds), Mathematical models in finance, 93–99. Chapman & Hall, New York.
15. Motzkin, T.S. and Straus, E.G. (1965): Maxima for graphs and a new proof of a theorem of Turán. Canad. J. Math. **17**, 533–540.

16. Nesterov, Y.E. and Nemirovskii, A.S. (1994): Interior point methods in convex programming: theory and applications. SIAM, New York.
17. Nowak, I. (1999): A new semidefinite programming bound for indefinite quadratic forms over a simplex. J. Global Optimization **14**, 357 - 364.

Computational Sensitivity Analysis of State Constrained Control Problems

Dirk Augustin[1] and Helmut Maurer[1]

Westfälische Wilhelms-Universität Münster
Institut für Numerische Mathematik
Einsteinstr. 62 , 48149 Münster , Germany
email: maurer@math.uni-muenster.de

Summary. Stability analysis of parametric control problems has recently been extended to control problems subject to pure state constraints. This paper illustrates the numerical aspects of stability analysis via a specific numerical example: the optimal control of the state constrained Van-der-Pol oscillator. The multiple shooting method is used to determine a nominal solution of high accuracy that satisfies first order necessary conditions. Second order sufficient conditions can be checked by solving an associated Riccati equation. Finally, sensitivity differentials of optimal solutions with respect to parameters are computed.

Keywords and phrases: parametric nonlinear optimal control, nonlinear ordinary differential equations, first and higher order state constraints, differentiability of solutions, computational sensitivity analysis

1. Introduction

In the last years, stability and sensitivity analysis of *parametric control problems* subject to mixed control–state inequality constraints have been studied extensively. Lipschitzian properties of optimal solutions with respect to parameters are given in [1, 2, 5] while Fréchet–differentiability is established in [7, 9]. The difficulties with extending these results directly to parametric control problems subject to *pure* state constraints are connected with the regularity of the multiplier associated with the state constraint.

Recently, stability results for control problems with *pure* state constraints of order one have been obtained in [3], [8]. Lipschitzian stability is derived in [3] via an abstract implicit function theorem in nonlinear spaces. Conditions for full Fréchet–differentiability of optimal solutions are developed in [8]. These conditions are obtained from the classical implicit function theorem which is applied to the multipoint boundary value problem underlying Pontryagin's minimum principle. The results in [8] extend the more heuristic approach in [12, 13].

Stability and sensitivity results are based on second order sufficient conditions (SSC) for the unperturbed (nominal) solution. Apart from [8] no explicit numerical examples with a detailed sensitivity analysis may be found in the literature. The purpose of this paper is to present a practical numerical example that illustrates both sensitivity analysis and the use of (SSC).

The example is provided by a control model for the Van-der-Pol oscillator. The control model belongs to the following class of perturbed control problems $OC(p)$ depending on a parameter (perturbation) $p \in \mathbb{R}^q$: Minimize the functional

$$J(x, u, p) = g(x(T), p) + \int_0^T L(x(t), u(t), p)\, dt \qquad (1.1)$$

subject to

$$\dot{x}(t) = f(x(t), u(t), p) \quad \text{for a.e.} \quad t \in [0, T] \ , \qquad (1.2)$$

$$x(0) = \varphi(p) \ , \quad \psi(x(T), p) = 0 \ , \qquad (1.3)$$

$$S(x(t), p) \leq 0 \quad \text{for a.e.} \quad t \in [0, T] \ . \qquad (1.4)$$

The final time T is fixed. It is assumed that the functions $g : \mathbb{R}^n \times \mathbb{R}^q \to \mathbb{R}$, $L : \mathbb{R}^{n+m} \times \mathbb{R}^q \to \mathbb{R}$, $f : \mathbb{R}^{n+m} \times \mathbb{R}^q \to \mathbb{R}^n$, $\varphi : \mathbb{R}^q \to \mathbb{R}^n$, $\psi : \mathbb{R}^n \times \mathbb{R}^q \to \mathbb{R}^r$, $0 \leq r \leq n$, and $S : \mathbb{R}^n \times \mathbb{R}^q \to \mathbb{R}$ are sufficiently smooth on appropriate open sets.

The admissible class is that of piecewise continuous control functions. Later on, regularity conditions will be imposed such that the optimal control is *continuous* and piecewise of class C^1. The problem $OC(p_0)$ corresponding to a *reference parameter* $p_0 \in \mathbb{R}^q$ is considered as the *unperturbed* or *nominal* problem. Suppose that there exists a local solution (x_0, u_0) of the reference problem $OC(p_0)$ such that $u_0 \in C(0, T; \mathbb{R}^m)$. The reference solution $x_0(t)$, $u_0(t)$ and the associated adjoint function $\lambda_0(t)$, $0 \leq t \leq T$, satisfy a boundary value problem (BVP). The regularity conditions ensure that $x_0(t)$ is of class C^1 while $\lambda_0(t)$ and $u_0(t)$ are piecewise of class C^1. In [8] conditions are given such that the following stability result holds.

Solution Differentiability: The unperturbed solution $x_0(t), \lambda_0(t), u_0(t)$ can be embedded into a family of optimal solutions $x(t, p), \lambda(t, p), u(t, p)$ to the perturbed problem $OC(p)$ where $x(t, p)$ is of class C^1 and $\lambda(t, p)$ and $u(t, p)$ are piecewise of class C^1 with respect to both variables (t, p) in a neighborhood of the reference parameter p_0 .

Before discussing solution differentiability for the Van-der-Pol oscillator, we briefly review in the next section the main results on (SSC) obtained in [6, 7].

2. Second Order Sufficient Conditions for State Constrained Control Problems

In the following, partial derivatives with respect to x and u are denoted either by D_x and D_u or by subscripts. Let us first define the *order* $l \geq 1$ of the state constraint $S(x, p) \leq 0$. We introduce recursively the following functions (cf. [4]):

$$S^j(x, u, p) : I\!\!R^n \times I\!\!R^m \times I\!\!R^q \mapsto I\!\!R,$$
$$S^0(x, u, p) := S(x, p), \quad S^{j+1}(x, u, p) := D_x S^j(x, u, p) f(x, u, p).$$

The constraint $S(x, p)$ is called to be of order $l \geq 1$ with respect to the dynamics $\dot{x} = f(x, u, p)$, if

$$D_u S^j(x, u, p) \equiv 0, \quad (j = 1, 2, ..., l-1), \quad D_u S^l(x, u, p) \not\equiv 0.$$

It follows from this definition that the functions $S^j(x, u, p) \equiv S^j(x, p)$ for $j = 0, 1, ..., l-1$ are independent of u. It can be easily seen that if $S(x, p)$ is of order l and (x, u) is the solution of $\dot{x} = f(x, u, p)$, then

$$\frac{d^j}{dt^j} S(x(t), p) = \begin{cases} S^j(x(t), p), & j = 0, 1, ..., l-1, \\ S^l(x(t), u(t), p), & j = l. \end{cases} \tag{2.1}$$

Let the pair (x_0, u_0) be optimal for $\mathrm{OC}(p_0)$ with reference parameter p_0. For convenience, arguments of all functions evaluated at the reference point $(x_0(t), u_0(t), p_0)$ will be often denoted by $[t]$, e.g., $f[t] := f(x_0(t), u_0(t), p_0)$, $g[T] := g(x_0(T), p_0)$, etc. First or second order partial derivatives are denoted by subscripts or by the symbol D resp. D^2.

For simplicity, the discussion will be restricted to first order state constraints with $l = 1$ in (2.1) (cf. [8]) and to a *scalar* control with $m = 1$ in (1.1)–(1.4). Usually, first order state constraints exhibit only *boundary arcs*. Therefore, we assume that the active set I_a consists of one boundary arc:

$$I_a = \{ t \in [0, T] \mid S(x_0(t), p_0) = 0 \} = [t_1, t_2] \quad \text{with} \quad 0 < t_1 < t_2 < T.$$

Consider the following Hamiltonian function H and the *augmented* Hamiltonian $\tilde{H}$, where the asterisk denotes transpose:

$$H(x, u, \lambda, p) = L(x, u, p) + \lambda^* f(x, u, p), \qquad \lambda \in I\!\!R^n, \quad (2.2)$$

$$\tilde{H}(x, u, \lambda, \nu, p) = H(x, u, \lambda, p) + \nu S^1(x, u, p), \quad \nu \in I\!\!R. \quad (2.3)$$

Here the first derivative S^1 is adjoined to the unconstrained Hamiltonian; cf. [4]. Assume that there exist an adjoint function $\lambda_0 : [0, T] \to I\!\!R^n$ and a multiplier function $\nu_0 : [0, T] \to I\!\!R$ that are piecewise of class C^1, and multipliers $\rho_0 \in I\!\!R^r$ and $\sigma_0 \in I\!\!R$ such that the following first-order necessary conditions hold for a.e. $t \in [0, T]$; compare [4] :

$$\dot{\lambda}_0(t) = -\tilde{H}_x(x_0(t), u_0(t), \lambda_0(t), \nu_0(t), p_0)^*, \tag{2.4}$$

$$\lambda_0(T) = (g + \rho_0^* \psi)_x^* (x_0(T), p_0), \tag{2.5}$$

$$\tilde{H}_u(x_0(t), u_0(t), \lambda_0(t), \nu_0(t), p_0) = 0, \tag{2.6}$$

$$\nu_0(t) \geq 0 \quad \text{and} \quad \nu_0(t) = 0 \quad \text{for} \quad t \notin [t_1, t_2], \tag{2.7}$$

$$\lambda_0(t_1^+) = \lambda_0(t_1^-) - \sigma_0 S_x(x_0(t_1), p_0)^*, \quad \sigma_0 \geq 0. \tag{2.8}$$

The following *regularity assumptions* are supposed to hold; cf. [8], Assumptions (I.4) and (I.8):

$$D_u S^1(x_0(t), u_0(t), p_0) \neq 0 \quad \text{for all } t \in [t_1, t_2] , \qquad (2.9)$$

$$\tilde{H}_{uu}(t) \geq c > 0 \quad \text{for all } t \in [0, T] \quad \text{with some } \; c > 0 . \qquad (2.10)$$

The second order sufficient optimality conditions in [6], [8], Section 8, and [10], Theorem 4.2, are satisfied if there exists an $n \times n$ symmetric matrix function $Q(t)$ that is a bounded solution of the following Riccati equation in the intervall $[0, T]$:

$$\dot{Q} = -Q f_x - f_x^* Q - \tilde{H}_{xx} + (\tilde{H}_{xu} + Q f_u)(\tilde{H}_{uu})^{-1}(\tilde{H}_{ux} + f_u^* Q) . \qquad (2.11)$$

The matrix $Q(t)$ satisfies the jump relation at the entry point t_1,

$$Q(t_1^+) = Q(t_1^-) - \sigma_0 S_{xx}(x_0(t_1), p_0) , \qquad (2.12)$$

and is subject to the boundary conditions

$$Q(T) \leq D_{xx}^2(g + \rho_0^* \psi)[T] \quad \text{on} \quad \ker(D_x \psi[T]) . \qquad (2.13)$$

3. Optimal Control of the Van-der-Pol Oscillator

Consider the following optimal control model for the Van-der-Pol oscillator with state constraints; compare [14]:

Minimize
$$\int_0^5 \left(u^2(t) + x_1^2(t) + x_2^2(t) \right) dt \qquad (3.1)$$

subject to
$$\dot{x}_1 = x_2, \quad \dot{x}_2 = (1 - x_1^2)x_2 - x_1 + u, \quad \text{for } t \in [0, 5], \quad (3.2)$$
$$x_1(0) = 1, \; x_2(0) = 0, \qquad (3.3)$$
$$p \leq x_2(t), \quad \text{for} \quad t \in [0, 5] . \qquad (3.4)$$

The state variable $x_1(t)$ represents the voltage in an electric circuit with capacity, resistance, inductivity and diode. The state inequality constraint (3.4) is of the form (1.4) with the function $S(x, p) := -x_2 + p \leq 0$ containing the scalar parameter p; we choose the *nominal parameter* $p_0 = -0.4$.

By studying first the unconstrained problem (3.1)–(3.3) omitting the state constraint (3.4), one realizes that the constrained solution has one boundary arc with $x_2(t) \equiv p$ for $t \in [t_1, t_2]$ and $0 < t_1 < t_2 < 5$. The function S^1 is computed from (2.1) as $S^1(x, u, p) = -\dot{x}_2 = -(1 - x_1^2)x_2 + x_1 - u$. From (2.4)–(2.7) we find the following adjoint equations, the control u and the multiplier ν associated with the state constraint:

$$\dot{\lambda}_1 = -2x_1 + 2x_1 x_2 \lambda_2 + \lambda_2 - \nu(2x_1 x_2 + 1), \quad \lambda_1(5) = 0, \quad (3.5)$$
$$\dot{\lambda}_2 = -2x_2 - \lambda_1 + \lambda_2(x_1^2 - 1) + \nu(1 - x_1^2), \quad \lambda_2(5) = 0, \quad (3.6)$$

where

while λ_1 is continuous at t_1. The multipoint boundary value problem (3.2), (3.3) and (3.5)–(3.11) can be solved with the code BNDSCO of [11]. Using the nominal parameter $p_0 = -0.4$ we obtain the following results for the initial values of the adjoint variables and the junction points:

$$
\begin{array}{llll}
\lambda_1(0) &=& 5.66576692 & , \\
t_1 &=& 0.62752939 & , \\
x_1(5) &=& -0.07420849 & , \\
\sigma_0 &=& 1.73011588 & ,
\end{array}
\qquad
\begin{array}{lll}
\lambda_2(0) &=& 0.82392266 \;, \\
t_2 &=& 1.46908215 \;, \\
x_2(5) &=& 0.02929680 \;, \\
J(x_0, u_0, p_0) &=& 2.95370134 \;.
\end{array}
$$

The nominal optimal solution is shown in Figure 1.

Next, we shall verify that the computed nominal solution satisfies all regularity conditions and the second-order sufficient conditions in Section 2. It is trivial to check the regularity conditions (2.9) and the strict Legendre condition (2.10). Moreover, note that the *non-tangential* junction condition holds; cf. [8], Assumption (I.9):

$$
\frac{d}{dt} S^1(x(t_1), u(t_1^-), p) = -1.62175826 \neq 0,
$$

$$
\frac{d}{dt} S^1(x(t_2), u(t_2^+), p) = -0.52365061 \neq 0.
$$

Additionally, note also that the strict complementarity condition holds (cf. [8], Assumption (I.11)) since the multiplier $\nu(t)$ satisfies $-d\nu/dt > 0$ for $t \in [t_1^+, t_2^-]$, cf. Figure 1. In order to check second-order sufficient conditions we have to find a bounded solution of the Riccati equations (2.11)–(2.13). Since the Van-der-Pol problem has dimension $n = 2$, we consider a symmetric 2×2 matrix in the form

$$
Q(t) = \begin{pmatrix} q_1(t) & q_2(t) \\ q_2(t) & q_4(t) \end{pmatrix}
$$

and obtain the explicit Riccati equations from (2.11) for $t \in [0, T]$:

$$
\dot{q}_1 = (4x_1 x_2 + 2)q_2 - 2 + 2x_2(\lambda_2 - \nu) + \frac{1}{2}q_2^2,
$$

$$
\dot{q}_2 = -q_1 - (1 - x_1^2)q_2 + (2x_1 x_2 + 1)q_4 + 2x_1(\lambda_2 - \nu) + \frac{1}{2}q_2 q_4,
$$

$$
\dot{q}_4 = -2q_2 - 2(1 - x_1^2)q_4 - 2 + \frac{1}{2}q_4^2.
$$

The functions q_i are continuous at the entry point t_1 due to the relations (2.12) and $S_{xx} \equiv 0$. By virtue of (2.13), the boundary conditions are $q_1(5) = q_2(5) = q_4(5) = 0$ because the final state is not given. Numerical integration along the nominal solution with $p_0 = -0.4$ shows indeed that there exist

$$u = \begin{cases} -\frac{\lambda_2}{2} & : \ t \in [0, t_1] \cup [t_2, 5], \\ (x_1^2 - 1)x_2 + x_1 & : \ t \in [t_1, t_2], \end{cases} \tag{3.7}$$

$$\nu = \begin{cases} 0 & : \ t \in [0, t_1] \cup [t_2, 5], \\ 2(x_1^2 - 1)x_2 + 2x_1 + \lambda_2 & : \ t \in [t_1, t_2]. \end{cases} \tag{3.8}$$

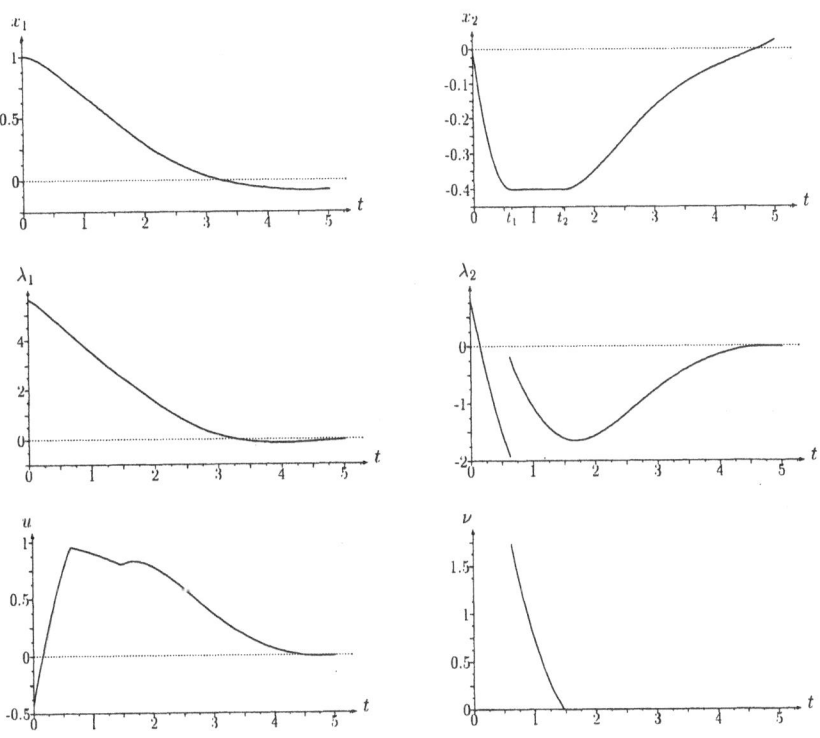

Figure 1: Nominal states $x_i(t)$, adjoint variables $\lambda_i(t)$, $i = 1, 2$, optimal control $u(t)$ and multiplier $\nu(t)$.

The junction conditions at the entry and exit point t_1 resp. t_2 are:

$$x_2(t_1) + 0.4 = 0, \tag{3.9}$$

$$\left(x_1^2(t) - 1\right) x_2(t) + x_1(t) + 0.5\lambda_2(t) = 0, \quad \text{at } t = t_1^- \text{ and } t = t_2^+. \tag{3.10}$$

In the last equation, we have used the control law $u = -\lambda_2/2$ valid on the interior arcs $[0, t_1] \cup [t_2, 5]$ and also the continuity of the control at the junction points. According to relation (2.8), the adjoint variable λ_2 has a jump at the entry point t_1,

$$\lambda_2(t_1^+) = \lambda_2(t_1^-) + \sigma, \quad \sigma \geq 0, \tag{3.11}$$

a bounded solution with $|q_i(t)| \le 6$, cf. Figure 2. Thus we arrive at the conclusion that the solution shown in Figure 1 is a local minimum.

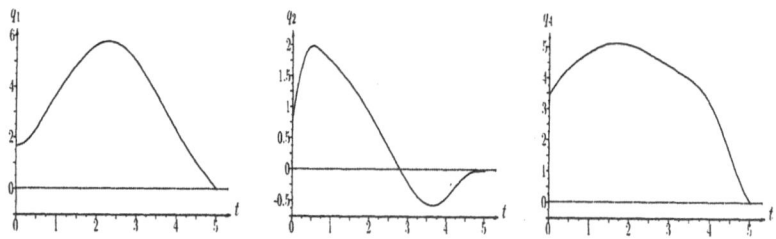

Figure 2: Solutions $q_i(t)$, $i = 1, 2, 4$, of the Riccati equation.

4. Computational Sensitivity Analysis

In the preceding sections we have verified numerically that the *unperturbed (nominal)* solution with $p_0 = -0.4$ meets all assumptions for *solution differentiability* in Theorem 8.3 of [8]. Thus the unperturbed solution can be embedded into a family of perturbed solutions

$$x_i(t, p), \quad \lambda_i(t, p) \ (i = 1, 2), \quad t_j(p) \ (j = 1, 2), \quad \nu(t, p), \quad \sigma(p),$$

of the perturbed problem $OC(p)$. The functions $x_i(t, p)$, $t_j(p)$, $\sigma(p)$ are of class C^1 while $\lambda_i(t, p)$, $\nu(t, p)$ are piecewise of class C^1. The perturbed solution satisfies the boundary value problem (3.2), (3.3) and (3.5)–(3.11) identically for all parameter p near p_0.

A quantitative *sensitivity analysis* consists in computing the *sensitivity differentials*

$$z_i(t) = \frac{\partial x_i}{\partial p}(t, p_0) \ , \quad \gamma_i(t) = \frac{\partial \lambda_i}{\partial p}(t, p_0) \ , \ (i = 1, 2) \ , \ p_0 = -0.4 \ .$$

The linear variational ODE's for $z_i(t)$ and $\gamma_i(t)$ are obtained by formally differentiating the nonlinear equations (3.2), (3.3) and (3.5)– (3.11) with respect to the perturbation p. This yields the linear ODE's:

$$\dot{z}_1 = z_2 ,$$
$$\dot{z}_2 = -2x_1 x_2 z_1 + (1 - x_1^2)z_2 - z_1 + u_p ,$$
$$\dot{\gamma}_1 = -2z_1 + 2(x_2 \lambda_2 z_1 + x_1 x_2 \gamma_2 + x_1 \lambda_2 z_2) + \gamma_2 - 2\nu(x_2 z_1 + x_1 z_2) -$$
$$\qquad -\nu_p(2x_1 x_2 + 1) ,$$
$$\dot{\gamma}_2 = -2z_2 - \gamma_1 + \gamma_2(x_1^2 - 1) + 2x_1 \lambda_2 z_1 - 2\nu x_1 z_1 + \nu_p(1 - x_1^2) ,$$

where

$$u_p = \frac{\partial u}{\partial p} = \begin{cases} -\frac{\gamma_2}{2} & , \ t \in [0, t_1] \cup [t_2, 5], \\ 2x_1 x_2 z_1 + (x_1^2 - 1)z_2 + z_1 & , \ t \in [t_1, t_2], \end{cases}$$

$$v_p = \frac{\partial v}{\partial p} = \begin{cases} 0 & , \ t \in [0, t_1] \cup [t_2, 5], \\ 4x_1 x_2 z_1 + 2(x_1^2 - 1)z_2 + 2z_1 + \gamma_2 & , \ t \in [t_1, t_2]. \end{cases}$$

Since the initial state is fixed and the final state is left free, we get the boundary conditions

$$z_i(0) = 0, \quad \gamma_i(5) = 0, \quad \text{for } i = 1, 2.$$

Moreover, differentiation of the junction condition $x_2(t_1(p), p) \equiv p$ yields the junction condition

$$z_2(t_1) = 1.$$

The jump relations for the sensitivity differentials γ_1, γ_2 at the entry point t_1 are obtained in the following way. Relation (3.11) holds identically in p and yields the conditions

$$\lambda_1(t_1(p)^+, p) \equiv \lambda_1(t_1(p)^-, p), \quad \lambda_2(t_1(p)^+, p) \equiv \lambda_2(t_1(p)^-, p) + \sigma(p).$$

If we differentiate these relations and use the fact that nominal derivatives $\dot{\lambda}_i, i = 1, 2$, are *continuous* at t_1, then we find the jump relations

$$\gamma_1(t_1^+) = \gamma_1(t_1^-), \quad \gamma_2(t_1^+) = \gamma_2(t_1^-) + \frac{d\sigma}{dp},$$

where $d\sigma/dp$ is the differential of the jump variable $\sigma(p)$.

Finally, the differentials of the junction points are obtained by differentiating the junction conditions $S^1(x(t_j(p), p), p) = 0$. This leads to the formulas

$$\frac{dt_j}{dp} = -\frac{(4x_1 x_2 + 2)z_1 + 2(x_1^2 - 1)z_2 + \gamma_2}{4x_1 x_2^2 - 2(x_1^2 - 1)^2 x_2 + 2(1 - x_1^2)x_1 - \lambda_1}\bigg|_{t=t_j} \quad \text{at } t_1^- \text{ and } t_2^+.$$

We get the following numerical results using the code BNDSCO in [11]:

$$\begin{array}{llll} \gamma_1(0) & = & 9.88324529 & , & \gamma_2(0) & = & -0.00482702 & , \\ dt_1/dp & = & -2.03284608 & , & dt_2/dp & = & 6.03498889 & , \\ z_1(5) & = & -0.26590753 & , & z_2(5) & = & -0.55722168 & , \\ d\sigma/dp & = & 22.0100099 & , & dJ/dp & = & 1.73011588 & . \end{array}$$

The sensitivity differentials are shown in Figure 3.

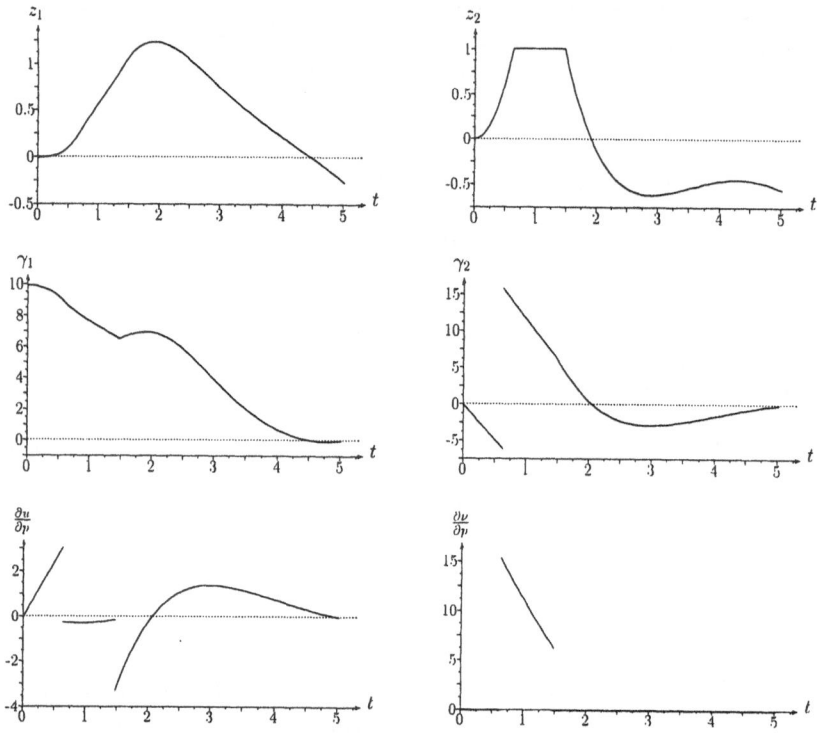

Figure 3: Sensitivity differentials $z_i = \partial x_i / \partial p$, $\gamma_i = \partial \lambda_i / \partial p$, $\partial u / \partial p$, $\partial \nu / \partial p$.

5. Conclusion

In this paper, a stability and sensitivity analysis has been carried out for the optimal control of the Van-der-Pol oscillator subject to a state constraint. The analysis provides an illustrative numerical example for the recent stability analysis of state constrained control problems in [8]. It has been shown that all theoretical conditions for solution differentiability given in [8] can be checked numerically. In particular, second order sufficient conditions for state constrained control problems in [6, 8] have been verified via bounded solutions of Riccati equations. A computation of sensitivity differentials has been performed which gives the possibility of applying real-time control techniques along the lines of [7, 9, 12, 13].

References

1. Dontchev, A.L., Hager, W.W. (1993): *Lipschitz Stability in Nonlinear Control and Optimization*, SIAM J. Control and Optimization **31** 569–603.
2. Dontchev, A.L., Hager, W.W., Poore, A.B. Yang, B. (1995): *Optimality, Stability and Convergence in Nonlinear Control*, Applied Mathematics and Optimization **31**, 297–326.

3. Dontchev, A.L., Hager, W.W. (1998): *Lipschitzian Stability for State Constrained Nonlinear Optimal Control*, SIAM J. Control and Optimization **36**, 698–718.

4. Hartl, R.F., Sethi, S.P., Vickson, R.G. (1995): *A Survey of the Maximum Principles for Optimal Control Problems with State Constraints*, SIAM Review **37**, 181–218.

5. Malanowski, K. (1995): *Stability and Sensitivity of Solutions to Nonlinear Optimal Control Problems*, Applied Mathematics and Optimization **32**, 111–141.

6. Malanowski, K. (1997): *Sufficient Optimality Conditions for Optimal Control Subject to State Constraints*, SIAM J. on Control and Optimization **35**, 205–227.

7. Malanowski, K., Maurer, H. (1996): *Sensitivity Analysis for Parametric Control Problems with Control-State Constraints*, Computational Optimization and Applications **5**, 253–283.

8. Malanowski, K., Maurer, H. (1998): *Sensitivity Analysis for State Constrained Optimal Control Problems*, Discrete and Continuous Dynamical Systems **4**, 241–272.

9. Maurer, H., Pesch, H.J. (1994): *Solution Differentiability for Parametric Nonlinear Control Problems with Control-State Constraints*, Control and Cybernetics **23**, 201–227.

10. Maurer, H., Pickenhain, S. (1995): *Second Order Sufficient Conditions for Optimal Control Problems with Mixed Control-State Constraints*, Journal of Optimization Theory and Applications **86**, 649–667.

11. Oberle, H.J., Grimm, W. (1989): *BNDSCO - A Program for the Numerical Solution of Optimal Control Problems*, Institute for Flight Systems Dynamics, DLR, Oberpfaffenhofen, Germany, Internal Report No. 515-89/22.

12. Pesch, H.J. (1989): *Real-Time Computation of Feedback Controls for Constrained Optimal Control Problems, Part 1: Neighbouring Extremals*, Optimal Control Applications & Methods **10**, 129–145.

13. Pesch, H.J. (1989): *Real-Time Computation of Feedback Controls for Constrained Optimal Control Problems, Part 2: A Correction Method Based on Multiple Shooting*, Optimal Control Applications & Methods **10**, 147–171.

14. Vassiliadis, V.S., Sargent, R.W.H., Pantelides, C.C. (1994): *Solution of a Class of Multistage Dynamic Optimization Problems, Part 2: Problems with Path Constraints*, Ind. Eng. Chem. Res. **33**, **No.9**, 2123-2133.

Rough Stability of Solutions to Nonconvex Optimization Problems

Hoang Xuan Phu[1], Hans Georg Bock[2], and Sabine Pickenhain[3]

[1] Institute of Mathematics, P.O.Box 631 Bo Ho, 10000 Hanoi, Vietnam
[2] IWR, University of Heidelberg, D–69120 Heidelberg
[3] Department of Mathematics, Technical University of Cottbus, D–03044 Cottbus

Summary. The optimal solution set $M(t)$ to some parametric optimization problem

$$\text{minimize } f(t, x) \quad \text{subject to } x \in D(t)$$

is said to be roughly stable w.r.t. the roughness degree $r > 0$ at $\bar{t} \in T$ if for all $\epsilon > 0$ there is a neighborhood $V(\bar{t}) \subset T$ of $\bar{t}$ such that diam $\left(\bigcup_{t \in V(\bar{t})} M(t) \right) < r + \epsilon$. This paper states some sufficient conditions for this kind of generalized stability. One of the most important assumptions is that f is strictly roughly convexlike w.r.t. the roughness degree r. The result is applied to some optimal control problems, in particular, to a shipping problem.

1. Introduction

Let T be a topological vector space with a countable local base and X be a subset of a linear normed space with the norm $\|\cdot\|$. Assume $\emptyset \neq D(t) \subset X$ for all $t \in T$. We consider a family of optimization problems

$$(P_t), \ t \in T : \qquad \text{minimize } f(t, x) \quad \text{subject to } x \in D(t),$$

where $f : T \times X \to \mathbb{R}$. Denote

$$v(t) := \inf_{x \in D(t)} f(t, x) \quad \text{and} \quad M(t) := \{ x \in D(t) : f(t, x) = v(t) \}$$

the optimal value or the optimal solution set, respectively. Suppose that $M(t)$ is nonempty for all $t \in T$. Our aim is to investigate the stability behavior of the optimal solution set $M(t)$.

Actually, this interesting problem was already investigated in many papers, see for instance [1–8].

Some important applications of stability research can be found in the theory of optimal control. The Pontryagin maximum principle states a parametric optimization problem, whose stability just means the continuity of the optimal control. To obtain this continuity one needs the uniqueness of the solution of the maximum condition, which is normally guaranteed by the local or global strict concavity of the Hamilton function w.r.t. the control variables (see for instance [9–11]). If the dynamic equation is affine w.r.t. the control variables, then the performance index is often required to be strictly convex w.r.t. to the control variables.

Of course, the requirement of strict convexity is sometimes too hard for practical problems. So this assumption has to be weakened. Well known generalizations using some growth conditions outside the optimal solution set deliver interesting results which are in some sense similar to the case with isolated local minimizers. But, they are not quite suitable for saying something about the stability behavior of optimal controls. Therefore, in this paper, we choose another approach, where the notion of strictly roughly convexlike functions introduced in [12]. is used to ensure the rough stability of optimal solution set which is defined as follows.

Definition 1.1. For a given $r > 0$, the optimal solution set $M(t)$ to the parametric optimization problem (P_t) $(t \in T)$ is said to be *roughly stable w.r.t. the roughness degree r* (or shortly, r-stable) at $\bar{t}$ if

$$\forall \epsilon > 0 \; \exists \text{ a neighborhood } V(\bar{t}) \text{ with } \operatorname{diam}\left(\cup_{t \in V(\bar{t})} M(t)\right) < r + \epsilon. \quad (1.1)$$

In Section 2 we recall some basic properties of strictly roughly convexlike functions In Section 3, some sufficient conditions for the rough stability of the optimal solution set to parametric optimization problem are stated. This result is applied to optimal control problems in Section 4. In particular, a shipping problem is considered in Section 5.

2. Strictly Roughly Convexlike Functions

Let f be a function which maps a convex subset D of a linear normed space $(X, \|\cdot\|)$ into $\mathbb{R}$ and $r > 0$.

Definition 2.1. f is said to be *strictly roughly convexlike (on D) with respect to the roughness degree r* (or shortly, *strictly r-convexlike*) provided that for all $x_0, x_1 \in D$ satisfying $\|x_0 - x_1\| > r$ there exists a $\lambda \in]0,1[$ such that

$$f(x_\lambda) < (1 - \lambda)f(x_0) + \lambda f(x_1), \quad \text{where } x_\lambda := (1 - \lambda)x_0 + \lambda x_1. \quad (2.1)$$

This notion was introduced in [12]. The reason of our interest in this kind of generalized convexity is the following.

Theorem 2.1. Let f be strictly r-convexlike (w.r.t. the roughness degree r) and let $\arg\min f$ denote the set of global minimizers of f. Then $\operatorname{diam}(\arg\min f) \leq r$. If f is additionally lower semicontinuous, then each r-local minimizer x^* defined by

$$f(x^*) \le f(x) \text{ for all } x \in D \text{ with } \|x - x^*\| \le r$$

is a global minimizer of f.

The first property mentioned above is corresponding to the uniqueness of the minimizer of a strictly convex function, which takes an important part for the stability of solutions to parametric optimization problems. The proof of this theorem is quite similar to that of Propositions 2.2–2.3 in [12]

Other properties of this kind of generalized convexity can be found in [12]. For our use in this paper, we only mention one sufficient condition for strictly roughly convexlike functions.

Theorem 2.2. Suppose that f is absolutely continuous on every segment $[x_0, x_1]$ of a convex set $D \subseteq \mathbb{R}^n$ and that

$$\langle \nabla f(y + (t+r)s) - \nabla f(y + ts), s \rangle > 0$$

holds for all $y \in D$, $\|s\| = 1$, and for almost all $t \in \{t \in \mathbb{R} : y + ts \in D, y + (t+r)s \in D\}$. Then f is strictly r-convexlike.

In fact, this theorem was proved in [12] as a sufficient condition for so-called strictly γ-convex functions introduced in [13–14]. But every strictly γ-convex function w.r.t. the roughness degree $r_\gamma > 0$ is strictly r-convexlike w.r.t. any roughness degree $r \ge r_\gamma$ ([12]). Therefore, it is also a sufficient condition for strict r-convexlikeness.

3. Sufficient Conditions for Rough Stability

First, let us mention some terms stated in [15]:

A (set-valued) map $F : T \to 2^X$ is called *open at* $\bar{t} \in T$ if $\{t_k\} \subset T$, $t_k \to \bar{t}$, and $\bar{x} \in F(\bar{t})$ imply the existence of an integer m and a sequence $\{x_k\} \subset X$ such that $x_k \in F(t_k)$ for $k \ge m$ and $x_k \to \bar{x}$.

F is named *closed at* $\bar{t}$ if $\{t_k\} \subset T$, $t_k \to \bar{t}$, $x_k \in F(t_k)$, and $x_k \to \bar{x}$ imply that $\bar{x} \in F(\bar{t})$.

F is said to be *continuous at* $\bar{t}$ if it is both open and closed at $\bar{t}$.

Actually, the terms open and closed are similar to lower and upper semi-continuous which were introduced by Bouligand ([16]), and by Kuratowski ([17]). Compare [15] and [18–19] for more details.

For our purpose, we also need the following notion: F is said to be *uniformly compact near* $\bar{t}$ if there is a neighborhood V of $\bar{t}$ such that the closure of the set $\bigcup_{t \in V} F(t)$ is compact.

Lemma 3.1. Suppose that $f(\bar{t}, .)$ is strictly r-convexlike $(r > 0)$, and that M is closed at $\bar{t} \in T$ and uniformly compact near $\bar{t}$. Then (1.1) holds true.

Proof. Assume the contrary that there exists an $\epsilon > 0$ with

$$\mathrm{diam}\left(\bigcup_{t \in V(\bar{t})} M(t)\right) \geq r + \epsilon$$

for each neighborhood $V(\bar{t})$ of $\bar{t}$. That means that each neighborhood $V(\bar{t})$ contains t' and t'' with

$$\exists x' \in M(t'), \, x'' \in M(t'') : \|x' - x''\| > r + \epsilon/2.$$

Consequently, we can extract four sequences $\{t_i'\}$ and $\{t_i''\}$ in T, and $\{x_i'\}$ and $\{x_i''\}$ in X with

$$t_i' \to \bar{t}, \quad t_i'' \to \bar{t}, \quad x_i' \in M(t_i'), \quad \text{and} \quad x_i'' \in M(t_i'').$$

Since M is uniformly compact near $\bar{T}$, we can assume without loss of generality that $\{x_i'\}$ and $\{x_i''\}$ converge to $\bar{x}'$ or $\bar{x}''$, respectively. Consequently, we have $\|\bar{x}' - \bar{x}''\| \geq r + \epsilon/2$. On the other hand, since the map M is closed at $\bar{t}$, $\bar{x}'$ and $\bar{x}''$ belong to $M(\bar{t})$, which implies by the strict r-convexlikeness of $f(\bar{t}, \cdot)$ that $\|\bar{x}' - \bar{x}''\| \leq r$ (Theorem 2.1), in contradiction with the preceding inequality. Hence, (1.1) must hold true. □

Of course, one cannot directly apply the above lemma, because the properties of M are usually unknown a priori. But, we can use it to prove some useful sufficient conditions for rough stability as follows.

Theorem 3.1. Suppose that the set-valued map D is continuous at $\bar{t}$ and uniformly compact near $\bar{t}$, f is continuous on $\bar{t} \times D(\bar{t})$, and $f(\bar{t}, .)$ is strictly r-convexlike on $D(\bar{t})$. Then (1.1) holds true, i.e., M is r-stable at $\bar{t}$.

Proof. Theorem 8 in [15] shows that M is closed at $\bar{t}$ if D is continuous at $\bar{t}$ and f is continuous on $\bar{t} \times D(\bar{t})$. Since $M(t) \subset D(t)$, the assumption yields that M is uniformly compact near $\bar{t}$. Hence, (1.1) follows from Lemma 3.1. □

In the next theorem, the uniform compactness of D is not assumed a priori.

Theorem 3.2. Suppose that

(a) X is a subset of a finite-dimensional normed space;
(b) D is closed on a neighborhood of $\bar{t}$ and open at $\bar{t}$, and the set $D(t)$ is convex for each t in a neighborhood of $\bar{t}$;
(c) f is continuous on $T \times X$, $f(t, .)$ is quasiconvex for fixed t, and $f(\bar{t}, .)$ is strictly r-convexlike on $D(\bar{t})$;
(d) $M(\bar{t}) \neq \emptyset$.

Then (1.1) holds true, i.e., M is r-stable at $\bar{t}$.

Proof. By Theorem 2.1, the boundedness of $M(\bar{t})$ follows from the strict r-convexlikeness of $f(\bar{t}, \cdot)$. Since $M(\bar{t})$ is closed and $\dim X < +\infty$, $M(\bar{t})$ is compact. Therefore, Corollary 9.1 in [15] implies that M is uniformly compact near $\bar{t}$. Moreover, due to Theorem 8 in [15], M is closed at $\bar{t}$. Hence, the conclusion follows from Lemma 3.1. $\qquad\qquad\square$

4. Rough Stability of Optimal Solutions to Control Problems

In this section, we consider the following optimal control problem

$$
\begin{aligned}
&\int_{t_0}^{t_f} L(t, x(t), u(t))dt \to \inf! \\
&\dot{x}(t) = \psi(t, x(t), u(t)), \quad u(t) \in U \subset \mathbb{R}^m, \\
&g_i(t, x(t)) \le 0 \quad \text{for} \quad t \in [t_0, t_f], \; 1 \le i \le k, \\
&h_0(x(t_0)) = 0, \quad h_f(x(t_f)) = 0.
\end{aligned}
\tag{4.1}
$$

Here, the functions

$$
\begin{aligned}
L : \mathbb{R} \times \mathbb{R}^n \times \mathbb{R}^m \to \mathbb{R}, \quad \psi : \mathbb{R} \times \mathbb{R}^n \times \mathbb{R}^m \to \mathbb{R}^n, \\
g_i : \mathbb{R} \times \mathbb{R}^n \to \mathbb{R}, \quad h_0 : \mathbb{R}^n \to \mathbb{R}^{s_0}, \quad h_f : \mathbb{R}^n \to \mathbb{R}^{s_f}
\end{aligned}
$$

are continuous. $L(t, ., u)$, $\psi(t, ., u)$, g_i, h_0 and h_f are supposed to be continuously differentiable (for all $t \in [t_0, t_f]$, $u \in U$, and $1 \le i \le k$). Each state function x is absolutely continuous and each control function u is bounded and measurable.

Denote

$$
H(t, x, u, p, \lambda_0) := p^T \psi(t, x, u) - \lambda_0 L(t, x, u).
$$

Due to [20-21], we have the following Pontryagin maximum principle.

Theorem 4.1. Let (x^*, u^*) be an optimal solution to (4.1). Then there exist a nonnegative number λ_0, two vectors $l_0 \in \mathbb{R}^{s_0}$ and $l_f \in \mathbb{R}^{s_f}$, a function $p : [t_0, t_f] \to \mathbb{R}^n$, and k nonnegative regular measures μ_i supported on $\{t \in [t_0, t_f] : g_i(t, x^*(t)) = 0\}$ $(1 \le i \le k)$, not all zero, such that

$$
\begin{aligned}
p(t) = \;& h_0'(x^*(t_0))l_0 - \int_{t_0}^{t} H_x(\tau, x^*(\tau), u^*(\tau), p(\tau), \lambda_0)d\tau \\
&+ \sum_{i=1}^{k} \int_{[t_0, t[} g_{ix}(\tau, x^*(\tau))d\mu_i, \\
p(t_f) = \;& h_f'(x^*(t_f))l_f,
\end{aligned}
\tag{4.2}
$$

and

$$
H(t, x^*(t), u^*(t), p(t), \lambda_0) = \sup_{u \in U} H(t, x^*(t), u, p(t), \lambda_0) \quad \text{a.e. in} \quad [t_0, t_f].
\tag{4.3}
$$

An optimal solution to (4.1) is said to be *normal* if it satisfies the maximum principle with $\lambda > 0$. In this case, one can set $\lambda_0 = 1$.

Let $I(z, x) := \{i : g_i(z, x) = 0\}$. To ensure the continuity of the solution p to the adjoint equation (4.2) at $z \in [t_0, t_f]$ we need the following assumption

$$g_{it}(z, x^*(z)) + \psi^T(z, x^*(z), u) g_{ix}(z, x^*(z)) \neq 0$$
$$\text{for} \quad u \in \partial U, \; i \in I(z, x^*(z)). \tag{4.4}$$

Furthermore we suppose that, for all $u \in \text{int } U$, there exists a $j = j(u) \in \{1, 2, \ldots, m\}$ such that

$$\text{either} \quad \psi_j^T(z, x^*(z), u) g_{ix}(z, x^*(z)) > 0 \; \text{ for all } \; i \in I(z, x^*(z)),$$
$$\text{or} \quad \psi_j^T(z, x^*(z), u) g_{ix}(z, x^*(z)) < 0 \; \text{ for all } \; i \in I(z, x^*(z)). \tag{4.5}$$

In [10], we proved the following

Theorem 4.2. Suppose that U is compact, (x^*, u^*) is an optimal solution to (4.1), and p is the correspondent solution to (4.2). If (4.4)–(4.5) hold at some $z \in]t_0, t_f[$, then p is continuous at z. If (4.4)–(4.5) hold for all $z \in]t_0, t_f[$, then p is continuous.

This result is now used for showing the rough stability of the optimal control u^*.

Theorem 4.3. Suppose that

(a) U is compact and convex,
(b) $L(t, x, .)$ is strictly r-convexlike and $\psi(t, x, .)$ is affine for all t and x,
(c) (4.4)–(4.5) holds true for all $z \in]t_0, t_f[$.

If (x^*, u^*) is a normal optimal solution to (4.1), then for all $z \in [t_0, t_f]$ and for all $\epsilon > 0$ there exists a $\delta > 0$ such that

$$\|u^*(t') - u^*(t'')\| < r + \epsilon \; \text{ whenever } \; |z - t'| < \delta, \; \text{ and } \; |z - t''| < \delta. \tag{4.6}$$

Proof. Theorem 4.2 implies the continuity of p on $[t_0, t_f]$. Therefore, (4.3) can be regarded to hold everywhere. Let

$$f(t, u) := -H(t, x^*(t), u, p(t), 1).$$

Then f is continuous on $z \times U$ and $f(z, .)$ is strictly r-convexlike on U for each $z \in [t_0, t_f]$. Since U is compact and independent of t, Theorem 3.1 yields that

$$M(t) := \arg\min_{u \in U} f(t, u)$$

is r-stable at each $z \in [t_0, t_f]$. Due to (4.3), $u^*(t) \in M(t)$. Hence, according to Definition 1.1, for each $z \in [t_0, t_f]$ and each $\epsilon > 0$ there exists a $\delta > 0$ such that (4.6) holds. $\qquad \square$

5. Application to a Shipping Problem

5.1 Mathematical Model

A transportation company has to carry goods from the warehouse of a factory to other places. Let d be the output (per time unit) of the factory, and u be the transportation rate. Then the change of the storage stock x is described by the differential equation

$$\dot{x}(t) = d(t) - u(t).$$

The warehouse capacity and the maximal transportation rate are given by α and β, respectively. With x_0 and x_f as initial and end states, we have the following constrained dynamic system.

$$\dot{x}(t) = d(t) - u(t), \ 0 \leq u(t) \leq \beta,$$
$$0 \leq x(t) \leq \alpha, \ x(t_0) = x_0, \ x(t_f) = x_f . \tag{5.1}$$

Suppose

$$\beta \in \mathbb{N}, \ 0 < d(t) < \beta \quad \text{for all} \quad t \in [t_0, t_f] \tag{5.2}$$

and

$$x_0 + \int_{t_0}^{t_f} d(t)\, dt > \alpha, \quad \text{and} \quad x_0 + \int_{t_0}^{t_f} d(t)\, dt < \beta(t_f - t_0). \tag{5.3}$$

The first condition implies that one cannot keep the stock at some constant level by extremal control values 0 or β, while the second one excludes the trivial cases $u \equiv 0$ and $u \equiv \beta$. Our aim is to minimize the total cost which consists of the storage cost $L_1(t, x)$ and the transportation cost $L_2(t, u)$, i.e.

$$\int_{t_0}^{t_f} \big(L_1(t, x(t)) + L_2(t, u(t)) \big)\, dt \rightarrow \min! \tag{5.4}$$

Assume that L_1 is continuous and $L_1(t, .)$ is continuously differentiable.

Let us state the transportation expenses. Assume the company only uses trucks of the same kind, each of which can carry maximal one unit of goods. The expense of one truck in the ideal situation depends on its load w and is given by some continuously differentiable function C defined on the interval $]0, 1]$ with

$$C(+0) = \lim_{w \downarrow 0} C(w) > 0, \ C'(w) > 0 \ \text{for} \ 0 < w < 1. \tag{5.5}$$

Let $R(u)$ denote the round off of u, i.e.

$$R(u) := \min\{z \in \mathbb{Z} : u \leq z\}.$$

To carry a good's amount of u one has to use $R(u)$ trucks. For the sake of simplicity, almost all trucks are loaded maximal, and the last one carries the remainder. In such a way, the total expense is $(R(u) - 1)C(1) + C(u - R(u) + 1)$.

In consequence of waiting time and traffic density, the speed is decreased, which causes an increasing factor F of cost. It is fair to assume that F depends on the number $R(u)$ of the trucks being in use and

$$1 = F(0) < F(1), \quad 2F(i) \leq F(i-1) + F(i+1) \quad \text{for} \ i \in \mathbb{N}. \tag{5.6}$$

This condition implies immediately

$$0 < F(1) - F(0) \leq \cdots \leq F(i) - F(i-1) \leq F(i+1) - F(i) \leq \cdots$$

and

$$1 = F(0) < F(1) < \cdots < F(i-1) < F(i) < F(i+1) < \cdots$$

Let a continuous function $E(t) \geq 1$ describe the time depending factor of cost, then the total transportation cost at each time unit is given by

$$L_2(t, u) = E(t)Q(u), \tag{5.7}$$

where

$$Q(u) := F(R(u)) \Big((R(u) - 1) \, C(1) + C(u + 1 - R(u)) \Big) \tag{5.8}$$

is the pure transportation cost.

Let (x^*, u^*) be an optimal solution to (5.1)–(5.4). Due to Theorem 4.1, there exist three real numbers $\lambda_0 \geq 0$, l_0, and l_f, a function $p : [t_0, t_f] \to \mathbb{R}$, and two nonnegative regular measures μ_1 and μ_2 supported on $\{t \in [t_0, t_f] : x^*(t) = 0\}$ or on $\{t \in [t_0, t_f] : x^*(t) = \alpha\}$, not all zero, such that

$$p(t) = l_0 + \lambda_0 \int_{t_0}^{t} L_{1x}(\tau, x^*(\tau)) d\tau + \mu_2[t_0, t[\, -\mu_1[t_0, t[\, , \atop p(t_f) = l_f, \tag{5.9}$$

and

$$p(t) \, u^*(t) + \lambda_0 L_2(t, u^*(t)) = \min_{u \in U} (p(t) \, u + \lambda_0 L_2(t, u)) \quad \text{a.e. in} \ [t_0, t_f]. \tag{5.10}$$

Lemma 5.1. Each optimal solution to Problem (5.1)–(5.4) is normal, i.e., it satisfies the Pontryagin maximum principle stated in (5.9)–(5.10) for $\lambda_0 = 1$.

Proof. Assume the contrary that some optimal solution (x^*, u^*) to (5.1)–(5.4) satisfies (5.9)–(5.10) for $\lambda_0 = 0$. Then

$$p(t) = l_0 + \mu_2[t_0, t[\, -\mu_1[t_0, t[\, , \ p(t_f) = l_f, \atop u^*(t) = \begin{cases} 0 & \text{a.e. in} \ \{t : p(t) < 0\} \\ \beta & \text{a.e. in} \ \{t : p(t) > 0\}. \end{cases}$$

If $p(t_0 + 0) > 0$, (5.1)–(5.3) implies that $u^*(t) = \beta > 0$ and x^* is strictly increasing in the following time period, until $z < t_f$ where x^* reaches the level α ($x^*(z) = \alpha$). Since $\mu_1 = 0$ when $x^* > 0$ and $\mu_2 \geq 0$, we have $p(t) > 0$

and $u^*(t) = \beta$ in a neighborhood on the right of z. Clearly, $\dot{x}^*(t) > 0$ in this neighborhood, i.e., the constraint $x(t) \leq \alpha$ is violated.

Analogously, the case $p(t_0 + 0) < 0$ is impossible. Hence, $p(t_0 + 0) = 0$ and $l_0 = 0$. That means that $p(t) = 0$ until some point $z < t_f$ where μ_1 or μ_2 attains some positive value. This z must exist, otherwise $\mu_1 \equiv 0$, $\mu_2 \equiv 0$, $p \equiv 0$, and $l_f = 0$, i.e., all multipliers are zero, which contradicts the Pontryagin maximum principle. But, similarly as above, one can show that the state constraint $0 \leq x(t) \leq \alpha$ is violated after such a point z. Hence, $\lambda_0 > 0$, and we can set $\lambda_0 = 1$. □

Note that Lemma 5.1 holds true for an arbitrary function L_2.

5.2 An Auxiliary Problem with Continuous Data

We will use Theorem 4.3 for pointing out the rough stability of optimal controls to Problem (5.1)–(5.4). Observe that all functions in (4.1) must be continuous. But the function L_2 defined by (5.7)–(5.8) is not continuous in u. To overcome this hindrance, we now introduce an auxiliary function $\tilde{Q}$ for the pure transportation cost function Q as follows. Since C is continuous on $]0,1]$ and $\lim_{w\downarrow 0} C(w)/w = +\infty$ (by (5.5)) and $F(1) > 1$ (by (5.6)), there is an $\omega \in]0, 1[$ such that

$$\omega < \min\left\{0.5, \frac{C(\omega)}{F(1)\, C(1)}\right\}, \quad \frac{C(w)}{w} > C(1) \quad \text{if} \quad 0 \leq w \leq \omega. \tag{5.11}$$

Define now

$$
\begin{aligned}
&\tilde{Q}(u) := \tilde{F}(u)\, \tilde{W}(u), \quad \text{where} \\
&\tilde{F}(u) := \begin{cases} \left(1 - \frac{u-i}{\omega}\right) F(i) + \frac{u-i}{\omega} F(i+1) & \text{if } i \leq u < i+\omega, \\ F(i+1) & \text{if } i+\omega \leq u < i+1, \end{cases} \\
&\tilde{W}(u) := \begin{cases} iC(1) + \frac{u-i}{\omega} C(\omega) & \text{if } i \leq u < i+\omega, \\ iC(1) + C(u-i) & \text{if } i+\omega \leq u < i+1, \end{cases} \\
&i = 0, 1, \ldots, \beta - 1.
\end{aligned} \tag{5.12}
$$

Lemma 5.2. The function $\tilde{Q}$ is absolutely continuous on $[0, \beta]$.

Proof. By definition, $\tilde{F}$ and $\tilde{W}$ are already continuous on $[i, i+\omega[$ and on $[i+\omega, i+1[$ for all $i = 0, 1, \ldots, \beta - 1$. In addition,

$$
\begin{aligned}
\tilde{F}(i+\omega - 0) &= F(i+1) = \tilde{F}(i+\omega), \\
\tilde{F}(i+1-0) &= F(i+1) = \tilde{F}(i+1),
\end{aligned}
$$

and

$$
\begin{aligned}
\tilde{W}(i+\omega - 0) &= iC(1) + C(\omega) = \tilde{W}(i+\omega), \\
\tilde{W}(i+1-0) &= (i+1)C(1) = \tilde{W}(i+1).
\end{aligned}
$$

Hence, $\tilde{F}$, $\tilde{W}$, and $\tilde{Q}$ are continuous on $[0, \beta]$. Moreover, $\tilde{Q}$ is absolutely continuous on $[0, \beta]$ because $\tilde{F}$ and $\tilde{W}$ are piecewise affine. □

For brevity, let us call the optimal control problem given by (5.1)–(5.6) with

$$L_2(t, u) = E(t)\, Q(u) \quad \text{or} \quad L_2(t, u) = E(t)\, \tilde{Q}(u)$$

(according to (5.8), or to (5.12)) by (P) or $(\tilde{P})$, respectively.

Lemma 5.3. Both Problems (P) and $(\tilde{P})$ have the same optimal solutions.

Proof. a) Consider an arbitrary integer i with $0 \le i \le \beta - 1$. For any $u \in \,]i, i+w[\,$, (5.6) and (5.11) imply $F(i+1) > F(i)$ and $C(u-i) > (u-i)C(1)$, which yields

$$
\begin{aligned}
Q(u) \; &= \; F(i+1)\big(iC(1) + C(u-i)\big) \\
&> \; \Big((i+1-u)F(i) + (u-i)F(i+1)\Big)iC(1) + F(i+1)(u-i)C(1) \\
&= \; (i+1-u)F(i)iC(1) + (u-i)F(i+1)(i+1)C(1),
\end{aligned}
$$

i.e.,

$$Q(u) > (i+1-u)Q(i) + (u-i)Q(i+1). \tag{5.13}$$

Therefore, due to (5.7), (5.10), and Lemma 5.1, the optimal control to (P) cannot attain values in $\,]i, i+w[\,$. That means that its values belong to

$$U' = \bigcup_{i=0}^{\beta-1} [i+w, i+1] \cup \{0\}.$$

b) (5.6) and (5.12) imply for $u \in \,]i, i+w[$

$$
\tilde{Q}'(u) = \\
\tfrac{1}{w}\Big(\big(F(i+1) - F(i)\big)iC(1) + C(w)\big(2\tfrac{u-i}{w}F(i+1) + \big(1 - 2\tfrac{u-i}{w}\big)F(i)\big)\Big)
$$

and

$$\tilde{Q}''(u) = \frac{2C(w)}{w^2}\big(F(i+1) - F(i)\big) > 0. \tag{5.14}$$

Consequently,

$$\tilde{Q}'(i+0) = \frac{1}{w}\Big(\big(F(i+1) - F(i)\big)iC(1) + C(w)F(i)\Big).$$

If $i = 0$, then it follows from $F(0) = 1$ and $C(w)/w > F(1)C(1)$ that

$$\tilde{Q}'(+0) = \frac{C(w)}{w} > F(1)C(1) = \tilde{Q}(1) - \tilde{Q}(0).$$

If $i \ge 1$, (5.11) implies

$$\frac{C(w)}{w} > C(1) \quad \text{and} \quad w < 0.5 \le \frac{i}{i+1}$$

and therefore

$$\tilde{Q}'(i+0) - (\tilde{Q}(i+1) - \tilde{Q}(i))$$
$$= \tfrac{1}{\omega}\big((F(i+1) - F(i))iC(1) + C(\omega)F(i)\big) - \big(F(i+1)(i+1) - F(i)i\big)C(1)$$
$$> \Big(\tfrac{i}{\omega}(F(i+1) - F(i)) + F(i) - F(i+1)(i+1) + F(i)i\Big)C(1)$$
$$= \tfrac{i-(i+1)\omega}{\omega}(F(i+1) - F(i))$$
$$> 0.$$

In all cases, we have $\tilde{Q}'(i+0) > \tilde{Q}(i+1) - \tilde{Q}(i)$, which together with (5.14) yields
$$\tilde{Q}'(u) > \tilde{Q}(i+1) - \tilde{Q}(i) \quad \text{for} \quad u \in \,]i, i+\omega[.$$

By integration we obtain
$$\tilde{Q}(u) - \tilde{Q}(i) > (\tilde{Q}(i+1) - \tilde{Q}(i))(u - i),$$

i.e.,

$$\tilde{Q}(u) > (i+1-u)\tilde{Q}(i) + (u-i)\tilde{Q}(i+1) \quad \text{for} \quad u \in \,]i, i+\omega[.$$

Hence, (5.10) and Lemma 5.1 imply that the optimal control to $(\tilde{P})$ only attains values in U', too.

c) Since $Q(u) = \tilde{Q}(u)$ for $u \in U'$, a) and b) imply that both Problems (P) and $(\tilde{P})$ must have the same optimal solutions. $\qquad\square$

This lemma explains why we can consider Problem $(\tilde{P})$ having continuous data instead of (P).

5.3 Rough Stability of Optimal Transportation Amount

In order to use Theorem 4.3 to prove the rough stability of the optimal control, we need the following.

Lemma 5.4. The function $\tilde{Q}$ is strictly r-convex w.r.t. $r = 1$.

Proof. In what follows, i is an arbitrary integer satisfying $0 \le i \le \beta - 2$. (5.5) and (5.12) imply
$$\tilde{W}(u+1) = \tilde{W}(u) + C(1) > \tilde{W}(u) \quad \text{for} \quad 0 \le u \le \beta - 1$$

and

$$\tilde{W}'(u+1) = \tilde{W}'(u) = \begin{cases} C(\omega)/\omega > 0 & \text{for } i < u < i+\omega \\ C'(u-i) > 0 & \text{for } i+\omega < u < i+1. \end{cases}$$

Due to (5.6) and (5.12), we have
$$\tilde{F}(u+1) - \tilde{F}(u)$$
$$= \begin{cases} F(i+1) - F(i) + \tfrac{u-i}{\omega}(F(i+2) - 2F(i+1) + F(i)) > 0, & i < u < i+\omega \\ F(i+2) - F(i+1) > 0, & i+\omega < u < i+1, \end{cases}$$

and

$$\tilde{F}'(u) = \frac{1}{\omega}\left(F(i+1) - F(i)\right) > 0,$$
$$\tilde{F}'(u+1) - \tilde{F}'(u) = \frac{1}{\omega}\left(F(i+2) - 2F(i+1) + F(i)\right) \geq 0, \ i < u < i+\omega,$$
$$\tilde{F}'(u) = \tilde{F}'(u+1) - \tilde{F}'(u) = 0 \ \text{if} \ i+\omega < u < i+1.$$

Using these results, we can deduce from (5.12)

$$\tilde{Q}'(u+1) - \tilde{Q}'(u)$$
$$= \tilde{F}(u+1)\tilde{W}'(u+1) - \tilde{F}(u)\tilde{W}'(u) + \tilde{F}'(u+1)\tilde{W}(u+1) - \tilde{F}'(u)\tilde{W}(u)$$
$$> \left(\tilde{F}(u+1) - \tilde{F}(u)\right)\tilde{W}'(u) + \left(\tilde{F}'(u+1) - \tilde{F}'(u)\right)\tilde{W}(u+1)$$
$$> 0 \quad \text{for} \quad u \neq j, \ u \neq j+\omega, \ j = 0, 1, \ldots, \beta - 1.$$

Consequently, Theorem 2.2 implies that $\tilde{Q}$ is strictly r-convex w.r.t. $r = 1$. □

Let us come to the main result of this section.

Theorem 5.1. Let (x^*, u^*) be an optimal solution to Problem (P). Then, for all $z \in [t_0, t_f]$ and $\epsilon > 0$, there exists $\delta > 0$ such that

$$|u^*(t') - u^*(t'')| < 1 + \epsilon \quad \text{whenever} \quad |t' - z| < \delta \ \text{and} \ |t'' - z| < \delta. \quad (5.15)$$

Proof. To obtain this claim, we apply Theorem 4.3 to Problem $(\tilde{P})$. Let

$$L(t, x, u) = L_1(t, x) + E(t)\tilde{Q}(u) \quad \text{and} \quad \psi(t, x, u) = d(t) - u.$$

Lemma 5.2 and Lemma 5.4 imply that L is continuous and $L(t, x, .)$ is strictly r-convexlike w.r.t. $r = 1$. For all $z \in [t_0, t_f]$, (4.4)–(4.5) follow from (5.2). By Lemma 5.1, every optimal solution to $(\tilde{P})$ is normal. Hence, if (x^*, u^*) is optimal to $(\tilde{P})$, Theorem 4.3 yields that for all $z \in [t_0, t_f]$ and $\epsilon > 0$ there exists a $\delta > 0$ such that (5.15) holds. By Lemma 5.3, optimal solutions to (P) also have this stability property. □

Note that for showing the continuity of the adjoint function p we do not need the continuity of d. Actually, L_1 need not be continuous w.r.t. t either.

(5.15) implies that

$$|R(u^*(t')) - R(u^*(t''))| \leq 2.$$

Therefore, Theorem 5.1 yields that, near any point of time, the optimal number of the trucks to be used changes at most by two vehicles. With an additional assumption, we are able to obtain the following stronger result.

Theorem 5.2. Let (x^*, u^*) be an optimal solution to Problem (P). Assume

$$\frac{C(w)}{w} > C(1) \quad \text{for all} \quad w \in \,]0, 1[. \quad (5.16)$$

Then u^* only attains integer values, and for all $z \in [t_0, t_f]$, there exists $\delta > 0$ such that

$$|u^*(t') - u^*(t'')| \in \{0, 1\} \quad \text{whenever} \quad |t' - z| < \delta \text{ and } |t'' - z| < \delta.$$

Proof. (5.16) implies hat (5.13) holds for all $u \in]i, i + 1[$. Therefore, an optimal control to (P) only attains integer values. Hence, by choosing $\epsilon = 0.5$, (5.15) implies our claim. □

(5.16) holds, for instance, if C is affine and $C(+0) > 0$. Theorem 5.2 says that, for the optimal solution to the shipping problem (P), all vehicles are always fully loaded and, at any point of time, the number of the vehicles to be used changes at most by one. In this sense, the optimal control u^* is roughly stable.

6. Concluding Remarks

This paper should present an example for stability investigation of solutions to nonconvex optimization problems. Some assumptions can actually be weakened. For instance, Section 5 shows that some continuity requirements in Sections 3–4 are not essential, because one can sometimes replace a discontinuous function by a continuous auxiliary function without changing the optimal solution.

7. References

[1] Bank, B., Guddat, J., Klatte, D., Kummer, B., Tammer, K. (1982): Non-Linear Parametric Optimization. Akademie-Verlag, Berlin
[2] Brosowski, B. (1976): Zur stetigen Abhängigkeit der Menge der Minimalpunkte bei gewissen Minimierungsaufgaben. Lecture Notes in Mathematics, vol. 556, 63–72, Springer-Verlag, Berlin Heidelberg New York
[3] Dantzig, G. B., Folkman, J., Shapiro, N. (1967): On the Continuity of the Minimum Set of a Continuous Function. Journal of Mathematical Analysis and Applications **17**, 519–548
[4] Evans, J. P., Gould, F. J. (1970): Stability in Nonlinear Programming. Operations Research **18**, 107–118
[5] Greenberg, H. J., Pierskalla, W. P. (1975): Stability Theorems for Infinitely Constrained Mathematical Program. Journal of Optimization Theory and Applications **16**, 409–428
[6] Klatte, D. (1994): On Quantitative Stability for Non-Isolated Minima. Control and Cybernetics, **24**, 183–200
[7] Kummer, B. (1977): Global Stability of Optimization Problems. Optimization **8**, 367–383

[8] Malanowski, K. (1987): Stability of Solutions to Convex Problems of Optimization. Lecture Notes in Control and Information Sciences, vol. 93, Springer-Verlag, Berlin Heidelberg New York

[9] Robinson, S. M. (1975): Stability Theory for System of Inequalities. SIAM Journal on Numerical Analysis **12**, 754–769

[10] Phu, H.X. (1984): Zur Stetigkeit der Lösung der adjungierten Gleichung bei Aufgaben der optimalen Steuerung mit Zustandsbeschränkungen. Zeitschrift für Analysis und ihre Anwendungen **3**, 527–539

[11] Pickenhain, S., Tammer, K. (1991): Sufficient Conditions for Local Optimality in Multidimensional Control Problems with State Restrictions. Zeitschrift für Analysis und ihre Anwendungen **10**, 397–405

[12] Phu, H. X. (1995): Strictly Roughly Convexlike Functions. Preprint 95-02, IWR, University of Heidelberg

[13] Phu, H. X. (1993): γ-Subdifferential and γ-Convexity of Functions on the Real Line. Applied Mathematics & Optimization **27**, 145–160

[14] Phu, H. X. (1995): γ-Subdifferential and γ-Convexity of Functions on a Normed Space. Journal of Optimization Theory and Applications **85**, 649–676

[15] Hogan, W. W. (1973): Point-to-Set Maps in Mathematical Programming. SIAM Review **15** (3), 591–603

[16] Bouligand, G. (1932): Sur la Semi-Continuité d'Inclusions et Quelques Sujets Connexes. Enseignement Mathématique, **31**, 14–22

[17] Kuratowski, K. (1932): Les Fonctions Semi-Continues dans l'Espace des Ensembles Fermés. Fundamenta Mathematicae, **18**, 148–159

[18] Aubin J.-P., Frankowska H. (1990): Set-Valued Analysis. Birkhäuser, Boston-Basel-Berlin

[19] Berge, C. (1963): Topological Spaces. Oliver & Boyd, Edinburgh and London

[20] Feichtinger, G., Hartl, R. F. (1986): Optimale Kontrolle Ökonomischer Prozesse. Walter de Gruyter, West-Berlin

[21] Ioffe, A. D., Tichomirov, V. M. (1979): Theory of Extremal Problems. North-Holland Publishing Company, Amsterdam New York Oxford

Data Envelopment Analysis as a Tool for Measurement of Eco-Efficiency

Mikulas Luptacik

Vienna University of Economics and Business Administration, Department of Economics, Augasse 2-6, A-1090 Wien, Austria

Abstract. The new concept in the current state of the public discussion on environmental policy is the concept of eco-efficiency, in which not only efficiency with respect to input and output productivity but efficiency from an environmental point of view is taken into account simultaneously. We propose data envelopment analysis as an approach for measurement of eco-efficiency. First, we measure the eco-efficiency in two steps: We estimate the technical efficiency (as the relation of the desirable outputs to the inputs) and the so-called ecological efficiency (as the relation of desirable to the undesirable outputs) separately. Then taking the results of both models as the output variables for the new data envelopment analysis model, the indicator for eco-efficiency can be obtained. Second, we formulate a model by taking into account simultaneously the inputs, the desirable outputs (goods) and the undesirable outputs (bads). The bads are treated as the inputs in the sense that we wish to expand desirable outputs and reduce the undesirable outputs and inputs. For illustration and comparison of both approaches a numerical example is provided.

1. Introduction

The improvement of environmental quality is one of the most important objectives in the framework of economic and social policy nowadays. The new concept in the current state of the public discussion on environmental policy is the concept of eco-efficiency: It is no more claimed that some of the goals of economic policy are in a trade-off relation with the environmental goals, implying that a better level of one goal variable must, cet. par., be "paid for" by lower levels of another. "There is no trade-off between economy and ecology. It must be a common denominator called "eco-efficiency", said the Swiss entrepreneur Stephan Schmidheiny (translated from DER STANDARD, 4 July 1997). The chairman of the board of trustees of the company Landes & Gyr, Heinz Felsner formulated the need for new indicators of the economic performance of the firm and of the national economy in the following way: "We are looking for eco-efficient solutions such that the good and services can be produced with less energy and resources and with less waste and emissions" (translated from DIE PRESSE, 31 December 1997).

How to define and to measure "eco-efficiency" in an operational way in order to provide a decision support for the entrepreneuers and for the economic policy? The main problem in developing of the eco-efficiency indicators is the lack of evaluations like market prices for the waste and emissions (or the undesirable outputs as the by-products of the many production processes).

Charnes – Cooper- Rhodes (1978) proposed a new approach for evaluating the (relative) technical efficiency of decision making units (DMUs) in the case that some of the outputs (or inputs) do not have market prices and they are given only in the physical units. This approach – called Data Envelopment Analysis (DEA) – has become one of the most widely used methods for the efficiency measurement (see Charnes – Cooper – Lewin – Seiford (eds. 1997) and the aim of this paper is to extend it for the purposes of measuring the eco-efficiency.

The paper is organized as follows. In Section 2 the extension of the basic DEA model when some outputs are undesirable will be presented. In Section 3 the proposed methodology is applied to evaluate the eco-efficiency of the Austrian industry. The last Section 4 closes by briefly summarising the results and discussing the topics for a further research.

2. Data envelopment analysis with undesirable outputs

Assume we have n decision making units (denoted by $j = 1, 2, ..., n$) each using m inputs ($i = 1, 2, ..., m$) and producing k desirable (denoted by $r = 1, 2, ..., k$) and p undesirable outputs ($s = 1, 2, ..., p$). Let $x_{ij} \geq 0$ be the i-th input of the j-th DMU, $y_r^g \geq 0$ the r-th desirable output (good) and $y_{sj}^b \geq 0$ the s-th undesirable output (bad) of the j-th DMU. We denote by $X \in R^{m \times n}$, $Y^g \in R^{k \times n}$ and $Y^b \in R^{p \times n}$ the matrices, consisting of nonnegative elements, containing the observed input, desirable and undesirable output measures for the DMUs.

In the original model (without undesirable outputs) proposed by Charnes - Cooper - Rhodes (1978), the efficiency measure of a DMU is defined as a ratio of a weighted sum of (desirable) outputs to a weighted sum of inputs subject to the constraints that corresponding ratios for each DMU is less than or equal to one. The model chooses nonnegative weights for the inputs and outputs for a DMU (whose performance is being evaluated) in a way which is most favourable for it.

Denoting the evaluated DMU by '0', the following optimization problem is to solve:

$$\text{maximize}_{u,v} \quad h_0 = \frac{\sum_r u_r y_{r0}^g}{\sum_i v_i x_{i0}} \tag{1}$$

subject to
$$\frac{\sum_r u_r y_{rj}^g}{\sum_i v_i x_{ij}} \le 1 \quad (j = 1, 2,..., n) \tag{2}$$

$$\begin{aligned} u_r &\ge 0 \quad (r = 1, 2,..., k) \\ v_i &\ge 0 \quad (i = 1, 2,..., m) \end{aligned} \tag{3}$$

In the subsequent paper by Charnes - Cooper - Rhodes (1978) the nonnegativity condition (3) was replaced by the following condition:

$$\begin{aligned} u_r &\ge \varepsilon \quad (r = 1, 2,..., k) \\ v_i &\ge \varepsilon \quad (i = 1, 2,..., m) \\ \varepsilon &> 0 \quad (\text{"Non - Archimedean"}) \text{ constant} \end{aligned} \tag{4}$$

Fortunately, as proposed by Charnes - Cooper (1962), the nonlinear problem (1), (2) and (4) can be transformed to linear programming problem by the following substitution of the variables:

$$\mu_r = t\, u_r, \; v_i = t\, v_i \quad and \quad t = \frac{1}{\sum_i v_i x_{i0}}.$$

The resulting model of the linear programming for the evaluated DMU$_0$ has the form:

$$\text{maximize}_{\mu,v} \quad h_0 = \sum_r \mu_r y_{r0}^g \tag{5}$$

subject to
$$\sum_r \mu_r y_{rj}^g - \sum_i v_i x_{ij} \le 0 \quad (j = 1, 2,..., n) \tag{6}$$

$$\sum_i v_i x_{i0} = 1 \tag{7}$$

$$\begin{aligned} \mu_r &\ge \varepsilon \quad (r = 1, 2,..., k) \tag{8} \\ v_i &\ge \varepsilon \quad (i = 1, 2,..., m) \tag{9} \end{aligned}$$

The corresponding dual model or the envelopment model (the model (5) – (9) is called multiplier model) is:

$$\text{maximize}_{\theta, \lambda, s^g, s} \quad \theta - \varepsilon \left(\sum_r s_r^g + \sum_i s_i^- \right) \tag{10}$$

subject to

$$\sum_j \lambda_j y_{rj}^g - s_r^g = y_{r0}^g \qquad (r = 1, 2, ..., k) \tag{11}$$

$$\theta\, x_{i0} - \sum_j \lambda_j x_{ij} - s_i^- = 0 \qquad (i = 1, 2, ..., m) \tag{12}$$

$$\lambda_j \geq 0 \qquad (j = 1, 2, ..., n) \tag{13}$$

$$s_r^g \geq 0 \qquad (r = 1, 2, ..., k) \tag{14}$$

$$s_i^- \geq 0 \qquad (i = 1, 2, ..., m) \tag{15}$$

where λ denotes the weights on DMUs, s^- is a vector of input slacks and s^g a vector of (desirable) output slacks. A DMU is efficient if and only if the following two conditions are slacks satisfied:

a) $\theta^0 = 1$

b) $s_i^- = s_r^{g^0} = 0$ for all i and r, where subscript "0" denotes the optimal solution of problem $(10) - (15)$.

The (scalar) variable θ_0° gives us the proportion of all inputs of DMU$_0$ must be sufficient – compared with the efficient units – to achieve the given output levels. In other words $1 - \theta_0^0$ gives the necessary proportional reduction of all inputs of the evaluated DMU$_0$ in order to be efficient (input oriented model). The nonzero slacks and the value of $\theta^\circ < 1$ identify the sources and amounts of inefficiency in each input and output of the DMU being evaluated.

The constraint (12) imply that even after the proportional reductions of all inputs, the inputs of the evaluated DMU$_0$ cannot be lower than the inputs $\sum \lambda_j x_{ij}$ $(i = 1, 2, ..., m)$ of the composite unit. Similarly (due to the constraints (11)), the (desirable) outputs of the DMU$_0$ cannot be higher than the (desirable) outputs $\sum \lambda_j y_{rj}^g$ $(r = 1, 2, ..., k)$ of the composite unit. The DMU$_0$ will be efficient when it has proved impossible to construct a composite unit that outperforms DMU$_0$. The positive values of λ_j provide the linear combination of the DMUs on the efficiency frontier faced closest to the DMU$_0$ (the peer group for DMU$_0$). In this way the problem $(10) - (15)$ constructs the piecewise linear envelopment surface.

The problem we are facing now is how to include undesirable outputs into the model in order to get an indicator for the eco-efficiency? Two ways to approach the problem can be considered:

First, we decompose the problem into two parts and measure the eco-efficiency in two steps. Beside the standard DEA model (10) – (15) for measuring a **technical efficiency** we formulate another DEA model for measuring the so-called **ecological efficiency** which is defined as a ratio of a weighted sum of desirable outputs to the weighted sum of undesirable outputs. The model takes the form:

$$\underset{u,d}{\text{maximize}} \quad h_0^E = \frac{\sum_r u_r y_{r0}^g}{\sum_s d_s y_{s0}^b}$$

subject to

$$\frac{\sum_r u_r y_{rj}^g}{\sum_s d_s y_{sj}^b} \leq 1 \quad (j = 1, 2,..., n)$$

$$u_r \geq \varepsilon \quad (r = 1, 2,..., k)$$
$$d_s \geq \varepsilon \quad (s = 1, 2,..., p)$$
$$\varepsilon > 0 \quad (\text{"Non - Archimedean"}) \text{ constant}$$

Again, by substitution of the variables $\mu_r = t\,u_r, \delta_s = t\,d_s, t = \dfrac{1}{\sum_s d_s y_{s0}^b}$ we obtain the following linear programming problem:

$$\underset{\mu,\delta}{\text{maximize}} \quad h_0^E = \sum_r u_r y_{r0}^g \tag{16}$$

subject to $\sum_r \mu_r y_{rj}^g - \sum_s \delta_s y_{sj}^b \leq 0 \quad (j = 1, 2,..., n)$ (17)

$$\sum_s \delta_s y_{s0}^b = 1 \tag{18}$$

$$\mu_r \geq \varepsilon \quad (r = 1, 2,..., k) \tag{19}$$
$$\delta_s \geq \varepsilon \quad (s = 1, 2,..., p) \tag{20}$$

The envelopment model for the ecological efficiency is the following:

$$\underset{\theta^E, \lambda, s^g, s^b}{\text{maximize}} \quad \theta^E - \varepsilon \left(\sum_r s_r^g + \sum_i s_s^b \right) \tag{21}$$

subject to
$$\sum_j \lambda_j y_{rj}^g - s_r^g = y_{r0}^g \qquad (r = 1, 2, ..., k) \tag{22}$$

$$\theta^E y_{s0}^b - \sum_j \lambda_j y_{sj}^b - s_s^b = 0 \qquad (s = 1, 2, ..., p) \tag{23}$$

$$\lambda_j \geq 0 \qquad (j = 1, 2, ..., n) \tag{24}$$

$$s_r^g \geq 0 \qquad (r = 1, 2, ..., k) \tag{25}$$

$$s_s^b \geq 0 \qquad (s = 1, 2, ..., p) \tag{26}$$

The formulation (21) – (26) implies that in order to increase the ecological efficiency, the DMU first should reduce proportionally all undesirable outputs.

In this way we obtain for every DMU an indicator for technical efficiency and one for ecological efficiency. In order to get an indicator for eco-efficiency we formulate a new DEA model with the technical and the ecological efficiency as the output variables and with the inputs equal 1. For simplicity we used for measuring a technical and an ecological efficiency the input oriented model with constant returns to scale, but their approach can be generalized to other DEA-models as well.

Second, we formulate a model, which simultaneously takes into account the (desirable and undesirable) outputs and the inputs. We treat the pollutants as the inputs in the sense that we wish to expand desirable outputs and reduce undesirable outputs and inputs. Most approaches to incorporate undesirable outputs were based on the technology set introduced by Färe – Grosskopf – Lovell – Pasurka (1989), where bads are weakly disposable (see also Tyteca (1996)). The approach by Scheel (1998) assumes strong disposability for both desirable and undesirable outputs using the technology set where the undesirable outputs data are negative in sign. This idea leads to the following model:

$$\underset{u,v,d}{\text{maximize}} \quad h_0 = \frac{\sum_r u_r y_{r0}^g}{\sum_i v_i x_{i0} + \sum_s d_s y_{s0}^b}$$

subject to
$$\frac{\sum_r u_r y_{rj}^g}{\sum_i v_i x_{ij} + \sum_s d_s y_{sj}^b} \leq 1 \qquad (j = 1, 2, ..., n)$$

$$u_r \geq \varepsilon \qquad (r = 1, 2, ..., k)$$
$$v_i \geq \varepsilon \qquad (i = 1, 2, ..., m)$$
$$d_s \geq \varepsilon \qquad (s = 1, 2, ..., p)$$

The transformation yields the following multiplier problem:

$$\underset{\mu, v, \delta}{\text{maximize}} \quad h_0 = \sum_r \mu_r y_{r0}^g \tag{27}$$

subject to
$$\sum_r \mu_r y_{rj}^g - \sum_s \delta_s y_{sj}^b - \sum_i v_i x_{ij} \leq 0 \qquad (j = 1, 2, ..., n) \tag{28}$$

$$\sum_i v_i x_{i0} + \sum_s \delta_s y_{s0}^b = 1 \tag{29}$$

$$\mu_r \geq \varepsilon \qquad (r = 1, 2, ..., k) \tag{30}$$
$$v_i \geq \varepsilon \qquad (i = 1, 2, ..., m) \tag{31}$$
$$\delta_s \geq \varepsilon \qquad (s = 1, 2, ..., p) \tag{32}$$

and the following envelopment model:

$$\underset{\theta, s^g, s^b, s^-}{\text{maximize}} \quad \theta - \varepsilon \left(\sum_r s_r^g + \sum_s s_s^b + \sum_i s_i^- \right) \tag{33}$$

subject to
$$\sum_j \lambda_j y_{rj}^g - s_r^g = y_{r0}^g \qquad (r = 1, 2, ..., k) \tag{34}$$

$$\theta \, y_{s0}^b - \sum_j \lambda_j y_{sj}^b - s_s^b = 0 \qquad (s = 1, 2, ..., p) \tag{35}$$

$$\theta \, x_{i0} - \sum_j \lambda_j x_{ij} - s_i^- = 0 \qquad (i = 1, 2, ..., m) \tag{36}$$

$$s_r^g \geq 0 \qquad (r = 1, 2, ..., k) \tag{37}$$
$$s_s^b \geq 0 \qquad (s = 1, 2, ..., p) \tag{38}$$
$$s_i^- \geq 0 \qquad (i = 1, 2, ..., m) \tag{39}$$

In this model, the DMU reduces simultaneously the inputs and pollutants in order to increase eco-efficiency. Because of the property of DEA models yielding the best possible result for every decision making unit, the eco-efficiency measured by the models (27) – (32) and (33) – (39) cannot be lower than the eco-efficiency obtained by the composition of the technical and ecological efficiency. Moreover, the models (27) – (32) and (33) – (39) provide a deeper insight in the causes of the eco-inefficiency and indicate potential improvement with respect to the particular inputs, outputs and pollutants. For other types of DEA-models measuring an eco-efficiency see Korhonen – Luptacik (1999).

3. Eco-efficiency of Austrian industry

In order to illustrate our approach we use the data for the 1993 of the sixteen OECD countries as described by Table 1. The desirable output is the industry production and the undesirable output emissions of CO_2. As the inputs we consider the labour and the capital stock (for lack of data on capital stock for some countries in OEDC (1997), we used an approximation. These data are denoted by "*").

	Labour (in 1000 employers)	Capital Stock (in billions ATS)	CO2 (in 1000 tons)	Industry- Production (in billions ATS)
Canada	1.647,00	1.867,05	93,60	2.693,85
USA	16.875,00	* 30319,065	694,30	35.481,53
Japan	10.885,00	31.100,01	296,90	30.413,81
Australia	1.009,00	1.039,56	46,60	1.218,09
N. Zealand	233,80	* 220	5,80	247,28
Austria	564,10	* 927,407	11,90	1.020,15
Denmark	500,30	770,21	5,00	616,29
Finland	342,80	926,08	14,50	647,15
Germany	7.203,90	11.802,18	152,70	13.066,45
Greece	317,40	* 431,747	9,20	237,20
Italy	2.801,00	6.438,52	84,60	4.226,74
Netherlands	719,60	* 1493,169	34,70	1.642,49
Norway	245,30	645,67	6,00	464,61
Spain	1.945,50	* 3100,74	45,30	2.363,80
Sweden	587,50	1.100,39	12,00	964,73
UK	4.379,00	7.178,00	81,80	6.570,72

Table 1

Source: OECD (1997) and own estimation.

Solving the models (10) – (15) and (21) – (26) we obtained the measures of the technical and ecological efficiency respectively. The numerical results are given in Table 2. The model for ecological efficiency (16) – (20) or (21) – (26) in our case is a very simple DEA model with only one input (CO_2) and one output (industry production). The only one efficient unit is the industry of Denmark. The Austrian industry is ecological-inefficient, for a given level of industry production the emission of CO_2 should be reduced by approximately 30% in order to be ecological efficient.

	Technical -efficiency	Ecological- efficiency	Eco-efficiency (as a composition of technical and ecological efficiency	Eco-efficiency estimated by (34) - (40)	Eco-efficiency under VRS
Canada	100,00	23,35	100,00	100,00	100,00
USA	100,00	41,46	100,00	100,00	100,00
Japan	100,00	83,11	100,00	100,00	100,00
Australia	81,21	21,21	81,21	84,00	84,92
N. Zealand	77,90	34,59	77,90	91,51	100,00
Austria	90,64	69,55	90,64	99,72	100,00
Denmark	64,11	100,00	100,00	100,00	100,00
Finland	70,11	36,21	70,11	70,11	90,48
Germany	91,10	69,42	91,10	100,00	100,00
Greece	41,63	20,92	41,63	46,45	73,66
Italy	62,10	40,53	62,10	63,06	63,28
Netherlands	99,93	38,40	99,93	99,93	100,00
Norway	71,54	62,82	74,12	74,71	100,00
Spain	62,00	42,33	62,00	67,93	68,07
Sweden	76,29	65,22	77,68	85,14	89,55
UK	75,34	65,17	77,30	87,32	87,34

Table 2

Technical efficient are industries in Canada, USA and Japan; the Austrian industry should reduce – for a given level of industry production – both inputs employment and capital stock by approximately 10% in order to achieve technical efficiency.

None of the countries is – under the assumption of constant returns to scale-technical **and** ecological efficient.

Taking the results of technical and ecological efficiency as output variables for the new DEA model (with input equal 1) we obtain the measure for eco-efficiency. The results are given in column 3 of Table 2 and the eco-efficiency frontier is drawn at Fig. 1.

Eco-efficient are industries in Denmark and in Japan. As can be seen from Fig. 1

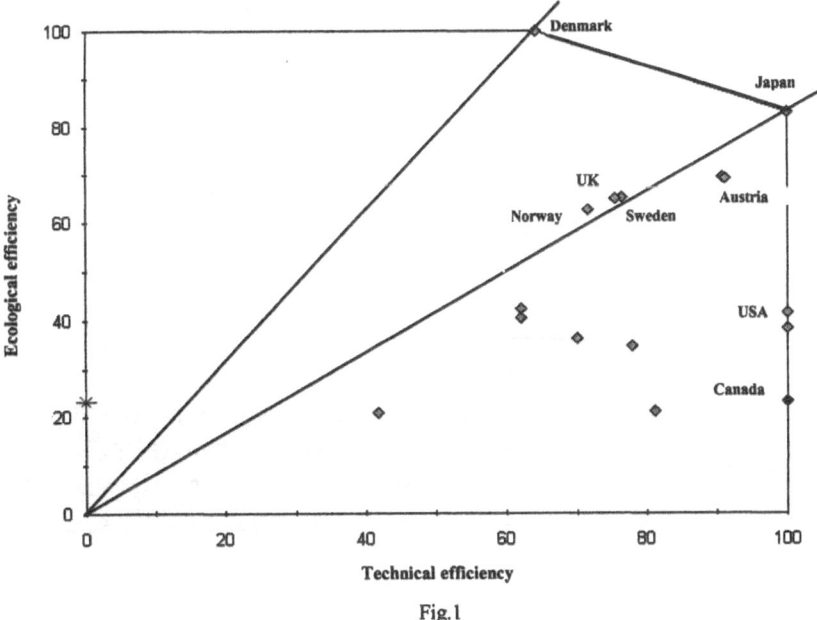

Fig.1

the industries in USA and Canada are weak eco-efficient because of the ecological inefficiency. Furthermore, for all eco-inefficient units lying outside the eco-efficiency cone (e.g. Austria), the indicator of the eco-efficiency is simple the better result from the models (10) – (15) and (21) – (26) respectively. For Norway, UK and Sweden lying inside of the cone the eco-efficiency indicator is higher than the indicators of the technical and ecological efficiency respectively.

Looking the solution of the corresponding multiplier or dual model the importance of technical and ecological efficiency in determining eco-efficiency can be seen. For most of the countries the reason for the eco-inefficiency lies in the ecological-inefficiency, only Denmark, Norway, Sweden and UK have their strength or advantages in the ecological efficiency. The contribution of ecological efficiency in determining the eco-efficiency of industry in Denmark was of 77% (the contribution of technical efficiency of 23%), in Norway of 65% (35%), in Sweden of 64% (36%) and in the United Kingdom of 65% (35%).

In order to get a deeper insight in the causes of the eco-efficiency we solve the model (34) – (40). The input variables are labour, capital stock and CO_2. The only output variable is the industry production. The results under the assumption of constant returns to scale are given in column 4, under the assumption of variable returns to scale in column 5 of Table 2 respectively.

Comparison of the columns 3 and 4 of Table 2 provides in tendency the same results. Because of the property of DEA models yielding the best possible result for every decision making unit, the eco-efficiency measured by model (34) – (40) cannot be lower than the eco-efficiency obtained as a composition of technical and ecological efficiency given in column 3. The eco-efficient countries Denmark and Japan are eco-efficient again and in the previous model weak – eco-efficient industries in Canada and USA and eco-inefficient industry in Germany are eco-efficient with respect to model (34) – (40). The solution of the multiplier problems (27) – (33) indicates the reasons for the eco-efficiency in the particular countries with respect to particular inputs and outputs. The input variable capital was given an importance rating of 53% and the variable CO_2 an importance rating of 47% in determining of the eco-efficiency in Denmark. The reason for the eco-efficiency in Japan, USA and Canada lies in the technical efficiency. The contribution of the input labour in determining the eco-efficiency was of 33% for Japan, of 40% in USA and 51% in Canada. An importance rating of 67% in Japan, of 60% in USA and of 49% in Canada was given to the input capital. In Germany an importance rating of 12% was given to labour, of 79% to capital and of 9% to undesirable output CO_2 in determining its industry eco-efficiency.

The Austrian industry is under the assumption of constant returns to scale eco-ineffient (it is eco-efficient under variable returns to scale). For given level of industry production it should reduce its labour input by 1% in order to be eco-efficient. The most important factor in determining the eco-efficiency is capital with rating of 56% and the undesirable output CO_2 with 44%. The solution of the envelopment model (34) – (40) characterizes the peer group for any eco-inefficient unit. For the Austrian industry is the peer group created by Germany and Japan. The linear combination of Germany (by approximately 98%) and of Japan (by 2%) provides the projection of the Austrian industry on the eco-efficiency frontier.

In similar way the results for other countries can be interpreted. An interesting result can be found for Norway. Comparing with the countries on the eco-efficiency frontier (Japan and Denmark) Norway can reduce the labour input by 26%, the capital input by 25% and CO_2 emission by 25%. The highest importance rating (of 56%) in determining the eco-efficiency is given the undesirable output CO_2.

4. Concluding remarks

In the paper DEA as a tool for measurement of eco-efficiency was presented.

First, we obtained the eco-efficiency as a composition of the technical and ecological efficiency.

Second, we extended the standard DEA model by taking into account simultaneously the inputs, the desirable outputs (goods) and the pollutants or undesirable outputs (bads).

Both approaches lead in tendency to the same results. However, the extended DEA model provides a deeper insight into the causes of the eco-inefficiency and indicates the potential improvements with respect particular inputs and (desirable and undesirable) outputs.

Concerning the application of our approach to the industry in sixteen OECD countries the following remark must be added. Obviously, the eco-efficiency of the industry in the particular countries depends on its structure, especially on the proportion of the steel production on the total industry production. According to the OECD data for 1993 (OECD, 1997) the proportion of the steel production on the industry production in Denmark was 0.013, in Austria 0.065, in Italy 0,072 however in eco-efficient Japan 0,062. Nevertheless, the more interesting analysis should be done for particular industrial branches rather than for the whole industry. The subject of our illustration in Section 3 was determined by the availability of the data.

As a topic for further research we intend to extend our analysis to the intertemporal comparisons. The aim is to decompose the change of eco-efficiency into the change of the technical efficiency and of the ecological efficiency respectively.

References

Charnes, A., Cooper, W. W., Rhodes, E. (1978): Measuring Efficiency of Decision Making Units. European Journal of Operational Research 2, 429 – 444

Charnes, A., Cooper, W. W., Lewin, A.Y., Seiford L. M. (1997): Data Envelopment Analysis Theory, Methodology and Applications. Kluwer

Charnes, A., Cooper, W.W., Rhodes, E. (1979): Short Communication: Measuring Efficiency of Decision Making Units. European Journal of Operational Research 3(4), 339

Charnes, A., Cooper, W. W. (1962): Programming with Linear Fractional Functionals. Naval Research Logistics Quarterly, 9(3/4), 181-185

Färe, R., Grosskopf, S., Lovell, K., Pasurka, C. (1989): Multilateral Productivity Comparisons When Some Outputs Are Undesirable: "A Non-parametric Approach". The Review of Economics and Statistics, 90-98

Korhonen, P., Luptacik, M. (1999): How to Measure Eco-Efficiency: An Application of Data Envelopment Analysis,. Paper presented at the 5th International Conference, Decision Sciences Institute, Athens, 4-7th July, 1999

Scheel, H.: Negative data and undesirable outputs in DEA. Paper presented at the EURO Summer Institute 1998: Data Envelopment Analysis – 20 years on

OECD (1997): Flows and Stocks of Fixed Capital", Edition 1997

OECD (1995): Environmental Data Compendium

Tyteca, D. (1996): On the Measurement of the Environmental Performance of Firms. A Literature Review and a Productive Efficiency Perspective. Journal of Environmental Management 46, 281-308

Evaluating the New Activity-Based Hospital Financing System in Austria

Margit Sommersguter-Reichmann[1], Adolf Stepan[2]

[1] Institute of Industrial Management, University of Graz, Universitätsstrasse 15, A-8010 Graz, Austria
[2] Institute of Industrial Engineering, Ergonomics and Business Economics, Technical University of Vienna, Theresianumgasse 27/II, A-1040 Wien

Abstract. This paper analyses the effects of the new activity-based hospital financing system on hospital performance and hospital costs in Austria. The research concentrates on differences in response among publicly-owned and privately-owned, not for profit hospitals by exploring hospital data from 1994 to 1997 with regard to credit point optimisation, shifts in performance from the inpatient to the outpatient care unit and from the outpatient care unit to physicians with their own practice, and changes in average length of stay. Techniques applied range from index figures to weighting figures and, at least, Data Envelopment efficiency scores.

1 Introduction

The new hospital financing system follows a system which was designed to finance inpatient care on the basis of the well observable figure of inpatient days. Whilst the old financing system set odd incentives (Smith et al., 1997), like maximising the number of inpatient days and length of stay, respectively, thereby increasing inpatient hospital capacity utilisation, the new system (so-called 'Leistungsorientierte KrankenanstaltenFinanzierung', LKF[1]) – based on the worse observable figure of credit points - completely reverses incentive structures. This leads to a minimisation of length of stay and thus inpatient hospital capacity utilisation should be reduced.[2]

The specific issue of interest presented here is a study of hospitals' reaction to changed incentives that mainly result from the combining of an activity-based financing system, which is similar to a system of Diagnoses Related Groups (DRG) with fixed funds. In addition, the research concentrates on the differences in response between publicly-owned (for example owned and run by state or fed-

---

[1] The new financing system is based on the so-called 'Vereinbarung gemäß Art. 15a B-VG über die Reform des Gesundheitswesens und der Krankenanstaltenfinanzierung für die Jahre 1997 bis 2000'.

[2] Unless the intended minimisation is not compensated by increasing the number of patients treated.

eral government) and privately-owned (in this case owned and run by different religious orders), not-for-profit hospitals. Privately-owned, not for profit hospitals (PoNFP hospitals) in Austria are hardly comparable to US-PoNFP hospitals since they are considerably subsidised by the government. There are negotiations between the PoNFP hospital owners and the local government concerning the coverage of total hospital costs. Apart from the negotiations the new financing system now provides a possibility of increasing the governmental share and of decreasing the residue to be covered by the hospital owners. The field of supplementary care is the only sector which is entirely privately financed in Austria. Supplementary care is offered both in publicly-owned and in PoNFP hospitals.

The reactions of hospitals to the new incentive system are analysed mainly with regard to the optimisation of credit points and shifts of performance from the inpatient care unit to the outpatient care unit and from the outpatient care unit to physicians with their own practice (general practitioners, specialists). We do not analyse shifts in funds between 1996 and 1997 since the bodies responsible to distribute funds among hospitals tried to guarantee that hospitals' 1997 funds do not differ significantly compared to their 1996 share by compensating for possible variations. In the first year of introduction hospital revenues which have been determined according to the number of credit points earned in 1997 were adapted to the 1996 share; for example if there were positive deviations between the 1996 and the 1997 share (i.e. the 1997 share increased compared to 1996) the 1997 share was reduced by 50 per cent of the positive deviation. On the contrary if there were negative deviations the 1997 share was raised by 50 per cent of the negative deviation. For this purpose particular variation funds were established in 1997. There has been no compensation for variations in subsequent years.

Techniques applied to perform the study range from index figures and weighting figures to Data Envelopment Analysis (DEA) efficiency measures. To preview particular results several hospitals, especially PoNFP hospitals, obviously registered as many credit points as possible to increase their share of funds. Additionally shifts in performance from inpatient care to outpatient care and from outpatient care to the field of physicians with their own practice were detected.

Accordingly, this paper is organised as follows: Section 2 briefly describes the new activity-based hospital financing system and then formulates hypotheses regarding the expected effects on hospital performance and costs. Section 3 introduces techniques to be employed in assessing the effects of the new financing system and the new incentive structures. Sample and parameters are defined in section 4. In section 5 the results derived from empirical observations are presented. The last section closes by briefly summarising the findings obtained and discussing necessary and possible extensions to the existing analysis.

2 The new activity-based hospital financing system in Austria

2.1 Characteristics of the new financing system

With effect of January 1st 1997, the 'LKF' was implemented in Austria. Following the DRG system, it is based on a system of credit points for 250 groups of diagnoses and 158 groups of surgical and non-surgical treatments to cover the spectrum of inpatient care. Besides, – and this is one of the most important characteristics of the new system – the activity-based hospital financing is combined with funds that consist of fixed and variable payments (see figure 1). So, to some extent funds are fixed.

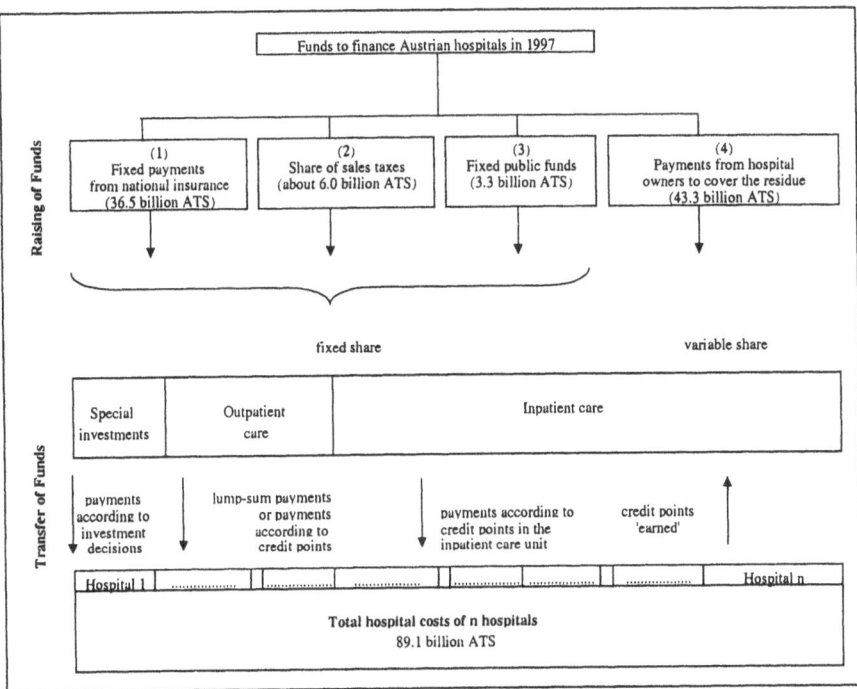

Figure 1: Raising and transfer of funds in 1997

The main share of funds consists of payments from national insurance (1), which correspond to the payments made by national insurance companies to hospitals and to the so-called 'Krankenanstaltenzusammenarbeitsfonds'[3] ('KRAZAF') in 1994, adjusted according to pre-arranged rules of valorisation. Combining the 'LKF' with a fixed payment constraint has been essential for the national insurance companies,

---

[3] One of the main tasks of the governmental body was the financing of hospitals and the stipulating of hospital capacity. In 1997 the 'KRAZAF' was replaced by nine bodies located in each province to allocate funds among hospitals.

because they feared an uncontrollable rise in hospital services. So they exclusively consented to the 'LKF' on condition that there would be fixed payments as far as national insurance payments to funds are concerned.

Furthermore, funds are endowed with a certain share of value-added taxes (VAT) (2) and further public funds (3). The funds (1) to (3) are distributed among the nine Austrian provinces according to pre-arranged rules of distribution.[4] In each Austrian province (x) there is one body, the so-called 'Krankenanstalten-finanzierungsfonds' ('x-KAFF') which is responsible to distribute funds among hospitals. The payments (1) to (3) represent the minimum payments that have to be placed to the disposal of the 'x-KAFF'. In several provinces (Lower Austria, the Tyrol, Vorarlberg), the payments from the owners of the hospital (governmental, private bodies) to cover the residue (4) are made available to the 'x-KAFF', too. (Dienesch et al., 1997). In the case of the hospitals that are analysed in this paper payments from hospital owners are not added to the funds which are distributed by the 'x-KAFF'.

The new process of transferring funds to hospitals can be characterised as follows: Each of the nine Austrian bodies administers its own funds for the financing of hospital care. Since there is no standardised procedure with regard to the transfer of funds, either, we have to distinguish two main possibilities: Firstly, funds may be decomposed into payments to finance inpatient care, outpatient care and special investments. Whilst special investments and outpatient care may be financed according to investment decisions and lump-sum payments, inpatient hospital care always has to be financed through the total number of credit points 'earned' by the hospitals. Secondly, special investments may be financed according to investment decisions and both outpatient care and inpatient care are financed through the total number of credit points registered in the inpatient care unit. In our analysis the first case applies. Consequently, payments to finance outpatient care (determined according to past cost and past revenues of treatment in the outpatient care unit) and special investments (determined according to political decisions) are subtracted from the total amount which is available to the 'x-KAFF'. The remaining sum is divided by the total number of credit points registered by the hospitals. Each hospital receives its share to finance inpatient care in accordance with the number of credit points it has 'earned'. Information from the Austrian Ministry of Health says that, originally, credit points per diagnosis corresponded to average full cost of inpatient treatment (BMAGS, 1998). But due to fixed funds, the more points that are 'earned' the smaller the corresponding value of one credit point, so that credit points apparently no longer correspond to average cost of inpatient treatment. The share may vary year after year, depending on the total amount of points registered for each year. Thus, if any consequences with regard to hospital performance are expected, they are more likely to be due to fixed funds than to the activity-based financing. Since the remaining residue (4) has to be covered by the owners of the

---

[4] See Art. 10 of the 'Vereinbarung gemäß Art. 15a B-VG über die Reform des Gesund-heitswesens und der Krankenanstaltenfinanzierung für die Jahre 1997 bis 2000.'

hospitals they may try to ensure that their hospitals maximise the number of credit points they can 'earn', i.e. to exploit any legal (and illegal?) possibility of increasing the amount of credit points. The Ministry of Health makes available a computer-based scoring program to determine how many credit points are to be assigned to particular diagnoses and particular treatments. This procedure of determining credit points per diagnosis and treatment respectively will be referred to as 'coding' in the course of the paper. Additionally, hospitals may buy privately distributed software packages that determine the *maximum* number of attainable credit points per diagnosis or treatment. Altogether, hospitals are now confronted with the fact that the risk of quantitative changes has been shifted from national insurance to hospitals and hospital owners, respectively.

2.2 Expected effects on hospital performance and hospital costs

The activity-based hospital financing system provides new incentives which may directly influence the performance of hospitals. We would expect the following reactions from hospitals:

When the system is first introduced we can expect that more careful registering will occur. Hospitals may be tempted to (legally?) optimise coding, for example by generously interpreting coding rules. In this case, which is known technically as 'creeping' (Stepan, 1997), changes in points do not correspond to changes in costs. To a great extent, creeping is unproductive and does not improve performance.

Secondly, increased points and increased costs may indicate an increase in the absolute number of patients treated or an increase in the frequency of treatment performed. An increased number of cases may be observed both in the inpatient or the outpatient care unit. A rising number of in-patients may be due to shifts in performance from the outpatient care unit to the inpatient care unit whenever this helps to improve a hospital's position regarding the 'earning' of credit points. On the other hand, less profitable (in terms of achievable credit points) diagnoses, if medically acceptable, may now be treated in the outpatient care unit. Performance shifts from the inpatient to the outpatient care unit are quite desirable if quality of treatment is guaranteed.

Moreover, the new financing system involves a strong incentive to discharge patients as early as possible. The subsequent reduction in average length of stay may be for two reasons: premature discharge and/or reduction of excess resource utilisation. Premature discharge may lead to an increase in treatment carried out by outpatients' departments and physicians with their own practice. However, we have to investigate whether the effects of reduced average length of stay are neutralised by raising the number of patients treated in the inpatient care unit.

Summing up we analyse the incentives of the new activity-based hospital financing system with regard to the optimisation of credit points 'earned', and shifts in performance from inpatient care to outpatient care and from the outpatient care unit to the field of physicians with their own practice. Furthermore, we examine whether average length of stay is reduced and whether there are changes in hospital capacity utilisation and in the number of patients treated in the inpatient and in the outpatient care unit.

3 Methods

To provide a comprehensive examination of possible reactions to the new incentives of the 'LKF', the following three methods will be used to test the assumptions formulated above.

At first, indices of time series of selected parameters[5] are designed to describe the development from 1994 to 1997. To receive a percentage change, that does not depend on each hospital's size, hospital data of each year are divided by their 1994 levels. Consequently, index figures represent the relative (i.e. related to the data level of 1994) change of parameter p_i, $i = 1,..., P$ in year t, t = 1994, 1995, 1996 or 1997, of hospital k, k = 1, ..., K.

By relating parameters to their 1994 level no statements about changes among hospitals can be made. For this reason, index figures of time series are divided by the very index figures of time series of the smallest[6] hospital. This way, we receive weighting figures that show the relative change in hospital performance compared to the smallest hospital. Consequently, weighting figures illustrate the extent of competitive behaviour and the moment of participation in competition among hospitals.

The analysis of a hospital's reaction to new incentives by means of index figures and weighting figures only allows analysis of single parameters and single factor productivities, it does not, however, reveal the effects on total factor productivity. Since analysing the effects of introducing the activity-based hospital financing system ultimately leads to analysing the effects on the relative performance of hospitals, Data Envelopment Analysis (DEA) is used to yield information of the effects on productive performance.

DEA is a non-parametric approach (for introduction to theory and applications see Charnes et al., 1997) which is used to construct an efficient production frontier from observed inputs and outputs by mathematical programming techniques. The inefficiency of each hospital is determined by its distance from the efficient frontier which is constructed from all observations of the sample. Furthermore, DEA allows decomposition of overall efficiency into technical and scale efficiency, i.e. it is possible to consider different returns to scale, thereby enabling us to discover,

---

[5] See page 8f.

[6] The size of hospitals is measured by the size of each parameter analysed, i.e. if the parameter of 'credit points' is going to be analysed, the smallest hospital is the hospital that shows the lowest amount of credit points registered.

in which specific segments of the production possibility set (constant, increasing, decreasing returns to scale) a hospital is performing.[7]

In this paper, hospital performance is analysed on the assumption of both constant returns to scale using the Charnes-Cooper-Rhodes (CCR) model (Charnes et al., 1978) and variable returns to scale using the Banker-Charnes-Cooper (BCC) model (Banker et al., 1984) to be able to distinguish between inefficiencies due to excessive use of scarce resources (technical inefficiency) and inefficiencies due to suboptimal hospital activity levels (scale inefficiency). The radially input-oriented measure used in this analysis has the advantage of being able to compare hospitals with a given input-output mix to other hospitals with a similar mix by means of a neutral total factor productivity measure in a multiple-input multiple-output framework. Consequently, the DEA efficiency measure combines earlier results of index and weighting figures.

4 Data and parameter specification

4.1 Data and sample

In 1994 it was formally announced that, with effect from January 1st 1997, a new hospital financing system would be implemented in Austria. Although being financed on the basis of inpatient days from 1994 to 1996, hospitals were obliged to apply the new credit point system and to report the total amount of credit points to the Ministry of Health during the same time period. This, however, did not have any effects on allocation of funds. The analysis of hospital data from 1994 to 1997 will include two different types of data: on the one hand data obtained from the 'cold phase' from 1994 to 1996 (the period in which hospitals familiarise themselves with the new activity-based financing system), and on the other hand data obtained from the 'hot phase', namely from 1997.

We used data from the bodies responsible for allocating funds to hospitals in each Austrian province, from hospital owners and the Austrian Ministry of Health.

The sample consists of 23 so-called 'Fondskrankenanstalten'. 'Fondskrankenanstalten' are hospitals which are eligible for financing by the new activity-based financing system.[8] The sample comprises 18 publicly-owned and 5 PoNFP hospitals.

It is important to mention that all publicly-owned hospitals are run by the same owner. This may be important with regard to the hospitals' response to the 'LKF'. If we consider that the 18 hospitals have a larger capacity and higher performance we can assume that the public owner has more negotiating strength than the different

---

[7] See the appendix for more details.

[8] See Art. 2 of the 'Vereinbarung gemäß Art. 15a B-VG über die Reform des Gesundheitswesens und der Krankenanstaltenfinanzierung für die Jahre 1997 bis 2000'.

private owners have. Consequently, different reactions of these hospitals compared to PoNFP hospitals are expected.

4.2 Parameter specification

In order to judge the effects of the new activity-based hospital financing system on hospital performance and costs the following cost and performance parameter will be analysed: full cost of hospital performance (COST)[9], 'total number of credit points' (PTS), 'credit points per hospital bed' (PTS/BED), 'credit points per inpatient day' (PTS/INDAY), 'credit points per inpatient case' (PTS/INPAT), 'credit points per unit of cost' (PTS/COST), 'frequency of treatment performed in the outpatient care unit' for every 'outpatient case' (FREQ/OUTPAT), 'frequency of treatment performed to in-patients in the outpatient care unit' (FREQOUT), 'average length of stay' (ALOS), 'hospital capacity utilisation' (CAPU), 'number of outpatient cases' (OUTPAT) and 'number of inpatient cases' (INPAT).

To discover whether the hospitals that are run by the public owner react differently to the introduction of the activity-based financing system the weighting figures 'full hospital cost' (COST) and 'credit points' (PTS) will be scrutinised.

To obtain comprehensive results, the effects of the new financing system on total factor productivity of each hospital will be examined by means of DEA. For that purpose, it is necessary to quantifiy hospital production (output) and utilisation of resources (input). Since the most appropriate measure of hospital output, namely improvement of health status, cannot be measured, it is necessary to include hospital output by the intermediate good of health services. Consequently, the amount of credit points (PTS) 'earned' by each hospital is chosen to indicate inpatient treatment performed.[10] No further differentiating is made since the amount of registered credit points includes both differences in treatment and the existence of additional diagnoses. The total number of outpatient cases treated (OUTPAT) is included to quantify performance of the outpatient care unit. The input vector comprises the full time equivalent of medical (MFTE) and non-medical labour (NMFTE) as well as the number of hospital beds (BED).

In order to judge the effects of the new financing system on total factor productivity hospital efficiency is examined by pooling data from 1994 to 1997 based on the output vector {PTS, OUTPAT} and the input vector {MFTE, NMFTE, BED}. At first the CCR Model is calculated to illustrate overall efficiency of the 92 'different' hospitals (n=23·4=92). Additionally, the BCC Model is run to allow decomposition of overall efficiency into technical and scale efficiency. This may be interesting because we assume that hospital management cannot directly influence

---

[9] Except for the cost of depreciation and the cost of non-productive cost centers.

[10] Besides, the number of credit points is chosen as indicator of hospital output because there are hardly data available to be included as output in the efficiency analysis. Consequently, if creeping occurs, it might be reflected in a seeming increase of total factor productivity.

hospitals activity levels in Austria, as major changes in hospital capacity and hospital treatment performed are the result of political decisions at regional level and have to be in line with the so-called 'Österreichische Krankenanstaltenplan' ('ÖKAP'). The 'ÖKAP' is a plan which stipulates hospital location and hospital capacity. It is published by the National Department of Public Health.

5 Results

5.1 Index figures

Since the quantity of results concerning index figures is relatively large we refrain from presenting each index figure calculated. We rather describe the main results concerning 1997 index figures as well as changes between 1996 and 1997 (see table1).[11]
The analysis of hospital costs and credit points indicates that - in fact - creeping occurred. The highest increase in credit points appeared from 1996 to 1997, whereas there was hardly a change in hospital costs during the same time period. The result is confirmed both by the rapid increase in PTS/BED, PTS/INPAT, and PTS/INDAY from 1996 to 1997. The increase in PTS/INDAY together with shortened ALOS points out careful coding and/or optimisation of coding by changing the strategy of discharge. The question of whether reduction of ALOS was due to premature discharge or to reduction of excess resource utilisation partly remains open. The decrease in ALOS contributed to the decrease in hospital capacity utilisation (CAPU) although the decrease partly was compensated by the increase in the number of in-patients treated (INPAT). The increase in the number of out-patients treated (OUTPAT) may indicate that necessary aftercare has been shifted from the inpatient care unit to the outpatient care unit. Additionally, the financing of the outpatient care unit via lump-sum payments obviously induced PoNFP hospitals to reduce the number of in-patients treated in the outpatient care unit (FREQOUT).
As far as index figures are concerned we rather suppose that the increase in credit points was more likely the result of credit point optimisation than of changes in performance structure. If changes in performance structure are organised correctly with regard to the requirements of the whole Austrian health care system, the changes will improve the performance of the whole system. Specialisation will be supported and positive effects on costs and quality can be expected. Ideally, each hospital will arrive at its own optimal performance program in accordance with regional requirements reconciled with the 'ÖKAP'. However, strategies of this kind take time and incur the cost of reorganisation. Consequently, it is hard to believe that such a reorganisation during the time of introduction, i.e. from 1996 to 1997, has been carried out successfully.

---

[11] Comprehensive results are presented in Sommersguter-Reichmann/Stepan (1999).

| | 1997 Index figure | | Changes between 1996 and 1997 | |
	Publicly-owned	PoNFP	Publicly-owned	PoNFP
PTS	1.19235	1.24631	11.4%	11.0%
COST	1.17269	1.14980	1.0%	0.0%
PTS/COST	1.01677	1.08394	9.5%	10.9%
PTS/BED	1.25157	1.24116	15.0%	15.2%
PTS/INDAY	1.32307	1.29928	18.6%	22.0%
PTS/INPAT	1.14376	1.16282	8.4%	8.8%
FREQ//OUTPAT	0.97689	0.97539	-1.1%	5.5%
FREQOUT	1.01742	0.86587	1.3%	-16.3%
ALOS	0.86447	0.89497	-8.7%	-10.8%
CAPU	0.94596	0.95527	-2.8%	-5.6%
OUTPAT	1.04880	1.15939	-0.5%	-0.3%
INPAT	1.04250	1.07180	2.8%	2.0%

Table 1: 1997 index figures and changes between 1996 and 1997

Moreover, it is obvious yet still a mistery, that – in isolated cases - performance shifts from the inpatient care unit to the outpatient care unit and vice versa occurred. The direction of shifts depends on how profitable it is to treat a patient in the inpatient care unit. In any case, it is important to guarantee quality of treatment. Concerning shifts of performance from the outpatient care unit to physicians (general practitioners, specialists) with their own practice, it will become important to include the extramural field in the process of reorganising the funding of health care services. This has not taken place up to now.

Overall, index figures indicate that PoNFP hospitals rather seemed to be anxious to guarantee and to increase their share of credit points in order to maximise the number of credit points.

5.2 Weighting figures

Weighting figures illustrate competition among hospitals and show quite different findings. The relative increase in credit points ranges from a minimum of 1.2% (publicly-owned hospital) to a maximum of 60.1% (privately-owned hospital) between 1994 and 1997. On the other hand, there was no continuous increase in the parameter of COST. With regard to the forthcoming introduction of the new financing system, PoNFP hospitals[12] have started competition for points earlier than publicly-owned. This may be an indication that PoNFP hospitals were more inclined to guarantee their share of funds by optimising coding (see table 2).

| Hospital | PTS | | | COST | | |
	1995	1996	1997	1995	1996	1997
1	0.99403	0.94205	1.16711	1.05884	1.31320	1.13601
2	1.15358	1.19381	1.37497	1.01553	1.15591	1.14676
3	1.07625	1.14964	1.29149	1.06790	1.18357	1.19901

[12] Marked with a star.

4	1.02552	0.97800	1.09220	1.09798	1.27140	1.33367
5	1.00517	1.01034	1.18894	1.02988	1.14217	1.20373
6	1.06161	1.11024	1.35740	1.08030	1.27237	1.25856
7*	1.11334	1.18484	1.45191	1.09205	1.30429	1.27040
8*	1.02487	1.03621	1.30121	1.01280	1.13735	1.09904
9*	1.02740	1.10928	1.22532	1.00570	1.12900	1.14117
10	1.02132	0.94292	1.24663	1.05997	1.15697	1.16675
11	0.98039	0.91533	1.04015	1.07155	1.15560	1.17214
12	0.97895	0.95524	1.32618	1.04473	1.18232	1.18369
13	1.03784	1.05784	1.23438	1.03781	1.14467	1.15315
14	1.00000	1.00000	1.00000	1.00000	1.00000	1.00000
15	0.97445	0.95535	1.21675	1.05520	1.13224	1.12978
16	1.00648	0.86372	1.39176	0.98184	1.08860	1.19737
17	0.97829	0.92329	1.12738	1.06486	1.13202	1.14338
18*	1.02797	1.00080	1.07260	1.01885	1.18888	1.15766
19	1.02303	0.99211	1.15087	1.04572	1.19691	1.20368
20	1.04579	0.95547	1.01155	1.05271	1.17835	1.16973
21*	1.16720	1.21182	1.60096	1.09957	1.31344	1.30963
22	0.99996	0.96940	1.08772	1.05082	1.13230	1.10691
23	1.06452	1.00834	1.12069	1.09371	1.17246	1.20326

Table 2: Weighting figures

Since PoNFP hospitals cannot fall back on owners that have good negotiating strength with financial backers, they obviously tried to increase their share of funds by increasing the number of credit points. Clarifying the increase in credit points is difficult (creeping, changes in performance structure, etc.). This information can be obtained in discussions with the hospital managers responsible.

5.3 DEA efficiency scores

Analysing changes in input efficiency (see table 3), calculated by pooling hospital data from 1994 to 1997, resulted in interesting findings. With respect to overall efficiency most hospitals (8) were rated efficient in 1997 whereas there were only 5 hospitals rated efficient during the time period 1994 to 1996, i.e. 8 out of 13 best-practicing hospitals are '1997 hospitals'. The '1997 hospitals' obtained an average efficiency score of 90% compared to 83% in 1996. The fact that many hospitals increased the amount of registered credit points from 1996 to 1997 considerably is reflected by the seemingly sharp increase in efficiency from 1996 to 1997 (e.g. hospital 8: 13 percentage points, hospital 10: 13 percentage points, hospital 12: 18 percentage points, hospital 15: 18 percentage points, hospital 16: 33 percentage points, hospital 21: 17 percentage points).[13] Overall, we suppose that the introduction of the activity-based financing system rather produced a shift in technology than an increase in productive efficiency.

---

[13] The increase in credit points influences the efficiency scores since credit points are used to quantify hospital inpatient output.

Hospital	Overall Efficiency				Technical Efficiency				Scale Efficiency			
	1994	1995	1996	1997	1994	1995	1996	1997	1994	1995	1996	1997
1	69	65	62	72	79	76	73	81	87	86	85	89
2	83	89	93	99	84	93	94	100	99	96	99	99
3	88	88	96	100	88	88	97	100	100	100	99	100
4	91	87	85	89	91	87	87	90	100	100	97	99
5	70	68	70	73	74	72	74	76	95	94	95	96
6	98	98	100	100	100	99	100	100	98	99	100	100
7*	85	95	100	100	89	95	100	100	95	100	100	100
8*	89	84	87	100	91	85	88	100	98	99	99	100
9*	91	95	100	100	92	95	100	100	99	100	100	100
10	78	74	71	84	78	74	71	84	100	100	100	100
11	73	73	77	79	75	76	81	83	97	96	95	95
12	81	77	76	94	83	80	79	96	98	96	96	98
13	84	83	87	97	89	88	92	100	94	94	95	97
14	66	68	68	65	99	100	100	100	67	68	68	65
15	76	72	74	92	80	75	78	94	95	96	95	98
16	73	71	67	100	78	76	72	100	94	93	92	100
17	72	68	66	77	73	68	66	79	99	100	100	97
18*	98	100	100	100	100	100	100	100	98	100	100	100
19	91	86	83	87	92	87	84	90	99	99	99	97
20	86	84	78	77	86	84	79	77	100	100	99	100
21*	80	79	82	99	88	85	86	100	91	93	95	99
22	97	97	98	100	98	98	98	100	99	99	100	100
23	81	80	80	81	82	80	80	81	99	100	100	100
Mean	83	82	83	90	86	85	86	93	96	96	96	97
Variance	90	116	155	131	69	94	123	83	50	50	50	55
efficient	0	1	4	8	2	2	5	12	4	9	8	11
inefficient	23	22	19	15	21	21	18	11	19	14	15	12

Note: The analysis comprises 92 'different' hospitals, i.e. hospital 1 (2, 3, ..., 23) is treated as a different hospital in different years of analysis. Consequently only one best-practice frontier is calculated.

Table 3: Efficiency scores in %

If we decompose overall efficiency into technical efficiency and scale efficiency[14] we see that the sharp increase in overall efficiency from 1996 to 1997 was mainly due to the increase in technical efficiency. On average technical efficiency scores were slightly higher than overall efficiency scores. Consequently, average scale efficiency was - except for hospital 14 and hospital 1 – very satisfying. Overall, average scale efficiency did not change very much from 1996 to 1997. This confirms the assumption that hospital management is not able to influence activity levels considerably. Although most hospitals were rated scale efficient in 1997 (11) there is only a slight difference in average scale efficiency between the years 1994 and 1997 (94-95: 0,4%; 95-96: 0,0%; 96-97: 0,9%). If we concentrate on the years 1996 and 1997 only, i.e. if we pool only 1996 and 1997 hospital data, we obtain exactly the same 1996 and 1997 efficiency scores as presented in table 3.

Comparing technical efficiency scores of PoNFP hospitals with efficiency scores of publicly-owned hospitals we see that all PoNFP hospitals were rated technically efficient in 1997. On average hospitals 21 and 8 were rated considerably inefficient in 1994, 1995 and 1996, but the rapid increase in credit points from 1996 to

---

[14] See the appendix.

1997 (hospital 21: 22%; hospital 8: 17%) resulted in 100% technical efficiency in 1997.

Among publicly-owned hospitals 7 out of 18 hospitals were rated technically efficient in 1997. But during the time period 1994 to 1996 only 2 hospitals were technically efficient twice, each. As far as the worse efficiency scores of publicly-owned hospitals are concerned it has to be taken into consideration that these hospitals are so-called 'public' hospitals[15] whereas this applies only to one PoNFP hospital. This fact may influence hospital efficiency since public hospitals have no possibility of refusing treatment to insured patients. By way of contrast so-called 'private' hospitals have the possibility of specialising in some fields of inpatient and outpatient care, thereby exploiting the advantages of specialisation.

6 Summary

Summing up, the results of this paper show that there are differences in response to the 'LKF' among hospitals. In most cases the increase in points does not correspond to the increase in cost of treatment. Moreover hospital strategies of reacting to reversed incentives in terms of participation in competition for points strongly differ. Index figures and efficiency scores confirm that optimisation of coding occurs. Additionally shifts of performance between the inpatient and the outpatient care unit occur, depending on the profitability (in terms of receivable credit points) of treating a patient with a particular diagnosis in the inpatient care unit.

However, further research has to be done. Firstly, including all Austrian 'Fondskrankenanstalten' in the analysis would be of considerable interest. Moreover, it would be interesting how hospitals will be rated in 1998 and 1999. Additionally, if data were available further differentiation of the total amount of credit points should be made to be able to weigh the case mix of the hospitals.

Altogether, this research has to be interpreted as a starting point for further discussion in Austria. The interpretation of results in terms of economics does not allow a comprehensive assessment of real occurrences. Cooperation among hospital managers and economists is of utmost importance. This requires the understanding on the part of hospital managers and physicians. Discussions concerning findings are imperative to realise 'what we are really finding' (Chilingerian, 1995).

Appendix

Input-oriented CCR model (overall efficiency):

$$\min_{\theta, \lambda} \quad \theta$$

---

[15] See article 15 B-KAG.

s.t.

$$\theta \, x_{ik} \geq \sum_{j=1}^{n} \lambda_j x_{ij} \quad i = 1,\ldots,m$$

$$\sum_{j=1}^{n} \lambda_j y_{rj} \geq y_{rk} \quad r = 1,\ldots,s$$

$$\lambda_j \geq 0, \quad \forall i, j, r; \quad \theta \ldots free$$

The input-oriented BCC model (technical efficiency) includes as additional constraint

$$\sum_{j=1}^{n} \lambda_j = 1 \quad .$$

Scale efficiency is calculated by $\theta_{CCR} / \theta_{BCC}$.

References

Banker, R.D., Charnes, A., Cooper, W.W. (1984): Some models for estimating technical and scale efficiencies in Data Envelopment Analysis, Management Science 30, 1078-1093

Banker, R.D. (1984): Estimating most productive scale size using Data Envelopment Analysis, European Journal of Operational Research 17, 35-44

Bundesministerium für Arbeit, Gesundheit und Soziales (BMAGS) (1998): Leistungsorientierte Krankenanstaltenfinanzierung – LKF – Systembeschreibung, Wien

Charnes, A., Cooper, W.W., Lewin, Y.A., Seiford, L.M. (1997): Data Envelopment Analysis. Theory, Methodology and Applications, 2nd Edition, Kluwer Academic Publishers, Boston/Dordrecht/London

Charnes, A., Cooper, W.W., Rhodes, E. (1978): Measuring the efficiency of decision making units, European Journal of Operational Research 2, 429-444

Chilingerian, J.A. (1995): Evaluating physician efficiency in hospitals: A multivariate analysis of best practices, European Journal of Operational Research 80, 548-574

Dienesch, S., Heitzenberger, G. (1997): Krankenanstaltenfinanzierung 9 mal anders. Ein Streifzug durch den Finanzierungsdschungel in den österreichischen Bundesländern, Österreichische Krankenhaus-Zeitung 38, 5-8

Kodex des österreichischen Rechts, Krankenanstaltengesetze (1998): Krankenanstaltengesetze des Bundes (B-KAG), Orac Verlag, Wien

Smith, P.C., Stepan A., Valdmanis V., Verheyen P. (1997): Principal-agent problems in health care systems: an international perspective, Health Policy 41, 37-60

Sommersguter-Reichmann, M., Stepan, A. (1999): Ein hierarchisches Informationssystem zur Analyse von Anreizwirkungen in Spitälern. Aufbau und praktische Anwendung am Beispiel der geänderten Spitalsfinanzierung in Österreich, Working Paper, forthcoming in Zeitschrift für Betriebswirtschaft

Stepan A., Sommersguter-Reichmann M. (1998): Gutachten zur Überprüfung der Auswirkungen des LKF-Abrechnungssystems auf die Kostenentwicklung der Steiermärkischen Fondskrankenanstalten. Unpublished research report. Wien/Graz

Stepan, A. (Ed.) (1997): Finanzierungssysteme im Gesundheitswesen. Ein internationaler Vergleich, Wien

Kodex des österreichischen Rechts, Krankenanstaltengesetze (1998): Vereinbarung gemäß Art. 15a Bundes-Verfassungsgesetz (B-VG) über die Reform des Gesundheitswesens und der Krankenanstaltenfinanzierung für die Jahre 1997 bis 2000, Orac Verlag, Wien

Similarities between Solutions of Discrete-Time (Non-)Linear-Quadratic Games [+]

Doris A. Behrens[1,2], Reinhard Neck[1]

[1] Department of Economics, University of Klagenfurt, Universitätsstraße 65-67, A-9020 Klagenfurt, Austria
[2] Department of Operations Research and Systems Theory, Vienna University of Technology, Argentinierstraße 8, A-1040 Vienna, Austria

Abstract. We briefly summarize the algorithm OPTGAME 1.0 to solve two-person discrete-time LQ dynamic games exactly and discrete-time non-linear quadratic games approximately by means of an appropriate linearization procedure. Furthermore, we consider the relations between different solutions of these dynamic games, in particular between open-loop and feedback Nash and Stackelberg equilibria and Pareto-optimal solutions, and derive conditions for differences between the results of these solution concepts for a given game to become very small.

1. Introduction

Several studies have used dynamic game theory as mathematical instrument to illustrate questions of strategic interdependences between decision- and policy-makers, in particular in the analysis of oligopolistic markets or of international policy coordination. Different solution concepts are distinguished in the literature on dynamic games according to the information patterns and the strategy spaces of the players, which correspond to the degree of commitment assumed for the players (see, e.g., Basar and Olsder (1995), Leitmann (1974), Feichtinger and Jørgensen (1983), Mehlmann (1991), Petit (1990), Dockner and Neck (1988)).

First, we distinguish between non-cooperative and cooperative solution concepts of dynamics games. In the former case, we exclude binding agreements between the players, while these agreements are assumed to be established and to hold in the latter case. Intermediate possibilities and the question how cooperation can evolve in a non-cooperative context are usually ignored. Among non-cooperative solutions, we distinguish between Nash equilibrium solutions, where no player can improve her (his) performance by one-sided deviations from the equilibrium strategy, and Stackelberg equilibrium solutions, where the players have asymmetric roles: In the case of a two-person game, one player is the leader and has the ability or power to declare her (his) strategy first and to impose it on the other player, the follower, who reacts rationally to the leader's strategy.

[+] This research was financed by the Austrian Science Foundation (FWF) under contract no. P12745-OEK. An earlier version of this paper has been presented at the SOR '99 Symposium in Magdeburg, Germany.

For the Nash as well as for the Stackelberg solution of a dynamic game, additional assumptions concerning the information pattern of the players can be made. These assumptions specify the informational basis of each player's decision. Here, we consider open-loop information patterns, where the players' strategies depend only on the initial state of the dynamic system, and feedback information patterns, where the strategies depend on the current state of the system (but not on the initial conditions). According to the interpretation of the distinction between open-loop and feedback Nash equilibrium solutions suggested by Reinganum and Stokey (1985), with the open-loop information pattern, each player can be imagined to commit herself (himself) at the start of the game a priori to all future actions she (he) will take ("path strategies"). In the feedback information pattern, players can be imagined to observe the current values of the state variable and to react upon them by choosing their actions according to a "decision rule", i.e., a strategy specified at the initial period which makes the values of the control variables dependent on the current values of the state variables at each (future) point in time.[1]

Apart from briefly summarizing the algorithm OPTGAME 1.0 (see also Hager (1994), Hager et al. (2000)) for solving discrete-time LQ games exactly and, more importantly, for approximately solving discrete-time non-linear quadratic games, this paper aims at showing the conditions for similarities between the feedback and the open-loop Nash equilibrium solution, the feedback Stackelberg equilibrium solution, and the cooperative Pareto-optimal solutions. As these different solution concepts are meant to model different behavioral and institutional environments, the question how close their outcomes are seems to be interesting both from a theoretical and an applied point of view. It turns out that it is mainly the weight of penalizing deviations from the desired control values in the objective functions of the players, or, alternatively, the strength of the impact of the controls on the states in the state equation, which are responsible for differences or similarities, respectively, with respect to the solutions of the dynamic game.

We introduce the class of discrete-time (non-)linear-quadratic dynamic games considered in Section 2, give a description of the solution algorithm in Section 3, illustrate the conditions for similarities among the solutions in Section 4, and conclude the paper with a short summary.

---

[1] The importance of feedback Nash and feedback Stackelberg equilibrium solutions derives from the fact that both have the desirable property of subgame perfection or strong time consistency (Basar (1989)). The open-loop Nash equilibrium is weakly time consistent but not subgame perfect. The open-loop Stackelberg equilibrium solution (which will not be discussed in this paper) is not even weakly time consistent and hence also not subgame perfect.

2. The (Non-)Linear-Quadratic Dynamic Game

For the calculation of the non-cooperative solutions, we regard the following discrete-time two-person game-theoretic problem with a finite planing horizon T: Find values $u_{i1}, u_{i2}, ..., u_{iT}$, such that the intertemporal quadratic loss function of each player $i = 1,2$,

$$J_i = \sum_{t=1}^{T} L_{it} = \sum_{t=1}^{T} \left(X_t - \tilde{X}_{it}\right)' \cdot Q_{it} \cdot \left(X_t - \tilde{X}_{it}\right), \tag{1a}$$

(which is additive in time) is either minimized subject to a nonlinear first-order (time-invariant) system of difference equations,

$$x_t = f\left(x_{t-1}, x_t, u_{1t}, u_{2t}, y_t\right), \qquad t = 1,...,T, \quad x_0 = x(0), \tag{2}$$

where $y_t, t = 1,...,T$, denotes a q-dimensional vector of known, non-controlled exogenous variables, and $f^\alpha(\ )$ the -th component of the vector-valued function $f(\ )$ with $\alpha = 1, ..., n_s$, or (1a) is minimized subject to a linear system of state equations with time-dependent coefficients,

$$x_t = A_t \cdot x_{t-1} + B_{1t} \cdot u_{1t} + B_{2t} \cdot u_{2t} + s_t, \qquad t = 1,...,T, \quad x_0 = x(0), \tag{2*}$$

which might be numerically derived from the nonlinear system (2) as described in Section 3.1. Note that the control variables $u_{i(T+1)}$, $i = 1,2$, are not considered in the optimization.

For $t = 1,...,T$, X_t denotes an r-dimensional stacked "state" vector consisting of n_s state variables, x_t, n_1 control variables determined by player 1, $u_{1(t+1)}$, and n_2 control variables driven by player 2, $u_{2(t+1)}$,

$$X_t = \begin{pmatrix} x_t \\ u_{1(t+1)} \\ u_{2(t+1)} \end{pmatrix}_{r \times 1}, \qquad r = n_s + n_1 + n_2. \tag{3}$$

$\tilde{X}_{it}, i = 1,2$, denotes the r-dimensional stacked "desired objectives" vector which contains the "ideal levels" of the state and control variables for player 1 and 2, respectively,

$$\tilde{X}_{it} = \begin{pmatrix} \tilde{x}_{it} \\ \tilde{u}_{i1(t+1)} \\ \tilde{u}_{i2(t+1)} \end{pmatrix}_{r \times 1}, \qquad r = n_s + n_1 + n_2, \quad i = 1,2. \tag{4}$$

Furthermore, the (positive definite) matrices Q_{it}^x, $Q_{i(t+1)}^{u_1}$, and $Q_{i(t+1)}^{u_2}$ describe the penalties for deviations of the state variables x_t in time period t, and of the control variables $u_{1(t+1)}$ and $u_{2(t+1)}$ in time period $t+1$, respectively, from their desired values. These penalty matrices for states and controls in the objective functions are gathered in larger-dimensional matrices which eases the notation and guarantees fast numerical computer calculations. Hence, these larger matrices are defined as

$$
Q_{it} = \begin{pmatrix} Q_{it}^x & 0 & 0 \\ 0 & Q_{i(t+1)}^{u_1} & 0 \\ 0 & 0 & Q_{i(t+1)}^{u_2} \end{pmatrix}_{r \times r}, \quad r = n_s + n_1 + n_2, \quad i = 1,2. \tag{5}
$$

Note that $Q_{i(t+1)}^{u_1}$ and $Q_{i(t+1)}^{u_2}$ are assumed to have full rank in order to guarantee the existence of a solution. For the calculation of the cooperative (Pareto-optimal) solution in a two-person dynamic game, we aggregate the objective functionals by application of weights, $\mu \in [0,1]$, to

$$
J = \mu \cdot J_1 + (1-\mu) \cdot J_2 = \sum_{t=1}^{T} \{ \mu \cdot L_{1t} + (1-\mu) \cdot L_{2t} \}. \tag{1b}
$$

3. The Algorithm OPTGAME 1.0[2]

3.1 Linearization of the System Equations

In most game-theoretical models, the system dynamics of the form (2) are a convenient starting point. For the calculation of the Nash equilibrium, Stackelberg equilibrium, and Pareto-optimal strategies, however, another state-space representation will be more adequate. This representation does not include x_t at the right-hand side of the implicit function (2). Following Chow (1975) and Neck and Matulka (1992), it is easy to show how to eliminate x_t in the course of a linearization of system (2).

- With the given non-controlled exogenous variables, y_t, tentative control paths, $\hat{u}_{it}$ (which are determined either by historical values or by previous optimization runs), and the lagged tentative state variables, $\hat{x}_{t-1}$ (starting with $x(0) = x_0$), the autonomous non-linear system (2) can be solved for all t using the Gauss-Seidel approximation, a well-known non-linear equation-solving

---

[2] See Hager (1994) and Hager et al. (2000).

method. That is, for known values of $\hat{x}_{t-1}$, $\hat{u}_{1t}$, $\hat{u}_{2t}$ and y_t, we can compute a value $\hat{x}_t$ such that

$$\hat{x}_t - f(\hat{x}_{t-1}, \hat{x}_t, \hat{u}_{1t}, \hat{u}_{2t}, y_t) = 0, \qquad t = 1,\ldots,T,$$

by straightforward application of the numerical Gauss-Seidel approximation technique.

- Then, we linearize the vector-valued system function $f(\ )$ numerically around the reference values $\hat{x}_{t-1}$, $\hat{x}_t$, $\hat{u}_{1t}$, $\hat{u}_{2t}$ and the given reference path y_t according to the first-order Taylor approximation and obtain the following approximate non-autonomous system equations:

$$\begin{aligned} x_t &= A_t \cdot x_{t-1} + B_{1t} \cdot u_{1t} + B_{2t} \cdot u_{2t} + s_t, \qquad t = 1,\ldots,T, \\ x_0 &= x(0), \end{aligned} \qquad (2^*)$$

where

$$\left.\begin{aligned} A_t &= \left(I - F_{x_t}\right)^{-1} \cdot F_{x_{t-1}} \\ B_{1t} &= \left(I - F_{x_t}\right)^{-1} \cdot F_{u_{1t}} \\ B_{2t} &= \left(I - F_{x_t}\right)^{-1} \cdot F_{u_{2t}} \\ s_t &= \hat{x}_t - A_t \cdot \hat{x}_{t-1} - B_{1t} \cdot \hat{u}_{1t} - B_{2t} \cdot \hat{u}_{2t} \end{aligned}\right\} \qquad t = 1,\ldots,T,$$

$$\left(F_{x_{t-1}}\right)_{ij} = \frac{\partial f^i(\ )}{\partial x_{t-1}^j}, \qquad \begin{aligned} i &= 1,\ldots,n_s \\ j &= 1,\ldots,n_s \end{aligned}, \qquad t = 1,\ldots,T,$$

$$\left(F_{x_t}\right)_{ij} = \frac{\partial f^i(\ )}{\partial x_t^j}, \qquad \begin{aligned} i &= 1,\ldots,n_s \\ j &= 1,\ldots,n_s \end{aligned}, \qquad t = 1,\ldots,T,$$

$$\left(F_{u_{1t}}\right)_{ij} = \frac{\partial f^i(\ )}{\partial u_{1t}^j}, \qquad \begin{aligned} i &= 1,\ldots,n_s \\ j &= 1,\ldots,n_1 \end{aligned}, \qquad t = 1,\ldots,T,$$

$$\left(F_{u_{2t}}\right)_{ij} = \frac{\partial f^i(\ )}{\partial u_{2t}^j}, \qquad \begin{aligned} i &= 1,\ldots,n_s \\ j &= 1,\ldots,n_2 \end{aligned}, \qquad t = 1,\ldots,T.$$

Here and in what follows, I denotes the identity matrix of dimension n_s, and it has to be assumed that the matrix $I - F_{x_t}$ is non-singular. The matrices and vectors defined above are time-dependent functions of the reference paths along which they have been evaluated. If these paths change, the matrices will change, too.

3.2 Computation of the Solutions

In a deterministic optimum control setting, the solutions obtained by the application of the minimum principle or the dynamic programming method coincide. Though deterministic dynamic game models can be solved by using essentially the same techniques, the choice of the solution technique depends on the information pattern and determines the qualitative results of the game. E.g., the application of the minimum principle generates an open-loop solution, while the application of the dynamic programming technique determines a feedback solution.

Using the appropriate optimization technique corresponding to the assumed information pattern of the game yields so-called Riccati equations – determined separately for each solution concept by equations (8a-c) – which can be solved by backward integration, i.e. for $t = T, T-1, \ldots, 2$. The terminal conditions for the Riccati matrices can be defined easily as already shown by Kendrick (1981) in control theory. By simple substitution of the Riccati matrices into equations (6a,b), we are able to derive values of the optimal control variables expressed in feedback form, u_{it}^*, as well as optimal state values, x_t^*, for each player $i = 1, 2$ by forward iteration, for $t = 1, \ldots, T$, using the initial values of the states,

$$u_{it}^* = G_{it} \cdot x_{t-1}^* + g_{it}, \tag{6a}$$

$$x_t^* = K_t \cdot x_{t-1}^* + k_t, \tag{6b}$$

where we need the following matrices (for $i - 1, 2$):

$$K_t := A_t + B_{1t} \cdot G_{1t} + B_{2t} \cdot G_{2t}, \tag{7a}$$

$$k_t := s_t + B_{1t} \cdot g_{1t} + B_{2t} \cdot g_{2t}, \tag{7b}$$

$$G_{it} := -M_{it} \cdot \left(B_{it} + U_{it}\right)' \cdot H_{it} \cdot E_{it} \cdot A_t + S_{it}, \tag{7c}$$

$$g_{it} := -M_{it} \cdot \left(B_{it} + U_{it}\right)' \cdot H_{it} \cdot E_{it} \cdot F_{it} + m_{it} + v_{it} + w_{it}. \tag{7d}$$

The specification of the matrices $M_i, m_i, U_i, S_i, v_i, w_i, F_i$, and E_i ($i = 1, 2$) follows separately for each solution concept below.

3.2.1 Special Features of the Nash Equilibrium Solutions

The calculation of the following matrices holds both for the feedback Nash equilibrium solution and the open-loop Nash equilibrium solution.

$$U_{1t} = U_{2t} := 0, \quad S_{1t} = S_{2t} := 0, \quad w_{1t} = w_{2t} := 0,$$

$$M_{it} := \left(Q_{it}^{u_i}\right)^{-1}, \quad m_{it} := \tilde{u}_{iit}, \quad i = 1, 2,$$

$$F_{1t} = F_{2t} := s_t + B_{1t} \cdot (m_{1t} + v_{1t}) + B_{2t} \cdot (m_{2t} + v_{2t}),$$

$$E_{1t} = E_{2t} := \left(I + B_{1t} \cdot M_{1t} \cdot B_{1t}' \cdot H_{1t} + B_{2t} \cdot M_{2t} \cdot B_{2t}' \cdot H_{2t}\right)^{-1}.$$

The calculation of the Riccati equations differs, however. The feedback Nash equilibrium solution is generated using the dynamic programming technique. Thus, starting with the value of the objective function for the terminal period $t=T$ and proceeding by minimization of the cost-to-go function step by step towards the initial period, the Riccati equations are defined recursively backwards in time, $t = T, T-1, \ldots, 2$,

$$H_{i(t-1)} = K_t' \cdot H_{it} \cdot K_t + Q_{i(t-1)}^x + \sum_{j=1}^{2} G_{jt}' \cdot Q_{it}^{u_j} \cdot G_{jt},$$

$$h_{i(t-1)} = K_t' \cdot (h_{it} - H_{it} \cdot k_t) + Q_{i(t-1)}^x \cdot \tilde{x}_{i(t-1)} + \sum_{j=1}^{2} G_{jt}' \cdot Q_{it}^{u_j} \cdot (\tilde{u}_{jit} - g_{jt}), \quad (8a)$$

$$H_{iT} = Q_{iT}^x,$$

$$h_{iT} = Q_{iT}^x \cdot \tilde{x}_{iT},$$

where

$$v_{it} := M_{it} \cdot B_{it}' \cdot h_{it}.$$

The solution of equations (8a) yields the Riccati matrices which define the feedback matrices determined by equations (7c-d). Then equations (6a) describe linear relations between the optimal control actions of the players and the previous state variables. With the Riccati matrices and the feedback matrices, the system equations can be solved by forward iteration, which determines the (approximate) equilibrium solution paths for state and control variables.

On the other hand, the open-loop Nash equilibrium solution is determined by using the minimum principle. By defining the (current-value) Hamiltonian function with the state equations and the objective function for each player and appropriate differentiation, one obtains adjoint equations and necessary conditions for the control variables. Under the assumption of a linear relation between co-states and states again Riccati equations are derived, which can be solved by iterating backwards in time,

$$H_{i(t-1)} = A_t' \cdot H_{it} \cdot E_{it} \cdot A_t + Q_{i(t-1)}^x, \qquad H_{iT} = Q_{iT}^x,$$

$$h_{i(t-1)} = A_t' \cdot (h_{it} + H_{it} \cdot E_{it} \cdot F_{it}) - Q_{i(t-1)}^x \cdot \tilde{x}_{i(t-1)}, \qquad h_{iT} = Q_{iT}^x \cdot \tilde{x}_{iT}, \qquad (8b)$$

where

$$v_{it} := -M_{it} \cdot B_{it}' \cdot h_{it}.$$

3.2.2 Special Features of the Feedback Stackelberg Equilibrium Solution

The feedback Stackelberg equilibrium solution is determined in a similar way as the feedback Nash equilibrium solution except for the additional consideration of the reaction of the follower (player 2) to the announced strategy of the leader (player 1) which results in an asymmetric game. Again, the principle of dynamic programming is used to derive feedback matrices and Riccati equations which are calculated backwards in time. The necessity of considering an additional so-called reaction-coefficient, determined by

$$R_t := \frac{\partial u_{2t}}{\partial u_{1t}} = -\left(B_{2t}' \cdot H_{2t} \cdot B_{2t} + Q_{2t}^{u_2}\right)^{-1} \cdot B_{2t}' \cdot H_{2t} \cdot B_{1t},$$

yields a number of additional terms in the feedback matrices, whereas the Riccati equations for both players are identical to the feedback Nash case, determined by equations (8a).

$$M_{1t} := \left(Q_{1t}^{u_1} + R_t' \cdot Q_{1t}^{u_2} \cdot R_t + \left(B_{1t} + U_{1t}\right)' \cdot H_{1t} \cdot \left(B_{1t} + U_{1t}\right)\right)^{-1},$$

$$M_{2t} := \left(Q_{2t}^{u_2} + B_{2t}' \cdot H_{2t} \cdot B_{2t}\right)^{-1},$$

$$v_{it} := M_{it} \cdot \left(B_{it} + U_{it}\right)' \cdot h_{it}, \quad i = 1,2,$$

$$U_{1t} := B_{2t} \cdot R_t, \qquad\qquad\qquad m_{it} := M_{it} \cdot Q_{it}^{u_i} \cdot \tilde{u}_{iit}, \quad i = 1,2,$$

$$U_{2t} := 0, \qquad\qquad\qquad\qquad\qquad w_{2t} := 0,$$

$$E_{1t} := I - B_{2t} \cdot M_{2t} \cdot B_{2t}' \cdot H_{2t}, \qquad F_{1t} := s_t,$$

$$E_{2t} := I, \qquad\qquad\qquad\qquad\qquad F_{2t} := s_t + B_{1t} \cdot g_{1t},$$

$$S_{1t} := M_{1t} \cdot R_t' \cdot Q_{1t}^{u_2} \cdot M_{2t} \cdot B_{2t}' \cdot H_{2t} \cdot A_t,$$

$$S_{2t} := -M_{2t} \cdot B_{2t}' \cdot H_{2t} \cdot B_{1t} \cdot G_{1t},$$

$$w_{1t} := -M_{1t} \cdot \left(B_{1t} + U_{1t}\right)' \cdot H_{1t} \cdot B_{2t} \cdot \left(M_{2t} \cdot B_{2t}' \cdot h_{2t} + m_{2t}\right) +$$
$$+ M_{1t} \cdot R_t \cdot Q_{1t}^{u_2} \cdot \left(M_{2t} \cdot B_{2t}' \cdot \left(H_{2t} \cdot F_{1t} - h_{2t}\right) - m_{2t} + \tilde{u}_{12t}\right).$$

3.2.3 Special Features of the Pareto-Optimal Solutions

Since the Pareto-optimal solutions describe strategies where the players cooperate, one has to solve the classical optimal control problem of minimizing (1b) subject to system (2). For $i = 1,2$, the matrices are given by

$$U_{1t} = U_{2t} := 0, \ S_{1t} = S_{2t} := 0, \ w_{1t} = w_{2t} := 0,$$

$$M_{it} := \left(\mu \cdot Q_{1t}^{u_i} + (1-\mu) \cdot Q_{2t}^{u_i}\right)^{-1},$$

$$m_{it} := M_{it} \cdot \left(\mu \cdot Q_{1t}^{u_i} \cdot \tilde{u}_{1it} + (1 - \mu) \cdot Q_{2t}^{u_i} \cdot \tilde{u}_{2it} \right),$$

$$v_{it} := M_{it} \cdot B_{it}' \cdot h_{it}, \quad i = 1,2,$$

$$F_{1t} = F_{2t} := s_t + B_{1t} \cdot (m_{1t} + v_{1t}) + B_{2t} \cdot (m_{2t} + v_{2t}),$$

$$E_{1t} = E_{2t} := \left(I + B_{1t} \cdot M_{1t} \cdot B_{1t}' \cdot H_{1t} + B_{2t} \cdot M_{2t} \cdot B_{2t}' \cdot H_{2t} \right)^{-1}.$$

The Pareto-optimal solution can be derived in a similar way as the feedback Nash equilibrium solution using the dynamic programming technique with the only difference that a convex combination of the players' objectives is to be optimized.

For $Q_t := \mu \cdot Q_{1t} + (1 - \mu) \cdot Q_{2t}$ and $P_t := \mu \cdot Q_{1t} \cdot \tilde{X}_{1t} + (1 - \mu) \cdot Q_{2t} \cdot \tilde{X}_{2t}$, the Riccati equations are determined by

$$H_{t-1} = K_t' \cdot H_t \cdot K_t + \begin{pmatrix} I \\ G_{1t} \\ G_{2t} \end{pmatrix}' \cdot Q_{t-1} \cdot \begin{pmatrix} I \\ G_{1t} \\ G_{2t} \end{pmatrix},$$

$$h_{t-1} = K_t' \cdot (h_t - H_t \cdot k_t) - \begin{pmatrix} I \\ G_{1t} \\ G_{2t} \end{pmatrix}' \cdot Q_{t-1} \cdot \begin{pmatrix} 0_{n_s} \\ g_{1t} \\ g_{2t} \end{pmatrix} + \begin{pmatrix} I \\ G_{1t} \\ G_{2t} \end{pmatrix}' \cdot P_{t-1}, \qquad (8c)$$

$$H_T = \mu \cdot Q_{1T}^x + (1 - \mu) \cdot Q_{2T}^x,$$

$$h_T = \mu \cdot Q_{1T}^x \cdot \tilde{x}_{1T} + (1 - \mu) \cdot Q_{2T}^x \cdot \tilde{x}_{2T}.$$

Note that we do not have to distinguish between H_{1t} and H_{2t} (h_{1t} and h_{2t}, respectively) when substituting the Riccati matrices into equations (6a).

4. Conditions for Similarities

- **Similarities between the Open-Loop and the Feedback Nash Equilibrium Solutions**

  One set of conditions which gives rise to similarities between the open-loop Nash equilibrium solution and the feedback Nash equilibrium solution is a small impact of the control variables and of the vector s on the dynamics of the state variables at each point in time, i.e., $B_{it}, s_t, t = 1, \ldots, T, \ i = 1,2$ have to be rather small, together with small desired values for the states, $\tilde{x}_i, \ i = 1,2$. While the conditions might be frequently fulfilled for the desired state values, control variables often exercise a strong influence on the development of the states over time.

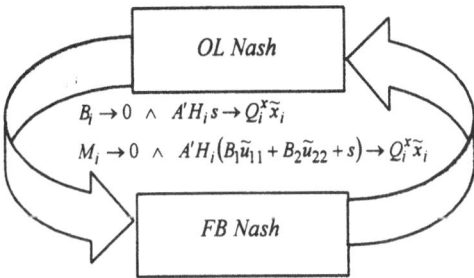

Fig.1. Visualization of the alternative similarity conditions for the open-loop and the feedback Nash equilibrium solutions for $i = 1,2$.

Another possibility to arrive at open-loop Nash equilibrium solutions which are close to feedback Nash equilibrium solutions is provided by a combination of small desired values, $\tilde{x}_i, \tilde{u}_{ii}, i = 1,2$, a small impact of s_t on the dynamics of the states at each point in time, and small matrices $M_i, i = 1,2$ which imply that the actual control values rapidly approach their desired values. E.g., the matrices, $M_i, i = 1,2$, become very small if one places large penalties for deviating from the desired values for the controls.

- **Similarities between the Feedback Stackelberg and the Feedback Nash Equilibrium Solution**
  A sufficient condition for the feedback Stackelberg equilibrium solution and the feedback Nash equilibrium solution to behave alike is given by a small impact of the control variables on the development of the states, i.e. by $B_i, i = 1,2$, being rather small. Note that small B-matrices dramatically decrease the reaction coefficient, R. In a well-specified control problem, however, this condition is not frequently met.

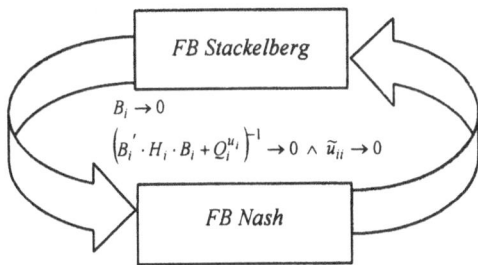

Fig.2. Visualization of the alternative similarity conditions for the feedback Stackelberg and the feedback Nash equilibrium solution for $i = 1,2$.

An alternative way to drive the reaction coefficient, R, towards zero is to place large penalties for the control variables to fail their desired values – relative to the impact of the control variables on the states, $B_i, i = 1,2$ – which causes very small matrices M_i, $i = 1,2$, together with small desired values $\tilde{u}_{ii}$, $i = 1,2$. To put it in non-mathematical terms, if one forces the controls to keep very close to their desired values, where the actual control values exercise an influence on the states which is quite small relative to the size of the penalties, the feedback Stackelberg equilibrium solution approaches the feedback Nash equilibrium solution.

- **Similarities between the Pareto-Optimal Solutions and the Feedback Nash Equilibrium Solution**
  If the weight in the objective functional (1b), μ, is supposed to be rather large, i.e. rather close to 1, the Riccati equations (8c) of the Pareto-optimal solution obtain the form of the Riccati equations (8a) of the feedback Nash equilibrium solution for player 1 – which is intuitively appealing. On the other hand, a small μ-value, i.e. close to 0, drives the Riccati equations of the Pareto-optimal solution towards those of the feedback Nash equilibrium strategy of player 2. I.e., for similarities between the solutions, one of the players must be regarded as "far more important" than the other one – which is expressed by the choice of the weight in the objective functional. Consequently, for $\mu = 1$ ($\mu = 0$) the Pareto-optimal solution equals the feedback Nash equilibrium solution of player 1 (2).

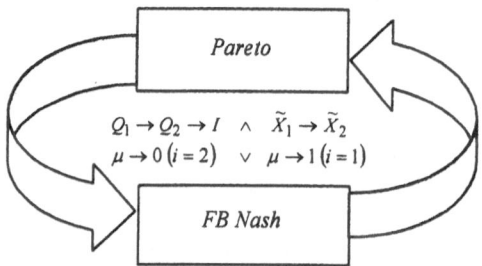

Fig.3. Visualization of the alternative similarity conditions for the Pareto-optimal solutions and the feedback Nash equilibrium solution.

Sufficient closeness and symmetry between the desired values of states and controls, $\tilde{x}_i, \tilde{u}_{ii}, \tilde{u}_{ij}, i, j = 1,2, j \neq i$, as well as between the penalties for deviations from the desired values, Q_i, $i = 1,2$, is another condition for similarities between the Pareto-optimal and the feedback Nash equilibrium solutions.

5. Conclusions

The algorithm OPTGAME 1.0 to solve two-person discrete-time LQ dynamic games exactly and discrete-time non-linear quadratic games approximately by means of an appropriate linearization procedure is briefly sketched in this paper. Furthermore, the relations between different solutions of these (non-)linear-quadratic dynamic games, in particular, open-loop and feedback Nash and Stackelberg equilibria and Pareto-optimal solutions, are considered, and conditions for differences between these solutions to become very small are given.

Basically, we are able to distinguish two main mechanisms (together with more distinct conditions mentioned in Section 4) which force different game-theoretic solutions approximated by OPTGAME 1.0 to behave almost identically. Firstly, the solutions approach each other whenever the impact of the controls on the state variables becomes rather small. Secondly, "large penalties" for failing the desired control values – as compared to the impact of the control variables on the development of the states – contribute towards achieving similar behavior by forcing the control variables to obtain their desired values as fast as possible in order to decrease the expected loss.

References

1. Basar, T. (1989): Time Consistency and Robustness of Equilibria in Non-Cooperative Dynamic Games. In: Dynamic Policy in Economics, eds. F. van der Ploeg and A. de Zeeuw. North-Holland, Amsterdam, 9–54.
2. Basar, T. and Olsder, G.J. (1995): Dynamic Noncooperative Game Theory. 2nd edition. Academic Press, London.
3. Chow, G.C. (1975): Analysis and Control of Dynamic Economic Systems. John Wiley, New York.
4. Dockner, E.J. and Neck, R. (1988): Time-Consistency, Subgame-Perfectness, Solution Concepts and Information Patterns in Dynamic Models of Stabilization Policies. Discussion Paper, University of Saskatchewan, Saskatoon.
5. Feichtinger, G. and Jørgensen, S. (1983): Differential Game Models in Management Science. European Journal of Operational Research 14, 137–155.
6. Hager, M. (1994): Non-Linear Dynamic Games: Theory, Numerical Solutions and Applications. Ph.D. Thesis, University of Technology, Vienna.
7. Hager, M., Behrens, D.A., and Neck, R. (2000): OPTGAME 1.0: An Algorithm to Derive Equilibrium Solutions of Two-Person Discrete-Time (Non-)Linear Dynamic Games. Discussion Paper, Department of Economics, University of Klagenfurt.
8. Kendrick, D. (1981): Stochastic Control for Economic Models. McGraw-Hill, New York.
9. Leitmann, G. (1974): Cooperative and Non-Cooperative Many Players Differential Games. Springer, Vienna.
10. Mehlmann, A. (1991): Applied Differential Games. Plenum Press, New York.
11. Neck, R. and Matulka, J. (1992): OPTCON: An Algorithm for the Optimal Control of Nonlinear Stochastic Models. Annals of Operations Research 37, 375–401.
12. Petit, M.L. (1990): Control Theory and Dynamic Games in Economic Policy Analysis. Cambridge University Press, Cambridge.
13. Reinganum, J.F. and Stokey, N.S. (1985): Oligopoly Extraction of a Common Property Natural Resource: The Importance of the Period of Commitment in Dynamic Games. International Economic Review 26, 161–173.

Optimization of an n-Person Game Under Linear Side Conditions

Werner Krabs, Stefan Pickl and Jürgen Scheffran

Department of Mathematics, Darmstadt University of Technology
Schlossgartenstr. 7, 64289 Darmstadt
pickl@mathematik.tu-darmstadt.de

Summary. This paper is concerned with n-person games which typically occur in mathematical conflict models [cf. [4], [7],[8]]. These games are so called cost-games, in which every actor tries to minimize his own costs and the costs are interlinked by a system of linear inequalities. It is shown that, if the players cooperate, i.e., minimize the sum of all the costs, they achieve a Nash Equilibrium. In order to determine Nash Equilibria, the simplex method can be applied with respect to the dual problem. An important special case is discussed and numerical examples are presented.

1. Introduction

The TEM-Model (Technology-Emissions-Means-Model) describes the dynamical interaction between several actors (players) who intend to minimize their emissions (E_i) caused by technologies (T_i) by means of expenditures of money (M_i) or financial means, respectively. The index stands for the i-th player, $i = 1, \ldots, n$. The players are linked by technical cooperations and the market, which expresses itself in the nonlinear time-discrete dynamics of the Technology-Emissions-Means-Model, in short: TEM-model.

The aim is to reach a steady state of the system by choosing the control parameters such that the emissions of each player become minimized. The focal point is the realization of the necessary optimal control parameters via a cost game, which is determined by the way of cooperation of the actors. In relation to the work of Leitmann, but not regarding solution sets as feasible sets, the τ-value of Tijs [9] is taken as a control parameter. This leads to a new class of problems in the area of 1-convex games. In this work the allocation problem is solved for a special case. With these equilibria a reasonable model for a Joint-Implementation process is developed, where its necessary fund is represented by the non-empty core of the analyzed game.

Using linear programming techniques and the simplex method Nash Equilibria are determined. Steering with these parameters of the feasible set, the TEM-model can be regarded as a useful tool to simulate and analyse an economical Joint-Implementation Program.

The obtained results can be used in a constructive way if we regard the following nonlinear time discrete system.

The TEM-model which was investigated in [7] and [8] looks as follows

$$E_i(t+1) = E_i(t) + \sum_{j=1}^{n} em_{ij}(t)(M_j(t) + u_j(t))$$

$$M_i(t+1) = M_i(t) -$$
$$\lambda_i(M_i(t) + u_i(t))[M_i^* - M_i(t) - u_i(t)]\{E_i(t) + \varphi_i \Delta E_i(t)\}$$

E_i emissions of actor i in percent

M_i financial means of actor i

em_{ij} describes the effect on the emissions of the i-th actor if the j-th actor invests money for his technologies

$u_j(t)$ control term of the j-th actor at the t-th time step

Considering that $\Delta M_i(t) = M_i(t+1) - M_i(t)$ we can construct a cost-game in the sense of cooperative game theory. The difference between the cooperative and the non-cooperative case is always positive. Now, the method is that at each time step, this amount is put into a central fund as feasible set for the control parameters.

Now we want to steer the system in a cost-minimal way in order to reach the states $M_i(\hat{t}) = 0$ and $E_i(\hat{t}) = 0$ $(i = 1,\ldots,n)$, for some $\hat{t} \geq 0$. This minimization problem leads directly to the following allocation problem which we will solve in the next section. Together with the basic theory [5] then we are able to simulate and analyse an economical Joint-Implementation Program [7] in the sense of Gustav Feichtinger [2], [3].

2. The Allocation Problem

In connection with the TEM-Model [7] which is based on a general conflict model [6] the following allocation problem is in the center of interest. In order to develop a Joint-Implementation Program we begin with the following formulation:

Given n players who pursue n goals which are given by an n-vector

$$b = (b_1,\ldots,b_n)^T \quad \text{with} \quad b_i \geq 0 \quad \text{for} \quad i = 1,\ldots,n.$$

In order to achieve these goals every player has to put in a certain amount of money, say $x_i \geq 0$ for the i-th player. The share of the player j at the goal b_i (where b_j is his own goal) when he spends one unit is assumed to be c_{ij} where (for good reasons)

$$c_{ii} > 0 \quad \text{for} \quad i = 1,\ldots,n . \tag{1}$$

If $i \neq j$, however, $c_{ij} \leq 0$ is also allowed for. In such a case player j can be considered as an opponent of player i. The requirement to achieve all the goals is expressed by the following system of linear inequalities

$$\sum_{j=1}^{n} c_{ij} x_j \geq b_i \quad \text{for} \quad i = 1, \ldots, n \ . \tag{2}$$

In the sequel we assume that there is a vector $x = (x_1, \ldots, x_n)$ with $x_i \geq 0$ for $i = 1, \ldots, n$ for which the inequalities (2) are satisfied.

Then, for every i, the i-th player, of course, is interested in minimizing his own contribution x_i. In general, this will not be possible simultaneously. So the players will have to cooperate. Let us assume that they choose $\hat{x} \in I\!R^n$ with (2) and $x_j \geq 0$, $j = 1, \ldots, n$, for a $x = \hat{x}$ such that

$$s(x) = \sum_{j=1}^{n} x_j \tag{3}$$

for $x = \hat{x}$, is as small as possible. Now let, for $i \in \{1, \ldots, n\}$, $x_i \geq 0$ be chosen such that

$$\sum_{\substack{j=1 \\ j \neq i}}^{n} c_{kj} \hat{x}_j + c_{ki} x_i \geq b_k \quad \text{for all} \quad k = 1, \ldots, n \ , \tag{4}$$

then

$$\sum_{j=1}^{n} \hat{x}_j \leq \sum_{\substack{j=1 \\ j \neq i}}^{n} \hat{x}_j + x_i \quad \Longrightarrow \quad \hat{x}_i \leq x_i \tag{5}$$

which implies that $\hat{x}$ is a Nash equilibrium.

3. On the Determination of Nash-Equilibria

The problem of minimizing (3) subject to (2) and

$$x_j \geq 0 \quad \text{for} \quad j = 1, \ldots, n \tag{6}$$

is a typical problem of linear programming whose dual problem consists of maximizing

$$t(y) = \sum_{i=1}^{n} b_i y_i \tag{7}$$

subject to

$$\sum_{i=1}^{n} c_{ij}\, y_i \le 1 \quad \text{for} \quad j = 1, \ldots, n \tag{8}$$

and

$$y_i \ge 0 \quad \text{for} \quad i = 1, \ldots, n . \tag{9}$$

If we put $y_i = 0$ for $i = 1, \ldots, n$, then we obtain a solution of (8) and (9). Under the above assumption that there exists a solution of (2) and (4) we can apply a well known duality theorem and conclude that there exists a solution $\hat{x} \in I\!\!R^n$ of (2) and (6) which minimizes (3) and a solution $\hat{y} \in I\!\!R^n$ of (8) and (9) which maximizes (7) and that $s(\hat{x}) = t(\hat{y})$ which is equivalent to

$$\hat{x}_j > 0 \quad \Longrightarrow \quad \sum_{i=1}^{n} c_{ij}\, \hat{y}_i = 1$$

and

$$\hat{y}_i > 0 \Longrightarrow \sum_{j=1}^{n} c_{ij}\, \hat{x}_j = b_i . \tag{10}$$

On introducing slack variables

$$z_j \ge 0 \quad \text{for} \quad j = 1, \ldots, n \tag{11}$$

the inequalities (8) can be rewritten as equations in the form

$$z_j + \sum_{i=1}^{n} c_{ij}\, y_i = 1 \quad \text{for} \quad j = 1, \ldots, n . \tag{12}$$

The dual problem is then equivalent to the minimization of

$$\sum_{j=1}^{n} 0 \cdot z_j + \sum_{i=1}^{n} b_i\, y_i \tag{13}$$

subject to (9), (11), and (12). This problem can be solved immediately by the simplex method starting with the basic solution

$$z_j = 1 \quad \text{for} \quad j = 1, \ldots, n \quad \text{and} \quad y_i = 0 \quad \text{for} \quad i = 1, \ldots, n . \tag{14}$$

Let us illustrate this by an example.

Choose

$$C = \begin{pmatrix} 1 & -0.8 & 0.1 \\ 0.2 & 1 & -0.8 \\ -0.1 & -0.5 & 1 \end{pmatrix} , \quad b = \begin{pmatrix} 0.2 \\ 0.2 \\ 0.2 \end{pmatrix} \tag{15}$$

The initial scheme of the simplex method applied to the dual problem is given by

		$-y_1$	$-y_2$	$-y_3$	
z_1	1	1	0.2	−0.1	1
z_2	1	−0.8	1	−0.5	*
z_3	1	0.1	−0.8	1	10
$\sum_i b_i y_i$	0	−0.2	−0.2	−0.2	

(16)

If we exchange y_1 and z_1 with the aid of a Jordan step, we obtain the scheme

		$-z_1$	$-y_2$	$-y_3$	
y_1	1	1	0.2	−0.1	5
z_2	1.8	0.8	**1.16**	−0.58	0.6896552
z_3	0.9	−0.1	−0.82	1.01	*
$\sum_i b_i y_i$	0.2	0.2	−0.16	−0.22	

(17)

If we exchange y_2 and z_2, we obtain the scheme

		$-z_1$	$-z_2$	$-y_3$	
y_1	0.6896552	0.862069	−0.1724138	0	*
y_2	1.5517241	0.6896552	0.862069	−0.5	*
z_3	2.1724138	0.4655172	0.7068966	**0.6**	
$\sum_i b_i y_i$	0.4482759	0.3103448	0.137931	−0.3	

(18)

If we exchange y_3 and z_3, we obtain the scheme

		$-z_1$	$-z_2$	$-z_3$
y_1	0.6896552			
y_2	3.3620689			
y_3	3.6206897			
$\sum_i b_i y_i$	1.5344828	0.5431034	0.4913793	0.5

(19)

From the last row of the last scheme we derive

$$\hat{x}_1 = 0.5431034 , \quad \hat{x}_2 = 0.4913793 , \quad \hat{x}_3 = 0.5 \tag{20}$$

and the first column gives

$$\hat{y}_1 = 0.6896552 , \quad \hat{y}_2 = 3.3620689 , \quad \hat{y}_3 = 3.6206897 \qquad (21)$$

which indeed implies

$$0.2 \, \hat{y}_1 + 0.2 \, \hat{y}_2 + 0.2 \, \hat{y}_3 = 1.5344828 = \hat{x}_1 + \hat{x}_2 + \hat{x}_3 . \qquad (22)$$

Further we also verify that

$$
\begin{aligned}
\hat{x}_1 - 0.8 \, \hat{x}_2 + 0.1 \, \hat{x}_3 &= 0.2 , \\
0.2 \, \hat{x}_1 + \hat{x}_2 - 0.8 \, \hat{x}_3 &= 0.2 , \\
-0.1 \, \hat{x}_1 - 0.5 \, \hat{x}_2 + \hat{x}_3 &= 0.2 .
\end{aligned}
\qquad (23)
$$

3.1 A Special Case

Now let us assume that for some $j \in \{1, \ldots, n\}$ it is true that

$$c_{ij} \leq 0 \quad \text{for all} \quad i = 1, \ldots, n , \quad i \neq j , \qquad (24)$$

i.e., the player j can be considered as an opponent of all the other players. If $\hat{x} \in I\!R^n$ is a solution of (2) and (6) that minimizes (3), it follows that

$$\sum_{k=1}^{n} c_{jk} \, \hat{x}_k = b_j . \qquad (25)$$

For otherwise $(\hat{x}_1, \ldots, \hat{x}_{j-1}, x_j^*, \hat{x}_{j+1}, \ldots, \hat{x}_n)$ with

$$x_j^* = \frac{1}{c_{jj}} \left(b_j - \sum_{\substack{k=1 \\ k \neq j}}^{n} c_{jk} \, \hat{x}_k \right) < \hat{x}_j$$

also solves (2) and (6) and it follows that

$$x_j^* + \sum_{\substack{k=1 \\ k \neq j}}^{n} \hat{x}_k < \sum_{k=1}^{n} \hat{x}_k$$

contradicting the assumption that $\sum_{k=1}^{n} \hat{x}_k$ is minimal.

Now let us assume that

$$c_{ij} \leq 0 \quad \text{for all} \quad i \neq j , \qquad (26)$$

i.e. all players can be considered as opponents to each other.

Then, for every solution $\hat{x} \in I\!R^n$ of (2) and (6) that minimizes (3), it follows that

$$\sum_{j=1}^{n} c_{ij}\, \hat{x}_j = b_i \quad \text{for all} \quad i = 1, \ldots, n \, . \tag{27}$$

If we assume (26) and

$$\sum_{j=1}^{n} c_{ij} > 0 \quad \text{for all} \quad i = 1, \ldots, n \, , \tag{28}$$

then (1) is satisfied and the matrix C is inverse monotone, i.e., the inverse C^{-1} exists and is positive (see [1]). In this case the solution of (27) is given by

$$\hat{x} = C^{-1}\, b\, (\geq \Theta_n) \, , \quad \Theta_n \quad n\text{-dimensional zero-vector} \, .$$

If $x = (x_1, \ldots, x_n)^T$ is any solution of (2) and (6), then it even follows that

$$x \geq C^{-1} b = \hat{x}, \quad \text{i.e.} \quad x_i \geq \hat{x}_i \quad \text{for all} \quad i = 1, \ldots, n \, .$$

This means that, if all players oppose each other but every player's contribution to achieving his own goal is larger than the negative sum of his opponents, then everybody can reach an absolutely minimal amount of money.
Let us demonstrate this by an example: Put

$$C = \begin{pmatrix} 1 & -0.8 & 0 \\ 0 & 1 & -0.8 \\ -0.1 & -0.5 & 1 \end{pmatrix} \quad \text{and} \quad b = \begin{pmatrix} 0.2 \\ 0.2 \\ 0.2 \end{pmatrix} \, .$$

Obviously (26) is satisfied but also (28). The solutions of (27) are given by

$$\hat{x}_1 = 0.761194 \, , \quad \hat{x}_2 = 0.7014925 \, , \quad \hat{x}_3 = 0.6268657 \, .$$

The two examples are special cases of the following situation.
    Let C be inverse monotone and let $C^* \geq C$. If $x \in I\!R^n$ is a solution of (2) and (4), then x also solves

$$C^* x \geq b \tag{29}$$

and, if $x^* \in I\!R^n$ solves (4) and (29) and minimizes (3), then

$$s(x^*) \leq s(\hat{x}) \, ,$$

if $\hat{x} \in I\!R^n$ solves (2) and (4) and minimizes (3).

Moreover, if

$$C^* x^* = b , \tag{30}$$

then

$$C\, x^* \leq C^*\, x^* = b \leq C\, \hat{x}$$

which implies $x^* \leq \hat{x}$. The last considerations can be interpreted as follows: If C is inverse monotone, then the players can achieve the best possible individual results by solving the linear system

$$C\, \hat{x} = b .$$

If they replace C by a matrix C^* with $C^* \geq C$ such that there exists $x^* \in \mathbb{R}^n$ with (6) and (30), then $x^* \leq \hat{x}$, i.e., every player gets a better result x^* which is a Nash equilibrium.

3.2 The General Case

We assume that there is a solution of (1) and (2) which implies that the dual problem has a solution. If this is obtained by $r \leq n$ steps of the simplex method, the result can be assumed to be of the following form

$$\begin{pmatrix} y_1 \\ \vdots \\ y_r \\ z_{r+1} \\ \vdots \\ z_n \end{pmatrix} = \begin{pmatrix} d_1 \\ \vdots \\ d_r \\ d_{r+1} \\ \vdots \\ d_n \end{pmatrix} + \tilde{D} \begin{pmatrix} z_1 \\ \vdots \\ -z_r \\ -y_{r+1} \\ \vdots \\ -y_n \end{pmatrix}$$

$$\tilde{D} = \begin{pmatrix} d_{11} & \cdots & d_{1r} & d_{1r+1} & \cdots & d_{1n} \\ \vdots & & & & & \\ d_{r1} & \cdots & d_{rr} & d_{r+r1} & \cdots & d_{rn} \\ d_{r+11} & \cdots & d_{r+1r} & d_{r+1r+1} & \cdots & d_{r+1n} \\ \vdots & & \vdots & \vdots & & \vdots \\ d_{n1} & \cdots & d_{n1} & d_{nr+1} & \cdots & d_{nn} \end{pmatrix}$$

$$\sum_{j=1}^{r} b_j\, y_j = \sum_{j=1}^{r} b_j\, d_j + \sum_{k=1}^{r} \left(\sum_{j=1}^{r} d_{jk}\, b_j \right) (-z_k) + \sum_{k=r+1}^{n} \left(\sum_{j=1}^{r} d_{jk}\, b_j \right) (-y_k)$$

where $d_j \geq 0$ for $j = 1,\ldots,n$ and $\sum_{j=1}^{r} d_{jk}\, b_j \geq 0$ for $k = 1,\ldots,n$.

The solution of the dual problem reads $\hat{y}_j = d_j$ for $j = 1, \ldots, r$ and $\hat{y}_j = 0$ for $j = r + 1, \ldots, n$.

Let us assume that

$$C_r^T = \begin{pmatrix} c_{11} & \cdots & c_{r1} \\ \vdots & & \vdots \\ c_{1r} & & c_{rr} \end{pmatrix}$$

is invertible. Then

$$D_r = \begin{pmatrix} d_{11} & \cdots & d_{1r} \\ \vdots & & \vdots \\ d_{r1} & \cdots & d_{rr} \end{pmatrix} = (C_r^T)^{-1} = (C_r^{-1})^T .$$

If we put

$$x^r = \begin{pmatrix} \sum\limits_{j=1}^r d_{j1} \, b_j \\ \vdots \\ \sum\limits_{j=1}^r d_{jr} \, b_j \end{pmatrix} = D_r^T \, b^r \quad \text{with} \quad b^r = \begin{pmatrix} b_1 \\ \vdots \\ b_r \end{pmatrix} ,$$

then $x^r \geq 0$ and $C_r \, x^r = b^r$. Further we obtain

$$d^r = \begin{pmatrix} d_1 \\ \vdots \\ d_r \end{pmatrix} = (C_r^T)^{-1} \, e^r \quad \text{with} \quad e^r = \begin{pmatrix} 1 \\ \vdots \\ 1 \end{pmatrix} ,$$

hence,

$$C_r^T \, d^r = e^r \quad \text{and} \quad (d^r)^T \, b^r = (e^r)^T \, C_r^{-1} \, b^r = (e^r)^T \, x^r .$$

All this implies that

$$\hat{x}_j = x_j^r \quad \text{for} \quad j = 1, \ldots, r ,$$
$$\hat{x}_j = 0 \quad \text{for} \quad j = r + 1, \ldots, n$$

minimizes $s(x)$ subject to (1) and (2).

Let us demonstrate this by an example which is taken from [4]. Put

$$C = \begin{pmatrix} 1.667 & -0.875 & 0.792 \\ -0.792 & 1.667 & -0.875 \\ -0.167 & -0.167 & 0.333 \end{pmatrix} \times 10^{-2}$$

and

$$b = \begin{pmatrix} -0.1 \\ -0.1 \\ -0.1 \end{pmatrix} - C \begin{pmatrix} 30 \\ 20 \\ 10 \end{pmatrix} = \begin{pmatrix} -0.3459 \\ -0.1083 \\ 0.0498 \end{pmatrix} .$$

The initial scheme for the simplex method applied to the dual problem reads

		$-y_1$	$-y_2$	$-y_3$
z_1	1	0.001667	−0.00792	−0.00167
z_2	1	−0.00875	0.01667	−0.00167
z_3	1	−0.00792	−0.00875	0.00333
$\sum_{i=1}^{3} b_i y_i$	0	0.3459	0.1083	−0.0498

The final scheme is given by

		$-y_1$	$-z_2$	$-z_3$
z_1	3.006206			
y_2	122.25319			
y_3	621.53617			
$\sum_{i=1}^{3} b_i y_i$	17.712486	0.2040928	1.8365155	15.87597

from which we deduce

$$\hat{y}_1 = 0, \quad \hat{y}_2 = 122.25319, \quad \hat{y}_3 = 621.53617, \text{ and}$$
$$\hat{x}_1 = 0, \quad \hat{x}_2 = 1.8365155, \quad \hat{x}_3 = 15.87597.$$

References

1. Collatz, L.: Funktionalanalysis und numerische Mathematik.
   Springer-Verlag, Berlin - Göttingen - Heidelberg 1964.
2. Feichtinger, G., Hartl, R.F.: Optimale Kontrolle Ökonomischer Prozesse.
   Anwendung des Maximumprinzips in den Wirtschaftswissenschaften.
   de Gruyter Verlag, Berlin 1986.
3. Leopold-Wildburger, U., Feichtinger, G., Kistner, K.-P. (Ed.),
   Modelling and Decisions in Economics: Essays in Honor of Franz Ferschl.
   Physica-Verlag, Wien 1999.
4. Jathe, M., Krabs, W. and Scheffran, J.: Control and Game-Theoretical
   Treatment of a Cost-Security Model for Disarmament.
   Math. Meth. Appl. Sc. 20, 653 - 666 (1997).
5. Krabs, W.: Mathematische Modellierung. B.G. Teubner, Stuttgart 1997.
6. Krabs, W., Pickl, S.: Controllability of a Time-Discrete Dynamical System with
   the Aid of the Solution of an Approximation Problem.
   Journal of Control Theory and Cybernetics 2000 (submitted).
7. Pickl, S.: Der τ-value als Kontrollparameter.
   Modellierung und Analyse eines Joint-Implementation Programmes mithilfe der
   kooperativen dynamischen Spieltheorie und der diskreten Optimierung.
   Dissertation TU Darmstadt 1998, Shaker-Verlag, Aachen 1999.
8. Scheffran, J.: Modelling Environmental Conflicts and International Stability. In:
   Huber, R., Avenhaus, R. (Ed.), Models for Security in the Post-Cold War Era.
   NOMOS, Baden-Baden 1996.
9. Tijs, S.H.: Bounds for the Core and the τ-Value. In: Moeschlin, O.
   and Pallaschke, D. (Ed.), Game Theory and Mathematical Economics.
   North-Holland Publishing Company, Amsterdam 1981.

The Dynamics of the Cobweb When Producers are Risk Averse Learners

Carl Chiarella and Xue-Zhong He

School of Finance and Economics, University of Technology, Sydney, PO Box 123 Broadway, NSW 2007, Australia

Abstract In this paper we investigate the dynamics of the traditional cobweb model where producers are risk averse and seek to learn the distribution of asset prices. We consider the subjective estimates of the statistical distribution of the market prices based on L-step backward time series of market clearing prices. With constant absolute risk aversion, the cobweb model becomes nonlinear. Sufficient conditions on the local stability of the unique positive equilibrium of the nonlinear model are derived and, consequently, we show that the local stability region is proportional to the lag length L. When the equilibrium loses its local stability, we show that, for $L = 2$, the model has a strong $1 : 3$ resonance bifurcation and a family of fixed points of order 3 becomes unstable on both sides of criticality. For general lag lengths, numerical simulations suggest that the model displays a variety of complex dynamics.

1. INTRODUCTION

In this paper, we investigate the dynamics of one of the simplest dynamic economic models—the well-known cobweb model with constant elasticity demand and supply functions

$$\begin{cases} p_{t-1,t}^e = bq_t^a, & \text{(supply)}, \\ p_t = \beta q_t^\alpha, & \text{(demand)}. \end{cases} \tag{1}$$

Here, q_t and p_t are quantities and prices, respectively, at period t, $p_{t-1,t}^e$ is the price expected at time t based on the information set at $t - 1$, and $a, b, \beta \, (> 0)$ and $\alpha < 0$ are constants.

It has been shown that non-linearities in the supply or demand curves may lead the cobweb model to exhibit both stable periodic and chaotic behaviour; we cite, in particular, Artsein [2], Boussard [3], Chiarella [4], Holmes and Manning [8], Hommes [9] and Jensen and Urban [11]. These authors consider a variety of backward looking mechanisms for the formation of the expectations $p_{t-1,t}^e$ ranging across the traditional naive expectation $p_{t-1,t}^e = p_{t-1}$, learning expectations (e.g., learning by arithmetic mean $p_{t-1,t}^e = (p_{t-1} + \cdots + p_{t-L})/L$) and adaptive learning expectation $p_{t-1,t}^e = p_{t-2,t-1}^e + w(p_{t-1} - p_{t-2,t-1}^e)$ with $0 \leq w \leq 1$. Our approach to the formation of expectations will be somewhat different in that we assume that the producer agents treat $\tilde{p}_t$ as a random variable drawn from a normal distribution whose mean and variance they are seeking to learn [1] . Let $p_{t-1,t}^e$

---

[1]It would of course be preferable (and more in keeping with models of asset price dynamics in continuous time finance) to treat $\tilde{p}_t$ as log-normally distributed. However this

and $v^e_{t-1,t}$ be, respectively, the mean and variance of the distribution of $\tilde{p}_t$ conditional on the information set at $(t-1)$.

One can rescale the equations by letting [2] $Q_t = q^a_t$. We then have

$$p^e_{t-1,t} = bQ_t \qquad (2)$$

and

$$p_t = \beta Q_t^{\alpha/a} \qquad (3)$$

With constant absolute risk aversion A, the marginal revenue certainty equivalent is [3]

$$p^{cq}_t = p^e_{t-1,t} - 2Av^e_{t-1,t}Q_t. \qquad (4)$$

Suppose a linear (*in terms of Q, not q*) marginal cost, as in (2), so that the supply equation, under marginal revenue certainty equivalent becomes

$$p^{cq}_t = bQ_t. \qquad (5)$$

Combining (4) and (5) and equating supply and demand gives the market clearing quantity in period t as a function of the subjective mean $p^e_{t-1,t}$ and variance $v^e_{t-1,t}$

$$p^e_{t-1,t} - 2Av^e_{t-1,t}Q_t = bQ_t, \qquad (6)$$

that is

$$[b + 2Av^e_{t-1,t}]Q_t = p^e_{t-1,t}. \qquad (7)$$

Equation (2) may then be used to obtain the corresponding market clearing price.

1.1. Forming Expectations.
Equation (7) shows clearly that, as we would expect the method of formation of expectations plays a crucial role in the dynamics of the cobweb models. Indeed this is also known from the references cited in the previous section. We assume here that agents form their subjective estimates of the mean and variance by considering past market clearing prices over a window of length L, that is

$$p^e_{t-1,t} = \frac{1}{L} \sum_{i=1}^{L} p_{t-i}, \qquad (8)$$

---

would then move us out of the mean-variance framework so we leave an analysis of this approach to future research.

[2] This rescaling is equivalent to a change in numeraire

[3] With constant absolute risk aversion A, we assume the certainty equivalent of the receipt $r = \tilde{p}Q$ is $R(Q_t) = p^e_{t-1,t}Q_t - Av^e_{t-1,t}Q_t^2$. Then maximisation of this function with respect to Q_t leads to the marginal revenue certainty equivalent $p^{cq}_t = \frac{\partial R}{\partial Q_t} = p^e_{t-1,t} - 2Av^e_{t-1,t}Q_t$. We recall that this objective function is consistent with producers having the utility of receipts function $U(r) = -e^{-Ar}$.

and

$$v^e_{t-1,t} = \frac{1}{L} \sum_{i=1}^{L} [p^e_{t-1,t} - p_{t-i}]^2.$$ (9)

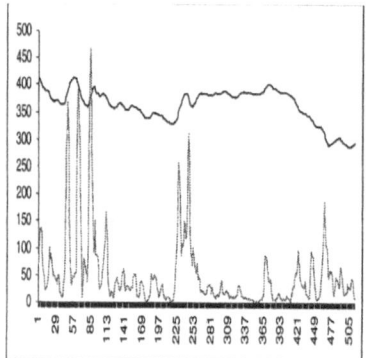 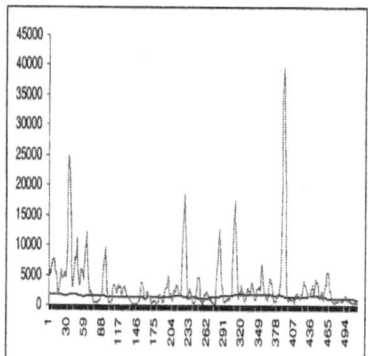

FIGURE 1. Running mean and variance of (a) gold ($/troy ounce) and (b) copper (£/tonne) prices from 11.1988 to 11.1998 with $L = 10$

Our choice of this mechanism has to some extent been motivated by the comments in Boussard [3] to the effect that a great deal of attention in the expectations formation literature has been devoted to schemes for the mean, but very little to schemes for the variance. In this connection it is of interest to consider the running mean and variance of actual commodity prices. Figure 1 displays such means and variances for weekly gold (a) and copper (b) prices over the period 11.1988–11.1998 with $L = 10$. In all figures the smoother curve is the running mean. A feature is the greater variability of the running variances compared to that of the running means. We have observed similar behavior for other commodity prices and lag lengths. We use this empirical observation as further support for our use of a scheme which allows both the subjective mean and variance to vary.

We substitute (8) and (9) into (7) and make use of (1) to obtain the market clearing quantity from

$$\left[b + \frac{2A\beta^2}{L} \sum_{i=1}^{L} \left(\frac{1}{L} \sum_{k=1}^{L} Q^{\alpha/a}_{t-k} - Q^{\alpha/a}_{t-i} \right)^2 \right] Q_t = \frac{\beta}{L} \sum_{i=1}^{L} Q^{\alpha/a}_{t-i}.$$ (10)

Using equation (3), we can rewrite equation (10) in terms of the price

$$p_t = \beta \left\{ \frac{\frac{1}{L}\sum_{i=1}^{L} p_{t-i}}{b + \frac{2A}{L}\sum_{i=1}^{L}[\frac{1}{L}\sum_{k=1}^{L} p_{t-k} - p_{t-i}]^2} \right\}^{\alpha/a}. \tag{11}$$

Let

$$u_{i,t} = p_{t-(i-1)}, \qquad i = 1, 2, \cdots, L. \tag{12}$$

Then equation (11) can be written as the following L dimensional system of first order difference equations

$$\begin{cases} u_{1,t+1} &= \beta \left[\frac{(1/L)\sum_{i=1}^{L} u_{i,t}}{b + (2A/L)\sum_{i=1}^{L}[(1/L)\sum_{k=1}^{L} u_{k,t} - u_{i,t}]^2} \right]^{\alpha/a} \\ u_{2,t+1} &= u_{1,t} \\ &\vdots \\ u_{L,t+1} &= u_{L-1,t}. \end{cases} \tag{13}$$

We note that this system can only generate positive price and quantity time series if we start from positive initial values.

2. LOCAL STABILITY

The system (13) has a unique positive equilibrium $u_1 = u_2 = \cdots = u_L = b(\beta/b)^{a/(a-\alpha)} \equiv p_o$. One can verify that, at the equilibrium point, the system (13) has the Jacobian matrix J given by the $L \times L$ matrix

$$J = \begin{pmatrix} \frac{\alpha}{aL} & \frac{\alpha}{aL} & \cdots & \frac{\alpha}{aL} & \frac{\alpha}{aL} \\ 1 & 0 & \cdots & 0 & 0 \\ 0 & 1 & \cdots & 0 & 0 \\ \vdots & \vdots & \ddots & \vdots & \vdots \\ 0 & 0 & \cdots & 1 & 0 \end{pmatrix}. \tag{14}$$

Set $\gamma = -\frac{\alpha}{aL}$ and $D_i(L) = \det(\lambda I - J)_{i \times i}$ with $i = 1, 2, \cdots, L$. We then have

$$D_1(\lambda) = \lambda + \gamma, \qquad D_2(\lambda) = \lambda D_1(\lambda) + \gamma = \lambda^2 + \gamma\lambda + \gamma,$$

and more generally,

$$D_i(\lambda) = \lambda D_{i-1}(\lambda) + \gamma, \qquad i = 1, 2, \cdots, L.$$

Thus

$$D_L(\lambda) = \lambda^L + \gamma\lambda^{L-1} + \cdots + \gamma\lambda + \gamma. \tag{15}$$

Using Jury's Test (see appendix A), we can derive the following local stability result. The proof of theorem 1 can be found in appendix A.

Theorem 1. *The unique steady state of the system (13) is locally asymptotically stable if and only if $-L < \frac{\alpha}{a}(< 0)$.*

It is interest to notice that both the steady state $(p_o, \cdots, p_o)$ and the local asymptotic stability condition of the system (13) are independent of the risk aversion parameter A. As pointed out by Boussard [3], this is a peculiarity of the particular expectation hypothesis chosen here. Under our assumption, $|\alpha/a|$, which is ratio of demand and supply elasticities, plays a key role on the local asymptotic stability of the positive equilibrium of the system (13). Furthermore, the local asymptotic stability condition $|\alpha/a| < L$ implies that the region of influence of the parameter α/a on the local asymptotic stability of the positive equilibrium of the system (13) is proportional to the lag length L. Theorem 1 tells us that larger time lags lead to larger regions of local asymptotic stability (in terms of the parameter α/a). This corresponds with a certain intuition that longer time delays stabilise otherwise unstable dynamics.

3. BIFURCATION ANALYSIS

The high dimensionality of the system (13) makes it difficult to undertake a bifurcation analysis for general lag length L. In this section we focus on the special case $L = 2$ and obtain some theoretical bifurcation results. We then use these results to guide our numerical investigations of the model for higher lag lengths in section 4.

Let us consider the simplest case $L = 2$ first so that the system (13) becomes

$$\begin{cases} u_{1,t+1} &= \beta \left[\dfrac{u_{1,t}+u_{2,t}}{2b+A(u_{1,t}-u_{2,t})^2} \right]^{\alpha/a} \\ u_{2,t+1} &= u_{1,t}. \end{cases} \tag{16}$$

Let $y_i = u_i - u^*$ with $u^* = b(\beta/b)^{a/(a-\alpha)}$ for $i = 1, 2$ and denote

$$g(y_1, y_2) = -u^* + \beta \left[\frac{y_1 + y_2 + 2u^*}{2b + A(y_1 + y_2)^2} \right]^{\alpha/a}. \tag{17}$$

Then the system (16) can be written as

$$\begin{cases} y_{1,t+1} &= g(y_{1,t}, y_{2,t}) \\ y_{2,t+1} &= y_{1,t}. \end{cases} \tag{18}$$

The Jacobian of the system (18) at the origin is then given by

$$J_2 = \begin{pmatrix} \frac{\alpha}{2a} & \frac{\alpha}{2a} \\ 1 & 0 \end{pmatrix}.$$

In the case $L = 2$, we have $\gamma = -\frac{\alpha}{2a}$. It follows from Theorem 1 that the equilibrium (p_o, p_o) of the system (16) loses its stability when γ passes through 1. Also, when γ is near 1, the Jacobian matrix J_2 has a pair of complex eigenvalues, say λ and $\bar\lambda$ with

$$\lambda = \lambda(\gamma) = \frac{1}{2}[-\gamma + i\sqrt{\gamma(4 - \gamma)}] = \sqrt{\gamma}e^{i\theta},$$

where θ satisfies

$$\sin\theta = -\frac{\sqrt{\gamma}}{2}, \qquad \cos\theta = \frac{\sqrt{4-\gamma}}{2}.$$

Let λ_o be the value of λ when $\gamma = 1$, that is,

$$\lambda_o = \frac{1}{2}[-1 + i\sqrt{3}].$$

We observe that $\lambda_o^3 = 1$. According to Kuznetsov [12] (page 350), for a map in $\mathcal{R}^2$, there is no "strong resonance" if there is an eigenvalue, say $\tilde{\lambda}$, satisfying $\tilde{\lambda}^q \neq 1$ for $q = 1, 2, 3, 4$. Otherwise, we say the map has a $1 : q$ **resonance** ($q = 1, 2, 3, 4$). Hence the map (18) has a $1 : 3$ resonance. As pointed out by Hale and Kocak [7] (page 481), *the dynamics of such maps – strong resonances – can be exceedingly complicated and the answer is not yet completely known.* The complexity of such maps is illustrated in one of the Examples 15.34 in Hale and Kocak [7] (pages 481-482). For more detailed discussion on the bifurcations of fixed points in discrete-time maps on $\mathcal{R}^2$ with both weak and strong resonances, we refer the reader to Iooss [10] when the maps involve one parameter and to Kuznetsov [12] when the maps involve two parameters.

The rest of this section is devoted to the study of generic bifurcations of the fixed points (p_o, p_o) of the map defined by (16). To keep the discussion simple, we will treat γ as the only parameter of the map.

Analysis of the bifurcation behavior of a map is aided by placing the map in standard normal form[4]. To perform a standard normal form calculation (see Arrowsmith et. al.[1]) for the system (18), we expand the function g as

$$g(y_1, y_2) - \sum_{j,k;j+k\geq 1} g_{jk} y_1^j y_2^k, \tag{19}$$

with

$$\begin{cases}
g_{10} &= g_{01} = \frac{\alpha}{2a} \\
g_{11} &= \frac{1}{4a^2 u^*}[\alpha^2 - a\alpha + 4aA\alpha(u^*)^2/b] \\
g_{20} &= g_{02} = \frac{1}{8a^2 u^*}[\alpha^2 - a\alpha - 4aA\alpha(u^*)^2/b] \\
g_{30} &= g_{03} = \frac{\alpha}{48a^3(u^*)^2}[\alpha^2 + 2a^2 - 3a\alpha - 12aA\alpha(u^*)^2/b] \\
g_{21} &= g_{12} = \frac{\alpha}{16a^3(u^*)^2}[\alpha^2 + 2a^2 - 3a\alpha + 4aA\alpha(u^*)^2/b].
\end{cases} \tag{20}$$

We now introduce complex coordinates

$$z = -iy_1 + iy_2\bar{\lambda},$$

from which, we have

$$y_1 = \frac{(-i)[\lambda z + \bar{\lambda}\bar{z}]}{\sqrt{\gamma(4-\gamma)}}, \qquad y_2 = \frac{z + \bar{z}}{\sqrt{\gamma(4-\gamma)}}. \tag{21}$$

---

[4]The standard normal form analysis essentially gives a polynomial approximation (usually up to order 3) of the stable and unstable manifolds of the equilibrium point of a dynamical system

Then the linear part of the map becomes

$$z_{t+1} = -i[-\gamma y_{1,t} - \gamma y_{2,t}] + i\bar{\lambda}y_{1,t} = -i[(\lambda + \bar{\lambda})y_{1,t} - \lambda\bar{\lambda}y_{2,t}] + i\bar{\lambda}y_{1,t} = \lambda z_t. \tag{22}$$

Hence the normal form of the map can be written in complex form as

$$z_{t+1} = \lambda z_t + \sum_{j+k\geq 2} \frac{1}{j!k!} a_{jk} z_t^j \bar{z}_t^k, \tag{23}$$

where

$$\begin{cases} a_{10} & = g_{10}\dfrac{-i\lambda}{\sqrt{\gamma(4-\gamma)}} \\ a_{01} & = g_{01}\dfrac{-i\bar{\lambda}}{\sqrt{\gamma(4-\gamma)}} \\ a_{11} & = \dfrac{1}{\gamma(4-\gamma)}[-i(\lambda+\bar{\lambda})g_{11} - 2\lambda\bar{\lambda}g_{20} + 2g_{02}] \\ a_{20} & = \dfrac{2}{\gamma(4-\gamma)}[-i\lambda g_{11} - \lambda^2 g_{20} + g_{02}] \\ a_{02} & = \dfrac{2}{\gamma(4-\gamma)}[-i\bar{\lambda}g_{11} - \bar{\lambda}^2 g_{20} + g_{02}]. \end{cases} \tag{24}$$

The following Lemma on the normal form of the map (18) (with 1 : 3 resonance) can be found in Kuznetsov [12] (p.382).

Lemma 2. *(Normal form map for a 1:3 resonance) The map (18) can be transformed by an invertible smooth and smoothly parameter-dependent change of variable, for all $\gamma = 1 + \epsilon$ with sufficiently small $|\epsilon|$, into the form*

$$\xi \mapsto \Gamma_\gamma(\xi) = \lambda(\gamma)\xi + B(\gamma)\bar{\xi}^2 + C(\gamma)\xi|\xi|^2 + O(|\xi|^4), \tag{25}$$

where

$$B(\gamma) = \frac{a_{02}(\gamma)}{2}, \tag{26}$$

and

$$C(\gamma) = \frac{a_{20}(\gamma)a_{11}(\gamma)[2\lambda(\gamma) + \bar{\lambda}(\gamma) - 3]}{2[\bar{\lambda}(\gamma) - 1][\lambda^2(\gamma) - \lambda(\gamma)]} + \frac{|a_{11}(\gamma)|^2}{1 - \bar{\lambda}(\gamma)} + \frac{a_{21}(\gamma)}{2}.$$

It can be verified that, when $\gamma = -\frac{\alpha}{2a} = 1$,

$$B(1) = \frac{1}{3}[(i - \sqrt{3})g_{11} + (3 - \sqrt{3}i)g_{02}].$$

Therefore, $Re(B(1)) = \frac{\sqrt{3}}{3}[-g_{11} + \sqrt{3}g_{02}]$. By using (20), one can see that $Re(B(1)) \neq 0$ provided $A(b\beta^2)^{\frac{1}{3}} \neq \frac{6-3\sqrt{3}}{4+2\sqrt{3}}$. This leads to the following bifurcation result.

Theorem 3. *For the system (16), if $A(b\beta^2)^{\frac{1}{3}} \neq \frac{6-3\sqrt{3}}{4+2\sqrt{3}}$, then there exists a single one-parameter family of fixed points of order 3 bifurcating from the positive equilibrium (u^*, u^*). The positive equilibrium, which is stable for $\gamma < 1$, becomes unstable for $\gamma > 1$, where the family of fixed points of order 3 which bifurcates on both sides of criticality is unstable on both sides.*

Theorem 3 indicates the dynamic structure near the positive equilibrium $P_o(p_o, p_o)$ and the hyperbolic periodic points bifurcating from P_o near the critical value $\gamma = 1$. We shall obtain more insight into these dynamics in the next section.

4. GLOBAL DYNAMICS

In the following discussion, some commonly used numerical simulation techniques, phase diagrams, time series and bifurcation diagrams, are employed in the study of the global dynamics of our nonlinear cobweb model. For more extensive simulations we refer the readers to Chiarella and He [5].

4.1. Pseudo-Phase plots for $L = 2$. It turns out that the phase plots of (Q_t, Q_{t-1}), instead of (p_t, p_{t-1}), give a clearer picture of the resonance bifurcation dynamics of the system. These plots are often called the **pseudo-phase plots**.

In the case of $L = 2$, the equation (10) can be written as the following 2-dimensional system

$$\begin{cases} x_{t+1} = \beta \dfrac{x_t^{\alpha/a} + y_t^{\alpha/a}}{2b + A\beta^2 [x_t^{\alpha/a} - y_t^{\alpha/a}]^2} \\ y_{t+1} = x_t, \end{cases} \tag{27}$$

where $x_t = Q_t, y_t = Q_{t-1}$.

Let $a = 1, b = 1, A = 0.005, \beta = 11$. Figure 2(a) is the pseudo-phase plot of the system when $\alpha = -1.9$ (and hence $\gamma < 1$), where we select three initial values $I_1(2.36, 2.4)$, $I_2(2.4, 2.4)$ and $I_3(1, 2.4)$. It shows that the solution with I_1, which is close to the equilibrium (say Q_o), converges to the fixed equilibrium and the solutions with I_2 and I_3 seem to converge to a bounded attractor, rather than Q_o. Based on Theorem 1 , when $\gamma < 1$, the unique equilibrium is locally asymptotically stable. Also, as indicated by Theorem 3, the nonlinear system has a single family of unstable fixed points of order 3 bifurcating from the equilibrium. In this simulation, the instability of the order 3 bifurcation drives the solutions with initial values I_2 and I_3 to the bounded attractor.

Figure 2(b) shows the case when $\alpha = -2.1$ (and hence $\gamma > 1$). It shows that the positive equilibrium is unstable. An unstable order 3 bifurcation can be observed and furthermore, when the iterations increase, the solution never settles down. It may indicate that the system has strange attractors, which may have different shape for different α.

For a much more detailed discussion and further simulations of the resonance structure of the model we refer the readers to Chiarella and He [5].

4.2. Time series plots for $L = 10$. In the case of $L = 10$, the time series plots of the price p can be used to identify stability, periodic and non-periodic oscillations. In Figure 3, we choose $a = 1, A = 0.05, b = 1, \beta = 11$ and $\alpha = -8, -11$ and -11.9, respectively. When $\alpha = -8$ (hence $\gamma < 1$), the positive equilibrium is asymptotically stable. For $\alpha = -11$ (hence $\gamma > 1$),

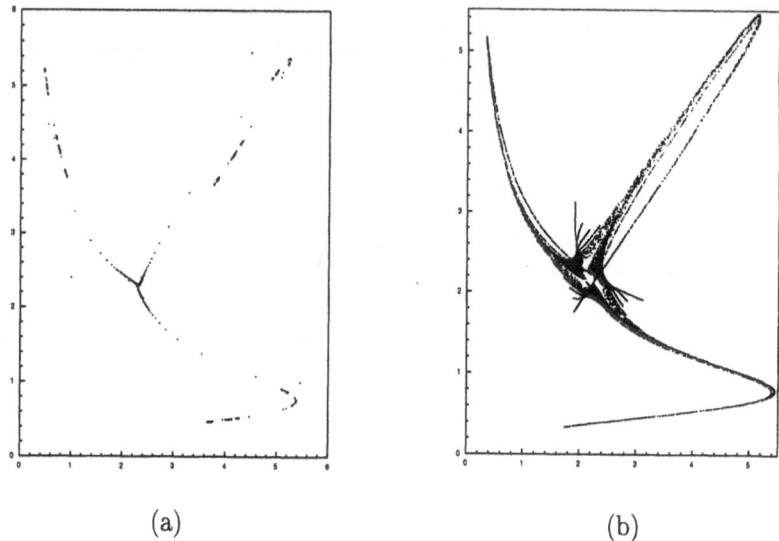

(a)   (b)

FIGURE 2.   Pseudo-phase plot with $\alpha = -1.9$ (a) and $\alpha = -2.1$ (b)

the solutions converge to a period-11 cycle; while for $\alpha = -11.9$, the time series plot does not indicate any pattern. In fact, when $\alpha = -11.9$, based on the Lyapunov exponent plot in Figure 4(b), the corresponding Lyapunov exponent is positive, which indicates chaotic behavior.

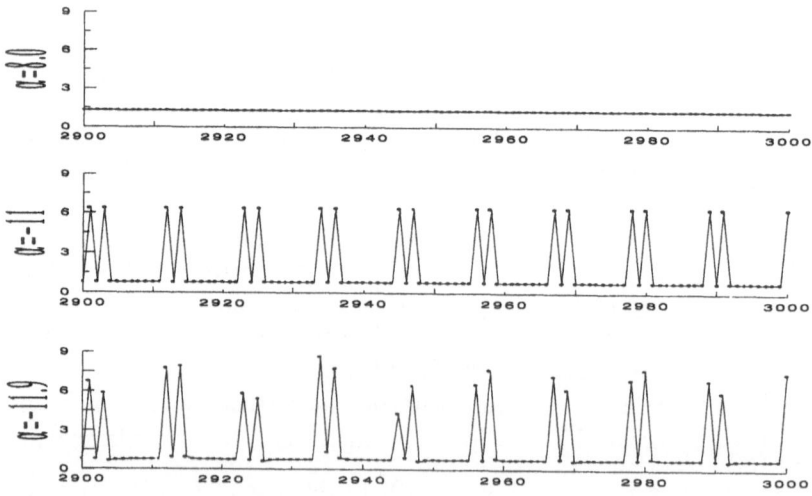

FIGURE 3.   Time series for $L = 10$ with $\alpha = -8, -11, -11.9$

4.3. **Bifurcation diagrams and Lyapunov exponent plots.** The bifurcation diagram and Lyapunov exponent of p as a function of α is shown in Figure 4 for both $L = 2$ (a) and $L = 10$ (b). Here we select $a = b = 1, A = 0.05$ and $\beta = 11$ in both cases. When $L = 2$, the positive equilibrium is stable for $\alpha > -2$. As α decreases, it bifurcates to a period 3 cycle; and as α decreases further, all types of cycles of order 6, 9, 12, ... appear and eventually a chaotic region appears where the corresponding Lyapunov exponent becomes positive. A similar dynamics can be observed in the case of $L = 10$. Obviously, an increase in the lag length stabilises the dynamics.

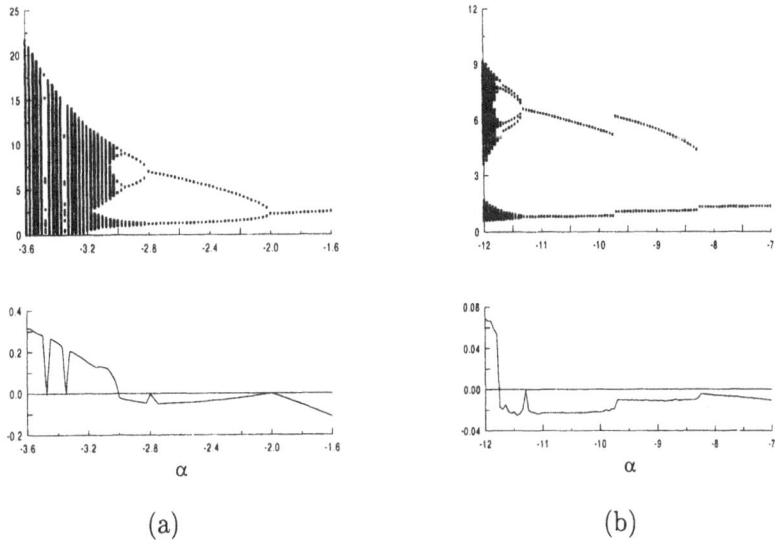

(a) (b)

FIGURE 4. Bifurcation diagrams and Lyapunov exponent
plots with respect to α for $L = 2$ (a) and $L = 10$ (b)

As indicated earlier, the risk aversion coefficient A does not appear in the local stability condition. However, it does affect the global dynamics of the nonlinear system. Figure 5 indicates the bifurcation diagrams and Lyapunov exponent plots as a function of A for both $L = 2$ (a) and $L = 10$ (b). For $L = 2$, we choose $\alpha = -3.2$ and for $L = 10$, $\alpha = -11.9$. The diagrams indicates that the complexity of the dynamics increases as risk aversion declines though the complexity is less for higher lag lengths.

5. CONCLUSION

We have formulated the cobweb model with constant elasticity supply and demand functions in a way that takes account of the risk aversion of producers and also allows them to learn about the distribution of the unknown price in the next period. The first and second moments of the price distributions are calculated from observations of past prices. The resulting cobweb dynamics form a complicated nonlinear expectations feedback structure whose

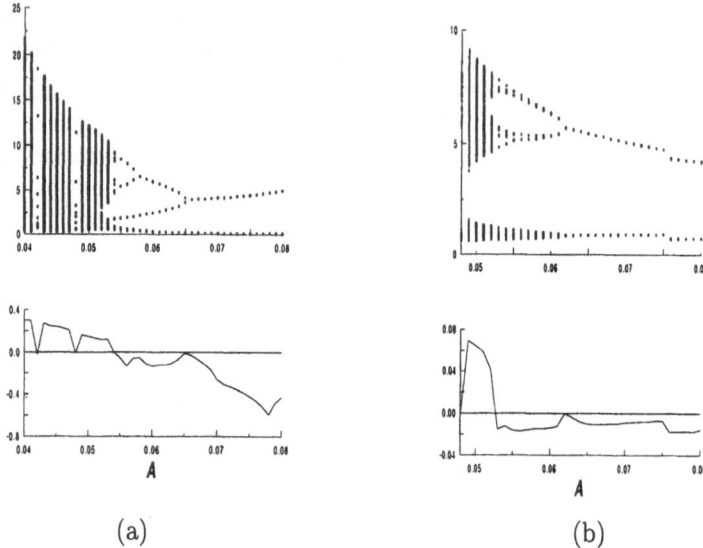

FIGURE 5. Bifurcation diagrams and Lyapunov exponent plots with respect to A for $L = 2$ (a) and $L = 10$ (b)

dimensionality depends upon the length of the window of past prices (the lag length) used to estimate the moments of the price distributions.

We have performed a local linear analysis of the dynamics for the case of general lag length and been able to determine the region of local stability as a function of the rate of supply and demand elasticities and the lag length. This local stability region turns out to be an increasing function of the lag length, which conforms the intuition obtained from earlier studies which use simpler adaptive expectations schemes.

We have carried out a detailed bifurcation analysis in the case of lag length 2 using the normal form of the dynamics. This analysis reveals that at the crossover from local stability to local instability the dynamics exhibits resonance behavior which is indicative of quite complicated dynamical behavior.

Our understanding of the dynamics obtained by the local linear and bifurcation analyses is then enhanced by a range of numerical simulations. In particular we focus on bifurcation diagrams and Lyapunov exponents. These indicate movement from stability to complex dynamics (including chaos) as the absolute value of the ratio of demand/supply elasticity increases, or as the coefficient of risk aversion decreases. We also find that all other things being equal, the dynamics are stabilised as the lag length increases.

The model developed here may be extended in a number of ways. First, it would be of interest to use supply and demand curves estimated from actual market data to see whether this type of cobweb model can reproduce at least the broad characteristics of real commodity price movements. Second, to base the producers' subjective distribution of next period's price on a more realistic dynamics for the commodity price as alluded to in footnote 1, here again it would be best to base this on an empirical study of the market under consideration. Some commodity price distributions might best be modelled as geometric Brownian motion but others may be drawn from mean-reverting processes. Finally, a richer set of learning mechanisms could be considered, such as Bayesian learning and various types of filtering techniques.

APPENDIX A.

Jury's Test

First we introduce the concepts of the *inners* of a matrix and the *positive innerwise* matrix, which can be found from the book by Elaydi [6] (pages 180–181).

Let $B = (b_{ij})_{n \times n}$ be a matrix. The *inners* of the matrix B are the matrix itself and all the matrices obtained by omitting successively the first and last rows and the first and last columns. A matrix B is said to be *positive innerwise* if the determinants of all its inners are positive.

Now consider the kth order scalar equation

$$x_{n+k} + p_1 x_{n+k-1} + p_2 x_{n+k-2} + \cdots + p_k x_n = 0, \tag{28}$$

where the p_i's are real numbers. Obviously, the characteristic equation of the equation (28) is given by

$$p(\lambda) = \lambda^k + p_1 \lambda^{k-1} + \cdots + p_k. \tag{29}$$

The Schur-Cohn criterion defines the conditions for the characteristic roots of equation (29) to fall inside the unit circle. More precisely, the following test will be used in the proof to Theorem 1.

Theorem 4. (**Jury's test**) *The zeros of the characteristic polynomial (29) lie inside the unit circle if and only if the following hold:*

- $p(1) > 0$
- $(-1)^k p(-1) > 0$,
- *the* $(k-1) \times (k-1)$ *matrices*

$$B_{k-1}^{\pm} = \begin{pmatrix} 1 & 0 & \cdots & 0 & 0 \\ p_1 & 1 & \cdots & 0 & 0 \\ \vdots & \vdots & \ddots & \vdots & 0 \\ p_{k-3} & p_{k-4} & \cdots & 1 & 0 \\ p_{k-2} & p_{k-3} & \cdots & p_1 & 1 \end{pmatrix} \pm \begin{pmatrix} 0 & 0 & \cdots & 0 & p_k \\ 0 & 0 & \cdots & p_k & p_{k-1} \\ \vdots & \vdots & \ddots & \vdots & \vdots \\ 0 & p_k & \cdots & p_4 & p_3 \\ p_k & p_{k-1} & \cdots & p_3 & p_2 \end{pmatrix}$$

are positive innerwise.

Proof of Theorem 1: Let $D_L(\lambda)$ be defined by (15). What needs to be shown is that all the zeros of the characteristic polynomial $D_L(\lambda)$ lie inside of the unit circle if and only if $0 \leq \gamma < 1$ or $-\frac{1}{L} < \gamma \leq 0$.

From $\gamma > 0$, it is easy to see that $D_L(1) = 1 + (L-1)\gamma > 0$ and $(-1)^L D_L(-1) = 1 - \gamma$ if L is odd and $(-1)^L D_L(-1) = 1$ if L is even.

Hence the first two conditions of Theorem 4 hold if and only if $0 \leq \gamma < 1$ or $-\frac{1}{L} < \gamma \leq 0$. To show the third condition is satisfied, it is enough to show that, for $k = 1, 2, \cdots, L-1$, the matrix $B_k^{\pm}$ with $p_1 = p_2 = \cdots = p_L = \gamma$ are positive if and only if $0 \leq \gamma < 1$ or $-\frac{1}{L} < \gamma \leq 0$.

Let $k = 2m$ be even. Then we have

$$
B_k^+ = \begin{pmatrix}
1 & 0 & \cdots & 0 & 0 & 0 & 0 & \cdots & 0 & \gamma \\
\gamma & 1 & \cdots & 0 & 0 & 0 & 0 & \cdots & \gamma & \gamma \\
\vdots & \vdots & \ddots & \vdots & \vdots & \vdots & \vdots & \mathinner{\mkern2mu\raise1pt\hbox{.}\mkern2mu\raise4pt\hbox{.}\mkern2mu\raise7pt\hbox{.}\mkern1mu} & \vdots & \vdots \\
\gamma & \gamma & \cdots & 1 & 0 & 0 & \gamma & \cdots & \gamma & \gamma \\
\gamma & \gamma & \cdots & \gamma & 1 & \gamma & \gamma & \cdots & \gamma & \gamma \\
\gamma & \gamma & \cdots & \gamma & 2\gamma & 1+\gamma & \gamma & \cdots & \gamma & \gamma \\
\gamma & \gamma & \cdots & 2\gamma & 2\gamma & 2\gamma & 1+\gamma & \cdots & \gamma & \gamma \\
\vdots & \vdots & \mathinner{\mkern2mu\raise1pt\hbox{.}\mkern2mu\raise4pt\hbox{.}\mkern2mu\raise7pt\hbox{.}\mkern1mu} & \vdots & \vdots & \vdots & \vdots & \ddots & \vdots & \vdots \\
\gamma & 2\gamma & \cdots & 2\gamma & 2\gamma & 2\gamma & 2\gamma & \cdots & 1+\gamma & \gamma \\
2\gamma & 2\gamma & \cdots & 2\gamma & 2\gamma & 2\gamma & 2\gamma & \cdots & 2\gamma & 1+\gamma
\end{pmatrix}
\tag{30}
$$

To evaluate the determinate of B_k^+, use (-1) to multiply the i-th columns and add to the $2m - (i-1)$-th columns, respectively, for $i = 1, \cdots, m$. Then we obtain

$$
|B_k^+| = \begin{vmatrix}
1 & 0 & \cdots & 0 & 0 & 0 & 0 & \cdots & 0 & \gamma-1 \\
\gamma & 1 & \cdots & 0 & 0 & 0 & 0 & \cdots & \gamma-1 & 0 \\
\vdots & \vdots & \ddots & \vdots & \vdots & \vdots & \vdots & \mathinner{\mkern2mu\raise1pt\hbox{.}\mkern2mu\raise4pt\hbox{.}\mkern2mu\raise7pt\hbox{.}\mkern1mu} & \vdots & \vdots \\
\gamma & \gamma & \cdots & 1 & 0 & 0 & \gamma-1 & \cdots & 0 & 0 \\
\gamma & \gamma & \cdots & \gamma & 1 & \gamma-1 & 0 & \cdots & 0 & 0 \\
\gamma & \gamma & \cdots & \gamma & 2\gamma & 1-\gamma & 0 & \cdots & 0 & 0 \\
\gamma & \gamma & \cdots & 2\gamma & 2\gamma & 0 & 1-\gamma & \cdots & 0 & 0 \\
\vdots & \vdots & \mathinner{\mkern2mu\raise1pt\hbox{.}\mkern2mu\raise4pt\hbox{.}\mkern2mu\raise7pt\hbox{.}\mkern1mu} & \vdots & \vdots & \vdots & \vdots & \ddots & \vdots & \vdots \\
\gamma & 2\gamma & \cdots & 2\gamma & 2\gamma & 0 & 0 & \cdots & 1-\gamma & 0 \\
2\gamma & 2\gamma & \cdots & 2\gamma & 2\gamma & 0 & 0 & \cdots & 0 & 1-\gamma
\end{vmatrix}
$$

$$
= (1-\gamma)^m \begin{vmatrix}
1 & 0 & \cdots & 0 & 0 & 0 & 0 & \cdots & 0 & -1 \\
\gamma & 1 & \cdots & 0 & 0 & 0 & 0 & \cdots & -1 & 0 \\
\vdots & \vdots & \ddots & \vdots & \vdots & \vdots & \vdots & \mathinner{\mkern2mu\raise1pt\hbox{.}\mkern2mu\raise4pt\hbox{.}\mkern2mu\raise7pt\hbox{.}\mkern1mu} & \vdots & \vdots \\
\gamma & \gamma & \cdots & 1 & 0 & 0 & -1 & \cdots & 0 & 0 \\
\gamma & \gamma & \cdots & \gamma & 1 & -1 & 0 & \cdots & 0 & 0 \\
\gamma & \gamma & \cdots & \gamma & 2\gamma & 1 & 0 & \cdots & 0 & 0 \\
\gamma & \gamma & \cdots & 2\gamma & 2\gamma & 0 & 1 & \cdots & 0 & 0 \\
\vdots & \vdots & \mathinner{\mkern2mu\raise1pt\hbox{.}\mkern2mu\raise4pt\hbox{.}\mkern2mu\raise7pt\hbox{.}\mkern1mu} & \vdots & \vdots & \vdots & \vdots & \ddots & \vdots & \vdots \\
\gamma & 2\gamma & \cdots & 2\gamma & 2\gamma & 0 & 0 & \cdots & 1 & 0 \\
2\gamma & 2\gamma & \cdots & 2\gamma & 2\gamma & 0 & 0 & \cdots & 0 & 1
\end{vmatrix}
\tag{31}
$$

Now for $i = 1, 2, \cdots, m$, first add the $2m - (i-1)$-the columns to the i-the columns, respectively. Then, multiply γ to the $2m - (i-1)$-th column and add to all the first $m-1$ columns. As a result, the upper left block matrix becomes a zero matrix and the lower left block matrix has 2γ as non-diagonal elements and $2\gamma + 1$ as diagonal elements. Correspondingly,

$$|B_k^+| = (-1)^m(1-\gamma)^m \begin{vmatrix} 2\gamma & \cdots & 2\gamma+1 \\ \vdots & \cdot^{\cdot^{\cdot}} & \vdots \\ 2\gamma+1 & \cdots & 2\gamma \end{vmatrix} \tag{32}$$

Multiply the first column by -1 and add it to all the rest of the columns. Then, multiply the columns 2 to k by -2γ and add them to the first column. As as result, a lower triangular matrix with $(1, 1, \cdots, 1, 2m\gamma + 1)$ as the diagonal elements is obtained. Therefore,

$$\det(B_k^+) = (1-\gamma)^m(L\gamma+1). \tag{33}$$

Similarly, one can show that

$$\det(B_k^-) = (1-\gamma)^m. \tag{34}$$

In conclusion, one has for $k = 2m$,

$$\det(B_k^+) = (1-\gamma)^m(2m\gamma+1), \qquad \det(B_k^-) = (1-\gamma)^m. \tag{35}$$

Next assume that $k = 2m + 1$. Then, using (35),

$$\det(B_k^-) = (1-\gamma)^{m+1}. \tag{36}$$

On the other hand,

$$B_k^+ = \begin{pmatrix} 1 & \cdots & 0 & 0 & 0 & \cdots & \gamma \\ \vdots & \ddots & \vdots & \vdots & \vdots & \cdot^{\cdot^{\cdot}} & \vdots \\ \gamma & \cdots & 1 & 0 & \gamma & \cdots & \gamma \\ \gamma & \cdots & \gamma & 1+\gamma & \gamma & \cdots & \gamma \\ \gamma & \cdots & 2\gamma & 2\gamma & 1+\gamma & \cdots & \gamma \\ \vdots & \cdot^{\cdot^{\cdot}} & \vdots & \vdots & \vdots & \ddots & \vdots \\ 2\gamma & \cdots & 2\gamma & 2\gamma & 2\gamma & \cdots & 1+\gamma \end{pmatrix} \tag{37}$$

Similarly, using row operations, we obtain that

$$\det(B_k^+) = (\gamma k + 1)(1-\gamma)^m. \tag{38}$$

Then from (36) and (38), for $k = 2m + 1$,

$$\det(B_k^+) = (\gamma k + 1)(1-\gamma)^m, \qquad \det(B_k^-) = (1-\gamma)^{m+1}. \tag{39}$$

Finally, it follows from (35) and (39) that $B_k^\pm$ are positive if and only if $0 \le \gamma < 1$ or $-\frac{1}{L} < \gamma \le 0$ and this completes the proof.

REFERENCES

[1] D. Arrowsmith, J. Cartwright, A. Lansbury, and C. Place. The Bogdanov map: Bifurcations, model locking, and chaos in a dissipative system. *Int. J. Bifurcation and Chaos*, 3:803–842, 1993.

[2] Z. Artstein. Irregular cobweb dynamics. *Economics Letters*, 11:15–17, 1983.

[3] J.-M. Boussard. When risk generates chaos. *Journal of Economic Behavior and Organization*, 29:433–446, 1996.

[4] C. Chiarella. The cobweb model, its instability and the onset of chaos. *Economic Modelling*, 5:377–384, 1988.

[5] C. Chiarella and X.-Z. He. *Learning about the Cobweb*, pages 244–257. 1998. Complex Systems'98, eds, Standish, R., Complexity Online Network, 1998, ISBN 0 7334 0537 1.

[6] S. Elaydi. *An Introduction to Difference Equations*. Springer, New York, 1996.

[7] J Hale and H. Kocak. *Dynamics and bifurcations*, volume 3 of *Texts in Applied Mathematics*. Springer-Verlag, New York, 1991.

[8] J. Holmes and R. Manning. Memory and market stability, the case of the cobweb. *Economics Letters*, 28:1–7, 1988.

[9] C. Hommes. Adaptive learning and roads to chaos: The case of the cobweb. *Economics Letters*, 1991:127–132, 1991.

[10] G. Iooss. *Bifurcations of maps and applications*, volume 36 of *Mathematics studies*. North-Holland, Amsterdam, 1979.

[11] R. Jensen and R. Urban. Chaotic price behaviour in a nonlinear cobweb model. *Economics Letters*, 15:235–240, 1984.

[12] Y.A. Kuznetsov. *Elements and applied bifurcation theory*, volume 112 of *Applied mathematical sciences*. SV, New York, 1995.

AS-AD Disequilibrium Dynamics and Economic Growth

Carl Chiarella,[1]   Peter Flaschel,[2]   Gangolf Groh,[2]   Willi Semmler[2]

[1] School of Finance and Economics, University of Technology, Sydney, Australia
[2] Faculty of Ecnomics, University of Bielefeld, Germany

Abstract. The paper reconsiders models of AS-AD growth from a disequilibrium perspective which in fact is easier to treat than AS-AD which admits disequilibrium in the labor market but not in the product market. We assume sluggish wage and price adjustments and a Metzlerian inventory dynamics. We obtain fluctuating rates of capacity utilization for both labor and capital. For exogenous productivity growth, the model implies six laws of motion. Stability, proved for sluggish adjustment speeds, gets lost through Hopf-bifurcations, implying the existence of attracting or repelling periodic motions close to such bifurcation values. We show that these results remain valid for endogenous growth, with new features concerning the production of technological change (which is subject to hysteresis).

Keywords. Integrated disequilibrium dynamics, growth, stability, cycles, endogenous technical change, hysteresis.

1. Introduction

Models of AS-AD growth, as the one in Sargent (1987, Ch.5), allow only for one type of disequilibrium, in the labor market, which is interpreted as being due to the assumed sluggish wage adjustment based on a conventional augmented money wage Phillips curve. The goods market is in equilibrium in a twofold way. Firms are on their supply or AS curve, operating at the desired level of capacity utilization. There is Keynesian goods and money market (or AD) equilibrium, since the price level, which is completely flexible, adjusts such that the Keynesian regime or quantity constraint becomes compatible with the profit maximizing choice of output of firms, but not compatible with labor market equilibrium. There has been an extensive discussion in the recent literature whether this is a sensible scenario for describing a Keynesian rationing of firms on the market for goods, see Chiarella, Flaschel, Groh and Semmler (1999, 7.10) for a brief survey.

We do not enter into this discussion here, but simply avoid the situation described above by positing that a Keynesian theory of fluctuations and growth should allow for both the stock of labor and the stock of capital to be under- or over-utilized. In line with Keynes (1936, p.4) we therefore should provide an explanation of the actual employment of all available resources, not only just labor. The traditional way to start the modeling of such a situation is to assume a sluggish adjustment of the price level as well – of a form that is still to be determined – and to study on this basis the evolution of IS-LM equilibria and the price level in time. One could follow thereby the example of the macroeconometric model in Powell and Murphy (1997) by assuming that prices sluggishly adjust towards competitive conditions as encapsulated in the AS curve.

Further reflection, however, shows that this is still too much equilibrium to start with, since firms always adjust towards IS-LM equilibrium with infinite speed. Assuming instead, and in line with price adjustment, also a somewhat sluggish quantity adjustment, indeed makes the Keynesian structure of the considered dynamics even more obvious and, as shown in Chiarella and Flaschel (1999a), also easier to analyze, though the number of dynamic variables is increased by two. In the place of ambiguous equilibrium descriptions we there have a clear sequence of causal events, which thus allows to understand AS-AD equilibrium growth from the perspective of disequilibrium growth to be presented below.

The following AS-AS model of disequilibrium growth is thus based on sluggish wage, price and quantity adjustment processes (including expectations) giving rise in particular to under- or over-utilized labor as well as capital in the dynamics it implies. We obtain a (minimally) complete and consistent structural form of a Keynesian monetary growth model with goods- as well as labor market disequilibrium, based on consistently formulated budget equations for households (workers and asset holders), firms and the government. We include Harrod neutral technical change of an exogenous rate n_l, supplementing natural growth n in the usual way, in order to approach the subject of the paper, the role of endogenously generated technical change in such models of AS-AD disequilibrium growth.

2. AS-AD disequilibrium growth: Exogenous technical change

The static and dynamic equations of this general continuous time model of disequilibrium growth are the following ones.[1]

---

[1] See also the appendix for the employed notation, and note that we use $\hat{x}$ to

First, households behavior (workers and asset holders) is described by:[2]

    1.   Households (workers and asset-holders):

$$C = (1 - \tau_w)\omega L^d + (1 - s_c)[\rho^e K + rB/p - T_c] \tag{1}$$
$$S_p = s_c[\rho^e K + rB/p - T_c] = (\dot{M}^d + \dot{B}^d + p_e \dot{E}^d)/p \tag{2}$$
$$\hat{L} = n = \text{const} \quad \text{growth rate of labor supply} \tag{3}$$
$$\omega = w/p \quad \text{real wage}, \quad u = \omega/x = wL^d/(pY) \quad \text{wage share} \tag{4}$$
$$\rho^e = (Y^e - \delta K - \omega L^d)/K \quad \text{expected real rate of profit} \tag{5}$$

Aggregate consumption C is based on classical saving habits, s_w, s_c, with the savings rate out of wages, s_w, set equal to zero for simplicity. We assume that wages are taxed with a uniform rate τ_w and that property income is subject to lump sum taxation T_c, again for reasons of simplicity. Allowing savings out of wages and other taxation schemes does not make much difference, see Chiarella and Flaschel (1999a,b) in this regard. We denote by L^d actual employment, by K the capital stock, and by L labor supply, which grows at the natural rate n. Bonds B are of the fixed price variety, with a money market determined nominal rate of interest r and with price set equal to 1. Real private savings S_p, here out of disposable interest income of asset holders solely, are allocated to desired changes in the stock of money, bonds and equities (E) held, the financial assets that exist in the considered economy.

    The production and investment behavior of firms is described next by the following set of equations:[3]

    2.   Firms (production-units and investors):

$$Y^p = y^p K, \; y^p = \text{const.}, \quad U = Y/Y^p, \tag{6}$$
$$L^d = Y/x, \; x = x_o \exp(n_l t) \; \hat{x} = n_l = \text{const.}, \quad V = L^d/L, \tag{7}$$
$$I = i_1(\rho^e - (r - \pi))K + i_2(U - \bar{U})K + (n + n_l)K, \tag{8}$$
$$S_f = Y_f = Y - Y^e = \mathcal{I}, \tag{9}$$
$$p_e \dot{E}/p = I + (\dot{N} - \mathcal{I}), \tag{10}$$
$$\hat{K} = I/K. \tag{11}$$

According to eq.s (6),(7), firms produce commodities in amount Y in the technologically simplest way possible, via a fixed proportions technology characterized by the potential output/capital ratio $y^p = Y^p/K$ and the ratio x between actual output Y and employment L^d needed to produce this output, where labor productivity x is assumed to rise with a constant rate of amount

---

denote a growth rate of a variable x.

   [2]See also Chiarella and Flaschel (1999a) for detailed presentations and analyses of such AS-AD disequilibrium growth models.

   [3]The case of smooth factor substitution is considered in Chiarella and Flaschel (1999a) and found to be of secondary importance with respect to the feedback structures contained in the model (and their implications).

n_l. This simple concept of technology allows for a straightforward definition of the rate of utilization U, V of capital as well as labor.[4]

In eq. (8) investment per unit of capital I/K is driven by three forces, the rate of return differential between the expected rate of profit ρ^e and the real rate of interest $(r - \pi)$, the deviation of actual capacity utilization U from the normal or non-accelerating-inflation rate of capacity utilization $\bar{U}$ and by an unexplained trend term $n + n_l$ which is determined such that capital widening in the steady state, where the first two terms are zero, is just sufficient to allow for 'full employment'. Savings S_f $(=$ income $Y_f)$ of firms, eq. (9), is equal to the excess of output Y over expected sales Y^e (and equal to planned inventory changes), since we assume in this model that expected sales are the basis of firms' dividend payments (after deduction of capital depreciation δK and real wage payments ωL^d).

The next equation shows the financial deficit of firms, due to the planned investment I and to unintended inventory changes $\dot{N} - \mathcal{I}$ (where N denotes the stock of inventories) which has to be financed by firms by issuing new equities. We assume here, as in Sargent (1987), that firms issue no bonds and retain no expected earnings. It follows, as expressed in eq. (10), that the total amount of new equities $\dot{E}$ issued by firms, valued at current share price p_e, must equal in value the sum of intended fixed capital investment and unexpected inventory changes, compare our later formulation of the inventory adjustment mechanism. Finally, eq. (11) states that (business fixed) investment plans of firms are always realized in this Keynesian (demand oriented) context, by way of corresponding inventory changes.

We now turn to a brief description of the government sector which in the present paper is not of central interest and is thus formulated in the simplest possible way in view of later steady state calculations.[5]

3. Government (fiscal and monetary authority):

$$T_w = (1 - \tau_w)\omega L^d, \quad \text{wage taxation,} \tag{12}$$

$$T_c \quad s.t. \quad (T_c - rB/p)/K = t_c^n = const, \quad \text{property income taxation,} \tag{13}$$

$$G = T_w + gK, \quad g = const, \quad \text{government expenditure} \tag{14}$$

$$S_g = T - rB/p - G, \quad T = T_w + T_c, \quad \text{budget surplus (or deficit),} \tag{15}$$

$$\hat{M} = \mu = const, \quad \text{growth rate of money supply} \quad M, \tag{16}$$

$$\dot{B} = pG + rB - pT - \dot{M}, \quad \text{accommodating debt financing.} \tag{17}$$

Government is here characterized by assuming as in Sargent (1987, Ch.5) that real (property income) taxes net of interest are constant and that the money supply grows at a constant rate μ. The consequences of these assumptions for government savings and debt financing are shown in equations (15), (17)

---

[4] Chiarella/Flaschel(1999a, Ch.5) show how such an approach can be extended to the case of smooth factor substitution without much change in its substance.

[5] A detailed presentation of the government sector is given in Chiarella and Flaschel (1999b).

in an obvious way. Much less restrictive fiscal and monetary policy rules are discussed in Chiarella, Flaschel, Groh, Köper and Semmler (1999).

The disequilibrium situation in the goods market is an important component driving the dynamics of the economy. This situation, as far as quantity adjustment processes are concerned, is described by the following equations:

4. Disequilibrium situation (goods-market adjustments):

$$S = S_p + S_g + S_f = p_e \dot{E}^d/p + \mathcal{I} = I + \dot{N} = p_e \dot{E}/p + \mathcal{I}, \tag{18}$$
$$Y^d = C + I + \delta K + G, \tag{19}$$
$$N^d = \beta_{n^d} Y^e, \quad \mathcal{I} = (n + n_l)N^d + \beta_n(N^d - N), \tag{20}$$
$$Y = Y^e + \mathcal{I}, \tag{21}$$
$$\dot{Y}^e = (n + n_l)Y^e + \beta_{y^e}(Y^d - Y^e), \tag{22}$$
$$\dot{N} = Y - Y^d = S - I. \tag{23}$$

Equation (18) of this disequilibrium block of the model describes various identities that can be related with the ex post identity of savings and investment for a closed economy. It is here added for accounting purposes solely. Equation (19) then defines aggregate demand Y^d which is never constrained in the present model.

In eq. (20) desired inventories N^d are assumed to be a constant proportion of expected sales Y^e and intended inventory investment $\mathcal{I}$ is then determined on this basis via the adjustment speed β_n multiplied by the current gap between intended and actual inventories $(N^d - N)$ and augmented by a growth term that integrates in the simplest way the fact that this inventory adjustment rule is operating in a growing economy. Output of firms Y in equation (21) is the sum of expected sales Y^e and planned inventory adjustments $\mathcal{I}$. Sales expectations are here formed in a purely adaptive way, again augmented by a growth term to take account of long run natural growth; see equation (22). Finally, in eq. (23), actual inventory changes $\dot{N}$ are given by the discrepancy between output Y and actual sales Y^d equal to the difference between total savings S and fixed business investment I.

We now turn to the wage-price module or the supply side of the model as it is often characterized in the literature.[6]

5. Wage-price adjustment equations and inflationary expectations):

$$\hat{w} = \beta_w(V - 1) + \kappa_w(\hat{p} + n_l) + (1 - \kappa_w)(\pi + n_l), \tag{24}$$
$$\hat{p} = \beta_p(U - 1) + \kappa_p(\hat{w} - n_l) + (1 - \kappa_p)\pi, \tag{25}$$
$$\dot{\pi} = \beta_\pi(\alpha\hat{p} + (1 - \alpha)(\mu - (n + n_l)) - \pi). \tag{26}$$

This 'supply side' description is based on fairly symmetric assumptions on the causes of wage- and price-inflation. Money wage inflation $\hat{w}$ according to

---

[6]A much more advanced wage price block is discussed in Chiarella, Flaschel, Groh and Semmler (1999).

eq. (24) is driven on the one hand by a demand pull component, given by the deviation of the actual rate of employment V from the NAIRU-rate $\bar{V}$, and on the other hand by a cost push term measured by a weighted average of the actual rate of price inflation $\hat{p}$, augmented by the rate of productivity growth n_l,[7] and a medium-run expected rate of inflation π, again augmented in the just described way. Similarly, in eq. (25), price inflation is driven by the demand pressure term $(U - \bar{U})$, where $\bar{U}$ denotes the non-accelerating-inflation rate of capacity utilization, and the weighted average of the actual rate of wage inflation $\hat{w}$ and a medium-run expected rate of inflation π, the former diminished by productivity growth, since this reduces the cost pressure that firms are experiencing. The latter rate of inflation, expected as average over the medium run, is in turn determined by a composition of backward-looking (adaptive) and forward-looking (regressive) expectations. It is easy to show, under suitable assumptions, that this amounts to an inflationary expectations mechanism as in (26) where expectations are governed in an adaptive way by a weighted average of the actual and the steady state rate of inflation. It is also easy to extend this mechanism to more refined backward looking procedures as well as to more refined price forecasting rules (so-called p^*-concepts and the like).[8]

Finally, we have the following set of equilibrium conditions with respect to the financial assets considered in this model:

6. Equilibrium conditions (asset-markets):

$$M = M^d = h_1 pY + h_2 pK(r_o - r) \quad [B = B^d, E = E^d], \tag{27}$$

$$r = \frac{\rho^e pK + \dot{p}_e E}{p_e E}, \tag{28}$$

$$\dot{M} = \dot{M}^d, \ \dot{B} = \dot{B}^d \quad [\dot{E} = \dot{E}^d]. \tag{29}$$

Asset markets are assumed to clear at all times, due to interest rate flexibility and the perfect substitute assumption as far as bonds and equities are concerned. The nominal interest rate r adjusts to clear the money market, (27), while the remainder of financial wealth is allocated to bonds and equities in a way that need not be considered explicitly in the following since asset holders are indifferent between these assets, since as stated, bonds and equities are assumed to be perfect substitutes, see equation (28). Finally, in eq. (29), it is assumed that wealth owners accept the inflows of money and bonds issued by the state for the current period, reallocating them only in the next period by adjusting their portfolios then anew. It is easy to check by means of the considered saving relationships that the assumed consistency of money and bonds flow supply and flow demand implies the consistency of the flow supply and demand for equities.

---

[7] This represents an augmented target in the wage negotiations of workers.

[8] See Chiarella and Flaschel (1999a) for further details on this wage-price block of the model.

Block 6. of the model provides us with a simple formula for the rate of interest on the intensive form level and is in all other respects irrelevant for the dynamical analyses that follow, since equities and bonds will not feed back into the dynamics of the private sector due to our choice of property income taxation and investment function, where the price of shares is not needed explicitly. Of course, the restrictive assumptions underlying this situation must be relaxed later on, see Köper and Flaschel (1999) in this regard.

Money demand, as well as all other behavioral and technological relationships, have been specified as simple linear functions. We use such linear relationships throughout since we want to formulate the dynamics on the basis of their intrinsic or unavoidable nonlinearities first. Behavioral nonlinearities may be important later on in order to get global boundedness in the case of instability, but they should then be introduced in a systematic fashion as a reflected response to the destabilizing feedback structures that may be obtained from the model in its present form.

It is obvious from this description of the model that it is, on the one hand, already a very general description of macroeconomic dynamics. On the other hand, it is still dependent on very special assumptions, in particular with respect to financial markets and the government sector. This can be justified at the present stage of analysis by observing that many of its simplifying assumption are indeed typical for macrodynamic models which attempt to provide a complete description of a closed economy, see in particular the model of Keynesian dynamics of Sargent (1987, Part I). We have considerably extended such a conventional model of a three sector / five market approach to economic dynamics in: 2. the sector of firms, 4. the disequilibrium adjustment process of the quantities produced and 5. the wage-price sector and the determination of inflationary expectations. Other extensions of this framework (a more plausible treatment of wealth W and less primitive policy rules) must here remain for future research. This is in line with the project begun in Chiarella and Flaschel (1999a), namely to develop a class of models of Keynesian variety (beginning with the supply-side oriented Keynes-Wicksell model considered in Chiarella and Flaschel (1995)) where each successor model removes at least one problematic feature of the directly preceding model type. The present model type is much further up in this hierarchy of Keynesian models. It provides a purely demand determined description of the macroeconomy with a particular emphasis on the behavior of firms and sluggish price as well as quantity adjustments. The stage meanwhile reached in the modeling of disequilibrium growth is surveyed in detail in Chiarella, Flaschel, Groh and Semmler (1999). At present it is however still the distorting lack of investigations of integrated models of this type – and not their specific form – which the research agenda of this paper seeks to address, here by incorporating the important issue of endogenous technological change in addition, see the section 4 of the paper.

3. The dynamics of the private sector

It is easy to reduce the extensive form dynamics of the preceding section to intensive form or state variable expressions, see Chiarella and Flaschel (1999a) for the details, giving rise to an integrated six–dimensional dynamical system in the wage share $u = w/x$, the full employment labor intensity in efficiency units $l^e = L \exp(n_l t)/K$, real balances per unit of capital $m = m/(pK)$, inflationary expectations π, sales expectations per unit of capital $y^e = Y^e/K$ and actual inventories per unit of capital $v = N/K$. There are further laws of motion (for bonds and equities per unit of capital) which however do not feed back into the core dynamics and are therefore here ignored for reasons of simplicity, see Chiarella and Flaschel (1999a) for their description. We denote in the following by β_{π_1} and β_{π_2} the expressions $\beta_\pi \alpha$, $\beta_\pi (1 - \alpha)$, respectively.

$$
\begin{aligned}
\hat{u} &= \kappa[(1 - \kappa_p)\beta_w(V - \bar{V}) + (\kappa_w - 1)\beta_p(U - \bar{U})] &&(30)\\
\hat{l}^e &= -i_1(\rho^e - r + \pi) - i_2(U - \bar{U}) &&(31)\\
\hat{m} &= \mu - (n + n_l) - \pi - (\kappa[\beta_p(U - \bar{U}) + \kappa_p \beta_w(V - \bar{V})]) &&\\
&\quad -(i_1(\rho^e - r + \pi) + i_2(U - \bar{U})) &&(32)\\
\dot{\pi} &= \beta_{\pi_1}\kappa[\beta_p(U - \bar{U}) + \kappa_p \beta_w(V - \bar{V})] + \beta_{\pi_2}(\mu - (n + n_l) - \pi) &&(33)\\
\dot{y}^e &= \beta_{y^e}(y^d - y^e) - (i_1(\rho^e - r + \pi) + i_2(U - \bar{U}))y^e &&(34)\\
\dot{v} &= y - y^d - (i_1(\rho^e - r + \pi) + i_2(U - \bar{U}) + n + n_l)v &&(35)
\end{aligned}
$$

These laws of motion are to be supplemented by the following algebraic equations, for output per unit of capital $y = Y/K$, aggregate demand per unit of capital $y^d = Y^d/K$ and the nominal rate of interest r that clears the money market, in order to make them an autonomous system of 6 differential equations in the six state variables enumerated above:

$$
\begin{aligned}
y &= (1 + n\beta_{n^d})y^e + \beta_n(\beta_{n^d}y^e - v)\\
y^d &= uy + (1 - s_c)(\rho^e - t_c^n) + i_1(\rho^e - r + \pi) + i_2(U - \bar{U})\\
&\quad + n + n_l + \delta + g\\
r &= r_0 + (h_1 y - m)/h_2.
\end{aligned}
$$

Note that these algebraic equations and the laws of motion are furthermore based on the following defining expressions for rates of capacity utilization, labor demand per unit of capital, and the expected rate of profit:

$$
\begin{aligned}
V &= l^{de}/l^e, \quad U = y/y^p, \quad l^{de} = L^d \exp(n_l t)/K = y/x_o\\
\rho^e &= y^e - \delta - uy
\end{aligned}
$$

Note finally that the above dynamical equations make use of the following solution of the wage-price block

$$\hat{w} - \pi - n_l = \kappa[\beta_w(V - \bar{V}) + \kappa_w\beta_p(U - \bar{U})], \quad \kappa = (1 - \kappa_w\kappa_p)^{-1}$$
$$\hat{p} - \pi = \kappa[\beta_p(U - \bar{U}) + \kappa_p\beta_w(V - \bar{V})], \quad \kappa = (1 - \kappa_w\kappa_p)^{-1}$$

in various places. Subtracting the second from the first equation furthermore implies

$$\hat{u} = \hat{w} - \hat{p} - n_l = \kappa[(1 - \kappa_p)\beta_w(V - \bar{V}) + (\kappa_w - 1)\beta_p(U - \bar{U})],$$

which gives the first of the differential equations shown.

In these reduced form equations, the interdependent wage and price Phillips curves defined in the preceding section have been separated from each other and now based on demand pressure expressions augmented by inflationary expectations in a seemingly conventional way, yet here based on demand pressure in the labor *and* the goods market, which now appear (with differing weights) in both of these equations for wage and price inflation.

Proposition 1
There is a unique interior steady–state solution or point of rest of the dynamics (30)–(35) fulfilling $u_0, l_0^e, m_0 \neq 0$ which is given by the following expressions:

$$y_0 = \bar{U}y^p, \quad \bar{V}l_0^e = l_0^{de} = y_0/x_o, \quad y_0^e = y_0^d = y_0/(1 + (n + n_l)\beta_{n^d}),$$
$$m_0 = h_1 y_0, \quad \pi_0 = \mu - (n + n_l), \quad r_0 = \rho_0^e + \pi_0, \quad \nu_0 = \beta_{n^d} y_0^e$$
$$\rho_0^e = \frac{g - t_c^n + n + n_l}{s_c} + t_c^n, \quad u_0 = \frac{y_0^e - \delta - \rho_0^e}{y_o},$$

Proof: See Chiarella and Flaschel (1999a, Proposition 6.4). ∎

Proposition 2
Consider the Jacobian J of the dynamics (30)–(35) at the steady state. The determinant of this $6 * 6$-matrix, $\det J$, is always positive. It follows that the system can only lose or gain asymptotic stability by way of a Hopf-bifurcation (if its eigenvalues cross the imaginary axis with positive speed).

Proof: The assertion is proved by exploiting appropriately linear dependencies between the rows of the Jacobian of the full dynamics at the steady state. See Chiarella and Flaschel (1999a, Proposition 6.4) for details. ∎

Proposition 3
This steady–state is locally asymptotically stable for all adjustment speeds $\beta_w, \beta_p, \beta_\pi, \beta_n$ chosen sufficiently low (and also β_{n^d} sufficiently low) and for sales expectations β_{y^e} that are revised sufficiently fast.

Proof: Freezing u, π, ν at their steady state values (and setting $\beta_n, \beta_{\pi_1}, \beta_{\pi_1}$ equal to zero) allows to apply to the remaining 3D dynamics (in l^e, m, y^e) the Routh Hurwitz theorem if h_2 is chosen sufficiently small and β_{y^e} sufficiently large, implying that the three eigenvalues of these reduced dynamics (at the steady state) must all have negative real parts in such a case. Next, one can show that the determinant of the Jacobian of the 4D dynamics (now including π) at the steady state, where β_{π_1}, β_{π_2} are now chosen positive (for all $\beta_{\pi_2} > 0$), must be positive which means that a small β_{π_1} will preserve the negativity of the real parts of the considered eigenvalues and add a further negative one, due to the continuity of eigenvalues with respect to the parameters of the dynamics. The same procedure applies to β_n which when made positive implies a negative determinant and thus gives rise to a negative real eigenvalue together with unchanged signs in the real parts of the other ones if the change in β_n is again sufficiently small. Finally, due to proposition 2, the determinant of the Jacobian of the full dynamics is always positive at the steady state and thus allows for another application of this procedure when the adjustment speeds of wages and prices are chosen sufficiently small. See also Chiarella and Flaschel (1999a, Ch.6) for further considerations of this type.[9] ∎

Proposition 4

The system will lose its local asymptotic stability if either β_w of β_p is chosen sufficiently large, if β_π is made large enough, or if β_n, β_{y^e} are chosen sufficiently large.

Proof: Assuming for example β_π sufficiently large will give rise to a negative principal minor of order 2 that dominates the sum of the other principal minors of order 2, which implies that one of the necessary conditions for local asymptotic stability of the Routh-Hurwitz criterion will be hurt in such a situation. ∎

We thus have that either wage or price flexibility must be destabilizing, that fast inflationary expectations are destabilizing as well as the Metzlerian inventory adjustment process when based on a sufficiently fast accelerator mechanism. Further, this proposition (when combined with the preceding ones) claims that such fast adjustments of prices, expectations or inventories will lead at certain critical parameter values to a cyclical loss of stability, either by the death of an unstable limit cycle as the Hopf-bifurcation value is approached or by the birth of a stable limit cycle when this bifurcation value is passed (since degenerate Hopf-bifurcations will generally not occur in these intrinsically nonlinear dynamics).

---

[9]The proof of this part of the proposition can also be obtained from Chiarella and Flaschel (1999a, 6.2) by setting β_n equal to zero and by applying the proof strategy there used to the then resulting 5D subdynamics. Positive, but small β_n combined with the positivity of the determinant of the Jacobian of the full dynamics then again imply the assertion 3.

4. Endogenous technical change

There exist various possibilities in the literature to model endogenous technical change, see Barro and Sala-i-Martin (1995) or Aghion and Howitt (1998) in this regard. We here follow Schneider and Ziesemer (1994, p.17) and use as representation of such technical change an approach based on Uzawa (1965) and Romer (1986), synthesized by Lucas (1988), called the URL approach in the following. Other representations of endogenous change will not significantly alter the conclusions of this section which therefore serves the purpose of illustrating the implications of an integration of the production of technological change into the AS-AD disequilibrium growth model of this paper.

The URL approach to endogenous technical change can be described by means of the following two equations, characterizing the productive activities of firms:

$$\dot{A} = \eta(L_2^d/L^d)A, \quad \eta' > 0, \quad \text{the research unit}, \tag{36}$$
$$Y = K^\beta(AL_1^d)^{1-\beta}A^\xi, \quad \xi > 0, \quad \text{the production unit}. \tag{37}$$

The activities of the employed workforce $L^d = L_1^d + L_2^d$ are here split between the production of output (37), described by a Cobb-Douglas production function based on the measure of labor productivity A and augmented by the Romer externality A^ξ, and the production of labor productivity growth as described by equation (36). The production function (37) is easily reformulated as follows:

$$Y = K^\beta(A^{\frac{1-\beta+\xi}{1-\beta}}L_1^d)^{1-\beta} = K^\beta(xL_1^d)^{1-\beta}$$

and shows in this way that it is of the usual type (Harrod neutral technical change), yet with a growth rate $\hat{x}$ of aggregate labor productivity x that exceeds the growth rate $\hat{A}$ produced by the firms due to the Romer externality: $\hat{x} = (1+\xi/(1-\beta))\hat{A}$. This approach to the production of technological change is considered in detail in Barro and Sala-i-Martin (1995, Ch.4).

In view of our approach of using linear relationships as much as possible in the initial formulation of our disequilibrium growth dynamics we reduce the above technological presentation to the case of fixed proportions in production and a linear production function for technical progress, which in the place of (36), (37) gives rise to

$$\dot{A} = \eta L_2^d/L^d, \quad \eta > 0 \tag{38}$$
$$Y = \min\{y^p K, AL_1^d A^\xi\} = \min\{y^p K, A^{1+\xi}L_1^d\} = \min\{y^p K, xL_1^d\} \tag{39}$$

in the notation used for our approach to disequilibrium growth in sections 2,3. The variable x of this earlier approach is now based on the relationship

$$x = A^{1+\xi}, \quad i.e., \quad n_l = \hat{x} = (1+\xi)\hat{A} = (1+\xi)\eta h$$

where $h = L_2^d/L^d$ denotes the proportion of employed workers that is devoted to the production of technological change. Depending on the variations of the scalar h there is thus now varying labor productivity growth n_l in the model in the place of the former assumption of a given rate of growth n_l of labor productivity x.

For illustrative purposes we assume as law of motion for the labor allocation ratio h the simple, but plausible rule

$$\dot{h} = \beta_h(V - \bar{V}), \; V = L^d/L \tag{40}$$

Firms therefore increase their efforts to increase labor productivity growth if the labor market gets tighter, and vice versa, since this signals how much buffer is available should for example the trend growth in labor demand exceed the growth rate of labor supply. This statement in fact implies that certain growth rates should also matter in the above law of motion for h, an extension of the model that will be considered in future extensions of this paper. We thus have now fluctuating employment in the aggregate and fluctuating allocation of the employed labor force between production proper and the production of technological change, producing fluctuations in labor productivity which in turn add to the fluctuations of the rate of employments V_1 in production as they are generated by the 6D dynamics.

We have now to distinguish between the overall rate of employment, $V = L^d/L$, and the one based on the production sector solely, $V_1 = L_1^d/L$ by way of the following revised algebraic relationships:

$$l^{de} = l_1^{de}/(1 - h), \; l_1^{de} = xL_1^d/K = y$$
$$V = l^{de}/l^e = (l_1^{de} + l_2^{de})/l^e = V_1 + V_2 = V_1 + hV = V_1/(1 - h), \quad i.e.,$$
$$V_1 = y/l^e, \quad V = y/((1 - h)l^e), \rho^e = y^e - \delta - ul^{de}.$$

We note that labor measured in efficiency units, for example labor supply, is now represented by xL in the place of $\exp(n_l t)L$, which however only means that all former expression are augmented by the given factor x_o. This does not change the form of the dynamics to a noteworthy degree, but is easier to read in the present situation. We observe that the formal structure of the model of the preceding sections is obtained, when $\beta_h = 0, h(0) > 0$ is assumed.

The first impression is that the dynamics have become more complex by the addition of endogenous technical change, since further intrinsic nonlinearities are introduced through the addition of a simple law of motion for the variable h, the ratio by which firms divide their workforce into productive and researching units. On the other hand, the wage price block gives rise to the same formal expressions as in the case of given technical change, due to the treatment of productivity increases in the formation of wage and price inflation. A second view however reveals that the state variable h, though it enters the initially considered dynamical system in various places, does so only via $V, n_l = \hat{x}$ and ρ^e. Furthermore, the rate of employment V is the only variable of the initial dynamical system that affects the new dynamical law for h.

The interior steady state solution for the above dynamics is no longer uniquely determined, since the variable h cannot be uniquely determined from setting the laws of motion to rest, due to the fact that an appropriate combination of the first six laws of motion generate the new law of motion (for h). The set of possible interior steady state solutions is thus now given by the following equations:

$$y_0 = \bar{U} y^p, \quad l_{10}^{de} = y_0, \quad \bar{V} l_0^e = l_0^{de} = l_{10}^{de}/(1 - h_0) \tag{41}$$

$$y_0^e = y_0^d = y_0/(1 + (n + (1 + \xi)\eta h_o)\beta_{n^d}), \tag{42}$$

$$m_0 = h_1 y_0, \quad \pi_0 = \mu - (n + (1 + \xi)\eta h_0), \tag{43}$$

$$r_0 = \rho_0^e + \pi_o, \quad \nu_0 = \beta_{n^d} y_0^e \tag{44}$$

$$\rho_0^e = \frac{g - t_c^n + n + (1 + \xi)\eta h_o}{s_c} + t_c^n, \quad \omega_0 = \frac{y_o^e - \delta - \rho_o^e}{l_o^d} \tag{45}$$

where h_0 can be any economically admissible number.

Proposition 5
There is a curve of interior steady–state solutions or points of rest of the dynamics which is given by the expressions (41) – (45) parameterized by an arbitrary choice of $h \in (0, 1)$.

Proposition 6
This steady–state is locally asymptotically stable for all adjustment speeds β_h sufficiently low in all cases where the 6D subdynamics is locally asymptotically stable.

This assertion follows directly from the fact that the dynamics with a given rate of labor productivity growth is qualitatively of the same type as the one where the variable h is of a given magnitude. Sufficiently sluggish wage, price and inflationary expectations adjustments coupled with fast sales expectations and a weak inventory accelerator mechanism, now also combined with slow shifts of employment between the production and research units of firms, will therefore be favorable for local asymptotic stability. Note in this context, that there is some kind of accelerator mechanism with regard to the research activities of firms which works as follows: In the case of a high employment rate V firms want to increase the growth of labor productivity by enlarging the number of workers in the R & D sector. Since, on the other hand, the number of employees concerned with the production of goods is already determined by aggregate demand, this results in a further increase of V, leading firms to chose again a higher value of h and so on. Thus, a destabilizing feedback mechanism emerges. Note furthermore, that the above stability assertions are here coupled with the situation of shock dependent convergence towards a continuum of steady states as the attractors of the dynamics. Should there be convergence to steady states it will be a path-dependent one, where history and shocks matter.

Proposition 7
Consider the Jacobian J of the dynamics at the steady state. The determinant of this 7 ⋆ 7-matrix, det J, is always zero, while the upper 6 ⋆ 6 principal minor is always positive (as in the 6D case of the preceding section). It follows that the system can only lose or gain asymptotic stability by way of a Hopf-bifurcation (if its eigenvalues cross the imaginary axis with positive speed).

Note for the considered situation that one eigenvalue must always be equal to zero, while no further eigenvalue can become zero in addition. Loss of stability therefore always takes place by the occurrence of two purely imaginary eigenvalues when the bifurcation point is reached. As before we get close to this situation either shrinking unstable limit cycle before the bifurcation point is reached or stable expanding ones after it has been passed.

Proposition 8
The system will lose its local asymptotic stability if β_h is made sufficiently large.

This proposition is proved in the same way as related ones for the system with exogenous technical change, here by simply showing that the parameter β_h appears in the trace of the Jacobian J of the dynamics at the steady state only once and there with a positive coefficient (due to the law of motion $\dot{h} = \beta_h(y/((1-h)l^e) - \bar{V}$ for the allocation of research workers h) which can be made arbitrarily large by means of the parameter β_h without change in the other elements in the trace of J.

The proofs for propositions 5 – 8 are thus not difficult to provide, since they are basically of the same type as the ones for the propositions of the preceding section. We thus in sum arrive at the conclusion that the attractors of the dynamics with endogenous technical change should not be very different from the ones of the dynamics considered in section 3, but are shock dependent, including the stable limit cycles that may be generated by the asserted existence of Hopf bifurcations. This means that the rate of productivity change of the economy will depend on average or in the limit on the history of the evolution of this economy.

5. Concluding remarks

In this paper we have investigated the dynamic properties of a Keynesian AS-AD disequilibrium model including not only sluggish wage adjustment but also sluggish prices as well as a Metzlerian inventory dynamics. Due to these elements the analysis, which is carried out first with exogenous Harrod neutral technical change and then with endogenous technological progress of the Uzawa-Romer type, is much easier than it would be on the basis of a conventional IS-LM-approach, see Flaschel, Groh, Semmler (1999). We have found that the dynamics are dominated by the AS-AD structure in both variants of technical progress. This means, that stability is supported by sluggish adjustments of wages, prices and inflationary expectations, by fast sales expectations and an inventory accelerator that is not too strong. If technical change is produced by the research activities of firms, stability furthermore requires, that the shifts of employment between the production of goods and the research sector are sufficiently slow. Thus, the inclusion of endogenous technical change adds realism to the model, but does not alter its behavior in a significant way.

6. Appendix: Notation

A. Statically or dynamically endogenous variables:

Y	Output
Y^d	Aggregate demand $C + I + \delta K + G$
Y^e	Expected aggregate demand
N	Stock of inventories
N^d	Desired stock of inventories
$\mathcal{I}$	Desired inventory investment
L^d	Level of employment
L_1^d	Employment in the production of goods
L_2^d	Employment in the production of technology
C	Consumption
I	Fixed business investment
I^p	Planned total investment $I + \mathcal{I}$
I^a	Actual total investment $= I + \dot{N}$ total investment)
r	Nominal rate of interest (price of bonds $p_b = 1$)
p_e	Price of Equities
S_p	Private savings
S_f	Savings of firms ($= Y_f$, the income of firms)
S_g	Government savings
$S = S_p + S_f + S_g$	Total savings
T	Real taxes
G	Government expenditure
ρ, ρ^e	Rate of profit (Expected rate of profit)
$V = L^d / L$	Rate of employment
Y^p	Potential output
$\Delta Y^e = Y^e - Y^d$	Sales expectations error
$U = Y / Y^p$	Rate of capacity utilization
K	Capital stock
w	Nominal wages
p	Price level
π	Expected rate of inflation (medium–run)
M	Money supply (index d: demand)
L	Normal labor supply

B	Bonds (index d: demand)
E	Equities (index d: demand)
W	Real Wealth
ω	Real wage ($u = \omega/x$ the wage share)
$\nu = N/K$	Inventory-capital ratio
$h = L_2^d/L^d$	Proportion of workers devoted to technology production

B. Parameters

$\bar{V} = 1$	1. NAIRU-type normal utilization rate concept (of labor)
$\bar{U} = 1$	2. NAIRU-type normal utilization rate concept (of capital)
δ	Depreciation rate
μ	Growth rate of the money supply
n	Natural growth rate
$i_{1,2} > 0$	Investment parameters
$h_{1,2} > 0$	Money demand parameters
$\beta_w \geq 0$	Wage adjustment parameter
$\beta_p \geq 0$	Price adjustment parameter
$\beta_\pi \geq 0$	Inflationary expectations adjustment parameter
$\beta_h \geq 0$	Adjustment parameter for the research activities of firms
$\alpha \in [0, 1]$	Weights for forward and backward looking expectations
$\beta_{n^d} > 0$	Desired Inventory output ratio
$\beta_n > 0$	Inventory adjustment parameter
$\beta_{y^e} > 0$	Demand expectations adjustment parameter
$\kappa_{w,p} \in [0, 1], \kappa_w \kappa_p \neq 1$	Weights for short– and medium–run inflation
κ	$= (1 - \kappa_w \kappa_p)^{-1}$
$y^p > 0$	Potential output–capital ratio ($\neq y$, the actual ratio)
$x > 0$	Output–labor ratio
$t(t^n = t - rb)$	Taxes (net of interest) per capital
$s_c \in [0, 1]$	Savings–ratio (out of profits and interest)
$s_w \in [0, 1]$	Savings–ratio (out of wages, $= 0$ here)
ξ	Parameter of the Romer externality

C. Mathematical notation

$\dot{x}$	Time derivative of a variable x
$\hat{x}$	Growth rate of x
l', l_w	Total and partial derivatives
$y_w = y'(l)l_w$	Composite derivatives
$r_o, etc.$	Steady state values
$y = Y/K, etc.$	Real variables in intensive form
$m = M/(pK), etc.$	Nominal variables in intensive form

7. References

AGHION, P. and P. HOWITT (1998): *Endogenous Growth Theory.* Cambridge, MA: MIT Press.

BARRO, R. and X. SALA-I-MARTIN (1995): *Economic Growth.* New York: McGraw Hill.

CHIARELLA, C. and P. FLASCHEL (1995): Real and monetary cycles in models of Keynes-Wicksell type, *Journal of Economic Behavior and Organisation*, 30, 327 – 351.

CHIARELLA, C. and P. FLASCHEL (1999a): *The Dynamics of Keynesian Monetary Growth. Macrofoundations.* Cambridge, UK: Cambridge University Press.

CHIARELLA, C. and P. FLASCHEL (1999b): Towards applied disequilibrium growth theory. I. The starting model. UTS Sydney: Discussion paper.

CHIARELLA, C., P.FLASCHEL, G. GROH and W. SEMMLER (1999): *Disequilibrium, Growth and Labor Market dynamics. Macroperspectives.* To appear: Springer Verlag.

CHIARELLA, C., P. FLASCHEL, G. GROH, C. KÖPER and W. SEMMLER (1999): Towards Applied Disequilibrium Growth Theory: VI. Substitution, money-holdings, wealth-effects and other extensions. UTS Sydney: Discussion Paper.

FLASCHEL, P., G. GROH and W. SEMMLER (1999): Investment of firms in capital and productivity growth. A macrodynamic analysis. University of Bielefeld: Discussion paper.

KEYNES, J.M. (1936): *The General Theory of Employment, Interest and Money.* New York: Macmillan.

KÖPER, C. and P. FLASCHEL (1999): Real – financial interaction. A Keynes-Metzler-Goodwin portfolio approach. Bielefeld University: Discussion paper.

LUCAS, R. (1988): On the mechanics of economic development. *Journal of Monetary Economics*, 22, 3 – 42.

POWELL, A. and C. MURPHY (1997): *Inside a Modern Macroeconometric Model. A Guide to the Murphy Model.* Heidelberg: Springer.

ROMER, P.M. (1986): Increasing returns and long-run growth. *Journal of Political Economy*, 94, 1002-37.

SARGENT, T. (1987): *Macroeconomic Theory.*, New York: Academic Press.

SCHNEIDER, J. and T. ZIESEMER (1994): What's new and what's old in new growth theory: Endogenous technology, microfoundation and growth rate predictions - A critical overview. Maastricht: Discussion paper.

UZAWA, H. (1965): Optimum technical change in an aggregate model of economic growth. *International Economic Review*, 6, 18-31.

Market Takeovers in Medieval Trade*

Herbert Dawid[1] and Michael Kopel[2]

[1] Department of Economics, University of Southern California, Los Angeles, CA
[2] Department of Managerial Economics and Industrial Organization, Vienna University of Technology, Austria

Summary. In this paper we focus on the emergence of trading patterns in a market where two groups of farmers compete. The members of each group share the transportation costs and we study the effect of differences in these costs on the market evolution. The agents do not have sufficient information to be able to always correctly judge whether it is advantageous to trade or not. Rather they use a simple decision rule where they observe the utility of some other farmer in the population and change their action with some probability if their own utility is smaller than the observed one. For the resulting market dynamics the basins of attraction of the coexisting equilibria are analyzed using the concept of critical curves.

1. Introduction

One of the important questions in the economics of trade is which factors can be identified to account for certain trading patterns and long run market outcomes. In particular from a historical and European perspective this is an interesting topic. European Economic Growth between the tenth and the fourteenth centuries was facilitated by the Commercial Revolution of the Middle Ages ", i.e. the reemergence of Mediterranean and European long-distance trade after an extended period of decline. This reemergence was caused by political and social events and was supported by the development of suitable institutions. Trade during this period was characterized by asymmetric information and limited contract enforceability, which made an exchange between the involved parties complex, and led to an organizational problem[1]. During the Commercial Revolution the evolving institutions governed the relations between merchants and their agents. The employment of agents was important during the Middle Ages, since goods were sold abroad only after being shipped to their destination. The merchants employed these agents to enable

---

* Helpful remarks by G.I. Bischi and M. Reimann are gratefully acknowledged.

[1] In recent years, based on historical evidence, several arrangements have been found and analyzed which were able to resolve this difficulty at this time (see Greif (1992, 1993)). A particularly interesting example is the system used by the Maghribi Traders. The institution which developed among the Maghribi traders was a coalition, where members of this coalition provided each other with agency services. The members of the coalition exchanged information on past behavior of other members providing agency services and involved the whole group in ostracizing cheaters. This made the reputation of each coalition member a valuable good and motivated each member to follow the contractual arrangements.

them to reduce the costs of trade by saving the time and risk of travelling and diversifying sales, and in this way they could achieve a large efficiency gain.

In this paper we focus on the emergence of trading patterns and long run market outcomes due to differences in the costs of trade of two merchant groups, but we abstract from the institutional and incentive issues. Our main interest will lie on the path-dependence of market outcomes, i.e. the question of lock-in behavior of the dynamic process which describes the interaction between the behavior of the merchants and the market. In particular we will analyze under which circumstances a trading group which is more cost effective than the rest can become dominant in a market where its initial market share is relatively small. To keep the model as simple as possible we consider a situation where there are only two competing groups of traders. Furthermore, we study the evolution of a small local market where local farmers offer their goods. The basic structure of such a local model captures the essence of a typical relationship between merchants and agents in medieval trade.

The decision of each local farmer whether to join a trading coalition has to be made under uncertainty about the gains of trade. The farmer does not know in advance how the supply-demand-relationship on the market is going to be and how many local farmers will join the trading group, i.e. how much of the transportation costs he has to cover. Given this, the farmers can in general not determine the utility maximizing action. Rather they use a simple decision rule where they observe the utility of some other farmer in the population and change their action with some probability if their own utility is smaller than the observed one. Thus this paper contributes to the fast growing stream of literature analyzing adaptive behavior of economic agents. In this respect our approach is somehow related to imitation type models (e.g. Schlag (1998)), aspiration level models (e.g. Karandikar et al. (1998)) or word of mouth learning (e.g. Ellison and Fudenberg (1995)). In our analysis we will use the concept of critical curves (see Mira et al. (1996)) which can be used in our model to characterize the shape and global properties of the basins of attraction of the stable equilibria. Exactly such a characterization is needed to address the questions mentioned above.

The paper is organized as follows. We introduce the model in section 2. In section 3 we focus on the impact of cost differences on the number, stability and basins of attraction of coexisting equilibria. In this section we also analyze the occurring global bifurcations and discuss the economic interpretation of the different changes of behavior. In section 4 we study the implications of an increase in the farmers' propensity to change strategy.

2. The Model

We consider a single good market where farmers from two different villages (we call them village A and village B) may offer their product via mediating agents in every period $t = 0, \ldots \infty$. Every farmer has also the option to consume all his produced goods himself without offering anything on the market. For reasons of simplicity we assume that the number of farmers in both villages is equal (denoted by $N > 0$) and that each farmer produces the same amount which is our unity. Consuming his product gives the farmer a utility of $u > 0$ where we again assume that this value is equal for all individuals. The farmers who have decided to offer their goods on the market together employ agents to transport the products to the marketplace and to sell it there[2]. The cost per farmer for this service is given by $C_i(n_{i,t}) = \frac{c_i}{1+\alpha_i n_{i,t}}$ for village $i \in \{A, B\}$. Here $n_{i,t}$ denotes the number of farmers in village i who decided to offer their product on the market in period t. Note that the transportation cost per farmer decreases with the number of participating individuals since the cost for necessary transportation means like a rack waggon are divided among them. We assume that on the market the price of the good is determined by a standard linear demand curve:

$$p_t = a - \tilde{b}(n_{A,t} + n_{B,t}).$$

For the following analysis it is more convenient to look at the fractions of farmers participating in the market rather than at the actual numbers. We denote this fraction by $x_{i,t} = \frac{n_{i,t}}{N}$ and now get the following net profit of a farmer who sells his goods at the market:

$$\pi_i(x_{A,t}, x_{B,t}) = a - b(x_{A,t} + x_{B,t}) - \frac{c_i}{1 + \beta x_{i,t}}, \qquad (2.1)$$

where $b = N\tilde{b}$ and we assume $\beta = N\alpha_1 = N\alpha_2$ to be equal in both villages. We assume that medieval farmers were risk neutral and thus utility is simply given by their net profit.

In order to make their decision whether to offer their products on the market or rather consume them the farmers follow a simple rule of thumb. Each period every farmer observes the profit of one randomly selected inhabitant of his village. In case the profit of the observed farmer is larger than his own, he changes his action with a probability which is increasing with the difference between the two payoffs and asymptotically goes to one if the payoff difference goes to infinity. We use the following analytical expression to model this switching probability:

$$\mathbb{P}(F \text{ changes behavior}) = \begin{cases} \frac{2}{\pi} \arctan(\lambda_i \Delta \pi) & \Delta \pi > 0 \\ 0 & \Delta \pi \le 0, \end{cases}$$

---

[2] We do not consider the hazards which threatened travelling traders at this time, in particular bandits and tax collectors of the ruling class. The interaction of these two groups with local farmers has been analyzed in detail e.g. in Feichtinger and Novak (1994).

where F denotes an arbitrary farmer, $\Delta\pi$ the difference between the observed and the own payoff and λ_i, $i \in \{A, B\}$ measures the general propensity to change behavior in village i.

We will analyze the evolution of the farmers' behavior using a dynamical system approach. Assuming that the number of farmers in both villages is large we consider the trajectory of the expected fraction of individuals in both populations who participate in the market. In order to do this we first calculate the expected number of individuals in village i who decide to trade in period $t + 1$ (action $a_t = 1$ means trading, $a_t = 0$ is consuming). Let us first assume that the utility of trading in period t was larger than the utility of consumption, i.e. $\pi_i(x_{A,t}, x_{B,t}) > u$ (T denotes a trading farmer, C a consuming farmer)

$$
\begin{aligned}
\mathbb{E}n_{i,t+1} & \\
&= N[\mathbb{P}(a_{t+1} = 1|a_t = 1)\mathbb{P}(a_t = 1) + \mathbb{P}(a_{t+1} = 1|a_t = 0)\mathbb{P}(a_t = 0)] \\
&= N[x_{i,t} + (\mathbb{P}(a_{t+1} = 1|a_t = 0, \text{F meets T})\mathbb{P}(\text{F meets T}) \\
&+ \mathbb{P}(a_{t+1} = 1|a_t = 0, \text{F meets C})\mathbb{P}(\text{F meets C}))(1 - x_{i,t})] \\
&= N[x_{i,t} + \frac{2}{\pi}\arctan(\lambda_i(\pi_i(x_{A,t}, x_{B,t}) - u))x_{i,t}(1 - x_{i,t})].
\end{aligned}
$$

Note that $\mathbb{P}(a_{t+1} = 1|a_t = 1) = 1$ and $\mathbb{P}(a_{t+1} = 1|a_t = 0, \text{F meets C}) = 0$. It is easy to see that for $\pi_i(x_{A,t}, x_{B,t}) < u$ we have

$$\mathbb{E}n_{i,t+1} = N[x_{i,t} - \frac{2}{\pi}\arctan(\lambda_i(u - \pi_i(x_{A,t}, x_{B,t})))x_{i,t}(1 - x_{i,t})].$$

Expressing these dynamical systems again in terms of $x_{i,t}$ we finally get the system

$$
\begin{aligned}
x_{A,t+1} &= x_{A,t} + x_{A,t}(1 - x_{A,t})\frac{2}{\pi}\arctan(\lambda_A(\pi_A(x_{A,t}, x_{B,t}) - u)) \\
x_{B,t+1} &= x_{B,t} + x_{B,t}(1 - x_{B,t})\frac{2}{\pi}\arctan(\lambda_B(\pi_B(x_{A,t}, x_{B,t}) - u)).
\end{aligned}
\tag{2.2}
$$

Obviously in our model we only have to consider the dynamic behavior on $[0, 1] \times [0, 1]$ and it is plain that the unit square is invariant under this dynamical system.

Such Word-of-Mouth learning dynamics have already been examined in the framework of Battle of the Sexes games in Dawid (1999). For recent analyses of learning under weak information see also Brock and Hommes (1997), Dawid (2000) or Gale and Rosenthal (1999).

3. Cost Differences, Market Dominance and Market Takeover

In what follows we will consider the question whether in the long run the market will be taken over by farmers from one of the two villages or whether the market will be shared. In particular we are interested in the impact of cost advantages and differences in the initial market shares on the long run evolution of the market.

It is easy to see that all corners of the unit square are fixed points of (2.2) and the edges are invariant. Furthermore, any interior fixed point has to be an equilibrium in a sense that each single farmer is indifferent between consuming or offering the products on the market. Fixed points on the edges are not necessarily Nash equilibria, however any fixed point which is stable with respect to (2.2) has to be a Nash equilibrium.

In order to study the equilibria and the dynamics of the system in such a case it is helpful to consider the conditions under which the profit of selling on the market exceeds that of consuming. Straight forward calculations show that a farmer in village i prefers to offer his good on the market iff

$$x_{j,t} < f_i(x_{i,t}) := \frac{a-u}{b} - x_{i,t} - \frac{c_i}{b(1+\beta x_{i,t})} \quad j \neq i.$$

The functions f_A, f_B will in what follows be used to determine the equilibria of the model and the direction of the dynamics. We start by considering a symmetric case where the costs of transportation to the market are equal for both villages (i.e. $c_A = c_B$). In our analysis we will focus on the impact of changes of the c_i. All other parameters will be assumed to be constant throughout the study. In order to guarantee that a single farmer has an incentive to enter an unserved market we always assume that $c_i < a - u$. The starting point of our analysis is a situation where the transportation costs c_i are such that it never pays off for farmers in village i to enter the market if all farmers in village j are already in the market. This means that $f_i(x_i) < 1 \ \forall x_i \in [0,1]$. The condition for this to hold is

$$c_i > c^{(1)} := \frac{b}{4\beta} \left(\beta \left(1 - \frac{a-u}{b} \right) - 1 \right)^2. \tag{3.1}$$

The equilibria and their stability can now easily be derived from figure 3.1 where we depict the unit square with the two functions f_A and f_B. We use the parameter values $a = 300$, $b = 100$, $c_A = c_B = 210, \beta = 1, u = 32, \lambda_A = \lambda_B = 0.1$. Note that we consider populations with very small propensities to change strategy here. We will however, discuss the effect of increasing the values of λ_i in the next section.

Two of the four corners – $(0,0)$ and $(1,1)$ – are no Nash equilibria and therefore unstable with respect to (2.2), the other two – $(0,1)$ and $(1,0)$ – are pure strategy equilibria and locally asymptotically stable. There is a single intersection point of f_A and f_B which is a mixed Nash equilibrium and a saddle point with respect to (2.2). The stable manifold of this saddle is given by the diagonal (which is invariant). In order to establish that this manifold divides the two basins of attraction it is sufficient to show that the diagonal is not only invariant with respect to the map but also with respect

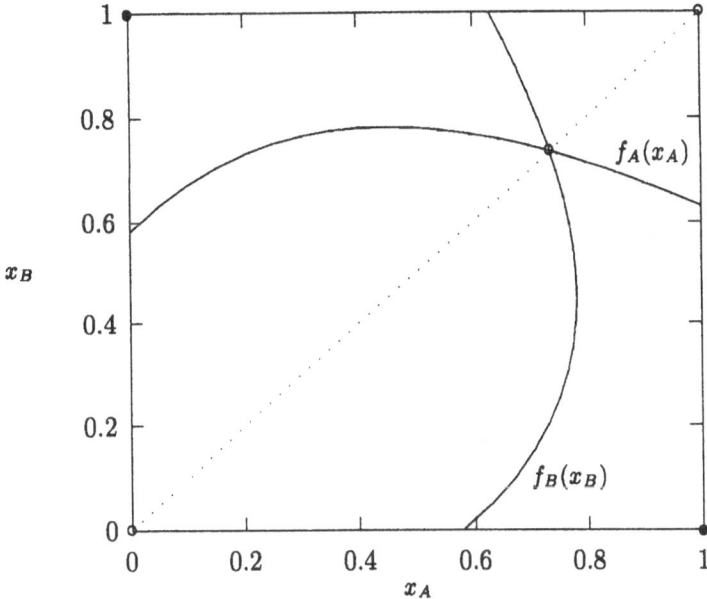

Fig. 3.1. The functions $f_A(x_A)$ and $f_B(x_B)$ and the stable manifold of the interior saddle point (dotted line) for symmetric cost parameters. Bullets mark stable fixed points of the dynamics, circles the unstable one.

to the inverse. This is trivially true if the inverse is single valued. In order to see whether the inverse is unique it is useful to consider the critical curves (LC) which separate the areas where the number of preimages of points coincide. They can be determined by applying the map to all points where the Jacobian of the map disappears (LC_{-1}); see Mira et al. (1996), Bischi and Kopel (1999) or Bischi et al. (1999) for more details on critical curves. In figure 3.2 we show the critical curve for this parameter constellation. It is obvious from this picture that in the largest part of the unit square points have only one preimage. There is only a small tongue around the saddle point equilibrium where points have three preimages (a so called 'Lip structure', see Mira et al. (1996) p. 156ff). However, since the eigenvalue of the Jacobian at the saddle point corresponding to the stable manifold is negative we know that at least locally around the saddle point each point on the diagonal has three preimages on the diagonal. Together this implies that the diagonal is also invariant with respect to the inverse map and accordingly separates the basins of attraction of the two pure strategy Nash equilibria. This is confirmed by our numerical calculations. In figure 3.2 we show the basins of attraction for the same parameter constellation as above[3].

We can clearly see that under these symmetric conditions the village which initially has the larger market share takes over the whole market in the long

---

[3] We would like to thank G.I. Bischi for providing us with the computer code used to plot the basins of attraction and the critical curves.

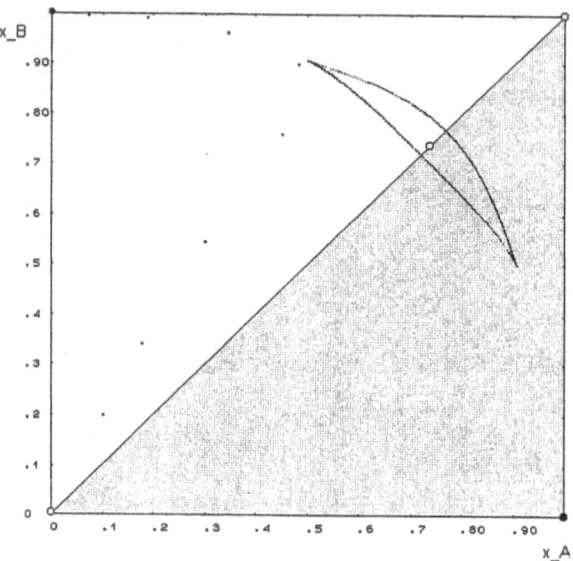

Fig. 3.2. The basins of attraction of the equilibria $(1,0)$ (light grey) and $(0,1)$ (white) and the critical curve surrounding the interior equilibrium for $c_A = c_B = 210$. All points inside the area surrounded by the curve have three preimages, all outside have only one.

run. However, it is interesting to look at the transient behavior inbetween. In figure 3.2 we also show a trajectory with initial conditions $x_{A,0} = 0.1$, $x_{B,0} = 0.2$. Initially, farmers in both villages prefer to offer their goods on the market and the trajectory evolves in the direction of the diagonal. However, at some point the price becomes so low that the small difference in transportation costs which is due to the different number of farmers selling on the market in the two villages results in utility which is smaller than u in village A but larger than u in village B. In other words, in village A consumption of the own goods becomes the preferred choice and farmers leave the market. However this does not lead to a price increase because the leaving farmers from village A are replaced by farmers from village B until eventually almost all farmers from village B offer their products on the market. The farmers from village A keep leaving until the whole market is taken over by individuals from village B. A symmetric picture would evolve if initially the farmers of village A have a slightly larger market share.

Now let us consider the impact of a reduction of transportation costs in one of the two villages, let us say village A. Such a cost reduction may result from innovations in the transportation technology, like larger rack waggons, faster horses or improvements in distribution systems (additional rest stations for the horses and the mediators). Note that the diagonal is no longer invariant if the two cost parameters differ. In a first step we consider a case where $c^{(1)} < c_A < c_B$. Here no additional fixed points of the dynamical sys-

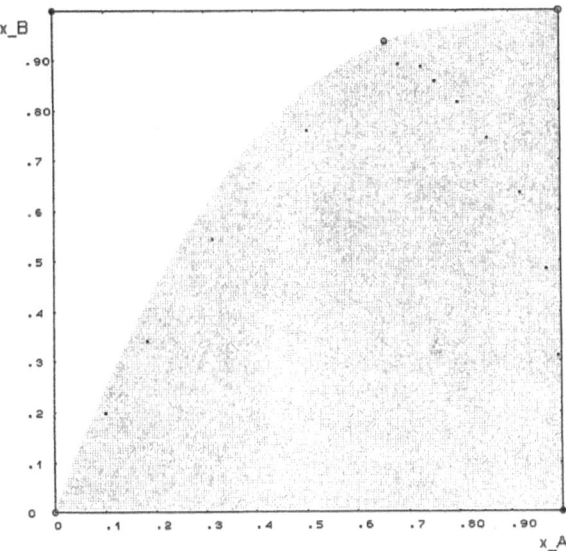

Fig. 3.3. The basins of attraction of the equilibria $(1,0)$ (light grey) and $(0,1)$ (white) for $c_A = 180$, $c_B = 210$.

tem appear, however the basins of attraction change. In figure 3.3 we show the basins for the cost values $c_A = 180$, $c_B = 210$ where all other parameters remain the same as in figure 3.2. We can clearly see that in this case the basin of attraction of the equilibrium $(1,0)$ enlarges. If we again consider the trajectory starting with $x_{A,0} = 0.1$, $x_{B,0} = 0.2$ we see that for a while the evolution is very similar to that in the case of symmetric cost parameters. The market share of village B remains larger than that of village A until a large fraction in both villages is on the market. Only at this stage the cost advantage of village A becomes crucial enabling the farmers from village A to stay profitably in the market whereas farmers from village B – although their number is larger – gain less utility than their peers who consume their own goods. Nevertheless for initial conditions in a rather large part of the unit square – if they have a substantial initial advantage – the farmers of village B are able to take over the market and drive out all individuals from village A although their transportation cost parameter is larger.

However, the picture changes instantaneously if village A is able to decrease their cost parameter below the level $c^{(1)}$. For our numerical example this parameter value is given by $c^{(1)} = 179.56$. If we assume that c_A can be reduced from the level of 180 by only 0.5 to 179.5 we suddenly have two more equilibria of the system. This can be easily seen in figure 3.4 where we depict the functions f_A, f_B for this parameter constellation.

The function f_A now has two intersection points with the line $x_B = 1$. Straightforward calculations show that the x_A-values of the two new equilib-

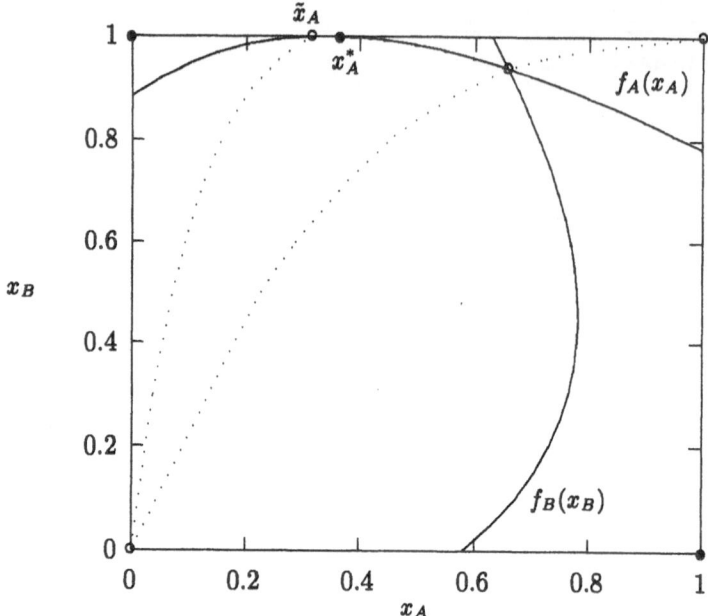

Fig. 3.4. The functions $f_A(x_A)$ and $f_B(x_B)$ and the stable manifold of the interior saddle point (dotted line) for the cost parameters $c_A = 179.5$, $c_B = 210$. Additionally the stable manifold of the saddle point $(\tilde{x}_A, 1)$ is shown. The area between the dotted lines is the basin of attraction of $(x_A^*, 1)$.

ria are given by

$$x_A = \frac{-\beta\left(1 - \frac{a-u}{b}\right) - 1 \pm \sqrt{\left(\beta\left(1 - \frac{a-u}{b}\right) - 1\right)^2 - 4\beta\frac{c_A}{b}}}{2\beta}.$$

Both of them are not only fixed points of (2.2) but also Nash equilibria where a certain fraction of farmers in village A are on the market but all of the farmers in village B. While the agents from village A which are on the market gain utility u and accordingly are indifferent between staying in the market and exiting, the utility of the farmers in village B is larger than u. Note however that their utility is smaller than in the equilibrium $(0, 1)$ which means that both new equilibria are Pareto dominated by $(0, 1)$ and thus inefficient. One of the two new equilibria – the one with the smaller fraction of individuals from village A in the market $((\tilde{x}_A, 1))$ – is unstable whereas the other one – which we denote by $(x_A^*, 1)$ – is, at least for small values of λ_A and λ_B, locally asymptotically stable. In figure 3.4 we also show the stable manifolds of the interior saddle point and of the saddle point on the upper edge. These two are crucial to understand the sudden changes in the basins of attraction which occur. Considering the critical curves (which we do not show in figure 3.4 in order to keep the picture as simple as possible) it is easy to see that the stable manifold of the saddle point on the upper edge lies entirely in an area with only one preimage. Thus it is invariant with respect to both the map and the inverse of the map. Together with the continuity of the map it follows that either no or all points are mapped from one side of this manifold to the other. Since the manifold is transversally repelling at the saddle point we conclude that a point and its image are always on the same side of this stable manifold. Similar arguments can be applied to the stable manifold of the interior saddle point[4]. Together this shows that all points between these two stable manifolds are attracted by the stable equilibrium $(x_A^*, 1)$. Thus, the previous basin of $(0,1)$ is now split into a basin of $(1,0)$ and a rather large basin of $(x_A^*, 1)$ (see figure 3.5). On the other hand, if we consider the basin of $(1,0)$ we see that it has remained almost unchanged.

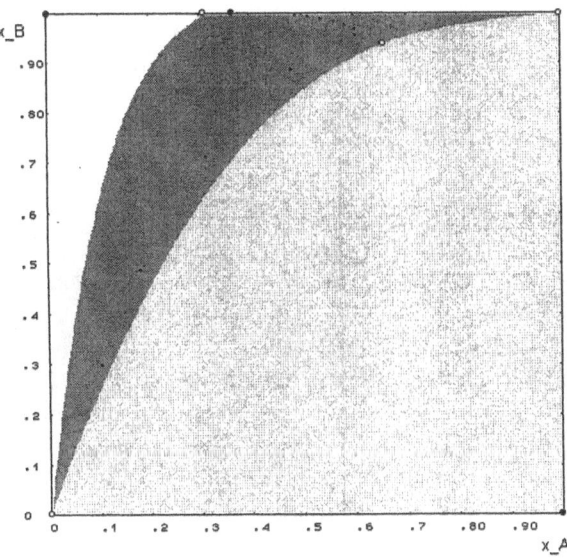

Fig. 3.5. The basins of attraction of the equilibria $(1,0)$ (light grey), $(0,1)$ (white) and $(x_A^*, 1)$ (dark grey) for $c_A = 179.5, c_B = 210$.

Let us consider for example a situation where initially 10% of farmers in village A and 30% of farmers in village B offer their goods on the market. For $c_A = 180$ village A would have been eventually driven out of the market. However for $c_A = 179.5$ the trajectory converges to the inefficient equilibrium $(x_A^*, 1)$. This means that by lowering the transportation costs village A is not able to increase the long run utility of its own individuals but only harms the other village.

---

[4] The situation is a little bit more intricate here since this manifold crosses a lip with three preimages. However, keeping in mind the symmetric case discussed above, continuity implies that these preimages result from overshooting of the equilibrium along the stable manifold.

If c_A is further decreased, the value x_A^* increases which means that for initial conditions in the corresponding basin of attraction the long run market share of village A farmers increases. However, their utility remains at u. The utility of village B farmers decreases however. From figure 3.4 we can further tell that the equilibrium $(x_A^*, 1)$ becomes unstable as soon as the value x_A^* crosses the intersection of f_B with the line $x_B = 1$. The value of c_A where this occurs is given by

$$c^{(2)} := \frac{c_B}{1+\beta} \left(\beta \left(\frac{a-u}{b} - 1 \right) + 1 \right) - \frac{\beta c_B^2}{b(1+\beta)^2}.$$

For our numerical example we have $c^{(2)} = 171.15$. If we consider the basins of attraction for $c_A = 172$ and $c_A = 171$ (see figure 3.6) we see that the basin of $(x_A^*, 1)$ still has a considerable extent for $c_A = 172$ but has completely vanished for $c_A = 171$. This means that by lowering the costs by one unit the farmers in village A are able to drive the farmers from village B out of the market and to gain utility larger than u for most of the initial conditions. The basin of attraction of $(0, 1)$ still exists, but now the number of farmers from village B which initially are in the market has to be much larger than the initial number of individuals from the other village in order to make it advantageous for individuals from village B to stay in the market.

Mathematically speaking the interior saddle point collides with the stable equilibrium on the upper edge resulting in a saddle point on the edge and another one outside the unit square (which is of course irrelevant for our model). The saddle point on the edge is attracting along the line $x_B = 1$ but transversally repelling. Accordingly, the previously stable manifold of the interior saddle point – which was the boundary between the basins of $(1, 0)$ and $(x_A^*, 1)$ – now disappears. Thus, now all points which lie below the stable manifold of $(\tilde{x}_A, 1)$ belong to the basin of $(1, 0)$.

Finally, if the transportation costs in village A are so small that a single individual from this village who enters a market where all farmers from village B are already present still gains a larger utility than u, then it is easy to see that the market is taken over by village A regardless of the initial condition. Formally this corresponds to the point where $f_A(0)$ crosses the line $x_B = 1$ and the equilibrium $(0, 1)$ becomes unstable. Obviously the cost threshold is given by

$$c^{(3)} = a - b - u.$$

For our parameter values we have $c^{(3)} = 168$.

Having discussed the structure of the basins in the presence of coexisting equilibria for different cost values in village A we end this section by summarizing the main implications of this analysis. If village A decreases its transportation cost parameter initially the effect is rather small with a slow increase of the size of the basin of attraction of $(1, 0)$. However, even if the costs are considerably below those of the competing agents, the farmers from

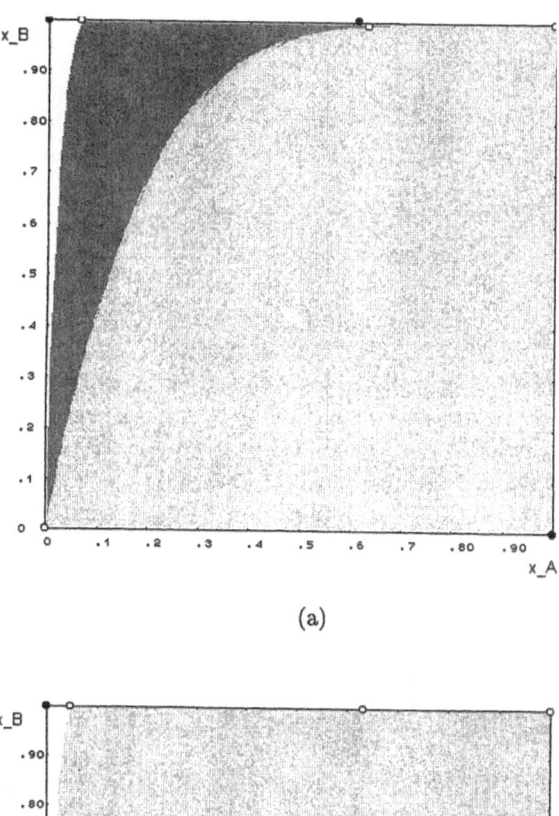

(a)

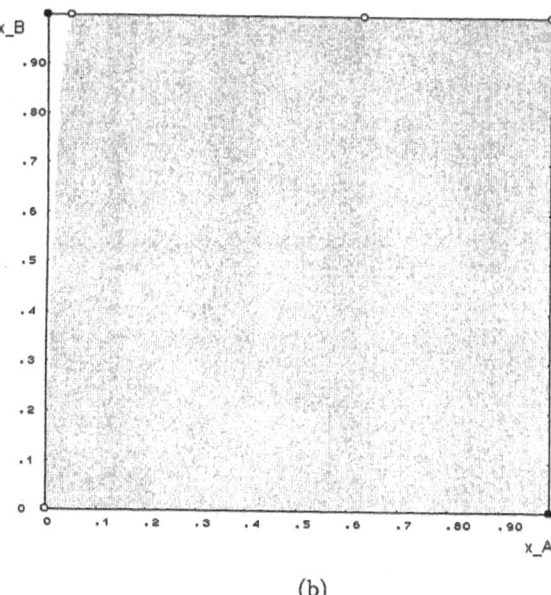

(b)

Fig. 3.6. The basins of attraction of the equilibria $(1, 0)$ (light grey), $(0, 1)$ (white) and $(x_A^*, 1)$ (dark grey) for (a) $c_A = 172, c_B = 210$ and (b) $c_A = 171$, $c_B = 210$.

village A are driven out of the market for most cases where the other village has initial advantages. This changes suddenly when the cost parameter crosses the value $c^{(1)}$. In this case, for most of the initial conditions with advantages for village B in the long run all its farmers are in the market and gain more utility, but at least a fraction of the individuals in village A are able to stay in the market. The next sudden change of behavior occurs for $c_A = c^{(2)}$. Here suddenly the basin of attraction of $(x_A^*, 1)$ disappears and all points which formerly belonged to this set now are part of the basin of attraction of $(1, 0)$. For all initial conditions where the farmers were able to just stay in the market they are now able to take over the market and drive the other individuals out of it. Once this barrier has been crossed a further decrease in the transportation costs leads to a continuous enlargement of the basin of $(1, 0)$ and for $c_A < c^{(3)}$ the farmers from village A have such a cost advantage that they are able to take over the market regardless of the initial conditions[5]. Before we continue with the analysis of the effect of differing levels of inertia in the populations we would like to point out that the inequalities $c^{(3)} < c^{(2)} < c^{(1)}$ which hold in our example do not hold in general. For different parameter constellations any of these inequalities may be violated leading to slightly different scenarios than those presented in this section. However, giving a complete characterization of all possible scenarios is beyond the scope of this paper.

4. The Impact of Inertia

In the previous section we have always assumed that the propensity to change strategy is very small even if the observed utility is smaller than the own. In other words we have analyzed the market evolution under the assumption of a high degree of inertia. In this section we will consider cases with less inertia by increasing the parameters λ_A and λ_B in our model. Since we have already documented the effect of increasing cost differences between the two populations, we will now restrict our attention to a single constellation, namely that where three stable equilibria coexist. All parameters are the same as in the previous section and we assume $c_A = 179.5$ throughout.

We know from the previous section that for $\lambda_A = \lambda_B = 0.1$ all three equilibria have basins of attraction with positive size which are simply connected sets bounded by smooth curves. If we increase the propensity to change we first observe that the interior saddle point becomes an unstable node via a flip bifurcation along the stable manifold. The following succession of local bifurcations leads to a 'wavelike' structure of the boundary between the basins of $(0, 1)$ and $(1, 0)$. The waves become more pronounced the larger the propensity to change.

---

[5] We neglect the set of initial conditions on the edges of the unit square, which however has measure zero.

Furthermore, the area of points with three preimages (the lip structure bounded by the critical curve) becomes larger with increasing $\lambda = \lambda_A = \lambda_B$. For a large set of values of λ this lip is entirely contained in the basins of attraction of $(1,0)$ and $(x_A^*, 1)$. However, if λ becomes sufficiently large eventually there is a contact between the lip and the stable manifold of $(\tilde{x}_A, 1)$. For this value a global bifurcation occurs marking the transition of the basin of $(0,1)$ from a simply connected to a multiply connected set, i.e. the basin of $(0,1)$ includes a large number of islands embedded in the basin of $(x_A^*, 1)$ after the bifurcation. We show the basins for $\lambda_A = \lambda_B = 3$ in figure 4.1a. The reason for the appearance of the islands is that now the area with three preimages intersects with the stable manifold of $(\tilde{x}_A, 1)$ (which was the boundary of the basin of $(0,1)$ before the bifurcation)[6]. This means that there is an open set I_0 of points which will eventually be attracted by $(0,1)$ and which has three preimages. As can be seen in figure 4.1b the set I_0 is very small in our case. One of the preimages of I_0 has to lie inside the closed curve LC_{-1}[7] and it can be numerically checked that this curve lies entirely to the right of the stable manifold of $(\tilde{x}_A, 1)$. Thus, at least one of the preimages of I_0 has to lie to the right of the stable manifold of $(\tilde{x}_A, 1)$ and, accordingly, we have an island of the basin of attraction of $(0,1)$ inside the previous basin of $(x_A^*, 1)$. We call this I_{-1}. Furthermore, at least one of the preimages of the set I_{-1} again lies to the right of the stable manifold of $(\tilde{x}_A, 1)$ and so on. This gives the complicated structure of the basins of $(0,1)$ and $(x_A^*, 1)$ shown in figure 4.1a.

Economically speaking we can observe several effects of a decrease in inertia. First, the basin of attraction of the equilibrium where the more cost effective population takes over the market gets smaller the less inertia there is. Furthermore, for a very low level of inertia the complicated structure of the basins of $(0,1)$ and $(x_A^*, 1)$ makes it extremely difficult to predict the long run outcome if the less cost effective traders initially have a larger market share. Given the intermingled basins it is not even clear whether an increase of the initial market share of the population with higher transportation costs is advantageous for its members. If we again look at figure 4.1a we can see that given an initial value of x_A a slight increase of x_B might lead to convergence to $(x_A^*, 1)$ rather than $(0,1)$ and therefore to a smaller utility in the long run for the members of village B. Thus we can observe some kind of sensitive dependence of the long run behavior on initial conditions. Note however that this notion of sensitive dependence on initial conditions differs from that usually used in the connection with chaotic dynamics. Whereas chaotic dynamics would lead to unpredictable fluctuations (on a chaotic attractor),

---

[6] Following standard notation we denote in figure 4.1b the set of points with three preimages – which lies to the right of LC – by Z_3, and the set of points with one preimage – left of LC – by Z_1.

[7] Recall that this is the set of points where the Jacobian vanishes. Note that this curve does not enter the region depicted in figure 4.1b.

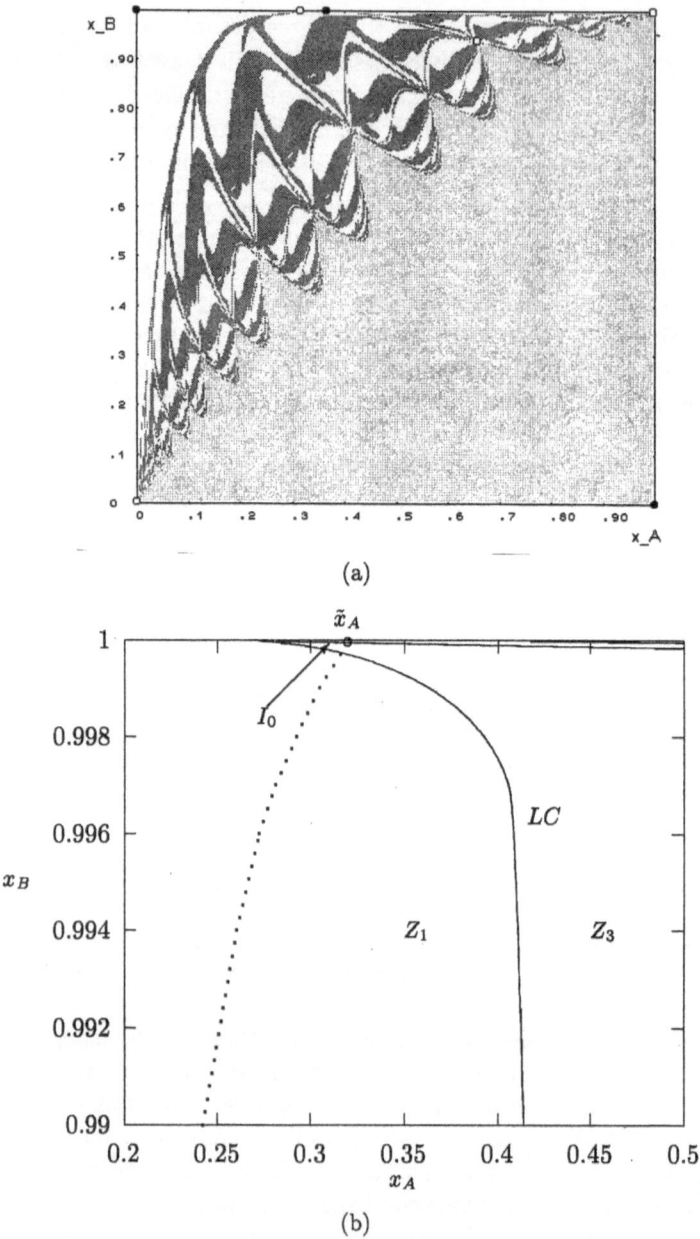

(a)

(b)

Fig. 4.1. (a) The basins of attraction of the equilibria $(1,0)$ (light grey), $(0,1)$ (white) and $(x_A^*, 1)$ (dark grey) for $c_A = 179.5, c_B = 210$ and $\lambda_A = \lambda_B = 3$. (b) The complicated structure of the basins is caused by the tiny set I_0 determined by the intersection of the critical curve with the stable manifold of $(\tilde{x}_A, 1)$.

here we always have convergence towards an equilibrium. However it is hard to predict from the initial conditions which equilibrium will be selected.

References

1. Bischi, G.I. and Kopel, M. (1999): Equilibrium Selection in a Nonlinear Duopoly Game with Adaptive Expectations, Working Paper, University of Urbino.
2. Bischi, G.I., Gardini, L. and Kopel, M. (1999): Analysis of global bifurcations in a market share attraction model. Journal of Economic Dynamics and Control, forthcoming.
3. Brock, W.A. and Hommes, C. (1997): A Rational Route to Randomness. Econometrica 65, 1059-1095.
4. Dawid, H. (1999): On the Dynamics of Word of Mouth Learning with and without Anticipations. Annals of Operations Research, 89, 273 - 295.
5. Dawid, H. (2000): The Emergence of Exchange and Mediation in a Production Economy. Journal of Economic Behavior and Organization, 41, 27 - 53.
6. Ellison, G. and Fudenberg, D. (1995): Word-of-Mouth Communication and Social Learning. Quarterly Journal of Economics CX, 93-125.
7. Feichtinger, G. and Novak, A. (1994): A Differential Game Model of the Dynastic Cycle - 3D Canonical System with a Stable Limit Cycle. Journal of Optimization Theory and Applications 80, 234-251.
8. Gale, D. and Rosenthal, R.W. (1999): Experimentation, Imitation, and Stochastic Stability. Journal of Economic Theory 84, 1 - 40.
9. Greif, A. (1992), Institutions and International Trade: Lessons from the Commercial Revolution. American Economic Review Papers and Proceedings 82, 128-133.
10. Greif, A. (1993): Contract Enforceability and Economic Institutions in Early Trade: The Maghribi Traders' Coalition. American Economic Review 83, 525-548.
11. Karandikar, R., Mookherjee, D. Ray, D. and Vega-Redondo, F. (1998): Evolving Aspirations and Cooperation. Journal of Economic Theory 80, 292 - 331.
12. Mira, C., Gardini, L., Barugola, A. and Cathala, J.-C. (1996): Chaotic Dynamics in Two-Dimensional Noninvertible Maps. World Scientific, Singapore.
13. Schlag, K. (1998): Why Imitate, and, if so, How? A Boundedly Rational Approach to Multi-Armed Bandits. Journal of Economic Theory 78, 130 - 156.

Cobweb Dynamics under Bounded Rationality

Cars Hommes

Center for Nonlinear Dynamics in Economics and Finance (CeNDEF), Department of Economics, University of Amsterdam, The Netherlands

Abstract This paper discusses recent work on expectation formation in the familiar cobweb 'hog cycle' model. A convenient feature of the model is the fact that it has a unique rational expectations equilibrium (REE). Attention will be focused on whether this unique REE can be learned under bounded rationality, with agents using simple habitual rule of thumb forecasting rules, or whether other 'approximate rational expectations equilibria' can arise, where forecasts are, although not perfect, approximately correct. In particular, the possibility of 'chaotic equilibria' under bounded rationality will be discussed. Both the homogeneous agent case and the heterogeneous case, with evolutionary competition between different forecasting rules, will be discussed. Recent experimental work on expectations formation in the cobweb economy suggests excess volatility and endogenous, possibly chaotic price fluctuations driven by the interaction of competing boundedly rational agents.

Keywords: expectations, adaptive learning, heterogeneity, nonlinear dynamics.

Acknowledgement. Financial support by the Netherlands Organization for Scientific Research (NWO) under Pionier grant 400-41-002 is gratefully acknowledged. I would also like to thank Joep Sonnemans, Henk van de Velden and Roy van der Weide for their help with some of the numerical simulations. Special thanks are due to Jan Tuinstra for his comments on an earlier draft.

1. Introduction

In Spring 1990 I gave a talk about the nonlinear cobweb model with adaptive expectations, in a special session on nonlinear economic dynamics at a big Operations Research conference in Vienna. I was still a Ph-D student, and it was one of my first international conferences. After 5 minutes a friendly looking man with a grey beard entered the room and sat down on the first row. He seemed to be particularly interested in my lecture and was asking several questions. His interest in nonlinear economic dynamics was a stimulating experience to me because as a Ph-D student, at different scientific meetings, I had met some general skepticism concerning applications of nonlinear dynamics in economics and finance.

The man on the first row was Gustav Feichtinger, the organizer of the conference. After the talk he invited me to visit his department in Vienna and we have met many times since then and became friends sharing a deep interest in nonlinear economic dynamics. It is therefore a pleasure to contribute to this *'Festschrift for Gustav Feichtinger'*, who has been a friendly supporter of research in nonlinear economic dynamics from the beginning, at the time when many others were still skeptical.

The main theme of my contribution will be the expectations hypothesis in economics and finance. How should we model expectations? Should we assume full rationality with agents deriving optimal forecasts from economic theory using underlying market equilibrium equations? Or is 'bounded rationality', with agents only observing time series and using simple habitual rule of thumb predictors, a more accurate description of expectation formation in a dynamic environment? Incomplete information about the model is only one aspect of bounded rationality. Another important issue is the possibility of heterogeneity. Can we use a *representative agent* model, or does *heterogeneity* in expectations contribute to excess price volatility? I will use the familiar demand-supply cobweb model to discuss these issues, because this model has the convenient feature of having a unique rational expectations equilibrium (REE). The main questions are then whether a representative agent could learn the REE from time series observations and whether in a heterogeneous world agents will be able to learn and coordinate on the unique REE.

In addressing these fundamental issues, I will discuss work that goes back to my thesis, as well as more recent work on cobweb dynamics that I have been doing over the years jointly with Buz Brock, with Gerhard Sorger, and more recently also some experimental work with Joep Sonnemans, Jan Tuinstra and Henk van de Velden, my co-workers from CeNDEF at the University of Amsterdam. In doing so, I will try not to repeat exactly the results from published papers, but instead focus on some relevant departures.

The paper is organized as follows. Section 2 discusses cobweb dynamics under various expectations rules, such as adaptive expectations. Section 3 discusses the notion of consistent expectations in the cobweb model, where forecasting rules are, although not perfect, correct in terms of sample averages and sample autocorrelations. Section 4 focuses on heterogeneity and evolutionary competition between different forecasting rules. Section 5 discusses recent experimental work on expectation formation in the cobweb model and finally section 6 concludes.

2. The Cobweb Model

The cobweb model describes fluctuations of equilibrium prices in an independent market for a non-storable good, that takes one time period to produce. Producers form price expectations one period ahead and derive their optimal supply from expected profit maximization:

$$S(p_t^e) = \text{argmax}_{q_t} \{p_t^e q_t - c(q_t)\} = (c')^{-1}(p_t^e). \qquad (2.1)$$

The cost function $c(\cdot)$ is assumed to be strictly convex so that the second order condition for profit maximization is satisfied. The marginal cost function is then invertible and supply strictly increasing in expected price. The simplest case arises when the cost function is quadratic, $c(q) = q^2/(2b)$, yielding a linear supply curve

$$S(p^e) = bp^e. \tag{2.2}$$

Here, we focus on a nonlinear, increasing, S-shaped supply curve derived from expected profit maximization with an increasing and convex cost function given by a third or higher order polynomial[1]

$$c(q) = \frac{1}{d+1}(q-1)^{d+1} + q, \tag{2.3}$$

where d is an odd integer. Optimal supply then becomes

$$q = S(p^e) = (p^e - 1)^{\frac{1}{d}} + 1. \tag{2.4}$$

Consumer demand D depends upon the current market price p_t. The demand curve D can be derived from consumer utility maximization, but for our purposes it is not necessary to specify these preferences explicitly. Throughout the paper we will simply work with a linear demand curve

$$D(p_t) = A - Bp_t + \epsilon_t, \tag{2.5}$$

where ϵ_t is an independently and identically distributed (IID) stochastic series representing an exogenous random demand shock. If beliefs are homogeneous, i.e., all producers have identical price expectations p_t^e, market equilibrium becomes

$$D(p_t) = S(p_t^e) \tag{2.6}$$

yielding the realized market price

$$p_t = D^{-1}(S(p_t^e)) = \frac{A + \epsilon_t - S(p_t^e)}{B}. \tag{2.7}$$

In order to complete the model, we have to specify how producers form price expectations. If producers have naive expectations, i.e. their prediction equals the latest observed price $p_t^e = p_{t-1}$, price dynamics becomes

$$p_t = D^{-1}(S(p_{t-1})). \tag{2.8}$$

According to the well known cobweb theorem (see e.g. Ezekiel (1938)), there are essentially two possibilities for the price dynamics under naive expectations, depending upon the ratio of marginal supply and marginal demand at the steady state p^*. When $-1 < S'(p^*)/D'(p^*) < 0$ the RE steady state p^* is (locally) stable, and prices converge to the REE. If on the other hand $S'(p^*)/D'(p^*) < -1$ the RE steady state p^* is (locally) unstable, and prices diverge from the REE. In the case of a nonlinear supply curve as in (2.4), if the REE is unstable, prices will converge to a stable 2−cycle.

It has been argued however, that simple forecasting rules such as naive expectations, lead to systematic forecasting errors. In particular, this argument seems to be valid, at least in theory, in the case the model generates

---

[1] For simplicity we have chosen a specification such that the point $(p,q) = (1,1)$ becomes the unique inflection point of the corresponding supply function.

a 2-cycle, even in the presence of exogenous shocks. When producers expect a high (low) price, they will supply a high (low) quantity and consequently the realized market price will be low (high). Along a 'hog cycle', expectations are thus systematically wrong, and forecasting errors are correlated. Rational agents would learn from these systematic forecasting errors and revise expectations accordingly, so the argument goes. Such considerations lead Muth (1961), using the cobweb model, to the introduction of rational expectations, where producers' subjective price expectations equal the objective mathematical expected value of the market price. Using market equilibrium (2.6) with the linear demand curve (2.5) rational expectations yields

$$p_t^e = E_t[p_t] = p^*, \tag{2.9}$$

where p^* is the unique price corresponding to the intersection of demand and supply. Given producers' rational forecast $p_t^e = p^*$, the actual law of motion (2.7) becomes

$$p_t = p^* + \frac{\epsilon_t}{B}. \tag{2.10}$$

The cobweb model therefore has a unique REE, given by an IID process with mean p^*. Along a REE expectations are self fulfilling and producers make no systematic forecasting errors; forecasting errors are uncorrelated. It is important to note that, in order to form rational expectations, perfect knowledge of underlying market equilibrium equations is required.

It is worthwhile to reconsider the issue of 'systematic forecasting errors' in the light of the recent discovery of chaotic dynamics in simple nonlinear deterministic systems and under the more plausible assumption of *bounded rationality*, where agents do not know underlying market equilibrium equations, but only use time series observations to forecast[2]. As an example, consider the cobweb model with *adaptive expectations*, i.e.,

$$p_t^e = (1 - w)p_{t-1}^e + wp_{t-1}, \qquad 0 \le w \le 1, \tag{2.11}$$

where w is the expectations weight factor. The expected price is a weighted average of yesterday's expected and realized prices, or equivalently, the expected price is adapted by a factor w in the direction of the most recent realization. Adaptive expectations may thus be seen as 'error learning' with a constant factor. Under adaptive expectations and, given the linear demand curve (2.5), the dynamics of expected prices in the cobweb model becomes

$$p_t^e = (1 - w)p_{t-1}^e + w\frac{A + \epsilon_t - S(p_{t-1}^e)}{B}. \tag{2.12}$$

Chiarella (1988) and Hommes (1991,1994) have shown that, without any random shocks ϵ_t, for *nonlinear*, but monotonic, demand and/or supply curves,

---

[2] At the time of the introduction of rational expectations by Muth (1961) and its introduction into macroeconomics by Lucas (1971) and others, the phenomenon of deterministic chaos was still unknown.

this nonlinear deterministic difference equation can easily generate chaotic fluctuations in expected prices, and therefore also in prices, quantities and forecasting errors. Fig. 2.1 shows a bifurcation diagram w.r.t. the expectations weight factor w, with the nonlinear, S-shaped supply curve (2.4). For high values of w prices converge to a 2-cycle, whereas for small values of w prices converge to the RE steady state. For intermediate w−values however, chaotic price oscillations arise.

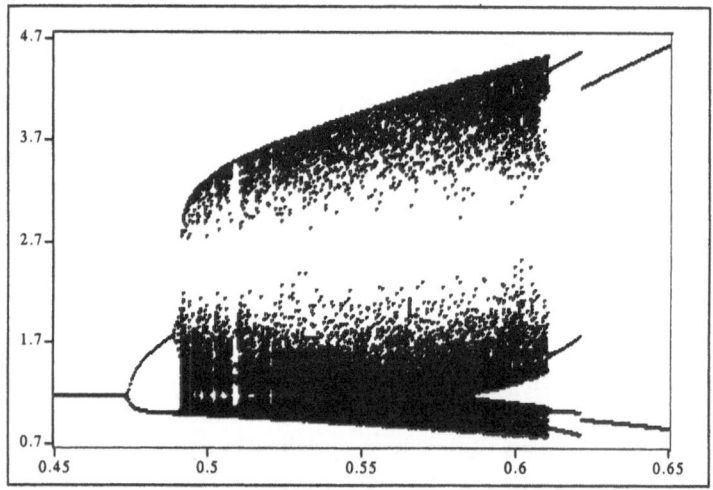

Fig. 2.1. Bifurcation diagram w.r.t. the expectations weight factor w, $0.45 \leq w \leq 0.65$, with the other parameters fixed at $A = 2$, $B = 0.25$ and $d = 5$.

When prices fluctuate chaotically, the corresponding forecasting errors will be highly unpredictable and the question arises whether boundedly rational agents would be able to detect structure in these chaotic forecasting errors and improve upon their simple adaptive forecasts. If patterns are indeed hard to discover, then such a chaotic situation might be a satisfactory (long run) equilibrium.

3. Consistent Expectations Equilibria

In order to detect structure in a chaotic time series generated by an unknown model, it seems natural that agents start to use linear time series techniques to detect patterns. Hommes (1998) investigated for various simple forecasting rules, including adaptive expectations and linear AR(p) forecasting rules, whether chaotic forecasting errors in the cobweb model are uncorrelated. In all chaotic examples the chaotic forecasting errors had strongly significant negative autocorrelation coefficient at the first lag. This suggests, that even in a nonlinear, chaotic cobweb economy, boundedly rational agents using simple linear time series techniques would be able to improve their forecasts. In related work, Sorger (1998) presented an example of an overlapping generations model with production, where agents believe that interest rates follow a stochastic IID process, and given this belief the actual interest rate follows a chaotic process without any autocorrelations. This is an example where beliefs, although not perfect, are self fulfilling in terms of mean and autocorrelations. Subsequently, Hommes and Sorger (1998) generalized this example and developed the notion of *consistent expectations equilibrium (CEE)*, which will be discussed below.

Assume that all agents believe that prices are generated by a stochastic AR(1) process[3]. Given this perceived law of motion and prices known up to p_{t-1}, the predictor for p_t minimizing the mean squared prediction error is

$$p_t^e = \alpha + \beta(p_{t-1} - \alpha), \tag{3.1}$$

where α and β are real numbers, $\beta \in [-1, 1]$. Given that agents use the linear predictor (3.1), the *implied actual law of motion* becomes

$$p_t = F_{\alpha,\beta}(p_{t-1}) := D^{-1}S(\alpha + \beta(p_{t-1} - \alpha)). \tag{3.2}$$

The empirical or sample average of a time series $(p_t)_{t=0}^{\infty}$ is defined as

$$\bar{p} = \lim_{T \to \infty} \frac{1}{T+1} \sum_{t=0}^{T} p_t \tag{3.3}$$

and the empirical or sample autocorrelation coefficients are given by

$$\rho_j = \lim_{T \to \infty} \frac{c_{j,T}}{c_{0,T}}, \qquad j \geq 1, \tag{3.4}$$

where

$$c_{j,T} = \frac{1}{T+1} \sum_{t=0}^{T-j} (p_t - \bar{p})(p_{t+j} - \bar{p}), \qquad j \geq 0. \tag{3.5}$$

A CEE is now defined as

Definition 3.1. A triple $\{(p_t)_{t=0}^{\infty}; \alpha, \beta\}$, where $(p_t)_{t=0}^{\infty}$ is a sequence of prices and α and β are real numbers, $\beta \in [-1, 1]$, is called a *consistent expectations equilibrium* (CEE) if

1. the sequence $(p_t)_{t=0}^{\infty}$ satisfies the implied actual law of motion (3.2),
2. the sample average $\bar{p}$ in (3.3) exists and is equal to α, and
3. the sample autocorrelation coefficients ρ_j, $j \geq 1$, in (3.4) exist and the following is true:
   a. if $(p_t)_{t=0}^{\infty}$ is a convergent sequence, then $\text{sgn}(\rho_j) = \text{sgn}(\beta^j)$, $j \geq 1$;
   b. if $(p_t)_{t=0}^{\infty}$ is not convergent, then $\rho_j = \beta^j$, $j \geq 1$.

---

[3] Hommes and Sorger (1998) emphasize that the definition of CEE can be generalized to higher order belief processes, e. g., AR(k) processes with $k \geq 2$.

A CEE is a price sequence together with an AR(1) belief such that expectations are self-fulfilling in terms of the *observable* sample average and sample autocorrelations. Along a CEE expectations are thus correct in a linear statistical sense. Given an AR(1) belief, there are at least three possible types of CEE:

- a *steady state CEE* in which the price sequence $(p_t)_{t=0}^{\infty}$ converges to a steady state p^*, with $\alpha = p^*$ and $\beta = 0$;
- a *2-cycle CEE* in which the price sequence $(p_t)_{t=0}^{\infty}$ converges to a period two cycle $\{p_1^*, p_2^*\}$, $p_1^* \neq p_2^*$, with $\alpha = (p_1^* + p_2^*)/2$;
- a *chaotic CEE* in which the price sequence $(p_t)_{t=0}^{\infty}$ is chaotic, with sample average α and autocorrelations β^j.

Which of these cases occurs in the cobweb model depends on the composite mapping $D^{-1}S$ in (3.2), determined by demand and supply curves, and will be discussed below.

So far, the notion of CEE involves a given AR(1) belief, with fixed parameters α and β. Now consider the more flexible situation of *adaptive learning* with agents changing their forecasting function over time within the class of AR(1) beliefs, and updating their belief parameters α_t and β_t, as additional observations become available. A natural learning scheme fitting the framework of CEE is based upon sample average and sample autocorrelation coefficients.

For any finite set of observations $\{p_0, p_1, \ldots, p_t\}$ the sample average is

$$\alpha_t = \frac{1}{t+1} \sum_{i=0}^{t} p_i, \qquad t \geq 1 \tag{3.6}$$

and the first order sample autocorrelation coefficient is

$$\beta_t = \frac{\sum_{i=0}^{t-1}(p_i - \alpha_t)(p_{i+1} - \alpha_t)}{\sum_{i=0}^{t}(p_i - \alpha_t)^2}, \qquad t \geq 1. \tag{3.7}$$

When, in each period, the belief parameters are updated according to (3.6) and (3.7) the (temporary) law of motion (3.2) becomes

$$p_{t+1} = F_{\alpha_t, \beta_t}(p_t) = D^{-1}S(\alpha_t + \beta_t(p_t - \alpha_t)), \qquad t \geq 0. \tag{3.8}$$

We call the dynamical system (3.6) - (3.8) the actual dynamics with *sample autocorrelation learning* (SAC-learning)[4]. The initial state for the system (3.6) - (3.8) can be any triple (p_0, α_0, β_0) with $\beta_0 \in [-1, 1]$.

Which type of CEE exist in the cobweb model, and to which of them will the SAC-learning dynamics converge? Hommes and Sorger (1998) show that

---

[4] Another adaptive learning process that has received much attention in the bounded rationality literature is ordinary least squares (OLS) learning, see Sargent (1993) for a survey and many references. Hommes and Sorger (1998) argue that the SAC- and the OLS-learning schemes are in fact closely related.

in the most relevant case, when demand is decreasing and supply is increasing, the *only* CEE is the RE steady state price p^*. This means that, even when underlying market equilibrium equations are *not* known, agents should be able to learn and coordinate on the REE price simply by looking at sample averages and sample autocorrelations. Although other simple forecasting rules, such as adaptive expectations, might lead to chaotic price fluctuations, these forecasting rules are *inconsistent* in terms of sample autocorrelations. Hence, in a nonlinear cobweb economy with monotonic demand and supply, boundedly rational agents should, at least in theory, be able to learn the unique REE from time series observations[5].

It is useful to discuss the possibility of more complicated, 2-cycle and chaotic CEE in the cobweb model. Hommes and Sorger (1998) present an example where demand is linear and supply is non-monotonic and piecewise linear. We slightly extend this example by adding IID demand shocks ϵ_t, uniformly distributed over the interval $[-0.25, 0.25]$. Let demand be given by

$$ D(p) = \begin{cases} max\{9 - p + \epsilon_t, 0\} & \text{if } 0 \le p \le 9, \\ max\{0, \epsilon_t\} & \text{if } p > 9 \end{cases} $$

and supply by the piecewise linear curve[6].

$$ S(p) = \begin{cases} (25/2)p & \text{if } 0 \le p \le 18/25, \\ 10 - (25/18)p & \text{if } 18/25 < p \le 18/5, \\ 5 + \epsilon[p - (18/5)] & \text{if } p > 18/5. \end{cases} $$

In the deterministic case without exogenous shocks ϵ_t, this example has a chaotic CEE with belief parameters $\alpha = 2$ and $\beta = 4/5$, as well as three steady state CEE (corresponding to the three intersection points of D and S, p_1^*, p_2^*, and p_3^*), and three 2-cycle CEE (corresponding to $\{p_1^*, p_2^*\}$, $\{p_1^*, p_3^*\}$, and $\{p_2^*, p_3^*\}$). To which of these CEE the SAC-learning process will converge depends upon the initial state of the learning process.

Fig. 3.1 shows an example of *learning to believe in chaos*, where the SAC-learning dynamics converges to a chaotic CEE. The key feature is that learning parameters converge to constants whereas prices do not converge but fluctuate chaotically on a strange attractor, with the correct sample average and sample autocorrelations[7]. A chaotic CEE may be seen as an example of

---

[5] Bray and Savin (1986) show that OLS learning also converges to the REE steady state in the cobweb model. Arifovic (1994) shows that agents using genetic algorithms can learn the REE steady state. See also Böhm and Wenzelburger (1999) for 'perfect prediction rules' in the cobweb and other models.

[6] At first sight, this non-monotonic supply curve may seem odd. However, Hommes and Rosser (1999) have recently shown that overfishing and logistic growth in an optimal fishery managment model leads to a smooth, non-monotonic supply curve similar to the piecewise linear curve. Chaotic CEE arise in this renewable resource model and they are robust w.r.t. noise.

[7] The notion *learning to believe in chaos* has been introduced in Hommes (1998, p.360), and the first examples have been given by Sorger (1998) and Hommes

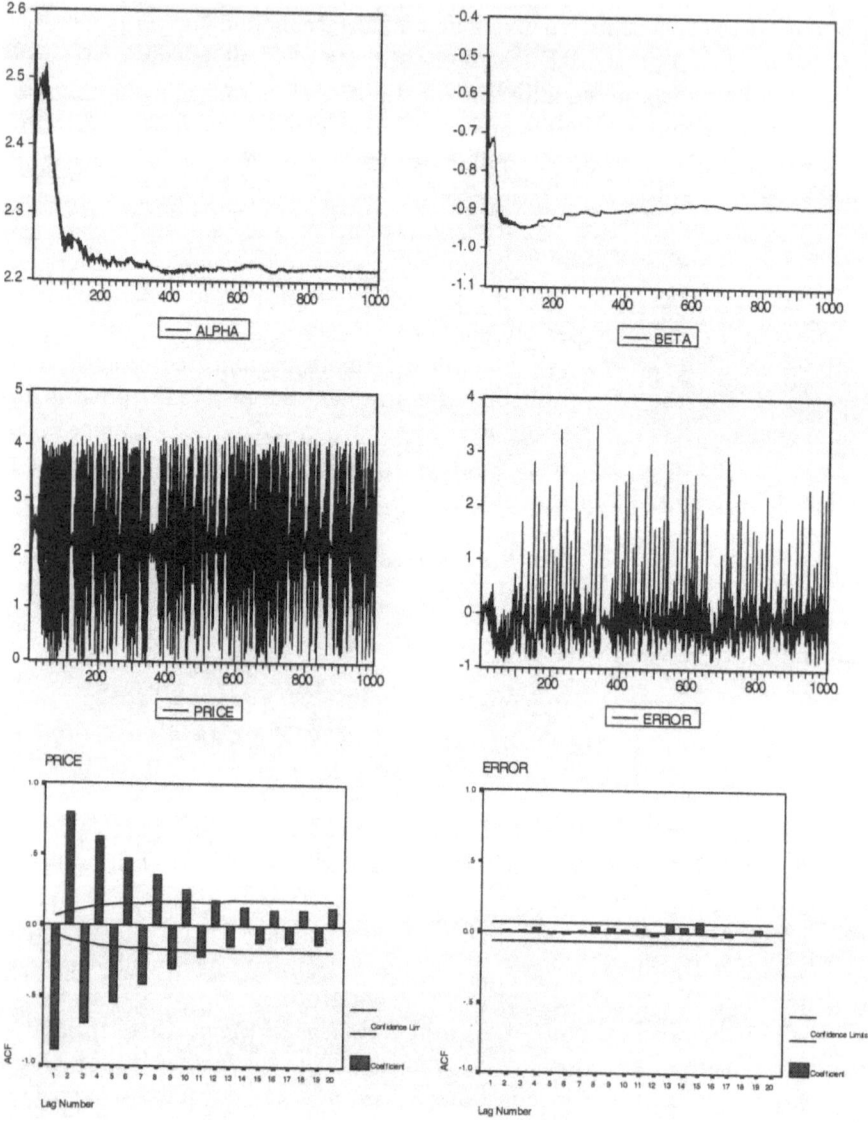

Fig. 3.1. Learning to believe in chaos. Belief parameters converge to constant values, i.e. $(\alpha_t, \beta_t) \to (\alpha^*, \beta^*) \approx (2.22, -0.89)$, but at the same time prices keep fluctuating chaotically with sample mean α^* and sample autocorrelations as for the corresponding AR(1) model (bottom left). Forecasting errors are chaotic and uncorrelated (bottom right). Statistical hypothesis testing would *not* reject the null hypothesis that prices follow a stochastic AR(1) process.

what Sargent (1998) calls an *approximate rational expectations equilibrium*, with optimal misspecified forecasts. A chaotic CEE is not a REE, because expectations do not coincide with the conditional mathematical expectations which could only be derived if underlying market equilibrium equations would be known. Along a CEE agents use time series observations only and arrive at an optimal linear predictor to forecast an unknown nonlinear actual law of motion. Along a chaotic CEE forecasting errors are uncorrelated.

4. Heterogeneous Expectations

So far we have focused on a representative agent cobweb model, where all producers have identical expectations. In this section we discuss *heterogeneous* expectations, with evolutionary competition between different forecasting rules, as introduced in Brock and Hommes (1997). Producers can choose between different forecasting rules H_j. The fractions $n_{j,t}$ of agents using predictor H_j at date t, will be updated over time based upon a publically available evolutionary fitness measure associated to each predictor.

BH focus on the special but important case of rational expectations, which can be obtained at costs C per time period, and naive expectations, which is freely available. Rational expectations represents a sophisticated forecasting rule, whereas naive expectations represents a rule of thumb forecasting rule. BH show the occurrence of a *rational route to randomness*, i.e. a bifurcation route to strange attractors and chaos as traders become more rational in the sense of using more frequently forecasting rules that performed well in the recent past.

Rational agents have perfect knowledge about market equilibrium equations and are aware of the fact that the market equilibrium price is affected by the presence of naive traders. Hence, rational agents have perfect knowledge about prices and quantities, but also about *beliefs* of *all* other traders. In a real market, this seems highly unrealistic. Therefore we will focus here on another two type case, with two simple, competing linear predictors. As a special case, the case of fundamentalists versus naive expectations is obtained. In contrast to rational agents, fundamentalists do not take into account the presence of other trader types, but instead predict that the price will always be equal to the REE steady state price p^*.

Consider the two linear AR(1) prediction rules

---

and Sorger (1998). For related work on the instability of OLS learning, see e.g. Bullard (1994) and Grandmont (1998). Schönhofer (1999) has recently employed the notion of learning to believe in chaos in a somewhat different context, namely when the entire OLS-learning process fluctuates chaotically. In Schönhofer's examples belief parameters of the OLS-learning scheme do *not* converge but keep fluctuating chaotically, and at the same time, due to inflation, prices diverge to infinity, so that agents are in fact running an OLS-regression on a non-stationary time series.

$$H_j(p_{t-1}) \quad = \quad \alpha_j + \beta_j p_{t-1}, \qquad j = 1, 2. \tag{4.1}$$

These predictors specialize to the case with fundamentalists versus naive expectations when $\alpha_1 = p^*$, $\beta_1 = 0$, $\alpha_2 = 0$ and $\beta_2 = 1$. Throughout this section we focus on the case where the supply curve is linear as in (2.2), with corresponding cost function $c(q) = q^2/(2b)$.

Market equilibrium prices in the cobweb model with two trader types, with linear predictors as in (4.1), is determined by[8]

$$A - Bp_t = n_{1,t-1}b(\alpha_1 + \beta_1 p_{t-1}) + n_{2,t-1}b(\alpha_2 + \beta_2 p_{t-1}), \tag{4.2}$$

where $n_{1,t-1}$ and $n_{2,t-1}$ denote the fractions of agents using respectively H_1 and H_2, at the beginning of period t. These fractions will be updated according to past realized profits. Net realized profit in period t for traders using predictor H_j is given by

$$\pi_{j,t} = bp_t H_j(p_{t-1}) - \frac{b}{2}(H_j(p_{t-1}))^2 - C_j, \tag{4.3}$$

where C_j represents the average costs per time period for obtaining predictor H_j. For a simple habitual rule of thumb predictor, such as naive or adaptive expectations, these costs C_j will be zero, whereas for more sophisticated predictors such as fundamentalists beliefs, based on fundamental analysis, costs C_j may be positive.

The fraction of agents using predictor H_j in period t are given by the discrete choice 'probabilities'

$$n_{j,t} = \frac{\exp(\beta \pi_{j,t})}{Z_t}, \qquad Z_t = \sum_{j=1}^{H} \exp(\beta \pi_{j,t}), \tag{4.4}$$

where Z_t is a normalization factor so that all fractions add up to one.

A key feature of this evolutionary predictor selection is that agents are *boundedly rational*, in the sense that predictors with higher evolutionary fitness attract more followers. The parameter β is called the *intensity of choice*, measuring how fast producers switch between different prediction strategies. For $\beta = 0$, all fractions are fixed over time and equal to $1/H$, whereas for the other extreme $\beta = \infty$, in each period *all* producers choose the optimal predictor. Hence, the higher the intensity of choice, the more rational agents are in choosing prediction strategies. Equivalently, the inverse $1/\beta$ is proportionally to the noise level in the fitness measure.

The timing of predictor selection in (4.4) is important. In (4.2) the old fractions $n_{t-1,1}$ and $n_{t-1,2}$ determine the new equilibrium price p_t. This new equilibrium price p_t is used in the fitness measure for predictor choice and

---

[8] In our simulations we will work in deviations $x_t = p_t - p^*$ from the fundamental RE steady state price p^*. This is equivalent to setting the parameter $A = 0$, so that the RE steady state $p^* = 0$.

the new fractions $n_{t,1}$ and $n_{t,2}$ are updated according to (4.4). These new fractions are then used in determining the next equilibrium price p_{t+1}, etc..

Fig. 4.1 shows the strange attractor and corresponding chaotic price fluctuations in the case of costly fundamentalists versus naive expectations. It is remarkable that both the attractor and the price time series are very similar to the case of rational versus naive expectations studied in BH[9].

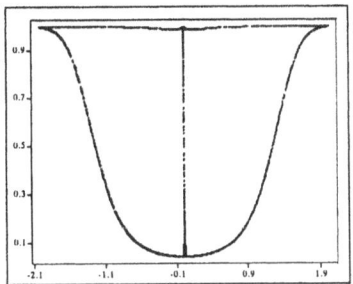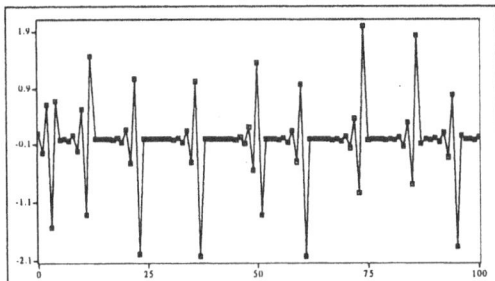

Fig. 4.1. Parameters: $A = 0$, $B = 0.5$, $b = 1.35$, $\beta = 3$, $\alpha_1 = 0$, $\beta_1 = 0$, $C_1 = 1$, $\alpha_2 = 0$, $\beta_2 = 1$ and $C_2 = 0$. Costly fundamentalists versus naive expectations. Strange attractor in the $(x_t, n_{1,t})$ phase space, where $x_t = p_t - p^*$ is the deviation from the fundamental steady state price, and chaotic time series x_t.

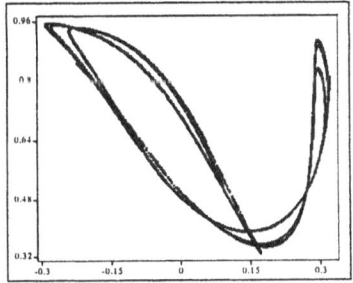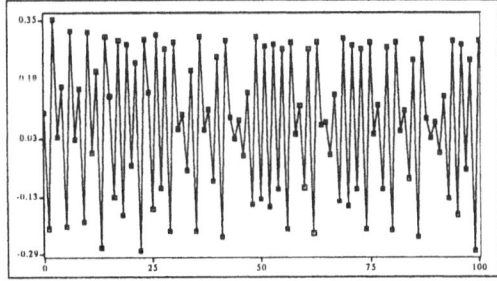

Fig. 4.2. Parameters: $A = 0$, $B = 0.5$, $b = 1.35$, $\beta = 18$, $\alpha_1 = -0.1$, $\beta_1 = 0$, $C_1 = 0$, $\alpha_2 = 0$, $\beta_2 = 1$ and $C_2 = 0$. Strange attractor and chaotic fluctuations due to evolutionary competition between downward biased forecast versus naive expectations, without any information costs.

Figure 4.2 illustrates an example without any information costs, with two competing linear forecasting rules. The first type is slightly downward biased, and predicts a constant price just below the (unknown) RE steady state, whereas the second type has again naive expectations. This example shows that, without any information costs, the evolutionary competition between simple, linear forecasting rules can generate chaotic price fluctuations.

---

[9] See the original working paper Brock and Hommes (1995) for more details concerning the case of fundamentalists versus naive expectations.

5. Laboratory Experiments

In joint work with my colleagues Joep Sonnemans, Jan Tuinstra and Henk van de Velden from CeNDEF, we have run individual as well as group experiments on expectation formation in the cobweb model. The results of these experiments are described in detail in Hommes, Sonnemans and van de Velden (1999a,1999b) and Sonnemans et al. (1999), and will be discussed here briefly.

The participants in the experiments were asked to predict next periods price of a certain, unspecified, good. The realized price p_t in the experiment was determined by the cobweb market equilibrium equation

$$D(p_t) = \frac{1}{K} \sum_{i=1}^{K} S(p_{i,t}^e), \tag{5.1}$$

where $D(p_t)$ is the demand for the good at price p_t, K is the size of the group, $p_{i,t}^e$ is the price forecast by participant i and $S(p_{i,t}^e)$ is the supply of producer i depending upon the forecast by participant i. Demand and supply curves D and S were fixed during all experiments (except for the random shocks to the demand curve) and unknown to the participants. In the group experiments $K = 6$, whereas in the individual experiments $K = 1$. Notice that on the righthand side of the market equilibrium equation (5.1) we take the average supplied quantity, in order to be able to compare the group experiments directly to the individual experiments. Solving (5.1) for the market equilibrium price, for the linear demand curve (2.5), yields

$$p_t = \frac{A_t + \epsilon_t - \frac{1}{K} \sum_{i=1}^{K} S(p_{i,t}^e)}{B}, \tag{5.2}$$

where ϵ_t are IID demand shocks, which are drawn from a normal distribution $N(0, 0.5)$ in the experiments discussed here. In the experiments we used the S-shaped supply curve

$$S(p_{i,t}^e) = tanh(\lambda(p_{i,t}^e - 6)) + 1, \tag{5.3}$$

which is geometrically equivalent to the S-shaped supply curve (2.4).

Expectation formation of the producers is the only part of the model that is affected by the participants in the experiments. Participants thus did not know underlying market equilibrium equations, nor were they informed about the distribution of any exogenous shocks to demand and/or supply. The participants were told that they were advisors of producers of an unspecified good and that the price was determined by market clearing. Based upon this information we asked the participants to predict next periods price. The predicted price had to be between 0 and 10 and the realized price was also always between 0 and 10. Participants' earnings in each period were a quadratic function of their squared forecasting error. The better their forecast, the higher their earnings. After every period the participants were informed about the realized price in the experiment. Also a time series of the participant's own prediction and a time series of the realized price in the experiment was shown on each computer screen.

In real markets, individual expectations usually only have a small effect upon realized market prices. Therefore, in the individual experiments, participants were not informed that the realized prices only depended upon their own expected price (and some small noise), and in the group experiments participants were not informed that the realized price depended upon their own price forecast and the forecasts of (five) other participants.

Participants in the experiments thus had very little information about the price generating process and had to rely mainly upon time series observations of past prices and predictions. The information in the experiment was thus similar to the information assumption underlying much of the bounded rationality literature, where agents form expectations based upon time series observations. Our setup thus enables us to test the expectations hypothesis in a controlled dynamic environment. The main question was whether agents can learn and coordinate on the unique REE, in a world where consumers and producers act as if they were maximizing utility and profits, but where they do *not* know underlying market equilibrium equations and only observe time series of prices and expected prices. Our choice for a nonlinear, S-shaped supply curve enables us to investigate whether agents can learn a REE steady state in a world where simple forecasting rules can lead to periodic cycles or even chaotic price fluctuations.

In the individual experiments, only about 35% of the participants were able to learn the unique REE steady state; in the remaining 65% prices kept fluctuating with large amplitude around the unstable REE steady state. In the group experiments, forecasting errors were significantly lower than in the individual experiments. Fig. 5.1 shows time series of the realized prices in two group experiments and Table 5.1 shows the mean and standard deviations over different subsamples in the experiments. Under RE the mean price would be 5.91 and the standard deviation (SD) would be 0.71 (equal to the SD of the demand shocks). For both time series, the sample mean over the full sample is fairly close to the RE mean, so that on average the price level is predicted correctly. In the first time series the amplitude of price fluctuations clearly decreases over time. Table 5.1 shows that the standard deviation in the last 25 time periods is close to the RE standard deviation. One might thus say that this group experiment converges to RE. In the second time series the amplitude of price fluctuations does *not* decrease much over time. The standard deviation in the last 25 periods is in fact close to the standard deviation in the first 25 periods, and about twice the SD under rational expectations. One might thus say that this group experiment exhibits excess price volatility due to endogenous expectations.

148

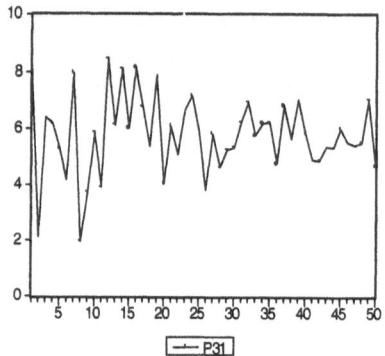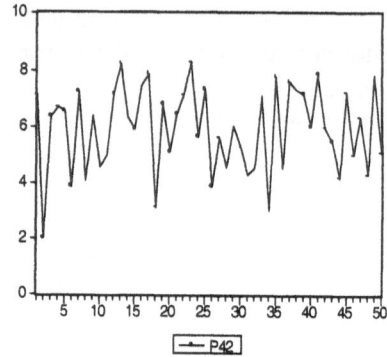

Fig. 5.1. Realized market prices in two different cobweb group experiments with exogenous demand shocks $\epsilon_t \sim N(0, 0.5)$. The first market settles down closely to REE, whereas the second one exhibits expectations driven excess price volatility.

Table 5.1. Statistics of realized market prices

statistic sample	mean 1-50	mean 1-25	mean 26-50	SD 1-50	SD 1-25	SD 26-50
rational expectations	5.91	5.91	5.91	0.71	0.71	0.71
group 31	5.78	5.92	5.64	1.41	1.83	0.81
group 42	5.95	6.12	5.77	1.52	1.61	1.43

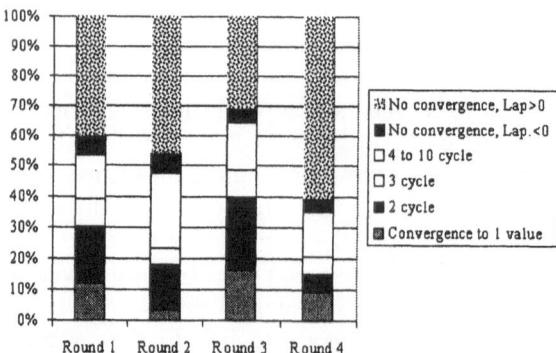

Fig. 5.2. Distribution of long run dynamics based upon 620 strategy simulations. Only 10% of all simulations converges to the RE steady state. In the final round, more than 60% appears to be chaotic.

Sonnemans et al. (1999) conducted a four round strategy experiment on expectation formation in the cobweb economy. Participants submitted a forecasting strategy and could revise their strategy after each round. Earnings were based upon the relative performance of the strategy. Over the four rounds, forecasting errors decreased, but at the same time the complexity of the price fluctuations also increased. Figure 5.2 shows the distribution of the long run dynamics, based upon 620 simulations, in each of the four rounds. Convergence to the unique REE steady state occurred in less than 10% of all cases. In the final round, more than 60% of the simulations appears to be chaotic, with a positive Lyapunov exponent. Strategy simulations with homogeneous agents typically show regular behaviour, with convergence to the REE steady state or to a stable periodic cycle. Heterogeneous interaction of simple prediction strategies thus seems to be the main source of the endogenous, possibly chaotic price fluctuations.

6. Concluding Remarks

In a cobweb economy with nonlinear, but monotonic demand and supply curves, it should be possible to learn the unique REE steady state price from time series observations. The steady state price forecast is the only forecast where sample averages and sample autocorrelations are correct. In theory, simply by looking at sample averages and sample autocorrelations, in particular trying to learn the negative first order autocorrelation so typical for the 'hog cycle', boundedly rational agents should be able to learn the unique REE. Experimental evidence shows however that this is not as easy as theory suggests. Even in individual experiments more than 65% is *not* able to learn the REE steady state. In heterogeneous agents experiments, although forecasting errors are significantly lower, typically prices do not settle down to the REE steady state. Interaction of simple, heterogeneous forecasting rules seems to generate complicated, possibly chaotic price fluctuations. Even in the simplest of all dynamic economic models, heterogeneity may thus cause expectations driven excess price volatility.

References

1. Arifovic, J., (1994) Genetic algorithm learning and the cobweb model, *Journal of Economic Dynamics and Control* 18, 3-28.
2. Böhm, V., and Wenzelburger, J., (1999) Expectations, forecasting and perfect foresight. A dynamical systems approach, *Macroeconomic Dynamics* 3, 167-186.
3. Bray, M.M., and Savin, N.E., (1986) Rational expectations equilibria, learning and model specification, *Econometrica* 54, 1129-1160.
4. Brock, W.A., and Hommes, C.H., (1995) Rational routes to randomness, SSRI working paper 9506, University of Wisconsin.

5. Brock, W.A., and Hommes, C.H., (1997) A rational route to randomness, *Econometrica*, 65, 1059-1095.
6. Bullard, J., (1994) Learning equilibria, *Journal of Economic Theory* 64, 468-485.
7. Chiarella, C., (1988) The Cobweb Model: Its Instability and the Onset of Chaos, *Economic Modeling*, 5, 377-384.
8. Ezekiel, M., (1938) The cobweb theorem, *Quarterly Journal of Economics* 52, 255-280.
9. Goeree, J.K. and Hommes, C.H., (1999), Heterogeneous beliefs and the nonlinear cobweb model, *Journal of Economic Dynamics and Control* 1999 forthcoming.
10. Grandmont, J.-M., (1998) Expectations Formation and Stability in Large Socio-Economic Systems, *Econometrica*, 66, 741-781.
11. Hommes, C.H., (1991) Adaptive learning and roads to chaos. The case of the cobweb, *Economics Letters* 36, 127-132.
12. Hommes, C.H., (1994) Dynamics of the cobweb model with adaptive expectations and nonlinear supply and demand, *Journal of Economic Behavior & Organization* 24, 315-335.
13. Hommes, C. H., (1998) On the Consistency of Backward-Looking Expectations: The Case of the Cobweb, *Journal of Economic Behavior and Organization*, 33, 333-362.
14. Hommes, C.H., and Barkley Rosser, Jr. J., (1999) Consistent Expectations Equilibria and Complex Dynamics in Renewable Resource Markets, Tinbergen Institute Discussion Paper TI 99-052/1, University of Amsterdam.
15. Hommes, C.H., Sonnemans, J., and van de Velden, H., (1999a) Expectations in an experimental cobweb economy: some individual experiments, In: Delli Gatti, M. et al. (eds.), *Market Structure, Aggregation and Heterogeneity*, Springer Verlag, Berlin.
16. Hommes, C.H., Sonnemans, J., and van de Velden, H., (1999b) Expectations driven price volatility in an experimental cobweb economy, working paper CeNDEF, University of Amsterdam.
17. Hommes, C. H., and Sorger, G., (1998) Consistent Expectations Equilibria, *Macroeconomic Dynamics* 2, 287-321.
18. Lucas, R.E., (1971) Econometric testing of the natural rate hypothesis, In: O. Eckstein (ed.) The econometrics of price determination Conference. Board of Governors of the Federal Reserve System and Social Science Research Council.
19. Muth, J.F., (1961) Rational expectations and the theory of price movements, Econometrica 29, 315-335.
20. Sargent, T.J., (1993) Bounded rationality in macroeconomics, Clarendon Press, Oxford.
21. Sargent, Thomas J., (1998) *The Conquest of American Inflation*, Princeton: Princeton University Press.
22. Schönhofer, M., (1999) Learning to Believe in Chaos, *Journal of Economic Behavior & Organization*, forthcoming.
23. Sonnemans, J., Hommes, C.H., Tuinstra, J. and van de Velden, H., (1999) The instability of a heterogeneous cobweb economy: a strategy experiment in expectation formation, working paper CeNDEF, University of Amsterdam.
24. Sorger, G. (1994) Imperfect foresight and chaos: an example of a self-fulfilling mistake, Journal of Economic Behavior & Organization 33, 363-383.

A Principal-Agent Problem in Continuous Time

Alfred Luhmer

Faculty of Economics and Management
Otto von Guericke University of Magdeburg

Summary: Kamien & Schwartz' famous maintenance model is taken as a basis to analyze an agency problem in continuous time. The owner of the system to be maintained delegates the maintenance job to an agent. The discount rate of the agent is assumed to be higher than that of the owner. The so-called first-best problem assumes the agent be able to commit himself to an agreed upon effort level for the whole contract period. The analysis shows that both parties can capitalize on the different discount rates. The optimal effort level turns out to be increasing during every given finite contract period. Both parties tend to choose the contract period as long as possible.

The "second-best" problem assumes the agent's effort level is unobservable, such that the owner has to rely on an incentive payment scheme to govern the agent's activity. The only piece of information that can serve as a basis for an incentive contract is the death of the system, triggering the end of compensation payments to the agent. Assuming the principal decides on and commits himself to the compensation scheme, the second-best problem seems to be a Stackelberg differential game. It degenerates, however, into a parameter optimization problem, due to the participation and incentive compatibility constraints. The higher discount rate of the agent requires a higher cost for the owner. Some analysis of the sensitivity of the solution with respect to the agent's characteristics concludes the article.

1. The Problem

Consider an individual or organization, termed the "principal" in what follows. The principal operates a system with a stochastic life. She relies on an agent to detect and take measures against disaster events that threaten the existence of the system. When a disaster event arrives and cannot be headed off by the agent, the system must be abandoned. This entails a loss to the principal. The contract with the agent stipulates that the agency ends on this condition and that payments from the principal to the agent stop immediately. Also the agent may suffer a loss from the disaster event. Since he is fired at once, he may be unemployed for a while or may lose reputation as a qualified and conscientious expert. An incentive contract of this kind might for instance be offered to

- free lance safety engineers to supervise a chemical or nuclear power plant, the maintenance of aircraft, high-speed trains or similar equipment,

- inspectors, auditors, accounting controllers,

- tax consultants, or

- family doctors.

Typically, the effort level of such an agent is not perfectly observable neither to the principal nor to an arbiter. Therefore a contract on a specified level of the agent's effort would not be enforceable. Since the only piece of publicly observable information is the disaster event itself, an incentive contract can only depend on this event. Possibly a contract could be written for a specified period of time at the end of which a compensation is paid on condition a disaster did not occur. This will lead to a maximal incentive effect, because any shirking of the agent would increase his risk to lose the entire compensation. An agent, however, will be ready to sign up a contract of this kind only if he is compensated not only for his effort but also for the high risk and low time value of money of such a compensation. Consequently, if the agent is more risk-averse and/or more impatient than the principal, an incentive payment will require a non-optimal allocation of risk and/or cost of capital between the parties. The resulting inefficiency appears as a higher present value of the compensation to be paid by the principal or as a lower level of the agent's performance than would be optimal if the agent could costlessly commit himself to a specified level of effort. This ineffiency is known as the "agency cost". The agency problem consists of determining a time pattern of payments during the contract period that minimizes the agency cost or, in other words, a "second-best" solution[1] for which the expected

---

[1] The solution is "second best" with respect to the situation where the agent is able to commit himself to a specified effort and gets a fixed payment at the beginning of the contract period. An arrangement of this kind would not require agency costs in terms of a risk premium or interest which is higher if shouldered by the agent. A first best contract of this kind, however, would not be enforceable.

total cost including the loss due to disaster and the required compensation of the agent is minimal.

In principle both parties could agree on a collateral or a penalty to be paid by the agent to the principal in case of disaster. These possibilities are excluded in what follows. Instead limited liability is assumed on the part of the agent.

2. Model Formulation

Suppose disaster events arrive according to independent Poisson processes. Let S denote the expected loss to the principal if a disaster occurs. Then the disaster process can be modeled as a compound Poisson process (see e.g. [5] p.23). Without any care of an agent, disaster events occur at a rate h per unit of time. The rate h can be reduced by the activity of an alert and circumspective agent. The agent's effort level u is measured by the proportion of the arrival rate h of disasters which is absorbed by his activities. Then $(1-u)h$ is the hazard rate after accounting for the agent's effort. If the agent's effort u is not time-constant, then the disaster process becomes nonhomogeneous. The hazard rate can be expressed in terms of the the distribution F of the waiting time until the next hazard event where $F(t) := \Pr\{$next disaster in the agent's area of responsibility occurs not later than $t\}$. Let $\dot{F} := \dfrac{dF}{dt} :=$ f denote the respective density function. The waiting time distribution can then be determined from the functional equation: $\dfrac{f(t)}{1-F(t)} = (1-u(t))h(t)$. This relation is in fact a differential equation and acts as the state equation in a dynamic model of preventive maintenance due to [3]. [2] ch. 10.1. analyze the model in depth by using their concept of state separability. A similar modelling approach was also used in a differential game context to model R&D races between duopolists ([4], [1]).

Let S denote the principal's loss from the disaster event and L the respective loss to the agent. The agent's effort to reduce the hazard rate imposes on him a disutility $hC(u)$, measured in monetary units. $C(u)$ increases at an increasing rate if he devotes himself to his job more intensively (see [2], p.288, (10.33a)):

$$C(0) = 0; \quad C'(0) = 0; \quad C''(u) > 0; \quad C'(1) = \infty. \tag{1}$$

The function C is common knowledge.

Both principal and agent use a discount rate to value expected future payments. The discount rates capture the time value of money and risk diversification opportunities. The principal is supposed to have access to the capital market at a given rate r. The agent, to the contrary, cannot sell future expected cash flow prospects on the anonymous capital market but he wants to invest in assets for business or consumptive purposes. His discount rate is therefore determined by the debit rate

$\rho > r$ offered by financial intermediaries. Suppose this to be the case for the entire contract period.

A first step of the analysis concerns the hypothetical situation that the agent could effectively commit himself to an effort level stipulated in the contract. Under these conditions the principal could pay the agent in advance and bear all the risk and capital cost. This would be the optimal allocation of these costs.

3. First-best Solution

Assume a predetermined contract period T. The principal's expected cost can be written as:

$$J^P = \int_0^T Se^{-rt}f(t)dt = \left(1 - x(T)e^{-rT}\right)S - \int_0^T rSe^{-rt}x(t)dt, \qquad (2)$$

where

$$x(t) := 1 - F(t)$$

denotes the reliability or survivorship probability of the system at time t. The agent will accept a contract for period $[0,T]$ only if he has no better opportunity (individual rationality or participation constraint). Assume for simplicity that the outside options to the agent yield a constant cash flow Y_0 known in advance with certainty. The cost on the part of the agent then amounts to

$$
\begin{aligned}
J^A &= \int_0^T \left(Le^{-\rho t}f(t) + (1 - F(t))\left[hC(u(t)) + Y_0\right]e^{-\rho t}\right)dt \\
&= (1 - x(T)e^{-\rho T})L - \int_0^T [\rho L - Y_0 - hC(u(t))]e^{-\rho t}x(t)dt. \qquad (3)
\end{aligned}
$$

Again for simplicity, assume $Y_0 = 0$ in what follows. There is no essential loss of generality from this assumption, since a positive Y_0 would play the same role as increasing the term ρL. The joint problem of both parties is to maximize

$$
\begin{aligned}
-J^P - J^A &= \int_0^T x(t)\left[e^{-rt}rS + e^{-\rho t}[\rho L - hC(u)]\right]dt \\
&\quad + x(T)\left(e^{-rT}S + e^{-\rho T}L\right) - S - L \qquad (4) \\
\text{subject to} \quad \dot{x} &= -(1 - u)hx, \quad x(0) = 1.
\end{aligned}
$$

This problem is a variant of the well-known Kamien-Schwartz maintenance model. The difference to the original, also intensively analyzed in [2], p. 283–291, is that the utility from maintaining the system is captured only in form of postponing the terminal loss and that the maintenance cost function is time dependent. Unfortunately, however, the well-known monotonicity result does not apply directly, because some underlying assumptions are violated.

The current value Hamiltonian of the joint optimization problem, the valuation based on the principal's discount rate, is

$$H(x,u,\lambda,t) = x\left[rS + e^{(r-\rho)t}(\rho L - hC(u)) - (1-u)h\lambda\right]$$

where λ, the variable adjoined to x, is determined by

$$\dot{\lambda} = [r + (1-u)h]\lambda - rS - e^{(r-\rho)t}(\rho L - hC(u)); \lambda(T) = S + e^{(r-\rho)T}L. \quad (5)$$

Maximizing the Hamiltonian w.r.t. the control u yields

$$C'(u)e^{(r-\rho)t} = \lambda. \quad (6)$$

(5) and (6) together imply

$$\dot{u}C''(u) = [\rho + (1-u)h]C'(u) - rSe^{(\rho-r)t} - \rho L + hC(u), \quad (7)$$

a differential equation for u which has no constant solution for $\rho > r$. Following a usual procedure in studying trajectories of differential equations, define a critical level $\hat{C}'(t)$ for the marginal cost $C'(u)$ at which $\dot{u} = 0$. From (7) one obtains

$$\hat{C}'(t) = \frac{rSe^{(\rho-r)t} + \rho L - hC(u(t))}{\rho + (1-u(t))h}. \quad (8)$$

Clearly, if $C'(u) > \hat{C}'(t)$ then $\dot{u}C''(u) > 0$ and vice versa, so that

$$\dot{u}(t) \gtreqless 0 \Leftrightarrow C'(u(t)) \gtreqless \hat{C}'(t)$$

follows. In particular, compare $\hat{C}'(T)$ to $C'(u(T)) = Se^{(\rho-r)T} + L$, i.e.

$$
\begin{aligned}
\hat{C}'(T) &= \frac{Se^{(\rho-r)T}}{\frac{\rho}{r} + (1-u(T))\frac{h}{r}} + \frac{L - \frac{h}{\rho}C(u(T))}{1 + (1-u(T))\frac{h}{\rho}} \\
&< \quad Se^{(\rho-r)T} \quad + \quad L \quad = C'(u(T))
\end{aligned}
$$

from which $\dot{u}(T) > 0$ follows. Therefore the first-best effort intensity is increasing at the end of the contract period. Consequently, due to continuity, there must exist an interval $(\tau, T]$ in which $C' > \hat{C}'$ and $\dot{u} > 0$. Suppose there is also an interval before τ in which $\dot{u} < 0$. Then, $C'(u(\tau)) = \hat{C}'(\tau)$, $\dot{u}(\tau) = 0$ and, consequently, $\frac{d}{dt}C'(u(\tau)) = 0$ must hold. But from (8) and $\dot{u}(\tau) = 0$ follows in fact

$$\frac{d}{dt}\hat{C}'(\tau) = \frac{(\rho - r)Se^{(\rho-r)\tau}}{\frac{\rho}{r} + (1-u(\tau))\frac{h}{r}} > 0, \quad (9)$$

a contradiction. This proves $\dot{u} > 0$ or in words:

Proposition 1 *If the agent's discount rate exceeds the principal's one, then the first-best optimal effort rate u* in the Kamien-Schwartz maintenance agency is increasing all the time during every finite contract period.*

The economic rationale behind this proposition may be explained by focusing on the time pattern of the marginal cost of the maintenance effort that the community of principal and agent has to bear: The marginal cost of a given effort level u is decreasing as time goes on, because the agent's disutility of effort becomes cheaper if the effort takes place later in time; the utility to the community of both, however, does not decline at the same rate. Therefore the later effort is relatively more rewarding[2].

For the same reason both parties will tend to choose the contract period as long as possible. This will enable them to make the most of their different valuations of transferable utility. Formally, this leads to the following

Proposition 2 *If the agent's discount rate exceeds the principal's one, then for the autonomous problem no optimal contract period exists, provided the system shall be kept alive indefinitely.*

A proof of proposition 2 considers the problem of optimal contract period as analogous to the optimal lifetime problem for an infinite chain of identical investments. The present safety equivalent of the whole chain of contracts is

$$J^A \sum_{n=0}^{\infty} x(T)^n e^{-n\rho T} + J^P \sum_{n=0}^{\infty} x(T)^n e^{-nrT} = \frac{J^A}{1 - x(T)e^{-\rho T}} + \frac{J^P}{1 - x(T)e^{-rT}}$$
$$= : \quad K^A(T) \quad + \quad K^P(T)$$

with J^A and J^P according to (3) and (2), respectively. It can be demonstrated by some straightforward but tedious algebra that the second order necessary condition

$$\ddot{K}^A(T) + \ddot{K}^P(T) \geq 0$$

is violated. The details are omitted here.∎

Note that proposition 2 does not mean that the optimal contract period is infinite. In fact, for the infinite contract period an optimal solution to the first-best problem does not exist. A formal proof of this conjecture is not attempted here.

---

[2] Note that usually $\dot{u} < 0$ holds in the Kamien-Schwartz model. The difference is due to the fact that the present formulation considers only the cost to keep the system alive without assuming a current utility that ceases at the end of the contract period.

4. Second-best Solution

Now drop the assumption that the agent is able to commit himself to an agreed upon effort level. The principal chooses the payment profile, consisting of an initial payment V_0, payable at $t = 0$ and a current compensation $v(\cdot)$ during the contract period which is assumed indefinite from now on. She wants to minimize total expected discounted cost consisting of

- the expected loss caused by disaster events
- the expected compensation of the agent

and is supposed to be able to commit herself effectively to the payment schedule for the whole contract period.

Once he has accepted the contract, the agent can revise his decision on his effort level every time instant. Even if he can commit himself to an infinite contract period he is able to change his mind on the effort level at any time. Therefore the principal, when she determines the compensation schedule, has to consider an incentive compatibility constraint and a participation constraint for every future time instant. The incentive compatibility constraint for time τ requires that the second-best effort level be optimal for the control problem:

$$\Upsilon(\tau, v) = \max_{u(\cdot)} \left[\int_\tau^\infty e^{-\rho(t-\tau)} \frac{x(t)}{x(\tau)} [v(t) + \rho L - hC(u(t))] dt \right] - L \quad (10)$$

$$\text{subject to } \dot{x} = (1 - u)hx. \quad (11)$$

Note that at time τ it is publicly known whether the system is alive or not. So the expectation is conditional on the system being up at τ. Therefore all the probabilities in (10) are fractions with denominator $x(\tau)$.

The participation constraint for time τ must ensure that the agent prefers the optimal utility $\Upsilon(\tau, v)$ to all his outside options. The agent will enter the contract only if

$$V_0 + \Upsilon(0, v) \geq 0. \quad (12)$$

(12) represents the participation constraint at the initial time 0. Once he has accepted, the agent can no longer avoid the loss L that occurs with the disaster event. Consequently, later on, he can only require

$$\Upsilon(\tau, v) + L \geq 0. \quad (13)$$

The agent's problem $\{(10), (11)\}$ can be decomposed into a finite horizon problem with an arbitrary time horizon T and an infinite horizon problem for the rest of the time:

$$\Upsilon(\tau, v) = \max_{u(\cdot)} \left\{ \int_\tau^T e^{-\rho(t-\tau)} \frac{x(t)}{x(\tau)} [v(t) + \rho L - hC(u(t))] dt \right.$$

$$\left. + e^{-\rho T} x(T) \cdot \Upsilon(T, v) - L \right\} \quad \text{subject to (11).} \quad (14)$$

The optimal solution (cf. [2], p. 284-291 for an analysis of the nonautonomous Kamien-Schwartz model) can be characterized using the current value Hamiltonian

$$H^A(y,\eta,t) = x[v + \rho L - hC(u) - \eta[\rho + (1-u)h]],$$

and must satisfy

$$C'(u) = \eta; \tag{15}$$
$$\dot\eta = [\rho + (1-u)h]\eta - v - \rho L + hC(u); \quad \eta(T) = \Upsilon(T,v). \tag{16}$$

Let u^* denote the optimal control for problem $\{(10),(11)\}$, given v and observe that differentiating $\Upsilon(\tau,v)$ with respect to time τ yields

$$\dot\Upsilon = \left[\rho - \frac{\dot x}{x}\right]\Upsilon - [v - hC(u^*) + \rho L]$$
$$= [\rho + (1-u^*)h]\Upsilon - v - \rho L + hC(u^*) \tag{17}$$

which coincides with the adjoint equation (16) of the agent's problem. Consequently, incentive compatibility requires

$$C'(u^*(\tau)) = \Upsilon(\tau,v) \tag{18}$$

for all τ. (18) is an special feature of the Kamien-Schwartz model. It means that the agent will choose his effort level $u^*(\tau)$ at any time τ such that the marginal cost of effort is equal to the maximal expected utility of outstanding payments net of cost, provided the participation constraint (13) is satisfied. $\Upsilon(t,v)$ is independent of $u^*(\tau)$ for all $\tau < t$. Only whether the process ends before t depends on the control implemented during $[0,t)$.

The principal's problem of determining the optimal compensation scheme is formally a Stackelberg differential game with the principal as the leader and the agent as the follower. According to (2), after accounting for the agent's compensation, it reads as follows:

$$Z_0 = \min_{v(\cdot),V_0} \int_0^\infty e^{-rt} x(t) \left[(1 - u^*(t,\Upsilon(t,v_{(t,\infty)})))hS + v(t)\right] dt + V_0 \tag{19}$$

subject to the participation constraints (12) and (13) for all τ,

and the incentive compatibility constraint (18).

Only the participation constraints (13) for $\tau > 0$ need consideration, because the constraint (12) for $\tau = 0$ can be fulfilled by adapting V_0 accordingly. The decomposition (14) leads to a short cut solution using a renewal-theoretic argument: Since the agent's problem is autonomous, it must remain the same irrespective of the initial time τ at which it is considered, provided $\tau > 0$. Therefore, an optimal incentive $\Upsilon(t,v)$ to the agent must be invariant to translations along the time axis, i.e.

$$\dot\Upsilon = 0$$

so that, due to (18), u^* must be constant while

$$
\begin{aligned}
v &= [\rho + (1 - u^*)h]\Upsilon - \rho L + hC(u^*) \\
&= [\rho + (1 - u^*)h]C'(u^*) - \rho L + hC(u^*)
\end{aligned}
\tag{20}
$$

follows from (17). Clearly,

$$
\frac{dv}{du} = C''(u)[\rho + (1 - u)h] > 0,
$$

i.e. a higher level of effort requires a higher compensation. Furthermore a high loss L imminent on the agent in case of the disaster reduces the necessary current compensation, but will increase the fixed salary component V_0. The optimal level of effort the agent should be induced to make can be determined by optimizing the principal's objective functional. Substituting (20) in (19), neglecting V_0, yields

$$
\begin{aligned}
Z(u) &= \int_0^\infty e^{-[r + (1-u)h](t-\tau)} \left[(1 - u)hS + (\rho + (1 - u)h)C'(u) + hC(u) - \rho L \right] dt \\
&= \left[(1 - u)h(S + C'(u)) + hC(u) + \rho(C'(u) - L) \right] \int_0^\infty e^{-[r + (1-u)h]t} dt \\
&= \frac{hC(u) + (1 - u)h[C'(u) + S] + (C'(u) - L)\rho}{r + (1 - u)h}.
\end{aligned}
\tag{21}
$$

Setting the derivative of Z equal to zero determines $\hat{u}$, a candidate for a cost minimizing solution:

$$
Z(\hat{u}) + \left[\frac{\rho}{h} + 1 - \hat{u} \right] C''(\hat{u}) - S = 0
\tag{22}
$$

But a solution to (22) between zero and one may not exist. For instance, if C is quadratic and S small, then the derivative of Z is always positive. This means that there must exist a ceiling $\bar{u} < 1$ for the effort level which cannot be crossed if the model of the system dynamics is not to lose its sense. The optimal effort level, then, coincides with the ceiling $\bar{u}$. Otherwise, if S is large enough for $Z' < 0$ to hold, then the second-best effort level will be zero. In the case of a quadratic C, one of these extreme solutions must be optimal. A solution $\hat{u}$ to (22) if it exists, cannot be optimal, since it would violate the second order necessary optimality condition as can be seen from

$$
\mathrm{sgn}\, Z''(\hat{u}) = \mathrm{sgn}\, \left[\left[\frac{\rho}{h} + 1 - \hat{u} \right] C'''(\hat{u}) - C''(\hat{u}) \right].
$$

The quadratic cost function, however, is not a satisfactory cost model in the present context, because it does not satisfy requirements (1). An example of a function that does comply with these requirements is

$$
C(u) = 1 - \sqrt{1 - u^2}.
\tag{23}
$$

Figure 1 displays this function along with its first three derivatives.

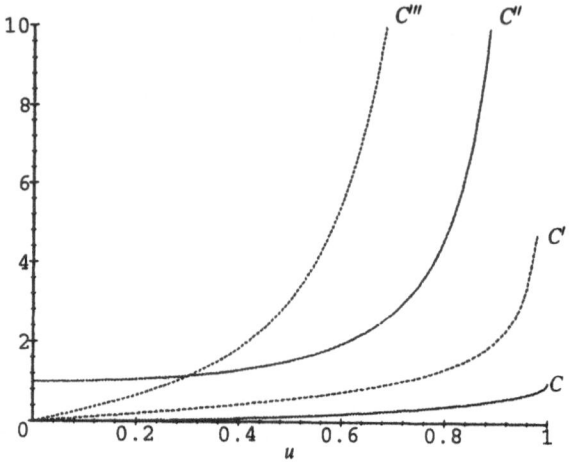

Figure 1: Cost function $C(u) = 1 - \sqrt{1 - u^2}$ and its derivatives

It may be interesting to analyze how ρ influences the second-best solution. While in the first-best case a high discount rate ρ of the agent turned out to be favorable for the two parties, the result for the second-best solution is ambiguous. The reaction of the objective function value $Z(u)$ for a fixed effort level u depends on $C'(u) - L$. The more severe the loss L expected by the agent from the disaster, the cheaper the effort level u for the principal; note, however, that the present certainty equivalent of this loss must be compensated by a higher V_0 in $t = 0$ as dictates the participation constraint (12). But this is cheaper for the principal than an equivalent incentive compensation, because of $\rho > r$.

The optimal effort level $\hat{u}$ depends on the agent's discount rate according to

$$\hat{u}_\rho \gtreqless 0 \Leftrightarrow C'(\hat{u}) + (\frac{r}{h} + 1 - \hat{u})C''(\hat{u}) - L \gtreqless 0. \tag{24}$$

(24) can be substantiated as follows. If the marginal cost $Z'(\hat{u}(\rho), \rho + \varepsilon)$ becomes positive, then the minimizer of $Z(\cdot, \rho + \varepsilon)$ must be situated to the left of the minimizer of $Z(\cdot, \rho)$ as shown in figure 2.

For the solid curve the optimal u is 0.8; the cost curve for a larger ρ is depicted as the dotted line. Its slope at $u = 0.8$ is positive. Therefore the minimum of this curve must be on the left of 0.8. (Actually, for the dotted line, $\hat{u} = 0.7$). A similar argument yields

$$\hat{u}_L > 0. \tag{25}$$

Proposition 3 *If the marginal maintenance cost $C'(u)$ at the optimal effort level $u = \hat{u}(\rho)$ exceeds the loss L to be borne by the agent in case of disaster at a given*

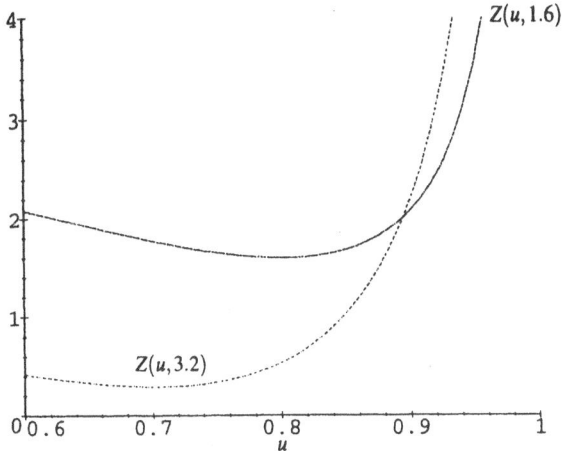

Figure 2: Dependence of the optimal effort level on the agent's discount rate

discount rate ρ of the agent, then the optimal effort level $\hat{u}(\rho)$ is locally decreasing if ρ increases. The converse, however, does not hold. Also with a loss L that is smaller than the marginal cost $C'(u)$ the optimal effort level $\hat{u}$ can be lower for higher ρ if the second derivative C'' of the cost function is sufficiently large. A larger loss L will always entail a higher optimal effort level.

Proof: Direct consequence of (24) and (25). ■

Interestingly, in the example of figure 2, the optimal cost for the principal is less for the higher ρ, i.e. for the lower level of effort. The conditions on which this occurs are spelled out by the following

Proposition 4 *If the marginal maintenance cost $C'(u)$ at the optimal effort level $u = \hat{u}(\rho)$ for a given discount rate ρ of the agent exceeds the loss L, then the principal's optimal total cost $Z^*(\rho)$ including the agent's compensation is locally increasing with ρ. If $C'(u) < L$ then it is locally decreasing.*

Proof: The proposition is a direct consequence of (21) together with the envelope theorem. ■

Figure 2 depicts a situation in which the optimal cost Z^* decreases while $\hat{u}$ decreases for increased ρ. Proposition 4 indicates that this situation occurs only if the loss L is relatively high w.r.t. the marginal cost $C'(\hat{u})$. This might be a less usual case. Usually one might expect that a lower ρ will not only induce a higher effort level $\hat{u}$ but also a lower cost Z^* for the principal. But the example makes clear that this is not generally true.

Propositions 3 and 4 may be relevant to the principal in selecting an agent from a set of applicants with different characteristics L and ρ. Furthermore there may be implications for third parties concerned about the disaster. They might be interested in adding to the risk L of the agent or in lowering his discount rate, because this enhances his effort level.

References

[1] Feichtinger, G.(1982), Optimal policies for two firms in a noncooperative research project. In: Feichtinger G., (Ed.): Optimal Control Theory and Economic Analysis. North Holland, Amsterdam.

[2] Feichtinger, G., Hartl, R.F.(1986): Optimale Kontrolle ökonomischer Prozesse. De Gruyter, Berlin

[3] Kamien, M.I., Schwartz, N.L. (1971): Optimal Maintenance and Sale Age of a Machine Subject to Failure. Management Science 17, B495-504.

[4] Kamien, M.I., Schwartz, N.L. (1982): Market Structure and Innovation. Cambridge University Press, Cambridge (Mass.).

[5] Ross, S.M. (1971), Applied Probability Models with Optimization Applications. Holden-Day, San Francisco.

A Note on Investment, Credit and Endogenous Cycles

Andreas J. Novak

Department of Statistics and Decision Support Systems, University of Vienna, Austria

Summary. In a recent paper Faria and Andrade (1998) present a model of two different representative agents, borrowers and lenders, and investigate conditions such that the borrower's problem results in a cyclical relationship between capital and loans by applying Hopf bifurcation theory. Nevertheless, the question about the existence of *stable* cycles is still unresolved. In this note a numerical example leading to *stable* cycles is presented.

Key words. Hopf bifurcation theory; Stable limit cycles; Investment

1. The Model

Faria and Andrade (1998) consider a model with borrowers and lenders as representative agents and financial intermediaries with inside and outside money. The model is set up as infinite time-horizon optimal control problems for the representative lender as well as for the representative borrower. These two control problems are linked together by additional equilibrium conditions for the financial intermediaries which have to hold in the equilibrium.

1.1 The lender's problem

The representative lender derives utility from the rate of consumption $c(t)$ as well as from money holdings $m(t)$. She chooses $c(t)$ and $m(t)$ as to maximize the current value of the discounted utility stream over an infinite horizon subject to a constraint on her savings. These savings are allocated in bank deposits $D(t)$ (inside money) and outside money $m(t)$ and their change is determined by the difference between interests paid by deposits and consumption and inflation. Thus the lender's problem can be written as [1]

$$\max_{c,m} \int_0^\infty e^{-rt} U(c,m)dt \tag{1.1}$$

subject to

$$\dot{D} + \dot{m} = (\phi - \pi)D - c - \pi m \tag{1.2}$$

where r denotes the positive rate of time preference which is assumed to be the same for lenders and borrowers, π denotes the inflation rate and ϕ is the nominal interest rate paid by deposits.

---

[1] Note that in the following the time argument is suppressed to simplify the notation.

Following Faria and Andrade (1998) the nominal interest rate ϕ is endogenously given by $\phi = r + \pi$ and the utility function is specified as $U(c, m) = c^{2/3}m^{1/3}$.

1.2 The borrower's problem

Setting up the borrower's problem this can be written in the following form according to Faria and Andrade (1998) :

$$\max_{I} \int_0^\infty e^{-rt}V(sL - I - C(I, k))dt \tag{1.3}$$

$$\dot{L} = (s + i - \pi)L + \delta k - f(k) \tag{1.4}$$
$$\dot{k} = I \tag{1.5}$$

where the control I denotes optimal investment to capital k, and L is the real amount of loans. The real amount of loans changes according to equation (1.4). Loans increase due to a constant fraction sL which is additionally raised to finance consumption, investments and adjustment costs $C(I, k)$ and due to interests iL paid for the loans. Also capital depreciation at the rate δ increases loans. On the other hand loans decrease by devaluation due to inflation πL and by production given by the production function $f(k)$. The objective functional is given as the discounted stream of utility derived by consumption $V(\tilde{c})$, where current consumption $\tilde{c}(t)$ is financed by newly raised loans sL reduced by investment I and adjustment costs $C(I, k)$.

Faria and Andrade (1998) specify the functions as

$$C(I, k) = \alpha I k^{-1} \tag{1.6}$$

$$V(\tilde{c}) = \frac{\tilde{c}^{1-\sigma}}{1 - \sigma} \tag{1.7}$$

$$f(k) = a_0 k^\beta \tag{1.8}$$

Making use of Pontryagin's Maximum Principle they derive the canonical system

$$\dot{L} = (s + i - \pi)L + \delta k - f(k) \tag{1.9}$$
$$\dot{k} = I \tag{1.10}$$
$$\dot{\lambda} = \lambda(r - s - i + \pi) - [sL - I(1 + \alpha k^{-1})]^{-\sigma}s \tag{1.11}$$
$$\dot{q} = rq - \lambda(\delta - f_k) - [sL - I(1 + \alpha k^{-1})]^{-\sigma}(I\alpha k^{-2}) \tag{1.12}$$
$$I = [sL - (1 + \alpha k^{-1})^{1/\sigma}q^{-1/\sigma}](1 + \alpha k^{-1})^{-1} \tag{1.13}$$

where λ and q denote the shadow prices of loans and capital, respectively.

1.3 Financial intermediaries

The borrower's problem is linked together with the lender's problem by two equilibrium conditions for the financial intermediaries that hold in the equilibrium point. First it is assumed that the borrower obtains deposits at a preparation rate θ and gets non-stochastic cash injections $Z(m)$ from the monetary authority, i.e. $L = Z(m) + (1 - \theta)D$, and secondly it is assumed that all deposits and loans clear up, i.e. $L = D$.

It is clear that these two equilibrium conditions can only hold when two of the model parameters are endogenously determined. Therefore we assume that additionally to ϕ also the interest rate i paid for the loans is endogenously given. The function $Z(m)$ is specified as $Z(m) = 1 + m$.

2. Endogenous cycles in the borrower's problem

2.1 Hopf bifurcation theorem

As Faria and Andrade (1998) now prove the existence of persistent oscillations between capital and loans by applying the Hopf bifurcation theory to the borrower's problem, we first restate the corresponding theorem. (See, e.g., Guckenheimer and Holmes, 1983, for a good introduction into the Hopf bifurcation theory.)

Theorem 2.1. *Suppose that a dynamical system* $\dot{z} = f_\alpha(z), z \in \mathbf{R}^n, \alpha \in \mathbf{R}$ *has an equilibrium* $(z_{crit}, \alpha_{crit})$ *at which the following property is satisfied:*

The Jacobian $D_z f_{\alpha_{crit}}(z_{crit})$ *has a simple pair of purely imaginary eigenvalues and no other eigenvalues with zero real parts.*

Then this implies that there is a smooth curve of equilibria $(z^\infty(\alpha), \alpha)$ *with* $z^\infty(\alpha_{crit}) = z_{crit}$. *The eigenvalues* $\zeta(\alpha), \bar{\zeta}(\alpha)$ *of* $D_z f_\alpha(z^\infty(\alpha))$ *which are imaginary at* $\alpha = \alpha_{crit}$, *i.e.,* $\pm i\omega$, *vary smoothly with* α.

If moreover,

$$\frac{d}{d\alpha}(Re\zeta(\alpha)) = D \neq 0, \ at \ \alpha = \alpha_{crit}, \tag{2.1}$$

then there is a unique three-dimensional center manifold passing through $(z_{crit}, \alpha_{crit})$ *in* $\mathbf{R}^n \times \mathbf{R}$, *and a smooth system of coordinates* $\xi = \xi(z), \eta = \eta(z)$, *for which the Taylor expansion of degree 3 on the center manifold is given by the following normal form:*

$$
\begin{aligned}
\dot{\xi} &= [D(\alpha - \alpha_{crit}) + A(\xi^2 + \eta^2)]\xi - [\omega + C(\alpha - \alpha_{crit}) + B(\xi^2 + \eta^2)]\eta \\
\dot{\eta} &= [\omega + C(\alpha - \alpha_{crit}) + B(\xi^2 + \eta^2)]\xi + [D(\alpha - \alpha_{crit}) + A(\xi^2 + \eta^2)]\eta
\end{aligned}
$$

$$\tag{2.2}$$

If $A \neq 0$, there is a surface of periodic solutions in the center manifold which has quadratic tangency with the eigenspace of $\zeta(\alpha_{crit}), \bar{\zeta}(\alpha_{crit})$ agreeing to second order with the paraboloid $\alpha = \alpha_{crit} - (A/D)(\xi^2 + \eta^2)$. Moreover, if $A < 0$, then the periodic solutions are stable limit cycles, while in the case $A > 0$, the periodic solutions are repelling.

Faria and Andrade (1998) show that the Jacobian of the canonical system evaluated at the equilibrium point possesses a pair of purely imaginary eigenvalues for a certain set of parameter values. According to the above theorem this is not enough to guarantee the existence of stable limit cycles as also unstable cycles may be generated by a Hopf bifurcation.[2] Unstable cycles, however, are economically irrelevant, as in this case the solution will spiral towards the stable equilibrium point in the long run, provided the starting point lies inside the cycle. Therefore it is essential to determine the stability of the cycles by additional analytical or numerical computations. Unfortunately, analytical expressions for at least the coefficients A and D can only be obtained for very simple models (for an example see Feichtinger et al., 1994). In most models the coefficients of the normal form can only be determined numerically.

As the borrower's problem is too complex to treat the question about stable cycles analytically we present a numerical example in the next section.

2.2 A numerical example

We specify the values of the exogenous parameters as follows:

$$r = 0.09, \pi = 0.46, a_0 = 0.27, \beta = 0.83,$$
$$s = 0.03, \theta = 0.99, \sigma = 1.4, \delta = 0.04. \tag{2.3}$$

Choosing α as bifurcation parameter the Jacobian of the above canonical system evaluated at the equilibrium point possesses a pair of purely imaginary eigenvalues for the critical value $\alpha_{crit} = 0.025274$. The endogenous parameters and the fixed point values are then given by $\phi = 0.55, i = 0.584314,$ $L^\infty = 1.072607, k^\infty = 0.663373, \lambda^\infty = -57.309455, q^\infty = 127.541602, I^\infty = 0$ and the corresponding equilibrium point values of the lender's problem are $D^\infty = L^\infty, m^\infty = 0.061881.$

The pair of imaginary roots implies the existence of cycles for values of α near α_{crit}. The stability of the cycles and the direction of the bifurcation (i.e. whether a subcritical or supercritical bifurcation occurs) are determined along with the sign of the coefficients A and A/D of the normal form (2.2). These coefficients are determined numerically by means of the computer code "BIFDD" (see Hassard et al., 1981) and we get $A = -0.010011, D =$

---

[2] Note that the example presented in Faria and Andrade (1998) leads to unstable cycles.

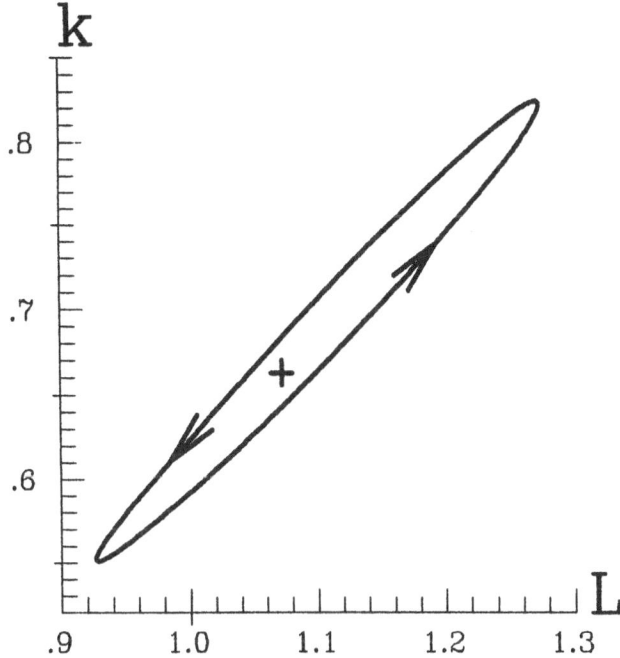

Fig. 2.1. Cycle in the state space ((L, k)−plane).

0.098660. Thus a supercritical Hopf bifurcation occurs at $\alpha = \alpha_{crit}$ and stable cycles exist for values of α slightly larger than α_{crit}. Note that the endogenously given parameter i also depends on the bifurcation parameter α which has to be taken into account when determining the coefficients A and D.

To illustrate the cyclical behaviour of optimal investment and the corresponding fluctuations of capital and loans we apply the boundary value problem solver COLSYS (see Ascher et al., 1978, and Steindl, 1981) to the canonical system for $\alpha = 0.026$ and find a closed stable orbit. Figure 2.1 shows the cycle in the state space, i.e. in the (L, k)−plane. The unstable equilibrium point is marked by a cross within the cycle. Figure 2.2 depicts the time paths of the states, of optimal investment and of the consumption rate $\tilde{c} = sL - I - C(I, k)$.

We now give a short description of the cycle (see Figure 2.2). Let us start at point P_1 characterized by the highest level of capital along the limit cycle and a high level of loans. As the capital is high, the loans are already decreasing. Because consumption is financed by part of the loans, the representative borrower starts to disinvest as this will on one hand reduce capital leading to a slower decay of the loans (compare equation (1.4)) and on the other hand disinvestment leads to higher consumption. From P_2 onwards the consumption rate will also start to decrease due to further decreasing loans.

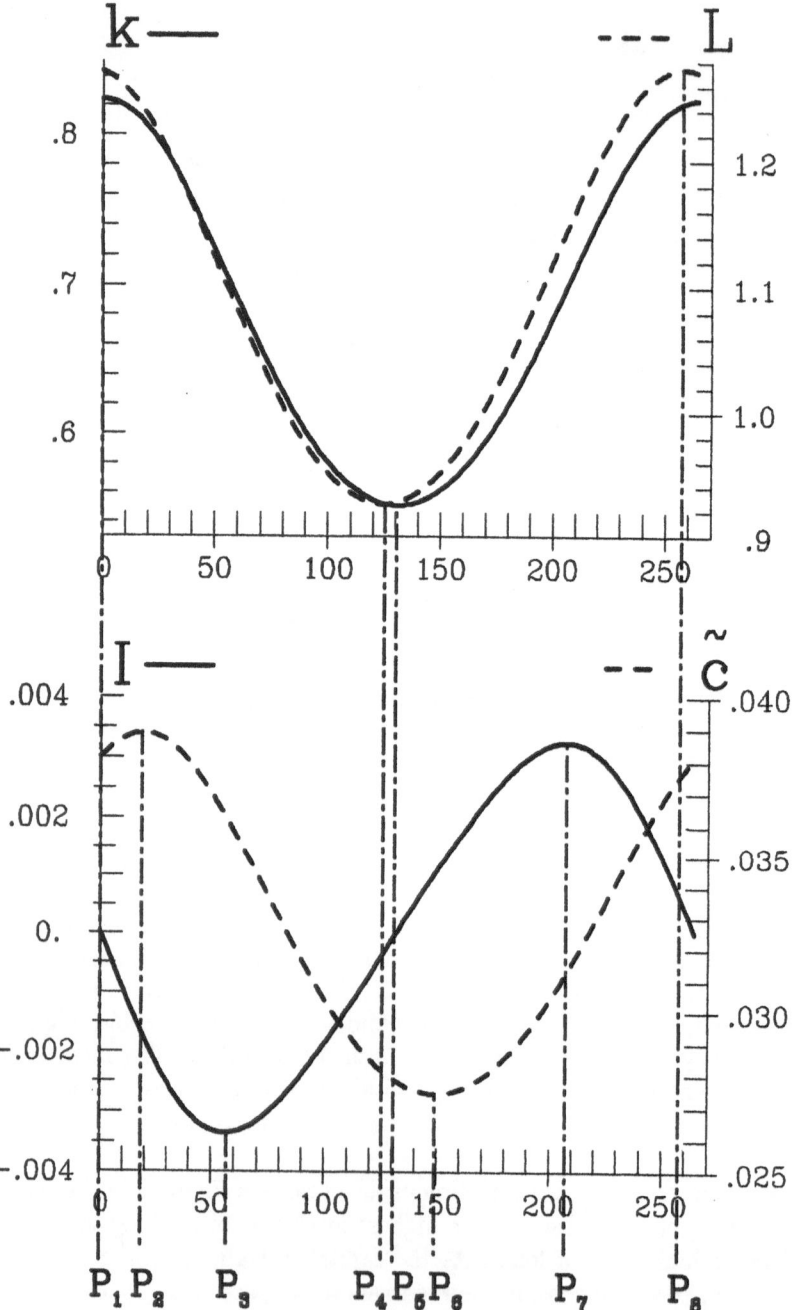

Fig. 2.2. Time paths of states L, k, of optimal investment I and of consumption $\tilde{c}$.

Becoming aware of the still decreasing capital the borrower reduces disinvestment (from P_3 onwards) but until he starts to invest at point P_5 capital is of course still decreasing. The stock of loans achieves its minimum value shortly before capital at point P_4, where production reduced by capital depreciation $f(k) - \delta k$ equals part of loans used for consumption, plus payment for debt minus inflation $(s + i - \pi)L$.

Although loans increase after P_4 consumption is still decreasing until P_6 due to high adjustment costs $C(I, k)$ caused by the low capital stock. Further increasing loans together with decreasing adjustment costs lead to a growing consumption rate although the borrower invests with a maximal investment rate at P_7.

The loans achieve its highest level at P_8, where the consumption is still growing due to a decreasing investment rate and decreasing adjustment costs as capital is growing until the end of the period, where the cycle is closed.

Moreover it can be seen from the figures that loans and capital are strongly correlated and that capital lags behind loans.

3. Summary

As discussed in Faria and Andrade (1998) the optimal investment of a representative borrower may lead to oscillations between capital and loans. Until now the existence of *stable* cycles has still been in question. We therefore present an example leading to such *stable* cycles in the present paper.

References

1. Ascher U., J. Christiansen and R.D. Russell, (1978), A collocation solver for mixed order systems of boundary value problems, Mathematics of Computation vol. 33, pp. 659-679.
2. Faria J.R. and J.P. de Andrade (1998), Investment, credit and endogenous cycles, Journal of Economics, 67, No. 2, pp. 135-143.
3. Feichtinger G., A. Novak and F. Wirl, (1994), Limit Cycles in Intertemporal Adjustment Models - Theory and Applications, Journal of Economic Dynamics and Control, 18, pp. 353-380.
4. Guckenheimer J. and P. Holmes, (1983), Nonlinear Oscillations, Dynamical Systems and Bifurcations of Vector Fields, Springer Verlag: New York.
5. Hassard B.D., N.D. Kazarinoff and Y.-H. Wan, (1981), Theory and Applications of Hopf Bifurcation, Mathematical Society Lecture Note Series 41, Cambridge: Cambridge University Press.
6. Steindl A., (1981), "COLSYS", ein Kollokationsverfahren zur Lösung von Randwertproblemen bei Systemen gewöhnlicher Differentialgleichungen, Diplomarbeit an der Technischen Universität Wien.

The Role of Extrinsic Motivation in the Dynamics of Creative Professions

Sergio Rinaldi and Francesco Amigoni

Dipartimento di Elettronica e Informazione, Politecnico di Milano, Milano, Italy

Summary. We analyze in this paper the dynamics of the production rates in creative professions from a purely theoretical point of view. The analysis is an extension of a previous work in which self-esteem was assumed to be the unique source of motivation. By contrast, in this paper we assume that the individual is stimulated by a mix of intrinsic and extrinsic motivation and we show that extrinsic motivation is a stabilizing factor. Our theory explains the fluctuating patterns of production frequently observed in artists and scientists and the absence of such fluctuations in other professions.

1. Introduction

This paper deals with the dynamic characteristics of the production rate in creative professions. The study of career trajectories of creative people has been undertaken almost two centuries ago (see Quetelet (1968), originally published in 1835). Up to now, the attempts to model creative production have tackled the evolution of mean productivity along the whole life of an individual, based on data averaged over many people (Lehman (1953), Dennis (1966), Stephan and Levine (1988)). Thus, most ups and downs have been filtered out and the dynamics have been related to the evolution of the productive ability due to learning and aging. In particular, the pattern of mean productivity was modelled by Simonton (1997) through a linear model, giving rise to a double exponential profile: productivity first reaches a maximum, and then declines, because of the exhaustion of the initial creative potential. As pointed out in Marchetti (1985), the basic pattern of the current average productivity can also be modelled by means of a logistic equation.

In this paper, by contrast, we model the medium term fluctuations observed in the career of many artists and scientists. Figure 1 reports, for example, a twenty years segment of the production rate of Mozart and Poincaré and suggests the existence of a limit cycle contaminated by random noise. A systematic analysis of the production of numerous artists has pointed out three main patterns: stationary, cyclic, and random. Ibsen is a typical example of the first kind: he wrote exactly one comedy every two years in the last 20 years of his life. Cyclic behavior (of course, not in a strictly mathematical sense) is quite common. Besides the two examples given in Fig. 1, it can be detected in many other famous artists and writers (e.g. Van Gogh, Dewey, and Lovecraft) but also in a great number of much more normal professions.

Readers in their fifties (or more) are invited to spend a few minutes to portrait their production rate in the last twenty years. This can be simply done in the following way: first partition the set of publications into three categories, say low quality (weight 1), standard quality (weight 3), and high quality (weight 5); then divide the value of each paper by the number of authors and finally sum up the scores, year by year, and plot them. The reader should not be surprised if his graph will point out almost regular ups and downs like those detectable in Fig. 2, which refers to Gustav Feichtinger, to whom this paper is dedicated, and to the first author. Random patterns, in which no mark of regularity can be (even qualitatively) detected are also possible. Stravinsky and Jung are examples of this last category.

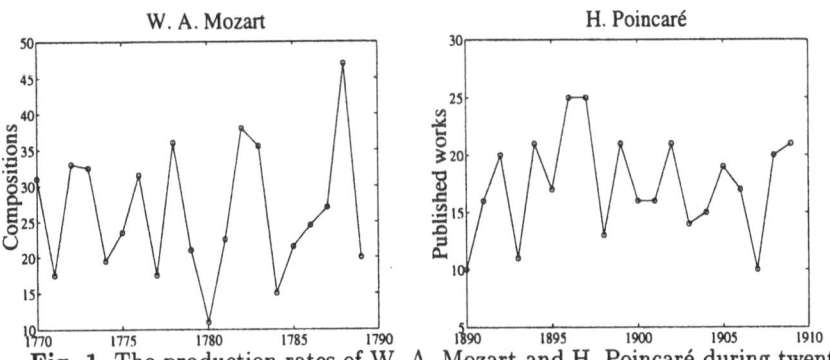

Fig. 1. The production rates of W. A. Mozart and H. Poincaré during twenty years (data taken from Lampson (1998) and Archives Henri Poincaré (1999))

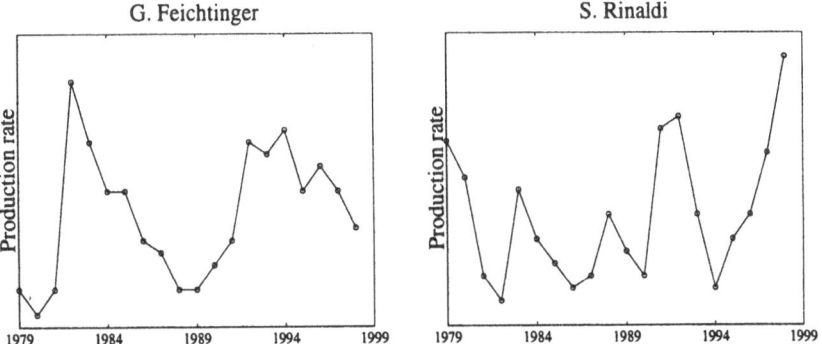

Fig. 2. Normalized production rates of Gustav Feichtinger and Sergio Rinaldi over a period of twenty years (see text for details on the evaluation of the production rate). Both diagrams point out recurrent ups and downs

The above spectrum of production dynamics suggests that the creativity of an individual has basically an endogenous origin and is characterized by a stationary or a cyclic regime, but that random exogenous events can hide these regular dynamics. Here we conjecture one potential reason for such an endogenous origin, by extending a previous contribution on the subject (see Rinaldi et al. (2000)).

In our opinion the dynamics of creative professions can be captured, at least qualitatively, by the interactions among three state variables. The first one, called *creativity*, is a measure of the fluency with which new ideas are conceived and, consequently, new results are achieved. The two others are *self-esteem* and *reputation*, which are a result of the past achievements. The current trend of a suitable combination of these two variables is assumed to stimulate creativity. This approach is in accordance with modern psychological research on motivation and creativity (Halphin and Halphin (1973), Kramer and Bayern (1984), Wells (1986)) and points out explicitly the dichotomy exisiting between intrinsic and extrinsic motivation (see, for example, Solomon and Corbit (1974), Landy (1978), Guastello (1987)). Of course the relative weights used in the combination of self-esteem and reputation can be rather different from case to case. People depending very much upon the opinion of the others (as, for example, managers and politicians) should be expected to be often motivated by reputation, while artists and scientists, or at least some of them, might be more sensitive to self-esteem.

The paper is organized as follows. In the next section we present our conjecture in the form of a third-order nonlinear dynamical model. The parameters of the model interpret psychological and behavioral characteristics of the individual and are assumed to be constant in time. This means that our conclusions should not be applied to the initial and final phases of a long career because adaptation, learning, and performance deterioration due to senescence might be relevant in those periods. Then, in the third section the model is analyzed and shown to have two possible asymptotic regimes, corresponding to stationary and cyclic production patterns. Finally, in the last section the results of the analysis are interpreted by focusing, in particular, on the stabilizing role of extrinsic motivation.

2. The Model

As already announced, our conjecture is now formulated in terms of a third-order continuous-time model. The three state variables, namely

$C = $ *creativity*
$S = $ *self-esteem*
$R = $ *reputation*

are non-negative: $C(t) = 0$ represents a state of complete catatonia at time t, while $S(t) = R(t) = 0$ only if no result has been achieved before time t.

The *flow of production*, indicated by $P(t)$, is instantaneously related to creativity, i.e.

$$P(t) = f(C(t)) \tag{1}$$

where the function $f(\cdot)$ satisfies the following properties (see Fig. 3a)

$$f(0) = 0 \qquad f'(C) > 0 \qquad f''(C) < 0 \qquad \lim_{C \to \infty} f(C) = P_{max} \qquad (2)$$

Such properties can be formally derived (see Rinaldi et al. (2000)) by subdividing the working time of each individual into time τ_c spent for conceiving new ideas and defining projects, and time τ_r spent for realizing such projects and by further assuming that τ_c is negatively correlated with creativity.

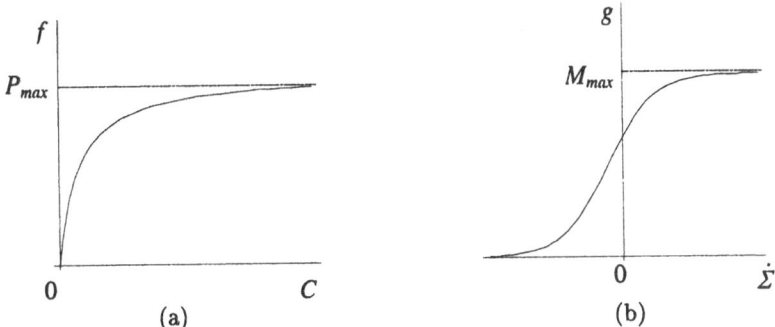

Fig. 3. The production rate P as a function $f(\cdot)$ of creativity C (a) and the motivation M as a function $g(\cdot)$ of the trend of satisfaction $\dot{\Sigma}$ (b)

Self-esteem, as well as reputation, are simply exponentially discounted integrals of the past achievements, i.e. the following two linear differential equations hold for them

$$\dot{S}(t) = -\lambda_s S(t) + \lambda_s P(t) \qquad (3)$$

$$\dot{R}(t) = -\lambda_r R(t) + \lambda_r P(t) \qquad (4)$$

where λ_s and λ_r are different forgetting coefficients. In the following, we will assume that

$$\lambda_s > \lambda_r \qquad (5)$$

since reputation raises and decays very slowly with respect to self-esteem. However, if the production rate $P(t)$ is constant, self-esteem and reputation coincide, after transient, to production rate. This means that at equilibrium self-esteem and reputation are two unbiased measures of productivity.

As far as creativity is concerned, we assume that it is an exponentially discounted integral of *motivation* $M(t)$, i.e.

$$\dot{C}(t) = -\lambda_c C(t) + M(t) \qquad (6)$$

Equation (6) is somehow an abstract definition of motivation, so that our conjecture is not fully specified until we do not specify how M depends upon the state variables. Loosely speaking, we could say that motivation M is related with the degree of satisfaction Σ that an individual feels as a mix of his self-esteem and reputation. In formulas, if we define *satisfaction* Σ as a convex combination of S and R, i.e.

$$\Sigma(t) = \mu R(t) + (1 - \mu) S(t) \tag{7}$$

we can then specify how M depends on Σ. This is, undoubtedly, the most delicate point of the model, since many assumptions reflecting different opinions are virtually possible. For example, one could imagine that motivation is simply proportional to satisfaction. But one could also assume (and we believe that this is more realistic) that motivation is mainly determined by the trend of satisfaction $\dot{\Sigma}$, i.e.

$$M(t) = g(\dot{\Sigma}(t)) \tag{8}$$

where the function $g(\cdot)$ is as shown in Fig. 3b, namely monotone increasing, first convex and then concave, and saturating to a maximum value M_{max}. A consequence of equations (6) and (8) is that when satisfaction rapidly increases, motivation is high and creativity has a sharp increase, while when satisfaction rapidly decreases the individual looses motivation. Since $\dot{\Sigma} = 0$ at equilibrium, the motivation in steady state conditions is simply $g(0)$. For a more detailed discussion on the relationship between satisfaction and motivation the reader might refer to Rinaldi et al. (2000), where a specific functional form of $g(\dot{\Sigma})$ is also derived through a microfounded model.

We summarize this section by showing the structure of the model in Fig. 4 and by rewriting the model equations in a more compact form (after substitution of equations (1), (7), and (8) into equations (3), (4), and (6))

$$\dot{S} = -\lambda_s S + \lambda_s f(C) \tag{9}$$

$$\dot{R} = -\lambda_r R + \lambda_r f(C) \tag{10}$$

$$\dot{C} = -\lambda_c C + g(\mu \dot{R} + (1 - \mu)\dot{S}) \tag{11}$$

Model (9-11) with the constraint (5) and with the functions $f(\cdot)$ and $g(\cdot)$ as in Fig. 3 is analyzed in the next section.

3. Analysis of the Model

First notice that model (9-11) is a positive dynamical system, since $S(0)$, $R(0)$, $C(0) \geq 0$ implies $S(t), R(t), C(t) \geq 0$ for all $t > 0$. Then, observe that there is only one equilibrium given by

$$\bar{S} = \bar{R} = f(\bar{C}) \qquad \bar{C} = \frac{g(0)}{\lambda_c}$$

and that such equilibrium does not depend upon λ_s, λ_r, and μ.

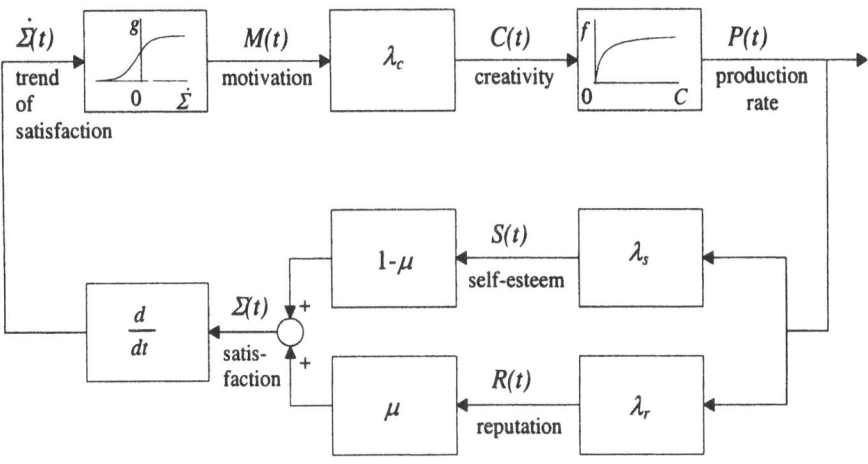

Fig. 4. The structure of the model pointing out the relationships among production rate, self-esteem, reputation, satisfaction, motivation, and creativity

The stability of this equilibrium can be studied through linearization, namely through the analysis of the Jacobian matrix J evaluated at the equilibrium. Such a matrix is given by

$$J = \begin{bmatrix} -\lambda_s & 0 & \lambda_s \bar{f}' \\ 0 & -\lambda_r & \lambda_r \bar{f}' \\ -\lambda_s(1-\mu)\bar{g}' & -\lambda_r\mu\bar{g}' & -\lambda_c + \bar{f}'\bar{g}'(\mu\lambda_r + (1-\mu)\lambda_s) \end{bmatrix}$$

where $\bar{f}'$ and $\bar{g}'$ are the derivatives of f and g evaluated at $\bar{C}$ and 0, respectively. The characteristic polynomial of J is a third order polynomial

$$\Delta(\lambda) = \det(\lambda I - J) = \lambda^3 + \alpha_1\lambda^2 + \alpha_2\lambda + \alpha_3$$

and the three coefficients α_i, $i = 1, 2, 3$, turn out to be given by

$$\alpha_1 = \lambda_s + \lambda_r + \lambda_c - \lambda_s(1-\mu)z - \lambda_r\mu z \tag{12}$$

$$\alpha_2 = \lambda_s\lambda_r + \lambda_r\lambda_c + \lambda_c\lambda_s - \lambda_s\lambda_r z \tag{13}$$

$$\alpha_3 = \lambda_s\lambda_r\lambda_c \tag{14}$$

where z is the product $\bar{f}'\bar{g}'$, i.e.

$$z = f'(\bar{C})g'(0)$$

The necessary and sufficient conditions for the asymptotic stability of the Jacobian matrix are

$$\alpha_1 > 0 \tag{15}$$

$$\alpha_1 \alpha_2 - \alpha_3 > 0 \tag{16}$$

$$\alpha_3 > 0 \tag{17}$$

Condition (17) is always satisfied (see (14)), while conditions (15) and (16) can be easily discussed because α_1 and α_2 depend linearly upon z (see (12) and (13)). Figure 5a shows, in particular, that the Jacobian matrix is asymptotically stable if and only if

$$z < z^* \tag{18}$$

where z^* is the lowest root of the equation

$$\alpha_1 \alpha_2 - \alpha_3 = 0 \tag{19}$$

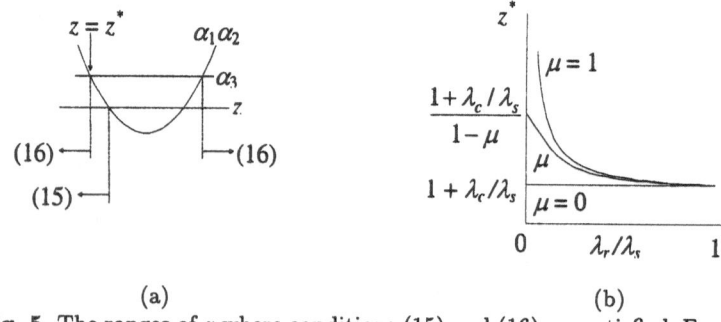

(a)   (b)

Fig. 5. The ranges of z where conditions (15) and (16) are satisfied. For $z < z^*$ both conditions are satisfied and the Jacobian matrix is asymptotically stable (a). The function $z^*(\mu, \lambda_r/\lambda_s)$ for a generic value of μ and for the two extreme values $\mu = 0$ and $\mu = 1$. The function satisfies the five properties (20-24) reported in the text (b)

Moreover, the well-known Routh criterion applied to third order polynomials says that whenever equation $z = z^*$ is satisfied, together with inequalities (15) and (17), two roots of the polynomial are purely imaginary and one is real and negative. Thus, when z is slightly smaller than z^* the Jacobian is asymptotically stable and has one negative eigenvalue and two complex conjugate eigenvalues with negative real part, while for z slightly bigger than z^* the Jacobian is unstable and has one negative eigenvalue and two complex conjugate eigenvalues with positive real part. In other words, when a parameter is varied in such a way that z becomes bigger than z^*, the equilibrium of the system (9-11) undergoes a Hopf bifurcation (Strogatz (1994)), i.e. from stable it becomes unstable and surrounded by a stable limit cycle (in order to fully support this statement we should also show that the Hopf bifurcation is supercritical). Thus, in conclusion, under condition (18) the equilibrium is asymptotically stable, while under the opposite inequality, i.e. for $z > z^*$, the attractor of system (9-11) is a limit cycle.

Of course z^* depends upon all parameters and this dependence could be explicitly specified solving equation (19) while taking equations (12-14) into account. Here we limit our discussion to the dependence of z^* upon μ and the ratio λ_r/λ_s, which vary in the unitary interval. In particular, we prove in the Appendix that the following five properties of $z^*(\mu, \lambda_r/\lambda_s)$ hold (see Fig. 5b)

$$z^*(0, \frac{\lambda_r}{\lambda_s}) = 1 + \frac{\lambda_c}{\lambda_s} \tag{20}$$

$$z^*(\mu, 1) = 1 + \frac{\lambda_c}{\lambda_s} \tag{21}$$

$$\frac{dz^*}{d\mu} > 0 \tag{22}$$

$$z^*(1, \frac{\lambda_r}{\lambda_s}) = 1 + \frac{\frac{\lambda_c}{\lambda_s}}{\frac{\lambda_r}{\lambda_s}} \tag{23}$$

$$z^*(\mu, 0) = \frac{1}{1 - \mu}(1 + \frac{\lambda_c}{\lambda_s}) \tag{24}$$

4. Interpretations of the Results and Conclusions

The formal results obtained in the previous section are now interpreted.

The first consequence of our conjecture on the relationship between satisfaction and motivation is that the production rate of an individual should be either stationary or cyclic. Indeed, the model we have analyzed has a unique global attractor which is either an equilibrium or a limit cycle. This means that our assumptions on the relationships between creativity, production rate, self-esteem, reputation, satisfaction, and motivation have not the power to explain chaotic production rates. Thus, if we believe that our model captures the main mechanisms involved in creative professions, we are forced to conclude that the deviations from pure cycles or equilibria that we observe by looking at the production profiles of many artists and scientists have exogenous sources. This is actually quite plausible since we know that social relationships, health, family life, and many other factors can have a great influence on motivation, and, hence, on creativity.

The condition for stationary production rates, namely $z < z^*$ (see (18)), and the opposite condition $z > z^*$ for fluctuating production rates can be interpreted in terms of characteristic psychological and behavioral parameters of the individual. For example, recalling that $z = f'(\bar{C})g'(0)$, we immediately recognize that a high sensitivity to variations of satisfaction (i.e., high $g'(0)$)

implies cyclic production patterns. By contrast, people rather insensitive to the variations of satisfaction should be expected to have stationary behaviors.

But more can be said by looking at z^*, which is the other term of the critical inequality $z \gtreqless z^*$. In fact, low values of z^* favor cyclic productivity, while high values favor stationary productivity. In this respect, Fig. 5b is rather helpful. It shows that a high value of μ, i.e. high extrinsic motivation, implies high values of z^*, and, hence, lower chances to have ups and downs of creativity. This is particularly true, if λ_r is small, i.e. if the environment reacts very slowly to the achievements of the individual. This situation describes quite well the case of people strongly dependent (in the success of their career) upon their reputation and working in slowly reacting environments. Politicians, directors of large public institutions, union leaders are perhaps classical stereotypes of this class. By contrast, low values of μ, i.e. great emphasis to intrinsic motivation (as opposed to extrinsic motivation), give higher chances to have ups and downs of performance. This should be typically expected in people who are only interested in pursuing their ideas, or in people who have already reached the top of their careers and are, therefore, inclined to look for intrinsic motivations. Poets, painters, and musicians are undoubtedly stereotypes of this class, but university professors can also easily fall in the same class.

In conclusion, our analysis has shown that our conjecture can explain typical production patterns of creative professions and support classical stereotypes explained until now by empirical observations. The most interesting property emerging from this study is that extrinsic motivation is a stabilizing factor.

References

Archives Henri Poincaré (1999): Henri Poincaré: List of Published Work. http://www.univ-nancy2.fr/ACERHP/bibliopk.html#Interrogation. [Accessed July 5, 1999].

Dennis, W. (1966): Creative Productivity Between the Ages of 20 and 80 Years. Journal of Gerontology **21**, 1-8.

Guastello, S. J. (1987): A Butterfly Catastrophe Modeling of Two Opponent Process Models: Drug Addiction and Work Performance. Behavioral Science **29**, 258-262.

Halphin, G., Halphin, G. W. (1973): The Effect of Motivation on Creative Thinking Abilities. Journal of Creative Behavior **7**, 51-53.

Kramer, D. E., Bayern, C. D. (1984): The Effects of Behavioral Strategies on Creativity Training. Journal of Creative Behavior **18**, 23-25.

Lampson, L. D. (1998): Köchel's Catalog of Mozart's Works. http://www.classical.net/music/composer/works/mozart/index.html. [Accessed July 5, 1999].

Landy, F. J. (1978): An Opponent Process Theory of Job Satisfaction. Journal of Applied Psychology **63**, 533-547.

Lehman, H. (1953): Age and Achievement. Princeton University Press, Princeton, NJ.

Marchetti, C. (1985): Action Curves and Clock-Work Geniuses. Technical Report WP-85-74, International Institute for Applied Systems Analysis, Laxenburg, Austria.

Quetelet, L. A. (1968): A Treatise on Man and the Development of his Faculties. Franklin, New York. [Original Work published in 1835].

Rinaldi, S., Cordone, R., Casagrandi, R. (2000): Instabilities in Creative Professions: A Minimal Model. Nonlinear Dynamics, Psychology, and Life Sciences, To appear.

Simonton, D. K. (1997): Creative Productivity: A Predictive and Explanatory Model of Career Trajectories and Landmarks. Psychological Review **104**, 66-89.

Solomon, R. L., Corbit, J. D. (1974): An Opponent Process Theory of Motivation I: Temporal Dynamics of Affect. Psychological Review **81**, 119-145.

Stephan, P. E., Levine, S. G. (1988): Measures of Scientific Output and the Age-Productivity Relationship. In Handbook of Quantitative Studies of Science and Technology, Elsevier-North Holland, Amsterdam, 31-80.

Strogatz, S. H. (1994): Nonlinear Dynamics and Chaos with Applications to Physics, Biology, Chemistery and Engineering. Addison-Wesley, Reading, Massachusetts.

Wells, D. H. (1986): Behavioral Dimensions of Creativity. Journal of Creative Behavior **20**, 61-65.

Appendix

Proof of property (20).

$$z^*(0, \frac{\lambda_r}{\lambda_s}) = 1 + \frac{\lambda_c}{\lambda_s}$$

The proof is very easy: put $\mu = 0$ in (12) and verify that $z = 1 + \lambda_c/\lambda_s$ is the lowest root of equation (19).

Proof of property (21).

$$z^*(\mu, 1) = 1 + \frac{\lambda_c}{\lambda_s}$$

Again the proof is just a check: put $\lambda_r = \lambda_s$ in (12), (13), and (14) and verify that $z = 1 + \lambda_c/\lambda_s$ is the lowest root of equation (19).

Proof of property (22).

$$\frac{dz^*}{d\mu} > 0$$

By definition, equation (19) gives $z = z^*(\mu)$. Thus, we can write

$$(\frac{\partial \alpha_1}{\partial \mu} + \frac{\partial \alpha_1}{\partial z}\frac{dz^*}{d\mu})\alpha_2 + \alpha_1(\frac{\partial \alpha_2}{\partial \mu} + \frac{\partial \alpha_2}{\partial z}\frac{dz^*}{d\mu}) - (\frac{\partial \alpha_3}{\partial \mu} + \frac{\partial \alpha_3}{\partial z}\frac{dz^*}{d\mu}) = 0 \quad \text{(I)}$$

But

$$\frac{\partial \alpha_1}{\partial \mu} = (\lambda_s - \lambda_r)z > 0$$

$$\frac{\partial \alpha_1}{\partial z} = -\lambda_s(1 - \mu) - \lambda_r \mu < 0$$

$$\frac{\partial \alpha_2}{\partial \mu} = 0$$

$$\frac{\partial \alpha_2}{\partial z} = -\lambda_s \lambda_r < 0$$

$$\frac{\partial \alpha_3}{\partial \mu} = 0$$

$$\frac{\partial \alpha_3}{\partial z} = 0$$

Thus, taking into account that α_1 and α_2 are positive for $z = z^*$, equation (I) gives $dz^*/d\mu > 0$.

Proof of property (23).

$$z^*(1, \frac{\lambda_r}{\lambda_s}) = 1 + \frac{\frac{\lambda_c}{\lambda_s}}{\frac{\lambda_r}{\lambda_s}}$$

As for properties (20) and (21), the proof is a simple check: put $\mu = 1$ in (12) and verify that $z = 1 + \lambda_c/\lambda_r$ is the lowest root of equation (19).

Proof of property (24).

$$z^*(\mu, 0) = \frac{1}{1 - \mu}(1 + \frac{\lambda_c}{\lambda_s})$$

Again, put $\lambda_r = 0$ in equations (12-14) and verify that $z = (1 + \frac{\lambda_c}{\lambda_s})/(1 - \mu)$ is the lowest root of equation (19).

Income Distribution and Endogenous Growth[*]

Gerhard Sorger

Department of Economics, Queen Mary & Westfield College, University of London, UK.

Summary. This paper studies a deterministic one-sector capital accumulation model with endogenous labor supply. Because of a production externality (learning-by-investing) the economy may experience endogenous growth. It is shown that the distribution of capital among the agents has an effect on the growth rate of per-capita output. There exists a continuum of balanced growth paths with different growth rates. A higher growth rate can be achieved when income inequality is greater, that is, when the income distribution is more strongly dispersed. The paper shows that countries with identical production technologies and identical preferences may have different GDP growth rates because wealth is distributed differently among their inhabitants. *JEL Classification Numbers:* O41, D31.

1. Introduction

This paper demonstrates that the distribution of wealth and income has an influence on the growth rate of the economy, provided that the labor supply of the households is endogenously determined. To prove this result formally we introduce an external effect in the form of learning-by-investing (as suggested by Arrow (1962) and Romer (1986)) into the model from Sorger (2000). The presence of this externality generates the possibility of permanent growth of per-capita output (endogenous growth). We first show that there exists a unique balanced growth path (BGP) if initial wealth is distributed uniformly among the households, and we determine the growth rate γ^* of per-capita output in this equilibrium. We then demonstrate that for non-homogeneous wealth distributions higher long-run equilibrium growth rates can occur. If there is no upper bound on the labor supply of households, any growth rate larger than γ^* can be achieved in a non-homogeneous BGP. Otherwise only a bounded interval of long-run growth rates (all of them larger than γ^*) is possible in equilibrium.

Since the model in this paper is very similar to the one in Sorger (2000), we refer to the latter paper for a motivation of studying the relation between aggregate output and income inequality and for a discussion of the relevant literature. In contrast to the present paper, the model in Sorger (2000) does not include any external effects and, thus, endogenous growth is ruled out. Therefore, the analysis in Sorger (2000) has focussed on stationary equilibrium paths. It is also worth pointing out that the model of the present paper uses a slightly different utility function than the model in Sorger (2000).

[*] This research is supported by grants P10850-SOZ and F010 from the Austrian Science Foundation.

Whereas in the latter model there can be either a positive or a negative relation between aggregate output and income inequality, there is an unambiguously positive relation between the growth rate of the economy and income inequality in the present model.

In Section 2 we describe the model and state the equilibrium conditions. The main results are presented in Section 3, where we also discuss their implications.

2. Model Formulation and Equilibrium Conditions

We consider a continuous-time model of a one-sector economy in which at each time $t \in [0, \infty)$ output $Y(t)$ is produced from capital $K(t)$ and labor $L(t)$ by the Cobb-Douglas technology

$$Y(t) = K(t)^{\alpha}[A(t)L(t)]^{1-\alpha}. \tag{2.1}$$

The variable $A(t)$ denotes the efficiency of labor at time t, which all agents in the economy take as given. Following the ideas developed by Arrow (1962) and Romer (1986) we assume that labor becomes more efficient if workers install new capital. This implies that the efficiency $A(t)$ increases if net investment $\dot{K}(t)$ is positive. If one assumes that the rate of change of $A(t)$ is proportional to net investment, then it follows that $A(t)$ must be proportional to $K(t)$. Choosing appropriate units of measurement, we may assume

$$A(t) = K(t). \tag{2.2}$$

Output is the numeraire good and we denote by $r(t)$ and $w(t)$ the rental rate of capital and the wage rate at time t. At every instant t, the representative firm maximizes profits

$$\Pi(t) = Y(t) - r(t)K(t) - w(t)L(t)$$

taking factor prices as given.

There exists a continuum of measure 1 (identified with the unit interval $I = [0, 1]$) of households.[1] At time t, household $i \in I$ consumes at the rate $c_i(t) \geq 0$ and supplies labor at the rate $\ell_i(t) \geq 0$ to the firms.[2] All households have identical preferences described by the utility functional

$$J_i = \int_0^{+\infty} e^{-\rho t} U(c_i(t), \ell_i(t)) \, \mathrm{d}t,$$

---

[1] When we use measure theoretic concepts we consider I as endowed with the Borel σ-algebra and Lebesgue measure.

[2] Note that a household's labor supply may become arbitrarily large. We discuss the effect of assuming an upper bound on $\ell_i(t)$ in Section 3 after presenting the results for the unconstrained case.

where $\rho > 0$ denotes the time preference rate. We assume that the instantaneous utility function has the form

$$U(c, \ell) = \begin{cases} \dfrac{c^{1-1/\theta} - 1}{1 - 1/\theta} - \dfrac{\ell^{1+\beta}}{1+\beta} & \text{if } \theta \in (0,1) \cup (1, +\infty), \\ \ln c - \dfrac{\ell^{1+\beta}}{1+\beta} & \text{if } \theta = 1. \end{cases}$$

The parameters are assumed to satisfy $\theta > 0$ and $\beta > 0$. Denoting by $\delta > 0$ the constant depreciation rate of capital, the interest rate is given by $r(t) - \delta$ and household i's flow of income at time t is

$$y_i(t) = [r(t) - \delta]k_i(t) + w(t)\ell_i(t), \tag{2.3}$$

where $k_i(t)$ is the wealth of household i at time t. Note that $k_i(t)$ can be negative or positive depending on whether the household is indebted or not. We denote the household's initial wealth at time 0 by k_{i0}. With these notations, the intertemporal budget constraint of household i can be written as

$$\dot{k}_i(t) = [r(t) - \delta]k_i(t) + w(t)\ell_i(t) - c_i(t), \ k_i(0) = k_{i0}, \tag{2.4}$$

$$\lim_{t \to +\infty} e^{-\int_0^t r(s) - \delta \, ds} k_i(t) = 0. \tag{2.5}$$

Equation (2.4) is the wealth accumulation equation and (2.5) is a no-Ponzi-game condition. Note that we assume that households are identical in terms of their preferences, but that they may have different initial endowments k_{i0}. The function $i \mapsto k_{i0}$ can be any measurable function such that aggregate initial wealth $K_0 = \int_0^1 k_{i0} \, di$ is positive.

The factor markets are in equilibrium if

$$K(t) = \int_0^1 k_i(t) \, di, \ L(t) = \int_0^1 \ell_i(t) \, di \tag{2.6}$$

holds for all t. The output market is in equilibrium if

$$Y(t) = \dot{K}(t) + \delta K(t) + C(t) \tag{2.7}$$

for all t, where $C(t) = \int_0^1 c_i(t) \, di$ denotes aggregate consumption.

Definition 2.1. A family of functions

$$E = (Y(\cdot), K(\cdot), L(\cdot), C(\cdot), r(\cdot), w(\cdot), \{k_i(\cdot), c_i(\cdot), \ell_i(\cdot) \mid i \in I\})$$

is called an *equilibrium* if there exists a function $A(\cdot)$ such that
(i) the functions $i \mapsto k_i(t)$, $i \mapsto c_i(t)$, and $i \mapsto \ell_i(t)$ are measurable for all $t \in [0, +\infty)$,
(ii) for all $t \in [0, \infty)$, the pair $(K(t), L(t))$ maximizes profit $\Pi(t)$ subject to the technological constraint (2.1) and non-negativity constraints on the inputs,

(iii) for all $i \in I$, $(k_i(\cdot), c_i(\cdot), \ell_i(\cdot))$ is an optimal solution to the problem of maximizing J_i subject to (2.4)-(2.5),

(iv) the market clearing conditions (2.6) and (2.7) and the externality condition (2.2) hold for all $t \in [0, +\infty)$.

An equilibrium is called a *balanced growth path* (BGP) if the variables $Y(t)$, $K(t)$, $L(t)$, and $C(t)$ grow at constant (but not necessarily identical) rates, and it is called *homogeneous* if the conditions $k_i(t) = K(t)$, $c_i(t) = C(t)$, and $\ell_i(t) = L(t)$ hold for all $t \in [0, +\infty)$ and all $i \in I$.

Because the population in the economy has the constant measure 1, we can interpret the variables $Y(t)$, $K(t)$, $L(t)$, and $C(t)$ either as aggregate variables or as per-capita variables. Note, however, that they need not coincide with the corresponding individual variables $y_i(t)$, $k_i(t)$, $\ell_i(t)$, and $c_i(t)$ for any particular agent i.[3]

In a homogeneous equilibrium, wealth and income are uniformly distributed among the agents. In this paper we are particularly interested in equilibria displaying non-homogeneous wealth and income distributions.

Profit maximization under perfect competition implies that capital and labor earn their marginal products, that is,

$$r(t) = \alpha A(t)^{1-\alpha}[K(t)/L(t)]^{-(1-\alpha)},$$
$$w(t) = (1 - \alpha)A(t)^{1-\alpha}[K(t)/L(t)]^{\alpha}. \tag{2.8}$$

Conditions (2.1), (2.4), (2.6), and (2.8) imply that (2.7) is satisfied (Walras' law). Thus, we may disregard condition (2.7). The first order optimality conditions for the optimization problem of household i are

$$c_i(t) = w(t)^{\theta} \ell_i(t)^{-\beta\theta} \tag{2.9}$$

and

$$\dot{c}_i(t)/c_i(t) = \theta[r(t) - \delta - \rho]. \tag{2.10}$$

Condition (2.9) shows that $c_i(t)$ is a strictly convex function of $\ell_i(t)$. In the following lemma we summarize the equilibrium conditions and draw a simple conclusion.

Lemma 2.1. *(i) A family of functions*

$$(Y(\cdot), K(\cdot), L(\cdot), C(\cdot), r(\cdot), w(\cdot), \{k_i(\cdot), c_i(\cdot), \ell_i(\cdot) \mid i \in I\})$$

is an equilibrium if and only if the conditions (2.1)-(2.2), (2.4)-(2.6), and (2.8)-(2.10) hold for all $t \in [0, +\infty)$ and all $i \in I$.

(ii) In every equilibrium and for every household $i \in I$ there exist constants $\mu_i > 0$ and $\nu_i > 0$ such that $c_i(t)/c_0(t) = \mu_i$ and $\ell_i(t)/\ell_0(t) = \nu_i$ for all $t \in [0, +\infty)$.

---

[3] Note also that $Y(t)$ is gross domestic product whereas $y_i(t)$ denotes individual income net of depreciation. Thus, one has $\int_0^1 y_i(t)\, di = Y(t) - \delta K(t)$. The other three aggregate variables are simply the integrals over I of the corresponding individual variables.

PROOF: Statement (i) is well known. From condition (2.10) follows

$$\dot{c}_i(t)/c_i(t) - \dot{c}_0(t)/c_0(t) = 0.$$

It is straightforward to see that this implies that $c_i(t)/c_0(t)$ is constant with respect to time. Together with (2.9) this shows that $\ell_i(t)/\ell_0(t)$ is constant, too.

3. Results and Discussion

From now on we focus on BGP. A stationary equilibrium is of course a special case of a BGP in which the growth rates of all variables are 0. We are mainly interested in BGP with strictly positive growth rates of $Y(t)$, $K(t)$, and $C(t)$. Part (ii) of the following lemma shows that such an equilibrium can exist only if utility of consumption is logarithmic, that is, if $\theta = 1$.

Lemma 3.1. *(i) In every BGP the aggregate labor supply $L(t)$ and the rental rate of capital $r(t)$ are constant and the growth rates of $Y(t)$, $K(t)$, $C(t)$, and $w(t)$ coincide.*
(ii) Denote the common growth rate of $Y(t)$, $K(t)$, $C(t)$, and $w(t)$ along a BGP by γ. Then $\gamma \neq 0$ is possible only if $\theta = 1$.

PROOF: (i) Conditions (2.2) and (2.8) imply

$$r(t) = \alpha L(t)^{1-\alpha}, \quad w(t) = (1-\alpha)K(t)L(t)^{-\alpha}. \tag{3.1}$$

Lemma 2.1(ii) implies that in every equilibrium the consumption paths of any two households are proportional. Consequently, it follows that $\dot{c}_i(t)/c_i(t) = \dot{C}(t)/C(t)$ for all t and all i. Along a BGP the growth rate of $c_i(t)$ must therefore be constant and independent of i. Using (2.10) it follows that the interest rate $r(t)$ must be constant along a BGP. Finally, using the first equation in (3.1) we see that labor supply along a BGP must be constant. From this property, (2.1), and (2.2) it is clear that output must be proportional to capital along a BGP and, consequently, the growth rate of output must coincide with the growth rate of capital. Dividing the aggregate resource constraint (2.7) by $K(t)$ it is easy to see that, along a BGP, consumption must be proportional to capital and it follows that the growth rate of consumption coincides with the growth rate of capital. These results together with the second equation in (3.1) imply that, along a BGP, the real wage $w(t)$ grows also at the same rate as the capital stock.

(ii) From the results proved in (i) and from condition (2.9) one can conclude that the growth rate of $\ell_i(t)$ is $(\theta - 1)\gamma/(\beta\theta)$ for all i. Integrating over $i \in I$ it follows that $L(t)$ must also be growing at the rate $(\theta-1)\gamma/(\beta\theta)$. Since we already know that $L(t)$ must be constant, it follows that $\theta = 1$ whenever $\gamma \neq 0$.

Our analysis of the model will be restricted to BGP. Thus, we assume $\theta = 1$ for the rest of this paper. The following lemma presents a necessary and sufficient condition for the existence of a BGP.

Lemma 3.2. *Assume $\theta = 1$ and let γ be any real number such that $\gamma > -(\delta + \rho)$. The following two statements are equivalent.*
(i) There exists a BGP with growth rate γ.
(ii) There exists a measurable function $\ell : I \mapsto (0, +\infty)$ such that the following two equations hold:

$$\int_0^1 \ell_i \, di = \left(\frac{\delta + \gamma + \rho}{\alpha} \right)^{1/(1-\alpha)}, \tag{3.2}$$

$$\int_0^1 \ell_i^{-\beta} \, di = \frac{\rho + (1 - \alpha)(\delta + \gamma)}{\alpha(1 - \alpha)} \left(\frac{\delta + \gamma + \rho}{\alpha} \right)^{\alpha/(1-\alpha)}. \tag{3.3}$$

PROOF: Assume that there exists a BGP with growth rate γ. It follows from (2.10) that $r(t) = \delta + \gamma + \rho$. Substituting this into (3.1) we obtain $L(t) = [(\delta + \gamma + \rho)/\alpha]^{1/(1-\alpha)}$ and

$$w(t) = K(t)(1 - \alpha) \left(\frac{\alpha}{\delta + \gamma + \rho} \right)^{\alpha/(1-\alpha)}. \tag{3.4}$$

The expression for $L(t)$ together with (2.6) yields (3.2). We know that along the BGP it must hold that $K(t) = \tilde{K}e^{\gamma t}$, $w(t) = \tilde{w}e^{\gamma t}$, $k_i(t) = \tilde{k}_i e^{\gamma t}$, and $c_i(t) = \tilde{c}_i e^{\gamma t}$, where $\tilde{K}$, $\tilde{w}$, $\tilde{k}_i$, and $\tilde{c}_i$ are constants satisfying $\int_0^1 \tilde{k}_i \, di = \tilde{K}$. Substituting this into (3.4), (2.9), and (2.4) one obtains

$$\tilde{w} = \tilde{K}(1 - \alpha) \left(\frac{\alpha}{\delta + \gamma + \rho} \right)^{\alpha/(1-\alpha)}, \quad \tilde{c}_i = \tilde{w}\ell_i^{-\beta}, \quad \tilde{k}_i = (\tilde{c}_i - \tilde{w}\ell_i)/\rho. \tag{3.5}$$

Combining the last two equations in (3.5) yields $\tilde{k}_i = (\tilde{w}/\rho)(\ell_i^{-\beta} - \ell_i)$. Integrating this equation over $i \in I$ we obtain

$$\tilde{K} = \frac{\tilde{w}}{\rho} \left[\int_0^1 \ell_i^{-\beta} \, di - \int_0^1 \ell_i \, di \right].$$

Substituting for $\int_0^1 \ell_i \, di$ from (3.2) and for $\tilde{w}$ from the first part in (3.5) one obtains equation (3.3). This proves that condition (ii) is necessary for condition (i). The sufficiency can easily be shown by constructing the equilibrium starting from values ℓ_i, $i \in I$, which satisfy (3.2) and (3.3). Note that this construction determines the growing variables (e.g. $K(t)$, $w(t)$, $c_i(t)$) only up to a common multiplicative constant. Thus, one can for example normalize by $\tilde{w} = 1$ and use (3.5) to compute $\tilde{c}_i$, $\tilde{k}_i$, and $\tilde{K}$. Details are omitted.

We can now prove the main result of this paper. The first part characterizes the unique homogeneous BGP whereas the second part shows that every growth rate that exceeds that of the homogeneous BGP can be achieved as the equilibrium growth rate in a non-homogeneous BGP.

Theorem 3.1. *Assume $\theta = 1$.*

(i) There exists a unique homogeneous BGP. The equilibrium growth rate in this BGP is the unique value $\gamma^* > -(\delta + \rho)$ which solves the equation

$$\frac{\rho + (1-\alpha)(\delta + \gamma^*)}{\alpha(1-\alpha)} \left(\frac{\delta + \gamma^* + \rho}{\alpha}\right)^{(\alpha+\beta)/(1-\alpha)} = 1. \qquad (3.6)$$

(ii) For every $\gamma > \gamma^*$ there exists a non-homogeneous BGP with growth rate γ.

PROOF: (i) In a homogeneous BGP we must have $\ell_i = L$ for all $i \in I$. Therefore, equations (3.2) and (3.3) yield

$$\left(\frac{\delta + \gamma + \rho}{\alpha}\right)^{-\beta/(1-\alpha)} = \frac{\rho + (1-\alpha)(\delta + \gamma)}{\alpha(1-\alpha)} \left(\frac{\delta + \gamma + \rho}{\alpha}\right)^{\alpha/(1-\alpha)}.$$

It is easy to see that this condition is equivalent to (3.6).

   (ii) The proof of this part is illustrated in Figure 3.1. The downward

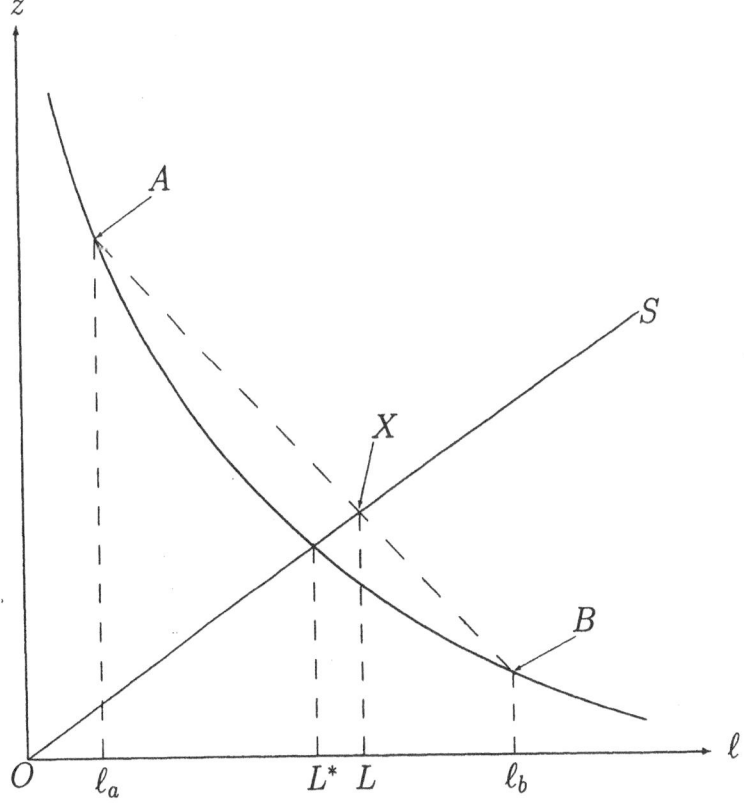

Fig. 3.1. Proof of Theorem 3.1(ii)

sloping hyperbolic curve in the figure is the graph of the function $\ell \mapsto \ell^{-\beta}$, the upward sloping curve OS is the locus $\{(\ell, z) \mid \ell = g_1(\gamma), z = g_2(\gamma), \gamma > -(\delta + \rho)\}$, where $g_1(\gamma)$ denotes the right-hand side of (3.2) and $g_2(\gamma)$ the right-hand side of (3.3). Both $g_1(\cdot)$ and $g_2(\cdot)$ are strictly increasing, $\lim_{\gamma \to -(\delta+\rho)} g_1(\gamma) = \lim_{\gamma \to -(\delta+\rho)} g_2(\gamma) = 0$, and $\lim_{\gamma \to +\infty} g_1(\gamma) = \lim_{\gamma \to +\infty} g_2(\gamma) = +\infty$. These properties imply that OS is strictly increasing, that it starts at the origin, and that it extends infinitely far into the north-east direction. The unique intersection of the two curves determines the aggregate labor supply in the homogeneous BGP, which we denote by $\ell = L^* := [(\delta + \gamma^* + \rho)/\alpha]^{1/(1-\alpha)}$. Now fix any $\gamma > \gamma^*$ and denote by L the corresponding value of the right-hand side of (3.2). Moreover, choose a value $\ell_a \in (0, L^*)$. Find the point on OS with $\ell = L$ (point X) and the point on the hyperbolic curve with $\ell = \ell_a$ (point A). By choosing ℓ_a sufficiently small one can always ensure that point A has a larger z-coordinate than point X. The intersection of the hyperbola and the straight line through A and X to the right of X is called B. Point B's existence and uniqueness is ensured by the hyperbolic shape of the curve. We denote the ℓ-coordinate of B by ℓ_b. Finally define $\lambda = (L - \ell_a)/(\ell_b - \ell_a)$. This construction ensures that $0 < \ell_a < L < \ell_b$, $\lambda \in (0, 1)$, $(1 - \lambda)\ell_a + \lambda \ell_b = g_1(\gamma)$, and $(1 - \lambda)\ell_a^{-\beta} + \lambda \ell_b^{-\beta} = g_2(\gamma)$. The function $\ell : I \mapsto (0, +\infty)$ defined by

$$
\ell_i = \begin{cases} \ell_a & \text{if } i \in [0, 1 - \lambda), \\ \ell_b & \text{if } i \in [1 - \lambda, 1], \end{cases}
$$

satisfies therefore the conditions stated in Lemma 3.1(ii) and the proof of the theorem is complete.

The main conclusion from the above theorem is that the model predicts a positive relation between income (or wealth) inequality and growth. The homogeneous BGP is the one with the smallest growth rate of per-capita output. Using the same methods as in Sorger (2000) one could show in a precise way that larger growth rates go along with higher income inequality. The reason for the positive relation between growth and income inequality in the present model is the interplay between the intertemporal saving decision and the labor/leisure choice of the households. For any given wage rate, the optimal consumption rate of a household is a strictly convex function of its labor supply. This implies that, for a given wage rate and a given aggregate labor supply, demand for consumption is higher if households supply different amounts of labor than if the labor supply is uniformly distributed.

One surprising aspect of Theorem 3.1 is that arbitrarily large growth rates can occur in non-homogeneous BGP. Scrutinizing the proof of the theorem shows that the potentially infinite labor supply of households is essential for this result to be true. If one would assume that households can only supply labor up to a certain limit, say $\bar{\ell}$, then it would only be possible to find BGP with growth rates $\gamma \in (\gamma^*, \bar{\gamma})$, where $\bar{\gamma}$ is determined by $\bar{\ell} = [(\delta + \bar{\gamma} + \rho)/\alpha]^{1/(1-\alpha)}$.

References

1. Arrow, K.J. (1962): The Economic Implications of Learning by Doing. *Review of Economic Studies* **29**, 155-173.
2. Romer, P.M. (1986): Increasing Returns and Long-Run Growth. *Journal of Political Economy* **94**, 1002-1037.
3. Sorger, G. (2000): Income and Wealth Distribution in a Simple Model of Growth, *Economic Theory* (forthcoming).

A Turnpike Theorem with Public Capital

Harutaka Takahashi*

Department of Economics, Meiji Gkuin University, Tokyo 158-8636, Japan

Summary. Most of the literature on global stability of optimal public capital often treats private and public capital as physically identical goods. As a result, one important characterlistic of public capital,"non-rivalness," will be neglected. This is clearly a serious analytical defect. To tackle the problem, a three-sector optimal growth model, where one of the sectors is a public sector and produces public capital, is set up, and the global stability of the optimal path of public capital will be demonstrated.

1. Introduction

A series of recent empirical studies of the influence of infrastructure on output and productivity in the private sectors have shown that infrastructure crucially affects output and productivity in private sectors. For example, Aschauer (1989) and Munell (1990) have reported that the estimated elasticities of production with respect to infrastructure for the US private sector are 0.36 and 0.15, respectively. Although many empirical studies have been done since then, the theoretical analysis of the infrastructure has not been sufficiently developed since the seminal paper by Arrow and Kurz (1970).

Arrow and Kurz (1970) have set up a continuous-time optimal growth model with two types of capital: private and public capital, and have shown that the optimal paths of private and public capital converge to the corresponding optimal steady states. This property is important for establishing the "stability of public instruments"; the proper policy parameters converge to finite values that achieve optimal paths of private and public capital independent of their initial capital stocks. To emphasize the insensitivity with respect to initial stocks, we may refer to this global convergence property as *Turnpike Theorem*. A study similar to that of Arrow and Kurz (1970) has also been done by Uzawa (1988).

Arrow and Kurz treat public capital as a public good that affects a person's utility. They also consider the case in which public capital increases the productivity of labor input: labor augmenting technical progress. Their model has been frequently used to analyze optimal policies in much of the literature of public finance. In spite of the popularity of the model, it has a serious drawback; the model neglects one important feature of public capital

---

* This work was supported by a grant from Tokyo Center for Economic Research. I also should like to thank an anonymous referee, Tohru Maruyama and seminar participants at Keio University for their valuable comments. Of course, any remaining errors are mine.

as the public good, namely "non-rivalness." In other words, they claimed that they are considering two different types of capital, but both types of capital are actually treated physically as the same. Therefore, one important question may emerge; if the private and public capital might be treated as physically different, how would the similar global convergence property shown by Arrow and Kurz (1970) hold?

To tackle this problem, I will set up an optimal growth model consisting of three sectors: a consumption-producing sector, a private capital- producing sector and a public sector that produces public capital. By doing so, private and public capital can be treated as physically different. To simplify the analysis, I may neglect the effect of the public capital on a person's utility. Therefore, I may regard public capital as special infrastructure: highways, railroads, and telecommunications. I will prove that under certain conditions, any optimal public capital will converge to a corresponding optimal steady state path of public capital. The turnpike theorem has been studied extensively by Brock and Scheinkman (1977) and McKenzie (1986) among others. I will apply their results to the model, especially the turnpike theorem proved by Brock and Scheinkman (1977).

The remainder of the paper is organized as follows. Section 2 explains the model and presents some assumptions. In Section 3, the model will be rewritten into a reduced model. The main result will also be proved in Section 3. Section 4 is assigned for the proof of Lemma 3. Section 5 discusses some possible extensions.

2. The Model

The model considered here is a slightly modified three-sector neoclassical optimal growth model where the sector denoted by the suffix 0 is a consumption-producing sector and the first and second sectors are producing private and public capital, respectively. All the three sectors use public capital as their input. A planner (or a government) will maximize a representative individual's utility. The problem will be then described as follows:

$$Max \sum_{t=0}^{\infty} \rho^{-t} y_0(t)$$

subject to:

$$y_0(t) = f(k_{01}(t), k_2(t), \ell_0(t)) \tag{2.1}$$
$$y_1(t) = g(k_{11}(t), k_2(t), \ell_1(t)) \tag{2.2}$$
$$y_2(t) = h(k_{21}(t), k_2(t), \ell_2(t)) \tag{2.3}$$
$$1 = \ell_0(t) + \ell_1(t) + \ell_2(t) \tag{2.4}$$

$$k_1(t) = k_{01}(t) + k_{11}(t) + k_{21}(t) \qquad (2.5)$$
$$k_1(t+1) = y_1(t) + k_1(t) \qquad (2.6)$$
$$k_2(t+1) = y_2(t) + k_2(t) \qquad (2.7)$$

$k_1(0)$ and $k_2(0)$ are historically given,

where the notation is as follows:

ρ	:	Subjective rate of discount,
$y_i(t)$	:	Per-capita net output of the i-th sector at the t-th period,
$k_{i1}(t)$	:	Private capital stock used in the i-th sector at the t-th period,
$\ell_i(t)$	:	Labor allocation per-capita to the i-th sector at the t-th period,
$k_1(t)$	:	Total supply of private capital at the t-th period,
$k_2(t)$	:	Total supply of public capital at the t-th period,
$f(.), g(.), h(.)$	:	Production functions of each sector.

Note that I make some assumptions in the model for simplifying the following argument. Three assumptions are important among others. First, I assume that the consumption will directly provide utility but public capital will not have any effect on a person's utility. Secondly, I ignore the depreciations of capital stocks. The first one may restrict our analysis of public capital to a narrower concept of public capital. This may be a severe restriction when we regard public capital as infrastructure, say roads or highways, which indeed directly affects a person's utility.

Assumption 1. $0 < \rho < 1$.

Assumption 2. i) All goods are produced non-jointly with production functions $f, g,$ and h, which are homogeneous of degree one, twice continuously differentiable for positive inputs. ii) They also are strictly quasi-concave and satisfy the second-order derivative condition: the n dimensional bordered determinant consists of the first and second derivatives of each sector's production function denoted by $\det B_n$ satisfies that $(-1)^n \det B_n > 0$ for $n = 1, 2, 3$. iii) No goods can be produced unless $k_{i1} > 0$, $k_2 > 0$ and $\ell_i > 0$.[1]

---

[1] Making the homogeneity assumption with respect to only private capital is a standard one in a neoclassical growth model. If public capital is introduced

3. Reduced Form Problem and Turnpike Theorem

We may first prove the following lemma which is similar to the one proved by Benhabib and Nishimura (1979) for positive arguments:

Lemma 1. *The constraints (2.1) through (2.5) can be summarized as the social production function $y_0 = T(y_1, y_2, k_1, k_2)$, which is twice continuously differentiable.*

Proof. To derive the social production function $T(y_1, y_2, k_1, k_2)$, we need to solve the following problem:

$$\begin{cases} y_0 = \max_{k_{01}, k_2, \ell_0} \quad f(k_{01}, k_2, \ell_0) \\ \\ \text{s.t. Equations (2.2) through (2.5) and } k_2. \end{cases}$$

Let us write the Lagrangian function of the problem as:

$$\begin{aligned} L \;=\; & f(.) + p_1[g(.) - y_1] + p_2[h(.) - y_2] + w_0[1 - (\ell_0 + \ell_1 + \ell_2)] \\ & + w_1[k_1 - (k_{01} + k_{11} + k_{21})] + w_2 k_2 \end{aligned}$$

Then the first order conditions of the problem are given as follows:

$$\partial f/\partial k_{01} - w_1 = 0$$
$$p_1 \partial g/\partial k_{11} - w_1 = 0$$
$$p_2 \partial h/\partial k_{21} - w_1 = 0$$
$$\partial f/\partial k_2 + p_1 \partial g/\partial k_2 + p_2 \partial h/\partial k_2 = w_2$$
$$\partial f/\partial \ell_0 - w_0 = 0$$
$$p_1 \partial g/\partial \ell_1 - w_0 = 0$$
$$p_2 \partial h/\partial \ell_2 - w_0 = 0$$
$$y_0 - f(k_{01}, k_2, \ell_0) = 0$$
$$y_1 - g(k_{11}, k_2, \ell_1) = 0$$
$$y_2 - h(k_{21}, k_2, \ell_2) = 0$$

---

as an input, then only empirical studies could justify this assumption: each sector's production function is homogeneous of degree one with respect to all inputs including public capital stock. Meade (1952) called it the case of "unpaid factors of production." Recent empirical works by Aschauer (1989) and Munell (1990) seem to suggest this case at least for the US data.

$$1 - (\ell_0 + \ell_1 + \ell_2) = 0$$

$$k_1 - (k_{01} + k_{11} + k_{21}) = 0$$

where the price of consumption good is normalized as 1 and, p_i and w_i are the Lagrange multipliers; the prices of outputs and inputs.

The Jacobian is the following:

$$
\begin{bmatrix}
f_{11} & f_{13} & 0 & 0 & 0 & 0 & f_{12} & 0 & 0 & -1 & 0 & 0 \\
f_{31} & f_{33} & 0 & 0 & 0 & 0 & f_{32} & 0 & 0 & 0 & -1 & 0 \\
0 & 0 & p_1 g_{11} & p_1 g_{13} & 0 & 0 & p_1 g_{12} & -g_1 & 0 & -1 & 0 & 0 \\
0 & 0 & p_1 g_{31} & p_1 g_{33} & 0 & 0 & p_2 g_{32} & -g_3 & 0 & 0 & 0 & -1 \\
0 & 0 & 0 & 0 & p_2 h_{11} & p_2 h_{13} & p_2 h_{12} & 0 & -h_1 & -1 & 0 & 0 \\
0 & 0 & 0 & 0 & p_2 h_{31} & p_2 h_{32} & p_2 h_{32} & 0 & -h_3 & 0 & 0 & -1 \\
f_{21} & f_{23} & p_1 g_{21} & p_1 g_{23} & p_2 h_{21} & p_2 h_{23} & \Delta & -g_2 & -h_2 & 0 & -1 & 0 \\
0 & 0 & -g_1 & -g_3 & 0 & 0 & -g_2 & 0 & 0 & 0 & 0 & 0 \\
0 & 0 & 0 & 0 & -h_1 & -h_3 & -h_2 & 0 & 0 & 0 & 0 & 0 \\
-1 & 0 & -1 & 0 & -1 & 0 & 0 & 0 & 0 & 0 & 0 & 0 \\
0 & 0 & 0 & 0 & 0 & 0 & -1 & 0 & 0 & 0 & 0 & 0 \\
0 & -1 & 0 & -1 & 0 & -1 & 0 & 0 & 0 & 0 & 0 & 0
\end{bmatrix}
$$

where $\Delta = f_{22} + p_1 g_{22} + p_2 h_{22}$. To simplify the proceeding argument, let us redefine the above Jacobian as

$$
Q = \begin{bmatrix} A & B \\ B^t & O \end{bmatrix}
$$

where A is a 9 by 9 matrix and B is a 3 by 7 matrix, and O is a 3 by 3 zero matrix.

Suppose that the Jacobian Q is singular. A non-zero vector X then exists such that $QX = 0$ where $X = (x, y)$. Then $X^t QX = x^t Ax = 0$. It follows that A must be singular. By applying the basic operations of determinant to det A, it follows

$$
\det A = \det(f_{ij}) + (p_1)^3 \det \begin{bmatrix} (g_{ij}) & (-g_i) \\ (-g_i)^t & O \end{bmatrix} + (p_2)^3 \det \begin{bmatrix} (h_{ij}) & (-h_i) \\ (-h_i)^t & O \end{bmatrix}.
$$

Due to ii) of Assumption 2 which implies the strict quasi-concavity of the production functions, the second and the third determinants are non-zero. So A is non-singular. This is a contradiction. Thus Q must be non-singular. By the implicit function theorem, k_{i1}, k_2, ℓ_i, p_i and w_1 are locally differentiable functions. Thus $y_0 = f(k_{01}(y, k), k_2(y, k), \ell_0(y, k))$ must be differentiable in $y > 0$ and $k > 0$. This completes the proof.

Let us define the following set **D**:

$$\mathbf{D} = \{(k(t), k(t+1)) \in \mathbf{R}_+^2 \times \mathbf{R}_+^2 : T(.) \geq 0\},$$

where $k(t) = (k_1(t), k_2(t))$.

Due to Lemma 1, we may rewrite our problem as the following reduced form problem:

$$\begin{cases} \max_{k(t)} \sum_{t=0}^{\infty} \rho^{-t} \mathbf{V}(k(t), k(t+1)) \\ \\ \text{s.t.} \quad (k(t), k(t+1)) \in \mathbf{D} \ for \text{ all } t \geq 0 \text{ and } k(0) \text{ is given} \end{cases}$$

where $V(k(t), k(t+1)) = T(k(t+1) - k(t), k(t))$.

I will introduce two important definitions now.

Definition 1. A path $\{k(t)\}_{t=0}^{\infty}$ of capital stocks is called a *regular path* if $\underline{k}$ and $\bar{k}$ $(0 < \underline{k} \leq \bar{k})$ exist such that $0 < \underline{k} \leq k(t) \leq \bar{k}$ for all $t \geq 0$.[2]

For the purposes of the paper, we have to restrict our discussion to sequences of capital that are bounded from above and below by certain positive vectors not equal to zero. Regularity of a path is a rather strong assumption but can be guaranteed if one requires production functions to satisfy generalized Inada conditions. The generalized Inada conditions have been introduced by Brock and Scheinkman (1977, Assumption 3) for a single firm adjustment cost model and could be easily adopted in my model. Also note that if $\{k(t)\}_{t=0}^{\infty}$ is a solution of the above problem, then it satisfies the following Euler equations:

$$\mathbf{V}_2(k(t), k(t+1)) + \rho \mathbf{V}_1(k(t+1), k(t+2)) = \mathbf{0},$$

where $\mathbf{V}_1 = \partial V/\partial k(t)$, $\mathbf{V}_2 = \partial V/\partial k(t+1)$ and $\mathbf{0}$ is a two-dimensional zero vector.

Definition 2. $k^\rho \in \mathbf{R}_+^2$ is an optimal steady state (abbreviated as O.S.S..) if $\mathbf{V}_2(k^\rho, k^\rho) + \rho \mathbf{V}_1(k^\rho, k^\rho) = \mathbf{0}$ holds.

As demonstrated by McKenzie (1986), under a fairly weaker set of assumptions than here, a unique O.S.S. exists denoted by k^ρ.

Define the Hessian matrix of the function V as

$$\begin{bmatrix} \mathbf{V}_{11} & \mathbf{V}_{12} \\ \mathbf{V}_{12} & \mathbf{V}_{22} \end{bmatrix},$$

where $\mathbf{V}_{11} = (\partial^2 V/\partial k^2(t))$, $\mathbf{V}_{12} = \mathbf{V}_{21} = (\partial^2 V/\partial k(t)\partial k(t+1))$, and $\mathbf{V}_{22} = (\partial^2 V/\partial k^2(t+1))$.

The following turnpike theorem is then a slightly modified version of the one proved in Theorem 1 of Brock and Scheinkman (1977). Also see Theorem 8.4 of McKenzie (1986).

---

[2] Let x and y be n-dimension vectors. Then, $x > y$ if $x_i > y_i$ for all i and $x \geq y$ if $x_i \geq y_i$ for all i.

Lemma 2. *If the Hessian matrix of V is negative definite along each regular path, there is a $\bar{\rho} > 0$ such that $\bar{\rho} \leq \rho < 1$ implies that all regular paths that satisfy the Euler equations satisfy $\lim_{t \to \infty} k(t) = k^\rho$.*

Proof. Taking $\bar{\rho}$ close enough to unity and applying Theorem 1 of Brock and Scheinkman (1977), the result follows.

From this lemma, all we need to show is that the Hessian matrix of V is negative definite. Furthermore, from the definition of V, the Hessian matrix of V is expressed by the Hessian matrix of the transformation function T as follows:

$$
\begin{bmatrix} \mathbf{V}_{11} & \mathbf{V}_{12} \\ \mathbf{V}_{12} & \mathbf{V}_{22} \end{bmatrix} = \begin{bmatrix} (\mathbf{T}_{11} - \mathbf{T}_{12} - \mathbf{T}_{21} - \mathbf{T}_{22}) & (-\mathbf{T}_{11} + \mathbf{T}_{21}) \\ (-\mathbf{T}_{11} + \mathbf{T}_{12}) & \mathbf{T}_{11} \end{bmatrix}
$$

$$
= \begin{bmatrix} -\mathbf{I}_2 & \mathbf{I}_2 \\ \mathbf{I}_2 & \mathbf{O}_2 \end{bmatrix} \begin{bmatrix} \mathbf{T}_{11} & \mathbf{T}_{12} \\ \mathbf{T}_{21} & \mathbf{T}_{22} \end{bmatrix} \begin{bmatrix} -\mathbf{I}_2 & \mathbf{I}_2 \\ \mathbf{I}_2 & \mathbf{O}_2 \end{bmatrix}
$$

where $\mathbf{T}_{11} = (\partial^2 T / \partial k^2(t))$, $\mathbf{T}_{12} = (\partial^2 T / \partial k(t) \partial k(t+1))$, $\mathbf{T}_{22} = (\partial^2 T / \partial k^2(t+1))$ and $\mathbf{I}_2$ is a two-dimensional unit matrix. $\mathbf{O}_2$ is a two-dimensional zero matrix.

Note that the first and third matrices of the last relation are clearly non-singular. Therefore, if the Hessian of T is negative definite, then the Hessian matrix of V is also negative definite. The following lemma will establish this result. Because the proof is rather long and tedious, so I relegate the proof to the next section.

Lemma 3. *The Hessian matrix of the social transformation function T is negative definite along each regular path.*

Proof. See Section 4.

Finally, combining Lemma 2 and Lemma 3 yields the following turnpike theorem:

Theorem. *Taking a $\rho \, (> 0)$ close enough to 1, then all the regular public and private capital accumulation paths that satisfy the Euler equations converge to each respective O.S.S. path. The convergence is insensitive to initial stocks.*

4. Proof of Lemma 3

Let us partition the Jacobian matrix $\mathbf{Q}$ differently as follows:

$$
\mathbf{Q} = \begin{bmatrix} \mathbf{D} & \mathbf{F} \\ \mathbf{F}^t & \mathbf{O} \end{bmatrix}
$$

where $\mathbf{O}$ is a 5 by 5 zero matrix and the matrix $\mathbf{D}$ consists of elements of all the Hessian matrices of sector's production functions.

The proof consists of two parts. The first one is to show that the matrix $\mathbf{D}$ is negative definite. We may then show that the Hessian matrix of social production function T is negative definite.

To show the first part, recall the argument by Lancaster (1968, Ch.8) and the comment by Kelly (1969) that the matrix consists of all the elements of the Hessian matrix of all sectors' production functions, denoted by $\mathbf{D}$ here, and is negative definite if factor usages in one production sector is a linear combination of factor usage in the other production sectors. We can show that this will never be the case if each production sector uses the infrastructure capital as its input. Suppose the consumption production sector changes all inputs at the constant rate $\alpha > 0$; the scalar change is expressed as $\alpha(k_{01}, k_2, \ell_0)$. Due to the factor constraints, the remaining production sectors must change their inputs. Let us define their scale change as $\beta \geq 0$ and $\gamma \geq 0$, respectively. The following must then hold simultaneously due to the factor constraints:

$$0 = \alpha\ell_0 + \beta\ell_1 + \gamma\ell_2,$$

$$0 = \alpha k_{01} + \beta k_{11} + \gamma k_{21},$$

$$\alpha k_2 = \beta k_2 = \gamma k_2.$$

Since the last constraint must hold for public capital; therefore, it is not difficult to see that the above relations hold simultaneously only if $\alpha = \beta = \gamma = 0$. Therefore, we cannot apply a scalar change of input to any sector. This means that the quadratic form of $\mathbf{D}$ is never zero and is negative definite based on the argument by Kelly (1969,pp.349-350).

We will show next that the social production function T has a negative definite Hessian matrix. Applying the chain rules to the first order conditions, the following relation will be established:

$$\begin{bmatrix} \mathbf{D} & \mathbf{F} \\ \mathbf{F}^t & \mathbf{O}_{5\times5} \end{bmatrix} \begin{bmatrix} \mathbf{G} \\ \mathbf{H} \end{bmatrix} = \begin{bmatrix} \mathbf{O}_{7\times4} \\ \mathbf{J}_{5\times4} \end{bmatrix}$$

where each of the submatrixes is defined as follows:

$$\mathbf{G} = \begin{bmatrix} \partial k_{01}/\partial y_1 & \partial k_{01}/\partial y_2 & \partial k_{01}/\partial k_1 & \partial k_{01}/\partial k_2 \\ \partial \ell_0/\partial y_1 & \cdots & \cdots & \partial \ell_0/\partial k_2 \\ \partial k_{11}/\partial y_1 & \cdots & \cdots & \partial k_{11}/\partial k_2 \\ \partial \ell_1/\partial y_1 & \cdots & \cdots & \partial \ell_1/\partial k_2 \\ \partial k_{21}/\partial y_1 & \cdots & \cdots & \partial k_{21}/\partial k_2 \\ \partial \ell_2/\partial y_1 & \cdots & \cdots & \partial \ell_2/\partial k_2 \\ \partial m/\partial y_1 & \partial m/\partial y_2 & \partial m/\partial k_1 & \partial m/\partial k_2 \end{bmatrix}'$$

$$\mathbf{H} = \begin{bmatrix} -\partial p_1/\partial y_1 & -\partial p_1/\partial y_2 & -\partial p_1/\partial k_1 & -\partial p_1/\partial k_2 \\ -\partial p_2/\partial y_1 & \cdots & \cdots & -\partial p_2/\partial k_2 \\ \partial w_0/\partial y_1 & \cdots & \cdots & \partial w_0/\partial k_2 \\ \partial w_1/\partial y_1 & \cdots & \cdots & \partial w_1/\partial k_2 \\ \partial w_2/\partial y_1 & \partial w_2/\partial y_2 & \partial w_2/\partial k_1 & \partial w_2/\partial k_2 \end{bmatrix}'$$

and

$$
\mathbf{J}_{5\times4} = \begin{bmatrix} 1 & 0 & 0 & 0 \\ 0 & 1 & 0 & 0 \\ 0 & 0 & 0 & 0 \\ 0 & 0 & 1 & 0 \\ 0 & 0 & 0 & 1 \end{bmatrix},
$$

where $\mathbf{O}_{5\times5}$ and $\mathbf{O}_{7\times4}$ are 5 by 5 and 7 by 4 zero matrixes, respectively. The matrix $\mathbf{Q}$ is non-singular, thus it follows

$$
\begin{bmatrix} \mathbf{G} \\ \mathbf{H} \end{bmatrix} = \begin{bmatrix} \mathbf{D} & \mathbf{F} \\ \mathbf{F}^t & \mathbf{O}_{5\times5} \end{bmatrix}^{-1} \begin{bmatrix} \mathbf{O}_{7\times4} \\ \mathbf{J}_{5\times4} \end{bmatrix}
$$

$$
= \begin{bmatrix} -\mathbf{D}^{-1} + \mathbf{D}^{-1}\mathbf{F}\mathbf{L}\mathbf{F}^{-1}\mathbf{D}^t\mathbf{D}^{-1} & \mathbf{D}^{-1}\mathbf{F}\mathbf{L} \\ \mathbf{L}\mathbf{F}^{-1}\mathbf{D}^{-1} & -\mathbf{L} \end{bmatrix} \begin{bmatrix} \mathbf{O}_{7\times4} \\ \mathbf{J} \end{bmatrix}
$$

Thus we have $\mathbf{H} = -\mathbf{L}\mathbf{J}_{5\times4}$. Recalling the definitions of matrices $\mathbf{J}_{5\times4}$, $\mathbf{F}$ and $\mathbf{D}$, we can establish the following:

$$
\mathbf{R} = \begin{bmatrix} (-\partial p_i/\partial y_j) & (-\partial p_i/\partial k_j) \\ (\partial w_i/\partial y_j) & (\partial w_i/\partial k_j) \end{bmatrix} = \mathbf{J}^t\mathbf{H} = \mathbf{J}^t\mathbf{L}\mathbf{J}
$$

$$
= \mathbf{J}^t\{-[-\mathbf{F}^t\mathbf{D}^{-1}\mathbf{F}]^{-1}\}\mathbf{J}.
$$

where $i, j = 1, 2$ and $\mathbf{J}_{5\times4}$ is denoted as $\mathbf{J}$ for simplicity.

Because the matrices $\mathbf{F}$ and $\mathbf{J}$ have rank 5 and rank 4, respectively, so the matrix $\mathbf{R}$ is negative definite due to the fact that $\mathbf{D}$ is negative definite. Because of the duality, $\partial T/\partial y_i = -p_i$ and $\partial T/\partial k_j = w_j$ hold. Thus the matrix $\mathbf{R}$ is the Hessian matrix of the social production function $T(y_1, y_2, k_1, k_2)$ and is negative definite. This completes the proof.

Tawada (1982) and a further study conducted by Manning and McMillan (1982) have demonstrated that, under the assumption that each sector's production function exhibits constant returns to scale with respect to all inputs, then will bring about the convex production possibility set with the strictly concave frontier to the origin. Note that because I need differentiability on the frontier, their result is not sufficient for our proof.

5. Concluding Remarks

We have generalized the turnpike theorem to the model which includes public capital as well as private capital. It is worth noting that, if no sector uses public capital as inputs and the production satisfies the constant returns to scale for all input, then the constructed social production function will not be strictly concave and will contain a flat segment in the production possibility frontier. This flat segment is often referred to as the *von Neumann facet*. In this case, the proof of the turnpike theorem will turn out to be more complicated, as discussed by McKenzie (1983,1986).

One important remaining task is to extend the model so that it explicitly contains the government budget constraint. By so doing, we may explicitly introduce a policy parameter into the model and can apply some comparative dynamic analysis.

6. References

1. Aschauer, D.(1989): Is public expenditure productive?, Journal of Monetary Economics 23,177-200.
2. Arrow K.J. and M. Kurz (1970): Public investment, the rate of return, and optimal fiscal policy (The Johns Hopkins University Press, Baltimore, Maryland).
3. Benhabib, J. and K. Nishimura (1979):The Hopf bifurcation and the existence and stability of closed orbits in multisector models of optimal economic growth, in Journal of Economic Theory 21,421-444.
4. Brock, W. and J.A. Scheinkman (1977): On the long-run behavior of a competitive firm, in Equilibrium and Disequilibrium in Economic Theory (G. Schwodiauer ed., Reidel Dordrecht).
5. Kelly, J. (1969): Lancaster vs Samuelson on the shape of the neoclassical transformation surface, Journal of Economic Theory 1,155-174.
6. Manning, R. and J. McMillan (1982): The scale effect of public goods on production possibility sets, in Production sets (M. Kemp ed.,Academic Press, New York, NY).
7. Lancaster, K.(1968):Mathematical Economics (Macmillan, New York).
8. Meade, J. E.(1952): Externalities and diseconomies in a competitive situation, Economic Journal 62,52-67.
9. Munnell, A. H.(1990):Why has productivity growth declined? Productivity and public investment, New England Economic Review, Federal Reserve Bank of Boston, January/February,3-22.
10. McKenzie, L.(1983): Turnpike theory, discounted utility, and the von Neumann facet, Journal of Economic Theory 30,330-352.
11. McKenzie, L.(1986):Optimal economic growth and turnpike theorems, in Handbook of Mathematical Economics Vol.III (K. Arrow and M. Intriligator eds., North Holland,Amsterdam)
12. Tawada, M.(1980):The production-possibility set with public intermediate goods, in Production sets (M. Kemp ed., Academic Press, New York, NY).
13. Uzawa, H.(1988): On the economies of social overhead capital, in Preference, Production, and capital (Cambridge University Press).

Wage Bargaining and Incentive Compatibility: Is Unemployment Optimal After All?

Franz Wirl[1]

[1] Faculty of Economics and Management, Otto-von-Guericke University of Magdeburg, P. O. Box 4120, D-39016 Magdeburg, Germany.

This paper investigates organised wage bargaining coupled with asymmetric private information. This attempts to rationalise the persistence of unemployment. The present strategy of (European) unions of continuing to demand high wages despite substantial unemployment may be rational, even if the unions account for the unemployed. In fact, a union accounting for the gains of all - those employed and those unemployed (who must receive unemployment benefits) - and not only for the employed workers raises its demands and thus unemployment! Other interesting features are that increasing the efficiency of workers at the 'top' increases unemployment, unless the unions reduce their demands. In contrast increasing efficiency at the 'bottom' mitigates unemployment, ceteris paribus.

1. Introduction

If certain economic phenomena persist, i.e. have passed the severe test of time, economists should start considering potential efficiency arguments that explain this very phenomenon. Fig. 1. documents the persistence of unemployment, in particular in Europe. Since unemployment is apparently hardly affected by the ups and downs of the economy - exceptions are the two oil price shocks 1973/74 and 1979-1981, while the price collapse 1986 seems of less importance - business cycle models seem insufficient to rationalise this evolution shown in Fig. 1. The United States with unemployment today similar to the 'golden' sixties seems an outlier from this trend.

The purpose of this paper is to develop an efficiency argument explaining the phenomenon of unemployment. While the efficiency wage hypothesis provides reasons why firms pay above market clearing wages, the following argument rests on collective wage bargaining (corresponding institutions exist in Germany, Austria and Scandinavia) and on the constraints imposed by an incentive compatible wage structure. Collective wage bargaining leads to high wages which bar certain fractions of labour from getting hired, because the (minimum) wage is too high considering the productivity of these workers. Yet a high wage can raise enough rents for those employed - their wage is essentially determined by an incentive compatibility constraint - so that they can compensate (again subject to an incentive compatibility constraint) the unemployed; re-distribution associated with (high) minimum wages is empirically investigated in Freeman (1996), Bell and

Wright (1996), Sloane and Theodossiou (1996) and Machin and Manning (1996). That is, the presently observed policy of (European) unions to demand high wages even for the price of unemployment and to resist vehemently any attempt to allow for low wage labour market segments (just watch the discussions in Germany) may be efficient for all workers (employed and unemployed). And given this potential efficiency, it might be difficult to uproot unemployment. In fact, this paper shows that a narrow minded union - only caring for those employed - actually mitigates unemployment considerably compared with a union that integrates the interest of the unemployed.

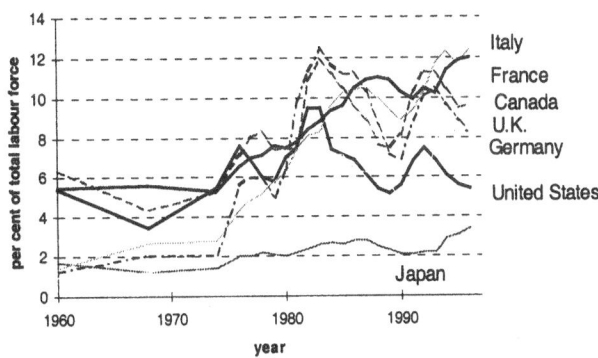

Fig. 1: Unemployment in major OECD countries

2. Workers

Although decision making runs from the unions over the firms to the workers, see Fig. 2, it is convenient (as often in economics) to start from the back, here with the workers. Workers differ in a single parameter characterising their type t, $t \in [\underline{t}, \ \bar{t}\ ]$, where $\underline{t}$ denotes the least and $\bar{t}$ the most efficient type. Although one could normalise to [0, 1], we keep the interval $[\underline{t}, \ \bar{t}\ ]$ but use instead the normalisation that the type $t = 0$, if existing, is only infinitesimally productive, i.e., can only be paid the wage 0. The number of all workers, whether employed or unemployed, is normalised to one. This efficiency parameter t is private information of each worker, yet the distribution $F(t)$ is common knowledge, $f(t)$ denotes the density $h(t) :=$ $f(t)/(1 - F(t))$ the hazard rate, and for simplicity given by the uniform distribution:

$$f(t) = \frac{1}{\bar{t} - \underline{t}} \ \text{for } t \in \left[\underline{t}, \bar{t}\right] \ F(t) = \frac{t - \underline{t}}{\bar{t} - \underline{t}} \ \text{and } h(t) = \frac{1}{\bar{t} - t}.$$

The characteristic t may determine a workers ability of producing a homogenous output $q = f(e, t)$ where e denotes the work effort and f is a conventional production function, $f(0,t) = 0$ for all t, increasing and concave in effort e. An alternative is that all workers are equally productive but differ in their work ethics: workers with a larger t have less disutility from work. All what matters in the following is that the 'payoff' of the workers for producing or carrying out the task q is given by a function $W(q,t)$. The following exposition chooses the second approach of workers having identical productivity:

$$q = f(e,t) = e, \tag{1}$$

but different degrees of work ethics expressed in the disutility $D(e, t)$, increasing and convex in the effort e and $D(0, t) = 0$ for all t. Without loss in generality we assume that a larger t corresponds to a more diligent type, $D_t < 0$; furthermore we assume that this incremental efficiency diminishes $D_{tt} > 0$. A worker's benefit from producing the output q and absent a wage, denoted W, is given by the negative of the disutility, using the production function to determine the effort contingent on output and type, $e = E(q, t) = q$:

$$W(q,t) = -D(E(q, t), t) = -D(q, t) = -\tfrac{1}{2}q^2/t. \tag{2}$$

Observe that the example specified on the right hand side of (2), which will be used for graphical presentations, is also implied by $f(e,t) = e\sqrt{t}$ and $D(e,t) = \tfrac{1}{2}e^2$, i.e., differences in productivity but not in the work ethics. Thus both characterisations, either referring to the abilities or to the effort exercised by workers, are compatible with W in (2). Maximising W with respect to effort implies $e = q = 0$, i.e. no output without monetary incentives. We assume in line with the related literature that W is concave in q (thus $D_{qq} > 0$), and that $W_{qt} = -D_{qt} > 0$ (Spence-Mirrlees' single crossing property); in addition, $W_{qqt} > 0$, and $W_{qtt} < 0$ are sufficient conditions for incentive compatibility of the so-called relaxed program, see Fudenberg-Tirole (1992). In the general case of differences in productivity and work ethics, these assumptions can be worked back in terms of assumptions imposed on f and D, yet for $f = e$, $D_{qt} < 0$, $D_{qqt} < 0$, and $D_{qtt} > 0$, while $W_q < 0$ and $W_t > 0$ follow directly from the assumptions about D.

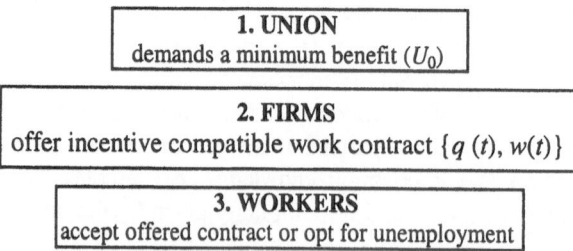

Fig. 2. Sequence of decisions

Given a work-compensation scheme $\{q(t),w(t)\}$, a worker of type t can, due to private information about t, pretend to be a type $\hat{t}$ and collect the net benefit:

$$U(\hat{t}, t) := w(\hat{t}) - D(q(\hat{t}), t). \tag{3}$$

Hence, a worker of type t facing the work – wage schedule will select that combination of work assignments and wages that maximises his payoff, i.e., he will pretend that type $\hat{t}$ that maximises $U(\hat{t}, t)$. However, the revelation principle allows to restrict the analysis to incentive mechanisms where each worker maximises $U(\hat{t}, t)$ for reporting the truth, $\hat{t} = t$. Therefore, we can re-define:

$$U(t) := U(t, t) \geq U(\hat{t}, t), \tag{4}$$

and $t = \arg\max_{i} U(\hat{t}, t)$. Hence, the corresponding first order condition must hold:

$$\frac{\partial U(\hat{t}, t)}{\partial \hat{t}} = \dot{w} - D_q \dot{q} = 0 \text{ for } \hat{t} = t, \tag{5}$$

where the dot denotes the (total) differentiation with respect to the type t. Differentiation of $U(t)$ with respect to the type t and substituting the above result that reporting the truth is optimal (revelation principle) yields:

$$\dot{U}(t) = \underbrace{\dot{w} - D_q \dot{q}}_{=0} - D_t = -D_t. \tag{6}$$

Although an individual worker would be willing to accept a job offer with the benefit $U(t) \geq W(0, t) = 0$, collective bargaining raises the reservation price to $U_0 \geq 0$. Consequently, hiring a worker requires that

$$U(t) = w(t) - D(q(t), t) \geq U_0. \tag{7}$$

Inequality (4), or respectively differential equation (6), is labelled the *incentive compatibility* constraint. Inequality (7) is the *individual rationality* constraint.

3. Firms

The firms make efficient use of the existing labour force accepting the union's demand, that the net benefit of each employed worker must exceed U_0. The price (= 1\$) taking firms design incentive mechanisms that maximise aggregate profits:

$$\pi = \int_{t_0}^{\bar{t}} [q(x) - w(x)] f(x) dx, \tag{8}$$

through prescribing (observable) outputs $q(t)$, offering the compensating wages $w(t)$ and pooling the workers, i.e. excluding inefficient types, $t \leq t_0$. This optimisation (8) is an optimal control problem and substituting $w = U + D$ from (3) eliminates one of the two controls:

$$\max_{\{q(x),t_0\}} \int_{t_0}^{\bar{t}} [q(x) - U(x) - D(q(x),x)]f(x)dx , \tag{9}$$

so that

$$\dot{U}(t) = -D_t(q(t),t), U(t_0) = U_0. \tag{10}$$

This formulation (9) – (10) follows, because the two constraints can be combined into a single differential equation constraint (incentive compatibility) including a boundary condition (due to the individual rationality constraint). What is unusual compared with the bulk of control theory applications in economics is that the control problem is 'static' and types play the role usually assigned to time, that 'starting', the choice of the 'marginal' type t_0 (but restricted to the manifold $U(t_0) = U_0$), instead of terminating is free.

Letting H denote the Hamiltonian and λ the costate associated with the state U,

$$H = [q(t) - D(q,t) - U]f - \lambda D_t , \tag{11}$$

the following first order optimality conditions result according to Feichtinger – Hartl (1986):

$$\dot{\lambda} = f, \lambda(\bar{t}) = 0, \lambda(t_0) = v, \lambda(\bar{t}) = 0, \tag{12}$$

$$H_q = (1 - D_q)f - \lambda D_{tq} = 0, \tag{13}$$

$$H(t_0) = 0. \tag{14}$$

Integrating the costate variable differential equation accounting for the transversality condition yields $\lambda(t) = F(t) - 1$. Substituting this solution into the maximum principle (13) and introducing with $h = (1 - F)/f$ the hazard rate gives:

$$1 - D_q = - D_{qt}/h = - D_{qt}(\bar{t} - t). \tag{15}$$

The last term in (15) follows from substituting the hazard rate h associated with the assumed uniform distribution. The solution of (15), denoted $q^*(t)$, is the so called relaxed program, compare Fudenberg-Tirole (1992). Observe that these production targets are independent of the characteristics of the least efficient types, $t = \underline{t}$. This relaxed program $q^*(t)$ is monotonically increasing,

$$\dot{q}^* = \frac{2D_{qt} - (\bar{t} - t)D_{qtt}}{(\bar{t} - t)D_{qqt} - D_{qq}} > 0, \tag{16}$$

for the assumptions imposed on W and thus satisfies the incentive compatibility constraint (4), see Fudenberg-Tirole (1992). Hence, not surprising, more diligent (or more productive) workers are asked to produce more. Another comparative static result of the relaxed program is

$$\frac{\partial q^*}{\partial \bar{t}} = \frac{D_{qt}}{D_{qq} - (\bar{t} - t)D_{qqt}} < 0. \tag{17}$$

Proposition 1: Increasing the efficiency of the already highly efficient types, i.e., raising $\bar{t}$, lowers the output of each type $t < \bar{t}$ and in particular of the least efficient types. In other words, productivity at the top will cast a shadow on the economic prospects of the less efficient types.

Solving (15) for the specification on the right hand side of (2) yields a simple analytical expression of the relaxed program:

$$q^*(t) = t^2/\bar{t} . \tag{18}$$

The relaxed program q^* is the optimal one in a world without collective bargaining and sufficiently low reservation prices so that full employment and second best[1] optimum results.

Suppose that we know the marginal type, t_0, employed. Then the wage $w(t)$ necessary to implement $q^*(t)$ at least for the types $t > t_0$ follows from integrating the state differential equation to determine the workers' net benefits

$$U(t) = \int_{t_0}^{t} -D_x(q^*(x), x)dx + U_0 = U_0 + (t^3 - t_0^3)/(6\bar{t}^2), \tag{19}$$

and from this, using the definition of U, the wage:

$$w(t) = U(t) + D(q^*(t), t) = U_0 + (4t^3 - t_0^3)/(6\bar{t}^2). \tag{20}$$

The explicit, analytical results on the right hand sides in (19) and (20) follow for the example specified in (2). If in addition $U_0 = 0$, which is sufficient but not necessary for full employment, the wage of a worker of type t follows from substituting $U_0 = 0$ in (19); the corresponding full employment solution is identified by upper bars, $\{\overline{U}(t), \overline{w}(t)\}$.

However, if union demands are binding, firms will exclude inefficient types and start hiring only types $t \geq t_0$. In order to determine the marginal type t_0 we have to use the condition (14), which becomes after proper substituting:

---

[1] Due to the incentive compatibility constraint. First best output follows from equating the left hand side of (15) to zero, $1 - D_q = 0$, which exceeds q^* except at $t = \bar{t}$, called *no distortion at the top.*

$$[q*(t_0)-D(q*(t_0),t_0)-\underbrace{U(t_0)}_{U_0}]f(t_0)-\underbrace{\lambda(t_0)}_{F(t_0)-1}D_t(q*(t_0),t_0)=0. \tag{21}$$

Hence, for an interior solution of the marginal type, $t_0 > \underline{t}$:

$$q*(t_0)=\underbrace{D(q*(t_0),t_0)+U_0}_{w_0}\underbrace{-D_t(q*(t_0),t_0)/h(t_0)}_{>0} \tag{22}$$

That is, the marginal product of the most inefficient of the hired workers must exceed his wage, $q(t_0) > w(t_0)$[2]. The reason for the strict inequality is that employing this type adds informational rents that must be paid to all more efficient types due to the incentive compatibility constraint. That is, employing an additional type raises the informational rents paid to all employed. In other words, employment creates a negative externality for the firm's payroll. The specification of W on the rhs of (2) simplifies (22) and yields:

$$\frac{t_0^2}{\underline{t}}=\frac{1}{2}\frac{t_0^2}{\underline{t}}+U_0 \text{ thus } t_0=\max\left\{\underline{t},\sqrt{2\underline{t}U_0}\right\}. \tag{23}$$

Fig. 3 shows an example for this incentive scheme for workers from the interval $t\in[4, 10]$ and for different union demands concerning the reservation price U_0. The prescription of output along the relaxed program $q*$ differs by a factor of 5 (exactly, ranging from 1.6 to 10). However, the union demands of reservation prices (the two choices of $U_0 > 0$ will be rationalised in section 4) eliminate the less efficient types causing unemployment. This in turn reduces the differences in observable output (to a factor of around 3 for $t_0 \approx 5.86$) and wages. Again this seems compatible with stylised facts concerning the differences between Europe and the United States. The union's demand of a minimum benefit U_0 shifts the associated wages (for those employed) considerably upwards compared with the reservation price $U_0 = 0$. Although wages are paid only to those employed, those unemployed receive a transfer payment equal to U_0, which leads to the discontinuity in wages (yet the benefit $U(t)$ is continuous) shown in Fig. 3, because the unemployed need no financial compensation for disutility from work.

---

[2]Note, however, that this inequality - the type must produce than more than he receives as a wage - needs hold for efficient types.

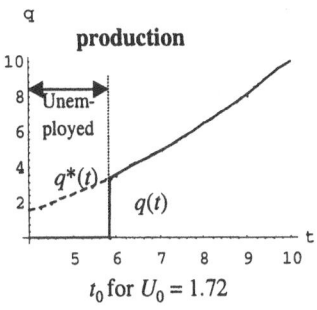 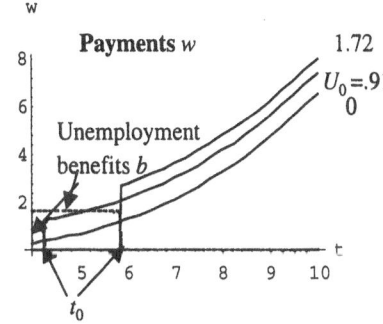

Fig. 3: Relaxed program, $t \in [4, 10]$, and wages calculated for $U_0 = 0$ (thus full employment), $U_0 \approx 0.91$ ($\Rightarrow t_0 \approx 4.26$, or 4.3% unemployed) and $U_0 \approx 1.72$ ($\Rightarrow t_0 \approx 5.86$, or 31% unemployed).

Collective wage bargaining[3], demanding $U(t) = w(t) - D(q^*(t), t) \geq U_0$, can lead to unemployment, because firms will hire only types, where the incremental product allows to pay the wage and the informational rents. There are two possible solutions, one at the boundary, $t_0 = \underline{t}$ and full employment, and the other in the interior, $t_0 > \underline{t}$ combined with unemployment. Full employment results if the least efficient type is sufficiently productive, thus $q^*(\underline{t}) > 0$ and U_0 is sufficiently small. However, for highly inefficient types, $t = \underline{t} = 0$ and thus according to the above convention $q^*(0) = 0$, any positive U_0 will restrict the set of workers actually finding a job. In this case of an interior solution - the choice of U_0 causes unemployment - the marginal type employed, denoted t_0, is determined by (11). Application of the implicit function theorem to (11) ensures the existence of a function T such that $t_0 = T(U_0)$ and this marginal type increases with respect to U_0, $T' > 0$, due to:

$$T' = \cfrac{1}{(\bar{t}-t)D_{tt} - 2D_t + \underbrace{(1 - D_q + (\bar{t}-t)D_{tq})\dot{q}}_{=0}} = \cfrac{1}{(\bar{t}-t)D_{tt} - 2D_t} > 0. \quad (24)$$

Proposition 2: A higher U_0 raises the marginal type and thus the number of unemployed, if the demanded level U_0 constrains indeed the firm's hiring.

This determination of the marginal type depends, in addition to the union's decision variable U_0, on other model parameters, e.g. on $\bar{t}$. This is straightforward for the specification of quadratic costs, $dt_0 / d\bar{t} = \frac{1}{2} t_0 / \bar{t} > 0$, but this qualitative property holds generally, because

---

[3]Indeed, unions usually have the net benefit in mind and bargain for a package of wages and fringe benefits, like full payment during sick leave (although there is no need to compensate for disutility from work). Nevertheless, this assumption that the union bargains for benefits rather than wages (although realistic since unions demand wage compensations for job nuisances and risks) is not crucial.

$$\frac{dt_0}{dt} = \frac{[D_q - D_{qt}(\bar{t}-t)-1]\frac{\partial q^*}{\partial \bar{t}} - D_t}{(\bar{t}-t)D_u - 2D_t} = \frac{-D_t}{(\bar{t}-t)D_u - 2D_t} > 0. \tag{25}$$

Proposition 3: An increase in productivity of the most efficient types increases unemployment, unless the unions accept a reduction of the minimum benefit.

4. Unions

While a higher U_0 might raise unemployment, this increase raises the entire wage and benefit schedule for all those workers finding a job, see Fig. 3. We consider two kinds of unions. The first maximises the aggregate benefit of all potential workers (whether employed or unemployed)[4]:

$$\max_{U_0} J := \int_{t_0}^{\bar{t}} U(t)f(t)dt = \int_{t_0}^{\bar{t}}\left[U_0 - \int_{t_0}^{t} D_x(q^*(x),x)dx\right]f(t)dt =$$

$$= \frac{(\bar{t}-t_0)}{\bar{t}-\underline{t}}(U_0 - \overline{U}(t_0)) + \frac{1}{\bar{t}-\underline{t}}\int_{t_0}^{\bar{t}}\overline{U}(t)dt. \tag{26}$$

The simplification in (26) results from substituting $(\overline{U}(t) - \overline{U}(t_0))$ for the integral over $W_x = -D_x$, since the relaxed program is independent of U_0. Maximising J amounts to a theoretical compensation test a la Hicks-Kaldor.

However, acquiring rents through setting $U_0 > 0$ need not be enough, because any unemployment benefit scheme must satisfy an incentive compatibility constraint too. That is, the unemployment benefits $\{b(t), t \in [\underline{t}, t_0]\}$ must satisfy (4) where the benefit b replaces the wage w and $q(t) = e(t) = 0$ for $t < t_0$. And there is only one solution satisfying this constraint: **All unemployed must receive the same net benefit as the marginal type employed, i.e., $b(t) = U_0 = U(t_0)$ for all $t < t_0$, (but less than the money wage, $b = U_0 < w(t_0)$).** The proof applies an indirect argument. First, $U(t)$ cannot be discontinuous across $t = t_0$. Assume that $U(t)$ jumps at $t = t_0$, then a worker of the type $(t_0 - \varepsilon)$ gains from pretending $\hat{t} = t_0$; a similar argument, but for $t = t_0 + \varepsilon$ applies if benefits increase discontinuously with unemployment. Thus continuity follows. Yet $U(t)$ cannot increase for $t < t_0$ because in this case all unemployed pretend $\hat{t} = t_{0-}$, the type that which is infinitesimally less efficient than the marginal type employed. And similarly $U(t)$ cannot decline for $t < t_0$, because in this case $\hat{t} = \underline{t}$ results. Therefore, all unemployed receive the payment $b = U_0$. An important political and empirical testable consequence is **that unemployment benefits correlate positively rather than negatively with un-**

---

[4]Of course, maximising J (without the following constraint) corresponds also to a union purely interested in benefits of those employed only, if someone else, using different means and channels (say property taxes), compensates those unemployed.

employment (of course, everything else equal). This result is a direct consequence from the incentive compatibility constraint and does not depend on the precise nature of wage bargaining. Moreover, this result seems compatible with the apparently joint rise in both, unemployment and unemployment benefits, during the seventies, and eighties.

As a consequence, it is conceivable that the Hicks-Kaldor test is passed but that the accrued benefits are insufficient to pay for the unemployment benefits in an incentive compatible manner. Thus not only firms, but also all employed face principal-agent slippage. As a consequence, a union that maximises J (thus includes the benefits of the unemployed) must ensure a Pareto-improvement for those employed (i.e., for the types $t \geq t_0$), because otherwise the employed will form their own union. This requires that the net gain to the employed (subtracting the transferred unemployment benefits) must exceed their gain under full employment:

$$J(U_0) - U_0 \underbrace{\int_{\underline{t}}^{t_0} f(x)dx}_{(t_0 - \underline{t})/(\bar{t} - \underline{t})} >$$

$$J(\tilde{U}_0) - \int_{\underline{t}}^{t_0}\left[[\tilde{U}_0 - \int_{\underline{t}}^{t} D_x dx]f(t)\right]dt = J(\tilde{U}_0) - \tilde{U}_0 \frac{t_0 - \underline{t}}{\bar{t} - \underline{t}} - \frac{1}{\bar{t} - \underline{t}}\int_{\underline{t}}^{t_0}\overline{U}(t)dt , \qquad (27)$$

where $\tilde{U}_0$ denotes the maximal benefit of the least efficient type that is compatible with full employment:

$$\tilde{U}_0 := \sup\{U_0| t_0 = T(U_0) = \underline{t}\} = q^*(\underline{t}) - D(q^*(\underline{t}),\underline{t}) - D_t(q^*(\underline{t}),\underline{t})(\bar{t} - \underline{t}). (28)$$

This magnitude $\tilde{U}_0$ is calculated in (28) from the optimal 'starting' condition supposing $t_0 = \underline{t}$ and using the relaxed program. This calculation simplifies considerably for the example of W specified on the rhs in (2): $\tilde{U}_0 = \frac{1}{2}\underline{t}^2/\bar{t} = \frac{1}{2}q^*(\underline{t})$.

Proposition 4: If $\underline{t} = 0$, then the union with the objective J chooses $U_0 > \tilde{U}_0 = 0$.

Therefore, a union maximising the aggregate benefits of the entire labour force, whether employed or unemployed, will cause unemployment.

Ignoring the constraint (27) for the moment, we will refer to this point in the proof of Proposition 5, it suffices to show that $J'(0)$ is positive such that $J(U_0) > J(0)$ at least in a local surrounding of $\tilde{U}_0 = 0$. Differentiating J yields:

$$J'(U_0) = \frac{1}{\bar{t} - \underline{t}}[(1 - \dot{\overline{U}} T')(\bar{t} - t_0) - (U_0 - \overline{U})T' - \overline{U} T' =$$

$$= \frac{1}{\bar{t} - \underline{t}}[(1 - \dot{\overline{U}} T')(\bar{t} - t_0) - U_0 T']. \qquad (28)$$

Since $\underline{t} = 0$, and setting $U_0 = 0$, thus $t_0 = 0$, implies finally substituting T' from (12) and $\ddot{U} = W_t = -D_{\underline{t}}$ from (9):

$$J'(0) = 1 - \ddot{U} \; T' = 1 + D_t T' = \frac{(\bar{t}-t)D_{tt} - D_t}{(\bar{t}-t)D_{tt} - 2D_t}, \tag{29}$$

i.e., $0 < J'(0) < 1$. Therefore, U_0 at least slightly above 0 increases J, but causes unemployment. However, this gain from $U_0 > 0$ may be insufficient to improve the lot of the actual workers. Since this is an immediate corollary from Proposition 5 no separate proof is here necessary. ◊

Proposition 4, which was proven for the assumption $\underline{t} = 0$, extends in a trivial manner to $\underline{t} > 0$ if $T'(0) = 0$ because in this case, raising the reservation price sufficiently moderate leads to a rent transfer from the firms to the workers without causing unemployment. This is always beneficial. However, it is less trivial, if the workers of type $t = \underline{t}$ are highly inefficient, $\underline{t} = 0$, such that any increase in U_0 (thus higher wages for those employed) must be paid with unemployment. In this case, $\underline{t} = 0$, the above proposition implies always the optimality of an unemployment outcome. This may explain the persistence of often very high unemployment in many developing countries considering the low levels of human capital.

In contrast, a union maximising the benefit of those employed only subtracts the (incentive compatible) transfer payments to the unemployed:

$$Z := J - U_0(t_0 - \underline{t})/(\bar{t} - \underline{t}). \tag{30}$$

Using the definition of Z in (30), $Z(U_0) > J(\tilde{U}_o)$ is a sufficient condition to satisfy the inequality (27). Therefore, the constraint (27) is redundant for a union maximising the benefits of those employed, i.e., when maximising Z.

Proposition 5: Consider a union concerned about the benefit of those employed only, but accounting for incentive compatible unemployment benefits, which are raised through lump sum contributions from those employed, i.e. maximising. Z. Then, an interior solution coupled with unemployment results, if the least efficient type is only infinitesimally productive, $\underline{t} = 0$. Furthermore, this union leads to a lower benefit demand U_0, thus to a lower minimum wage and hence to less unemployment compared with a union accounting explicitly for the interest and benefit of all including the unemployed, if both objectives, J and Z, are concave.[5]

Differentiation of Z with respect to U_0 yields:

$$Z' = J' - \frac{1}{\bar{t}-\underline{t}}(t_0 + U_0 T'), \tag{31}$$

---

[5]The concavity of J and Z seems plausible (almost a standard in economics) but is here assumed rather than derived, because this property depends on derivatives up to the fourth order, which escape an economic interpretation.

so that $Z' < J'$ for $U_0 > 0$ because $T' > 0$. Moreover, $Z' = J'$ at $U_0 = 0$ guarantees $Z' > 0$ for $U_0 = 0$ and thus the existence of an interior solution $U_0 > 0$ for $\underline{t} = 0$. Therefore, assuming concavity of Z and J implies that $Z' = 0$ at a lower level U_0 than $J' = 0$. This completes the proof for an interior solution of Z supposing that (27) does not bind. However, this constraint cannot bind at the maximum of Z, which by definition exceeds $Z(0)$ and thus satisfies (27) at the optimal level $U_0 > 0$ due to (19). And at this level of U_0, the constraint (27) cannot bind either, when maximising J. Therefore, even if (27) is binding when maximising J it must do so at a level greater arg max $Z(U_0)$, which ensures the claimed unconditional unemployment outcome of Proposition 1. $\Diamond$

The economic intuition behind this at first surprising finding is that a higher benefit bargained for all those employed must raise unemployment benefits too, due to the incentive compatibility constraint. The reason is that the unemployed become free riders of the wage bargaining process with the consequence that the casual conclusion reached in public debates - the union's strategy of high wages harms the less efficient and thus consequently unemployed workers - is wrong. The necessity to finance the transfers to unemployed reduces a union's demand for minimum benefits compared with the all inclusive union (maximising J). Consequently, this burden of paying for the unemployed plus the positive contribution of an additional member helps to reduce unemployment.

Fig. 4 compares the two objective functions J (referring to a union accounting for both employed and unemployed) and Z (corresponding to a union catering only for those employed) and shows the associated unemployment benefits. Indeed Z is just J minus the transfer to the unemployed. This figure highlights that a comprehensive union will substantially increase unemployment (of course, the actual magnitude depends on the specification of the production and disutility functions and the parameters such that the numerical results should be interpreted qualitatively only). Although a considerable rent transfer is possible retaining full employment (up to $U_0 = \tilde{U}_0 = \frac{1}{2}\underline{t}^2/\bar{t} = 9$), both kinds of unions opt for a higher $U_0 = 1.71673$ (for maximising J) and 0.907017 (for maximising Z) and thus for unemployment (31% and 4.3%, see Fig. 3).

These two propositions imply always unemployment, if $\underline{t} = 0$, which is crucial. What is the crucial point of assuming $\underline{t} = 0$? This assumption implies that this least efficient type $\underline{t}$ is 'asked' to produce nothing, $q^*(\underline{t}) = 0$, i.e., $\underline{t} = 0$ and thus only $U_0 = 0$ is compatible with full employment. However, if the least efficient type is quite productive, $\underline{t} > 0$ and $q(\underline{t}) > 0$, then $\tilde{U}_0 > 0$ and possibly 'large'. Hence, the union can bid up to $\tilde{U}_0$ without causing unemployment. In such circumstances, a full employment outcome is possible. For example, raising in our numerical example in Figs 2 and 3 the least efficient type $\underline{t}$ to 6 eliminates unemployment under both objectives. The economic reason is that compensating fairly efficient yet unemployed types at a rate $U_0 > \tilde{U}_0 > 0$ causes considerable dead-weight losses, which may deter unemployment. Therefore, **the more productive the least efficient workers are the better the chances are to establish full employment,** despite unionisation, due to the positive correlation between the productivity of the labour force at the bottom and $\tilde{U}_0$.

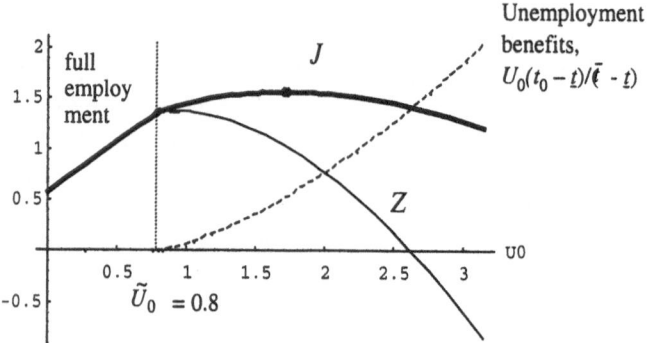

Fig. 4: Comparison of benefits, J, including those unemployed, and Z, the net gain to those employed after compensating the unemployed; $t \in [60, 200]$.

4. Final remarks

This paper is motivated by the persistence of unemployment in Europe and referred to efficiency to explain this phenomenon. The crucial point is that a union demanding high minimum benefits for employed workers risks unemployment but nevertheless increases the payoff of all employed. The income of those employed is determined by the union's demand for a minimum net compensation and an incentive compatibility constraint to induce maximum second best efforts of all employed workers. The efficient workers benefit from fixing the minimum requirements at the lower productivity end of the labour force spectrum. If the winners, those finding a job despite the high demands by the union, can compensate those becoming unemployed (for the same reason), then the entire labour force, whether employed or unemployed, gains. However, not only workers' compensations but also the unemployment benefits must satisfy an incentive compatibility constraint in order to support the high wage policy of the union. This in turn implies that those unemployed are not worse off than the least efficient, yet employed worker. This is a consequence of the incentive compatibility constraint faced by the union, because otherwise those unemployed start downward bidding and undermining the union's demand. Therefore, generous unemployment benefits are no accident and waste (from the union's perspective) but a conditio sine qua non to sustain a high wage policy with a reserve army (to use a Marxian phrase but with the opposite implication). This in turn suggests (ceteris paribus) a positive correlation between unemployment and unemployment benefits.

While the above objective - maximising the benefits to all workers employed - refers (maybe contrary to a first impression) to a union accounting for the benefit of those unemployed too (unemployed after all due to the union's strategy), a union concerned about the payoff of those employed only will subtract the unemployment benefits, because they must be raised from those employed. Now comparing these two kinds of unions, one caring for the entire labour force, whether employed or unemployed, and the other caring only about those employed, it is the first kind of a union that cares about the lot of the unemployed that leads to higher unemployment!

Furthermore, comparative statics implications of this framework are compatible with stylised facts underlying the change in regimes from full employment in the sixties to persistent unemployment in the eighties and the nineties. A possible explanation for this shift and compatible with the presented framework is that technological progress coupled with larger expenditures for education increased the productivity at the upper segment of the labour market and might have depreciated the productivity at the bottom, because both education and the ongoing revolution of the information technology favours people with high cognitive abilities and reduces the returns on manual work. The returns on manual work are further reduced by the increase in international competition and migration. While the entrance of more productive types depresses even under full employment wages and benefits of less efficient types (even if their productivity remains unchanged) it is the combination with collective wage bargaining that makes unemployment beneficial to all, those employed and those unemployed. Thus recent increases in unemployment are compatible with a rational union strategy. One possibility to reduce unemployment under these circumstances is to increase the productivity of lower ends of the labour force, since this raises the necessary transfer payments to the unemployed such that a full employment outcome may be compatible with the union's objective. How far such a public strategy is feasible (or even desirable) is open considering the at best mixed results about training. Another possibility is to weaken unions and other institutions that foster collective bargaining power. This might harm workers, but increases employment and thereby national output. This happened in the United States (experiencing an amazing employment boom) and to a lesser extent in the United Kingdom and this may explain the differences in the evolution of unemployment shown in Fig. 1.

References

Bell, D. N. F., Wright, R. W. (1996): The Impact of Minimum Wages on the Wages of the Low Paid: Evidence from the Wage Boards and Councils. Economic Journal 106, 650-656

Feichtinger, G., R. F. Hartl, R. W. (1986): Optimale Kontrolle ökonomischer Prozesse, deGruyter, Berlin.

Freeman, R. B. (1996): The Minimum Wage as a Redistributive Tool. Economic Journal 106, 639-649

Fudenberg, D., Tirole, J. (1992): Game Theory. 2nd, MIT Press Cambridge (Mass.)

Machin, S., Manning, A. (1996): Employment and Introduction of a Minimum Wage in Britain. Economic Journal 106, 650-656

Sloane, P. J., Theodossiou, I. (1996): Earnings Mobility, Family Income and Low Pay, Economic Journal 106, 657-666

Dynamic Investment Games

Engelbert J. Dockner[1] and Kazuo Nishimura[2]

[1] Institut für Betriebswirtschaftslehre, Universität Wien, Brünnerstraße 72, A-1210 Wien.

[2] Institute of Economic Research, Kyoto University, Yoshidahonmachi, Sakyoku Kyoto 606, Japan

Summary. Traditionally investment decisions of firms have been studied under the assumption of perfectly competitive markets. If firms, however, compete in oligopolistic product markets capacity choices and hence investment decisions of one firm influence the strategies chosen by the rival firms. Based on this observation Spence in his seminal paper introduced a class of dynamic games which are frequently referred to as capital accumulation or investment games. In this paper we present a survey of discrete time investment games. We concentrate on two different classes of games. In one class firms invest in a single capital stock that constitues a pure public good while in the second class each firm accumulates a private capital stock. For these classes of games we summarize existence and stability results for Open-Loop (OLE) and Markov Perfect Equilibria (MPE).

1. Introduction

One of the driving forces in a market economy is the growth of firms and industries. The investment strategy of a firm plays a crucial role for its growth. While traditionally economists have analyzed firm and industry growth under the assumption of perfectly competitive product markets (i.e. firms are assumed to be price takers in the output market) more recent research has focused on game theoretic models of growth and investment.

The aim of this paper is to present a survey of models and results for a class of dynamic games that is frequently referred to as capital accumulation or investment games. In these games two or more firms strategically invest in a physical capital stock when the output market is characterized by oligopolistic competition. In such a setting the dynamic evolution of a firm and that of an industry differs substantially from the predictions of perfectly competitive firm models. In particular the strategic interactions among rival firms give rise to a number of interesting conclusions such as *overaccumulation* of capital, preemption, entry (mobility) deterrence and complicated dynamics.

We begin our survey with the description of two classes of games. In the first class firms invest in a single capital stock which constitutes a pure public good. The public capital stock (like knowledge or infrastructure) is used in the production process and investment is costly. Firms choose their investment strategies in such a way so as to maximize the discounted stream of profits.

In the second class each firm invests in a private (firm specific) capital stock. In such a setting the reduced form profit function of each firm depends on all the capital stocks of the rival firms in the industry. This dependence is the result of strategic product market interactions (i.e. oligopolistic product markets). Again investment is costly and firms maximize the discounted stream of profits.

For both classes of dynamic investment games we discuss existence and stability of OLE and MPE. Based on these results we draw general conclusions for industry evolution and firm growth.

Our paper is organized as follows. In the next section we present the two classes of investment games, state their assumptions and discuss the concepts of open-loop and Markov (perfect) strategies. In Section 3 we present existence and qualitative properties of open-loop Nash equilibria. Section 4 is devoted to the discussion of existence and stability of Markov perfect Nash equilibria. Section 5 summarizes the results, discusses their implications for firm and industry growth and concludes the paper.

2. Alternative classes of dynamic investment games

The purpose of this section is to present two alternative classes of capital accumulation games. The first class was initiated by Fershtman and Nitzan (1991) and can be considered as a dynamic model of a pure public good. In this game two or more firms invest in a single public capital stock. Gross profits of each firm are functions of the public capital stock and investment is costly and results in adjustment costs.

The second class of investment games was initiated in the seminal paper by Spence (1979). In this class each firm strategically invests in a private capital stock. Profits of each firm depends on the firms own capital stock as well as on the capital stocks of all the rival firms and investment results in convex costs. While in the first class of games the main interaction of firms arises from a strategic externality in terms of a dynamic free rider problem, the second class of games is characterized by both strategic market interactions and a strategic externality.

2.1 Investment game with a single capital stock

Consider the case of two firms $i = 1, 2$ that invest in a single public capital stock. Let k_t be the public capital stock at time t and I_t^i be investment of firm i at time t. Under the assumption that capital depriciates at a constant rate β the accumulation equation becomes

$$k_{t+1} = I_t^1 + I_t^2 + (1 - \beta)k_t \qquad (2.1)$$

with $k_0 \geq 0$ as the given inital condition.

For each unit of investment firms have to pay a constant price given by p and they face adjustment costs given by $A^i(I_t^i)$ where A^i is a strictly convex function with $A_I^i > 0$[1]. Hence the cost functions are given by

$$C^i(I_t^i) \equiv pI_t^i + A^i(I_t^i).$$

Firms employ the public capital stock as the primary input so that their per period gross profits are given by $\pi^i(k_t)$ where π^i is a strictly concave function. Under the assumption of an infinite horizon the investment game of firm i becomes

$$\max \left\{ J^i \equiv \sum_{t=0}^{\infty} \rho^t \left[\pi^i(k_t) - C^i(I_t^i) \right] \right\} \tag{2.2}$$

subject to the constraint (2.1), where $0 < \rho < 1$ is the constant discount rate.

2.2 Investment game with multiple capital stocks

Consider two firms in an industry each investing in a private physical stock of capital k_t^i. The accumulation equations are given by

$$k_{t+1}^i = I_t^i + (1 - \beta^i)k_t^i \tag{2.3}$$

where β^i is a constant rate of depriciation. Investment is costly with the cost function given by $C^i(I_t^i) \equiv pI_t^i + A^i(I_t^i)$, where A^i is again assumed to be strictly convex. Gross profits of each firm now depend on the firm's own capital stock as well as on the capital stock of the rival. Hence, we get $\pi^i(k_t^i, k_t^j)$ where we assume that π^i is strictly concave with respect to k_t^i and that $\pi_{k^i k^j}^i \leq 0$ holds. If firms face an infinite planning horizon the investment problem of firm i becomes

$$\max \left\{ J^i \equiv \sum_{t=0}^{\infty} \rho^t \left[\pi^i(k_t^i, k_t^j) - C^i(I_t^i) \right] \right\} \tag{2.4}$$

subject to the constraint (2.3), where $0 < \rho < 1$ is again the constant discount rate.

These two investment games are the basis of our analysis. In order that the two games (2.2) and (2.4) are well defined we need to specify the strategy spaces. Throughout the paper we will discuss two contrasting games. One in which firms condition their current action only on the current calander time. In this case they choose a sequence of investment $\{I_t^i\}_{t=0}^{\infty}$ and ignore all other available information. Such a setting is equivalent to the case in which each firm chooses an investment sequence at the beginning of the game and than

---

[1] In what follows we use the following notation: $\frac{dA^i(I_t^i)}{dI_t^i} \equiv A_I^i$, $\frac{d^2 A^i(I_t^i)}{dI_t^i dI_t^i} \equiv A_{II}^i$, etc.

commits itself to stick to this preannounced path. We call this game the open-loop game and the corresponding equilibrium the open-loop equilibrium.

In a second type of game we allow firms to design their strategies as decision rules that depend on the current stock(s) of capital. This implies for the case of a public capital stock a Markovian decision rule given by $I^i(k_t)$ and in the case of two private capital stocks $I^i(k_t^i, k_t^j)$. In this type of game the strategies are choosen as feedback rules that associate a given level of investment with the current stocks of capital. Only if firms play Markovian decision rules can dynamic strategic interactions properly be taken into account.

3. Precommitment equilibria

As a first step we characterize precommitment equilibria for our games. It is easy to prove that both games admit an OLE. Hence, in what follows we will concentrate on the stability properties of the equilibrium. This is an interesting issue since it helps us to understand the growth process of firms and the corresponding evolution of industries.

3.1 OLE in case of a single capital stock

Consider first the case of the investment game with a single capital stock. If we make use of the accumulation equation (2.1) we can substitute for k_t in the objective functions to get

$$\max \left\{ J^i \equiv \sum_{t=0}^{\infty} \rho^t \left[\pi^i(I_{t-1}^1 + I_{t-1}^2 + (1-\beta)k_{t-1}) - C^i(I_t^i) \right] \right\}.$$

Therefore the Euler equations become

$$- C_I^i(I_t^i) + \rho \pi_k^i(I_t^1 + I_t^2 + (1-\beta)k_t) = 0. \tag{3.1}$$

Total differentiation of (3.1) results in

$$dI_t^i = \left[\frac{\rho(1-\beta)\pi_{kk}^i}{C_{II}^i - \rho\pi_{kk}^i} \right] dk_t + \left[\frac{\rho\pi_{kk}^i}{C_{II}^i - \rho\pi_{kk}^i} \right] dI_t^j.$$

Given our strict concavity assumptions this implies that we can apply the implicit function theorem and express $I_t^i = \phi^i(k_t, I_t^j)$, $i \neq j$. We therefore get

$$\phi_k^i = \frac{\rho(1-\beta)\pi_{kk}^i}{C_{II}^i - \rho\pi_{kk}^i}, \quad \phi_{I^j}^i = \frac{\rho\pi_{kk}^i}{C_{II}^i - \rho\pi_{kk}^i}.$$

Hence, from the two Euler equations we get

$$I_t^1 = \phi^1(k_t, I_t^2), \quad I_t^2 = \phi^2(k_t, I_t^1)$$

and after substitution

$$I_t^1 = \phi^1(k_t, \phi^2(k_t, I_t^1)), \quad I_t^2 = \phi^2(k_t, \phi^1(k_t, I_t^2))$$

which implies

$$I_t^1 = h^1(k_t) \quad I_t^2 = h^2(k_t).$$

Hence, the state dynamics are given by

$$k_{t+1} = (1 - \beta)k_t + h^1(k_t) + h^2(k_t).$$

This state equation characterizes the qualitative properties of the open-loop (precommitment) equilibrium. We therefore analyze this equation in detail now. First, we are interested in the existence of a steady state equilibrium and secondly, on the dynamics of the open-loop equilibrium paths. A steady state equilibrium is defined as $k_{t+1} = k_t = k^\infty$. A steady state must satisfy the system of equations

$$\beta k^\infty = I^1 + I^2 \tag{3.2}$$
$$\rho\pi_k^1(k^\infty) - C_I^1(I^1) = 0 \tag{3.3}$$
$$\rho\pi_k^2(k^\infty) - C_I^2(I^2) = 0. \tag{3.4}$$

From equations (3.3) and (3.4) we get

$$\frac{dI^1}{dk} = \frac{\rho\pi_{kk}^1}{C_{II}^1}, \quad \frac{dI^2}{dk} = \frac{\rho\pi_{kk}^2}{C_{II}^2},$$

so that we get

$$\frac{d(I^1 + I^2)}{dk} = \frac{\rho\pi_{kk}^1}{C_{II}^1} + \frac{\rho\pi_{kk}^2}{C_{II}^2} < 0. \tag{3.5}$$

Since βk is increasing in k and by (3.5) $I^1 + I^2$ is decreasing in k there exists a unique interior steady state.

To analyze the stability of the steady state we need to determine the characteristic roots of the dynamic equation

$$k_{t+1} = (1 - \beta)k_t + h^1(k_t) + h^2(k_t) \equiv H(k_t)$$

To get stability we need to show that $0 < H'(k_t) < 1$. From the Euler equations we have

$$I_t^1 = \phi^1(k_t, \phi^2(k_t, I_t^1)).$$

Total differentiation yields

$$(1 - \phi_I^1\phi_I^2)dI_t^1 = (\phi_k^1 + \phi_I^1\phi_k^2)dk_t$$

so that we get

$$\frac{dI_t^1}{dk_t} = h_k^1 = \frac{\phi_k^1 + \phi_I^1\phi_k^2}{1 - \phi_I^1\phi_I^2}.$$

After some algebraic manipulations we get

$$h_k^1 + h_k^2 = \frac{\rho(1-\beta)\left[\pi_{kk}^1 C_2'' + \pi_{kk}^2 C_1''\right]}{C_1'' C_2'' - \rho(\pi_{kk}^1 C_2'' + \pi_{kk}^2 C_1'')}.$$

Hence,

$$0 < H'(k_t) = (1-\beta)\left[\frac{C_1'' C_2''}{C_1'' C_2'' - \rho(\pi_{kk}^1 C_2'' + \pi_{kk}^2 C_1'')}\right] < (1-\beta) < 1$$

holds. So we are able to formulate our first result.

Theorem 3.1. *There exists an open-loop Nash equilibrium of the dynamic investment game with a single capital stock that is characterized by a unique interior steady state which is globally and asympthotically stable.*

3.2 OLE in case of multiple capital stocks

The case of a single capital stock implies that the capital stock is a pure public good. Hence, we have a dynamic free-rider problem incorporated into our game. In case of the multiple capital stocks, each firm considers its stock as a private good. In this section we are interested in the existence and stability of an OLE when each firm accumulates a single private stock of capital. The dynamic problem then becomes

$$\max\left\{J^i \equiv \sum_{t=0}^{\infty} \rho^t \left[\pi^i(k_t^i, k_t^j) - C^i(k_{t+1} - (1-\beta^i)k_t)\right]\right\}$$

which can be solved by using the Euler equations. For our game with two players the Euler equations become

$$\rho\left[\pi_{k_t^i}^i + C_{k_t^i}^i(k_{t+1} - (1-\beta^i)k_t)(1-\beta^i)\right] - C_{k_t^i}^i(k_t - (1-\beta^i)k_{t-1}). \quad (3.6)$$

From (3.6) it is clear that the dynamics of the capital stocks is governed by a system of nonlinear difference equations. In order to get qualitative insights into this system we linearize (3.6) around the steady state and look at the corresponding dynamics.

Assumption 1. The matrix of second order partial derivatives of the gross profit functions $\Pi \equiv (\partial \pi^i / \partial k^i \partial k^j) = (\pi_{k^i k^j}^i)$ have a dominant diagonal, i.e. there exist positive scalars $d_j \ j = 1, 2$ such that

$$d_j \mid \pi_{k^j k^j}^j \mid > d_i \mid \pi_{k^i k^j}^j \mid, \ i \neq j$$

hold.

The assumption of a dominant diagonal for the matrix of the second order partial derivatives has the following interpretation. A change of the firm's capital stock has a larger impact on its marginal profit than a change in the rival's capital stock.

With Assumption 4 we are now in a position to characterize the dynamics of an OLE in case of multiple capital stocks.

Theorem 3.2. *There exists a $\hat{\rho}$ such that for any ρ that satisfies $\hat{\rho} < \rho < 1$ there exists a unique steady state of the dynamic system (3.6) that is locally saddle point stable.*

The proof of this Theorem can be found in Dockner and Takahashi (1993) and will not be repeated here.

What we have shown in case of OLE can be summarized as follows. For both models there exists a open-loop Nash equilibrium with a unique steady state. This unique steady state is locally stable. Hence irrespective where the industry starts off there will be convergence towards a unique long run level of capital stocks.

4. Markov perfect equilibrium

In the preceeding section we have analyzed existence and stability properties of Nash equilibria in simple strategies. Simple or open-loop strategies are specified as time path. Hence, each player commits himself to choose a sequence of investment levels over the entire planning horizon and does not deviate from this predetermined levels. Clearly, such a behaviour puts a lot of structure on the dynamic game and hence, is rather restrictive. Therefore, we will discuss Markov perfect equilibria (MPE) now. When firms choose MPE they design their actions as decision rules, i.e. the current level of investment depends on the current level of the capital stocks. Depending on whether or not we discuss the single or the multiple capital stock models, investment is either a function of the current public capital stock or the current capital stocks of all rival firms.

4.1 MPE for the case of a single capital stock

In this section we derive MPE for our two classes of games and analyze their qualitative properties. We start out with the case of the single capital goods model. It is given by the equations (2.1) and (2.2). Before we are able to say anything specific about this game we need to make some assumptions.

Assumption 2. Adjustment costs for each firm are identically equal to zero, i.e. $A^i(I_t^i) \equiv 0$ and the price for a unit of investment is one, i.e. $p = 1$.

Assumption 3. For $i = 1, 2$ and for all k it holds that

$$0 \leq \pi^i(k) \leq (\delta + \beta)k.$$

We are now in a position to proof the existence of a MPE.

Theorem 4.1. *If Assumption 2 and 3 are satisfied, then there exists a pair of Markovian strategies given by*

$$I^i(k_t) = (\delta + \beta)k_t - \pi^j(k_t), \quad 1 \le i \ne j \le 2 \tag{4.1}$$

which constitutes a Markov perfect Nash equilibrium of the dynamic invest-ment game with a single capital good.

A proof of this Theorem is identical to the one given by Dockner et al. (1996) hence, there is no need to repeat it here. We therefore go on to discuss the stability properties of the game.

The dynamic equation that results from the MPE derived in the preceed-ing Theorem results in the following state dynamics

$$k_{t+1} = h(k_t) = (1 + \beta + 2\delta)k_t - \pi^1(k_t) - \pi^2(k_t) \tag{4.2}$$

Assumption 4. The profit functions are assumed to satisfy the following.

$$\pi^1(0) = \pi^2(0) = 0$$

$$\lim_{k \to 0} \frac{d\left(\pi^1(k) + \pi^2(k)\right)}{dk} > \beta + 2\delta$$

$$\lim_{k \to \infty} \frac{d\left(\pi^1(k) + \pi^2(k)\right)}{dk} < \beta + 2\delta$$

With this technical assumption we are in a position so say something about the existence and stability of a steady state.

Theorem 4.2. $k_1^\infty = 0$ is a trivial steady state, and there exists a nontrivial positive steady state k_2^∞. The trivial steady state is stable and the nontrivial steady state is unstable.

Proof: From (4.2)

$$h'(k_t) = (1 + \beta + 2\delta) - \frac{d\left(\pi^1(k_t) + \pi^2(k_t)\right)}{dk_t}$$

holds. Since $1 > \beta > 0$ we have

$$h'(k_t) > (2\beta + 2\delta) - \frac{d\left(\pi^1(k_t) + \pi^2(k_t)\right)}{dk_t}.$$

The right hand side of the inequality is positive since $\pi(0) = 0$, $\pi(k_t)$ is concave and Assumption 3 is satisfied. Hence, $h(k_t)$ is increasing. Since $\pi^1 + \pi^2$ is concave

$$h''(k_t) = -\frac{d^2\left(\pi^1(k_t) + \pi^2(k_t)\right)}{dk_t^2} > 0$$

and $h(k_t)$ is convex and increasing. Therefore if there exists a fixed point k^∞ of h it must be unique. By Assumption 4 $h(0) = 0$, $h'(k_t) < 1$ for sufficiently small k_t and $h'(k_t) > 1$ for sufficiently large k_t, and $h(k_t) < k_t$ for sufficiently small $k_t > 0$ and $h(k_t) > k_t$ for sufficiently large k_t, therfore $k^\infty > 0$ exists. It is now clear that 0 is stable and k^∞ is unstable. Q.E.D.

It is interesting to note that the stability properties of the MPE is totally different from that of the OLE. The economic reason for this is the strategic interaction that is taken into account when firms use Markovian decision rules rather than simple time paths as their strategies. If we have a closer look at the MPE we recognize that it is characterized by another strong property. It can be called an indifferent MPE which means that every response of firm i to the strategy (4.1) chosen by the rival firm is a best response. This property of MPE was first analyzed in detail by Dockner et al. (1996) and applied to the case of pollution control in Dockner and Nishimura (1999). It must be pointed out, however, that the indifferent MPE is not the only MPE that exists for the capital accumulation game with a single capital stock. In Dockner et al. (1996) it is shown that there exists a strict MPE which results in a most rapid approach dynamics and a bang bang investment strategy.

4.2 MPE for the case of multiple capital stocks

Next we are interested in the case where each firm accumulates a private stock of capital. As in the case with the public stock of capital we assume that the adjustment costs are identically equal to zero and that the price of a unit of investment is one. In this case the game is given by its reduced form

$$\max \left\{ J^i \equiv \sum_{t=0}^{\infty} \rho^t \left[\pi^i(k_t^i, k_t^j) - k_{t+1}^i + (1 - \beta^i)k_t^i \right] \right\} \qquad (4.3)$$

This class of games is analyzed in Dockner et al. (1999) where emphasis is put on three issues. The first one is the relationship between the long run capital stocks for the noncooperative Nash equilibrium and the collusive solution. The second one is the existence of an MPE and the third one is the dynamic properties of the MPE. Here we will not report all the results but will concentrate on the existence and stability of a MPE, where emphasis is put on so called indifferent MPE. Recall that an indifferent MPE implies that every response is a best response to the rivals indifferent strategy. As is demonstrated in Dockner et al. (1999) an indifferent MPE does not exist if the profit functions are not additively separable. This is, off course, the more interesting case from an economic point of view. If product markets are oligopolistic and firms compete in quantities it follows that the reduced form profit functions are not additively separable. To be, however, consistent with the discussion in the preceeding section we make the following assumption.

Assumption 5. The profit functions $\pi^i(k_t^i, k_t^j)$ are additively separable functions, i.e.

$$\pi^1(k_t^1, k_t^2) = \varphi^1(k_t^1) + \lambda_1 \psi^1(k_t^2)$$
$$\pi^2(k_t^2, k_t^1) = \varphi^2(k_t^2) + \lambda_2 \psi^2(k_t^1)$$

with $\lambda_i < 0$, φ^i and ψ^i increasing and $\varphi^i(0) = \psi^i(0) = 0$ and $\varphi_k^1(0) = \psi_k^1(0) = \varphi_k^2(0) = \psi_k^2(0) = \infty$

Theorem 4.3. *If a Markovian strategy profile* $(h^1(k_t^2), h^2(k_t^1))$ *is an indifferent MPE for a capital accumulation game satisfying Assumption 5, then it must have the following functional form*

$$
\begin{aligned}
h^1(k_t^2) &= \lambda_2^{-1}(\psi^2)^{-1}\left(\rho^{-1}(r+\delta_2)k_t^2 - \rho^{-1}\varphi^2(k_t^2) + d^1\right) \\
h^2(k_t^1) &= \lambda_1^{-1}(\psi^1)^{-1}\left(\rho^{-1}(r+\delta_1)k_t^1 - \rho^{-1}\varphi^1(k_t^1) + d^2\right)
\end{aligned}
$$

where d^i *are appropriately chosen constants.*

The proof of this Theorem can be found in Dockner et al. (1999).

To study the dynamic behaviour of indifferent MPE, we make use of specific functional forms of the profit functions. Because we are dealing with indifferent MPE the profit function must be as in Assumption 5 but on top of this we require $\lambda_1 = \lambda_2 = \lambda$, $\beta^1 = \beta^2 = \beta$ and $d^1 = d^2 = 0$ to hold and specify the functions ψ^i and φ^i as follows:

$$
\lambda\psi^1(k_t^2) \equiv k_t^2, \quad \lambda\psi^2(k_t^1) \equiv k_t^1 \tag{4.4}
$$

$$
\varphi^1(k_t^1) \equiv (k_t^1)^\alpha, \quad \varphi^2(k_t^2) \equiv (k_t^2)^\alpha \tag{4.5}
$$

where $\lambda < 0$ and α is constant and constrained by $0 < \alpha < 1$. With these specifications the Markov strategy profiles become

$$
k_{t+1}^1 = h(k_t^2); \quad t \geq 0
$$

$$
k_{t+1}^2 = h(k_t^1); \quad t \geq 0 \tag{4.6}
$$

where

$$
h(z) \equiv \frac{\beta+\delta}{\rho\lambda}z - \frac{1}{\lambda\rho}z^\alpha.
$$

Since $\rho^{-1} = 1 + \delta$, this becomes

$$
h(z) \equiv \frac{1+\delta}{\lambda}\left[(\beta+\delta)z - z^\alpha\right]. \tag{4.7}
$$

Given the specific functional form (4.7), the upper bounds for the capital stocks are given by

$$
\bar{k}^1 = \frac{1}{(\beta+\delta)^{\frac{1}{1-\alpha}}}, \quad \bar{k}^2 = \frac{1}{\beta+\delta)^{\frac{1}{1-\alpha}}}
$$

and the domain for the the equilibrium dynamics is given by $D = [0, \bar{k}^1] \times [0, \bar{k}^2]$. Finally, it should be noted that the equations for the equilibrium dynamics (4.6) form a two-dimensional system but with the specific property that the right-hand sides of the individual equations are decoupled in the sense that they only depend on the other firm's capital stock. Hence, a simple algebraic manipulation makes it possible that the two-dimensional system can be reduced to a one-dimensional one. We then get

$$z_{t+1} = h(h(z_{t-1})) = h^2(z_{t-1}) \tag{4.8}$$

where z_t is either k_t^1 or k_t^2. In case the initial conditions are identical $k_0^1 = k_0^2$, we get that for $t \geq 0$ $k_t^1 \equiv k_t^2 \equiv z_t$ holds, and the system becomes even simpler and reduces to the dynamic equation

$$z_{t+1} = \frac{1+\delta}{\lambda}\left[(\beta+\delta)z_t - z_t^\alpha\right]. \tag{4.9}$$

Now observe that the function (4.7) has a unique maximum given by

$$z_1 = \left(\frac{\alpha}{\beta+\delta}\right)^{\frac{1}{1-\alpha}}.$$

Assuming that the parameter restriction

$$\beta+\delta = \frac{\lambda}{(1+\delta)(1-\frac{1}{\alpha})\alpha^{\frac{1}{1-\alpha}}} \tag{4.10}$$

holds, implies that $h(z_1) = \bar{z}$, and, therefore the dynamical system (4.9) maps the domain $X = [0, \bar{z}]$ into itself. Hence, the dynamic equation (4.9) is defined as $h : X \to X$ which is a continuous map from the compact set X into itself. Hence the dynamical system we are going to analyze is fully characterized by the pair (X, h).

Theorem 4.4. *If (4.10) holds together with $\mid \alpha - \alpha^{\alpha/(1-\alpha)} \mid > 1$ then the dynamical system (X, h) has a period-3 cycle.*

The proof of this Theorem can be found in Dockner et al. (1999) and we do not repeat it here. What we see, however, is that our indifferent MPE does result in complicated dynamics. Contrary to the case with a single capital stock the MPE for the model with private capital does result in chaos and irregular behaviour.

5. Conclusion

In this paper we analyze two classes of capital accumulation games and derive open-loop and Markov-perfect equilibria. The two classes of games differ with respect to the capital stocks that are accumulated through investment. In the first case firms are investing in a public stock of capital and in the second case firms accumulate their private stock of physical capital. For both classes of games it is possible to derive the existence and stability of open-loop and Markov-perfect Nash equilibria. It turns out that while in the open-loop game there exists a unique interior steady state with simple dynamics, the MPE's are characterized by complicated dynamics with no convergence to a steady state. Hence, the industry dynamics of firms that are able to choose commitment strategies are easy to characterize while in the case in which strategic interactions are taken into account more complex behviour emerges.

Acknowledgement. Engelbert J. Dockner wishes to acknowledge financial support by the Austrian Science Foundation (FWF) under the grant SFB#010.

6. References

E.J. Dockner and H. Takahashi, *Further Turnpike Properties of Capital Accumulation Games*, Economics Letters 28 (1988), pp. 321–325.

E.J. Dockner and K. Nishimura, *Transboundary Pollution in a Dynamic Game Model*, The Japanese Economic Review 50 (1999), pp. 443–456.

E.J. Dockner, M. Plank and K. Nishimura, *Markov Perfect Equilibira for a Class of Capital Accumulation Games*, Annals of Operations Research 89 (1999), pp. 215–230.

E.J. Dockner and H. Takahashi, *Turnpike Properties and Comparative Dynamics of General Capital Accumulation Games*, in: BOLDRIN M. et al (eds), General Equilibrium, Growth and TradeII: The Legacy of L. McKenzie (1993), pp. 352–366.

E.J. Dockner, G. Sorger and N.V. Long, *Analysis of Nash Equilibria in a Class of Capital Accumulation Games*, Journal of Economic Dynamics and Control 20 (1996), pp. 1209–1235.

C. Fershtman and S. Nitzan, *Dynamic Voluntary Provision of Public Goods*, European Economic Review 35 (1991), pp. 1057–1067.

A.M. Spence, *Investment Strategy and Growth in a New Market*, Bell Journal of Economics 10 (1979), pp. 1-19.

Optimal Investments with Increasing Returns to Scale: A Further Analysis

Richard F. Hartl[1], Peter M. Kort[2]

[1] Institute of Management, University of Vienna, Brünnerstraße 8, A-1210 Vienna, Austria
[2] Department of Econometrics and CentER, Tilburg University, P.O.Box 90153, 5000 LE, Tilburg, The Netherlands

Abstract. This paper considers a capital accumulation model that was previously analyzed by Barucci (1998). The specific feature of the model is that revenue is a convex function of the capital stock. We extend Barucci's work by giving a full analytical characterization of the case where a saddle point with a positive capital stock level exists. Furthermore we also analyze the other cases.

1. Introduction

In this paper we study a standard capital accumulation model of the firm, where the objective is to maximize the discounted profit stream. The profit rate equals the difference between the revenue and the costs of investment. Revenue is obtained by selling goods on the market. The firm needs a capital stock to produce these goods. The higher the capital stock it owns , the more goods the firm produces, which in turn leads to a higher revenue. The firm can increase capital stock by investing. Technically spoken, this model is an optimal control model with one state variable, the capital stock, and one control variable, the investment rate.

The study of this framework goes back to the sixties, and started out with Eisner and Strotz (1963). In this contribution the revenue function was assumed to be concave and investment costs were convex. Using standard methods of control theory it is easily shown that optimal firm behavior describes convergence to a unique long run equilibrium at which marginal revenue equals marginal costs. Later it was recognized (Rothschild (1971)) that arguments could be found in favor of a (partly) concave shape of the investment cost function. The problems (chattering controls!) that then occur in the maximization problem were subject of study in Davidson and Harris (1981) and Jorgensen and Kort (1993).

On the other hand it can also be the case that the revenue function is convexly shaped for some intervals of capital stock values. Such a scenario was studied in Dechert (1983) and again Davidson and Harris (1981). From these contributions it can be concluded that partly convex revenue functions can lead to multiple equilibria. It then depends on the initial level of the capital stock to which of the equilibria it is optimal for the firm to converge to. In this sense we can speak of history dependent equilibria. Barucci (1998) studies the case where the revenue function is strictly convex throughout. He considered a framework where both the revenue function and the investment cost function are quadratic. As a result the isoclines, on which state, control, and co-state variables are constant, are linearly shaped, so that exactly one steady state exists. This means that multiple equilibria are ruled out. Barucci (1998) identifies the case where a saddle point equilibrium occurs for a positive level of the capital stock. He shows that convergence to this saddle point is the optimal policy.

Fascinated by the fact that such a simple optimal solution exists for the model with a fully convex revenue function, in this paper Barucci's framework is studied once again. We extend Barucci (1998) by (1) determining a full analytical characterization of the case with the saddle point with positive capital stock, and (2) by determining which other cases are also possible if the parameter values are different.

The contents of this paper is as follows. The model is formulated in Section 2. After establishing the necessary optimality conditions in Section 3, the equilibrium and its stability properties are studied in Section 4. In Section 5 all possible cases are studied, while some ideas for future research are outlined in Section 6.

2. Model formulation

Following Barucci (1998), the model we consider is the following:

$$\max_{u} \int_{0}^{\infty} e^{-\rho t} [r(k) - c(u)] dt , \tag{1}$$

$$\dot{k} = u - \mu k , \quad k(0) = k_0, \tag{2}$$

where k denotes the capital stock and u is investment. The revenue function is given by $r(k)$ while the investment costs are $c(u)$. The discount rate is ρ while μ denotes the depreciation rate.

Although Barucci did not impose this constraint, for economic reasons (see, e.g., Dixit and Pindyck (1996)) we assume that investments are irreversible:

$$u \geq 0 . \tag{3}$$

In order to be able to obtain a full analytical solution, like Barucci (1998) we assume quadratic revenue and cost functions:

$$r(k) = ak + bk^2, \quad c(u) = cu + du^2. \tag{4}$$

We require all parameters a, b, c, d, μ, and ρ to be positive. Hence, as already explained in the Introduction, the revenue function exhibits increasing returns to scale. To illustrate the importance of analyzing this framework, we can refer to, e.g., Hartl and Kort (1996). In this paper a variant of the current model was studied where the revenue function is concave and pollution was included. Also a second control in the form of abatement expenditures was added. For this model it turned out that, after solving for abatement expenditures, an optimal control model results for which in one particular case the objective is strictly convex in k. In Hartl and Kort (1996) this scenario was not analyzed because it seemed to complicated at that time.

Investment costs include costs of acquisition and adjustment costs. In the next section it turns out that the strictly convex shape of the investment cost function implies that u is continuous over time.

3. Necessary conditions

To obtain the necessary conditions for optimality we start out by presenting the current value Hamiltonian:

$$H = ak + bk^2 - cu - du^2 + q[u - \mu k].$$
(5)

From the maximum principle it is derived that:

$$\partial H / \partial u = 0 \quad \text{i.e.} \quad u = \frac{q - c}{2d}.$$
(6)

If (3) is imposed, then (6) holds only for $u > 0$ i.e. for $q > c$. Otherwise we have

$$u = 0 \quad \text{if} \quad q \le c.$$
(6a)

Since the Hamiltonian is strictly concave in u we know from, e.g., Feichtinger and Hartl (1986) that u is continuous over time. The adjoint equation is

$$\dot{q} = \rho q - \partial H / \partial k = (\rho + \mu) q - a - 2bk .$$
(7)

From (6), i.e., $q = 2du + c$ and (7) we get:

$$\dot{u} = \frac{\dot{q}}{2d} = (\rho + \mu) u + \frac{(\rho + \mu) c - a}{2d} - \frac{b}{d} k .$$
(8)

4. Equilibrium and its stability properties

The (unbounded) linear DE-system (2) and (8) has the following unique equilibrium:

$$\bar{k} = \frac{1}{2} \frac{c(\rho + \mu) - a}{b - d\mu(\rho + \mu)} \quad \bar{u} = \mu \bar{k} = \frac{\mu}{2} \frac{c(\rho + \mu) - a}{b - d\mu(\rho + \mu)}.$$
(9)

On the other hand, the original canonical system (2) and (7) has the unique equilibrium

$$\bar{k} = \frac{1}{2}\frac{c(\rho+\mu)-a}{b-d\mu(\rho+\mu)} \qquad \bar{q} = \frac{bc-ad\mu}{b-d\mu(\rho+\mu)} \tag{10}$$

which is the same as Barucci's result on p. 794 except for a sign error in the second formula.

For economic reasons only positive equilibria $\bar{k}$ make sense. Then also $\bar{u} = \mu\bar{k}$ is positive, which in turn implies that then also $\bar{q} = 2d\bar{u} + c$ is positive too.

Proposition 1: *The unique equilibrium is in the relevant region ($k > 0$, $u > 0$, first quadrant), iff the sign of $c(\rho+\mu)-a$ equals the sign of $b-d\mu(\rho+\mu)$.*

The Jacobian of the linear DE-system (2) and (8) is

$$J = \begin{bmatrix} -\mu & 1 \\ -\frac{b}{d} & \rho+\mu \end{bmatrix} \quad \text{and} \quad \det J = -\mu(\rho+\mu)+\tfrac{b}{d},$$

so that the equilibrium is a saddle point iff

$$\mu d(\rho+\mu) > b \tag{11}$$

as was also found by Barucci (1998); see (i) on p. 794.

The $\dot{k} = 0$-isocline is

$$u = \mu k \tag{12}$$

and the $\dot{u} = 0$-isocline is

$$u = -\frac{(\rho+\mu)c-a}{2d(\rho+\mu)} + \frac{b}{d(\rho+\mu)}k . \tag{13}$$

Comparing this with (11) it follows that the $\dot{k} = 0$-isocline is steeper than the $\dot{u} = 0$-isocline iff the equilibrium is a saddle point.

5. Solution in the four different cases

From Proposition 1 we obtain, that the signs of the expressions $c(\rho+\mu)-a$ and $b-d\mu(\rho+\mu)$ are crucial for the outcome of the model. Consequently we can distinguish four different cases.

5.1 Case 1: $c(\rho+\mu)-a < 0$ and $b-d\mu(\rho+\mu) < 0$

In this case, from (9) we get a positive equilibrium, which is, by (11), a saddle point. Figure 1 clearly illustrates the reverse accelerator feature of the stable investment path as expressed in Dechert (1983). This means that investment is lower the larger the difference between the steady state level and the current level of capital goods. The economic intuition behind this is that in this model marginal revenue increases with the capital stock so that investment is more profitable if the capital stock is large.

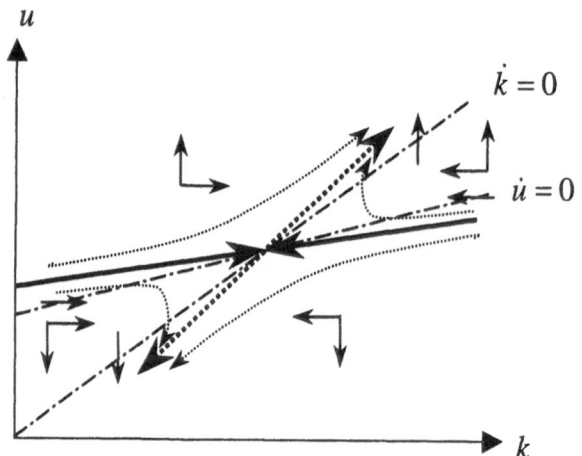

Fig. 1. The phase diagram in Case 1 where $c(\rho + \mu) - a < 0$ and $b - d\mu(\rho + \mu) < 0$.

We now compute the trajectories $u(t)$ and $k(t)$ along the saddle point path and evaluate the objective function. This is normally impossible, but here it can be done due to the fact that the functions $r(k)$ and $c(u)$ are quadratic (cf. (4)).

First, we have to obtain the eigenvalues associated with the Jacobian of the dynamic system (see section 4):

$$\lambda_1 = \frac{\rho - \sqrt{(\rho + 2\mu)^2 - \frac{4b}{d}}}{2}, \ \lambda_2 = \frac{\rho + \sqrt{(\rho + 2\mu)^2 - \frac{4b}{d}}}{2}. \tag{14}$$

It is easily obtained that λ_1 is negative while λ_2 is positive. Since the solution of Figure 1 is stable only λ_1 must be considered.

Then we need to compute the eigenvector $[k^*, 1]'$ associated with this negative eigenvalue λ_1 and get the solution:

$$\begin{bmatrix} k(t) \\ u(t) \end{bmatrix} = \begin{bmatrix} \bar{k} \\ \bar{u} \end{bmatrix} + \frac{k_0 - \bar{k}}{k^*} \begin{bmatrix} k^* \\ 1 \end{bmatrix} e^{\lambda_1 t}. \tag{15}$$

Taking once again the Jacobian of the dynamic system into consideration, the eigenvector is easily computed as follows:

$$\begin{bmatrix} -\mu - \lambda_1 & 1 \\ -\frac{b}{d} & \rho + \mu - \lambda_1 \end{bmatrix} \begin{bmatrix} k^* \\ 1 \end{bmatrix} = \begin{bmatrix} 0 \\ 0 \end{bmatrix},$$

which yields:

$$k^* = \frac{1}{\mu + \lambda_1}. \tag{16}$$

From (15) and (16) we generate the following expressions:

$$k(t) = \bar{k} + \left(k_0 - \bar{k}\right)e^{\lambda_1 t}, \tag{17}$$

$$u(t) = \bar{u} + \left(k_0 - \bar{k}\right)\left(\mu + \lambda_1\right)e^{\lambda_1 t}. \tag{18}$$

Then, with $\alpha = k_0 - \bar{k}$ and $\beta = \mu + \lambda_1$ the profit rate is

$$\pi(k,u) = ak + bk^2 - cu - du^2$$
$$= \left[a\bar{k} + b\bar{k}^2 - c\bar{u} - d\bar{u}^2\right] + e^{\lambda_1 t}\left[a\alpha + 2b\alpha\bar{k} - c\alpha\beta - 2d\alpha\beta\bar{u}\right] + e^{2\lambda_1 t}\left[b\alpha^2 - d\alpha^2\beta^2\right]$$

Evaluating the objective function (1) using this expression, we get:

$$\Pi(k_0) = \int_0^\infty e^{-\rho t}\left[\pi(k,u)\right]dt =$$

$$= \frac{a\bar{k} + b\bar{k}^2 - c\bar{u} - d\bar{u}^2}{\rho} + \frac{a + 2b\bar{k} - c\beta - 2d\beta\bar{u}}{\rho - \lambda_1}\alpha + \frac{b - d\beta^2}{\rho - 2\lambda_1}\alpha^2.$$

Using $\alpha = k_0 - \bar{k}$ we get the profit $\Pi(k_0)$ as a function of the initial state:

$$\Pi(k_0) = \left[\frac{a\bar{k} + b\bar{k}^2 - c\bar{u} - d\bar{u}^2}{\rho} - \frac{a + 2b\bar{k} - c\beta - 2d\beta\bar{u}}{\rho - \lambda_1}\bar{k} + \frac{b - d\beta^2}{\rho - 2\lambda_1}\bar{k}^2\right]$$
$$+ \left[\frac{a + 2b\bar{k} - c\beta - 2d\beta\bar{u}}{\rho - \lambda_1} - 2\frac{b - d\beta^2}{\rho - 2\lambda_1}\bar{k}\right]k_0 + \frac{b - d\beta^2}{\rho - 2\lambda_1}k_0^2.$$

Apparently, profit $\Pi(k_0)$ is a quadratic function of k_0. It is clear, that the net present value of the profit is higher the larger the initial capital stock is. This means, that the coefficient of k_0^2 is positive, and that the that the coefficient of k_0 is positive, i.e. the minimum of $\Pi(k_0)$ occurs for an infeasible $k_0 < 0$. This is shown in Appendix 1.

Extensive numerical experiments have shown, that in Case 1 the net present value of profit is always positive. Unfortunately it turned out to be too difficult to derive this result analytically. If one wants to determine the parameter values a, b, c and d such that $\Pi(0)$ is minimized, taking into account all the constraints that hold in Case 1 and the positivness of the parameters, then one can make $\Pi(0)$ a positive number arbitrarily close to zero. In this case d is very large compared to all the other parameters and $\bar{k}$ approaches zero. If this $\Pi(0)$ would not have been positive the solution that converges to the saddle point would have been dominated by a policy of zero investment throughout. The outcome of this exercise does not contradict Barucci's (1998) result (Proposition 4.1) that approaching the equilibrium is always optimal in Case 1.

5.2 Case 2: $c(\rho + \mu) - a < 0$ and $b - d\mu(\rho + \mu) > 0$

In this case the equilibrium is not in the first quadrant and it is not a saddle point:

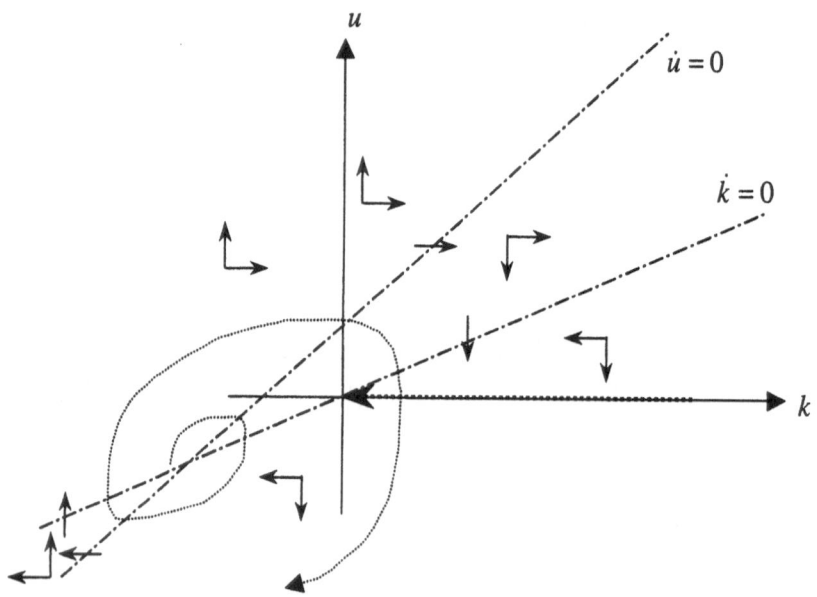

Fig. 2. The phase diagram in Case 2 where $c(\rho + \mu) - a < 0$ and $b - d\mu(\rho + \mu) > 0$.

The equilibrium with negative k and u is an unstable focus.

If (3) is imposed, $u = 0$ and $k \rightarrow 0$ could be expected to be optimal when looking at the figure. However, for economic reasons it is clear that this is not true. Case 2 is characterized by large values of the parameters a and b occurring in the revenue function. In this case, approaching $k = 0$ is certainly not optimal.

In fact, no optimal solution exists, since the objective is unbounded. This will now be verified by showing analytically that constant or proportional investment rates can yield arbitrarily high values of (1)

5.2.1 Constant investment

We first consider a constant investment policy

$$u(t) = u^* \text{ for all } t.$$

Then the capital stock develops according to

$$k(t) = \frac{u*}{\mu} + \left(k_0 - \frac{u*}{\mu}\right)e^{-\mu t}.$$

Evaluating the profit using these expressions, we get:

$$\pi(k,u) = ak + bk^2 - cu - du^2$$

$$= \left[a\frac{u*}{\mu} + b\frac{u*^2}{\mu^2} - cu* - du*^2\right] + e^{-\mu t}\left[a\left(k_0 - \frac{u*}{\mu}\right) + 2b\frac{u*}{\mu}\left(k_0 - \frac{u*}{\mu}\right)\right]$$

$$+ e^{-2\mu t}b\left(k_0 - \frac{u*}{\mu}\right)^2.$$

so that the objective function (1) becomes:

$$\Pi(k_0) = \int_0^\infty e^{-\rho t}\left[\pi(k,u)\right]dt =$$

$$= \frac{a\frac{u*}{\mu} + b\frac{u*^2}{\mu^2} - cu* - du*^2}{\rho} + \frac{a\left(k_0 - \frac{u*}{\mu}\right) + 2b\frac{u*}{\mu}\left(k_0 - \frac{u*}{\mu}\right)}{\rho+\mu} + \frac{b\left(k_0 - \frac{u*}{\mu}\right)^2}{\rho+2\mu}.$$

The terms with $u*^2$ are

$$\frac{\frac{b}{\mu^2} - d}{\rho} - \frac{2b\frac{1}{\mu^2}}{\rho+\mu} + \frac{b\frac{1}{\mu^2}}{\rho+2\mu} = \frac{2b - d(\rho+\mu)(\rho+2\mu)}{\rho(\rho+\mu)(\rho+2\mu)}.$$

Thus, the objective (1) can be made arbitrarily large, if the constant $u*$ is chosen large enough, provided that

$$b > \frac{d}{2}(\rho+\mu)(\rho+2\mu) = d\left(\mu^2 + \tfrac{3}{2}\rho\mu + \tfrac{1}{2}\rho^2\right).$$

Barucci shows that the objective is unbounded for

$$b > d\left(\mu^2 + \rho\mu\right)$$

which is a weaker condition.

5.2.2 Proportional investment

We now consider a constant investment policy

$$u(t) = [\mu + \varepsilon]k(t) \text{ for all } t.$$

Substitution of this expression into (2) yields that the capital stock develops according to

$$k(t) = k_0 e^{\varepsilon t} \quad \text{and} \quad u(t) = k_0 [\mu + \varepsilon] e^{\varepsilon t}.$$

Evaluating the profit function using these expressions, we get:

$$\pi(k,u) = ak + bk^2 - cu - du^2$$
$$= k_0 [a - c(\mu + \varepsilon)] e^{\varepsilon t} + k_0^2 [b - d(\mu + \varepsilon)^2] e^{2\varepsilon t},$$

which is again used to evaluate the objective function (1):

$$\Pi(k_0) = \int_0^\infty e^{-\rho t} [\pi(k,u)] dt = k_0 \frac{a - c(\mu + \varepsilon)}{\rho - \varepsilon} + k_0^2 \frac{b - d(\mu + \varepsilon)^2}{\rho - 2\varepsilon}$$

provided that $2\varepsilon < \rho$.

However (1) is infinite for $2\varepsilon \geq \rho$. In particular, it is $+\infty$ if $b > d(\mu + \varepsilon)^2$ and it is $-\infty$ if $b < d(\mu + \varepsilon)^2$.

So (1) is unbounded, provided that $b > d(\mu + \varepsilon)^2$ and $2\varepsilon \geq \rho$. Then it holds that

$$b > d\left(\mu + \frac{\rho}{2}\right)^2 = d\left(\mu^2 + \rho\mu + \frac{\rho^2}{4}\right).$$

This lower bound is better than that obtained for constant investment, but still above the Barucci boundary.

5.3 Case 3: $c(\rho + \mu) - a > 0$ and $b - d\mu(\rho + \mu) < 0$

In this case the equilibrium is not in the first quadrant and it is a saddle point:

This case is characterized by small values of the parameters a and b occurring in the revenue function. In this case, approaching $k = 0$ is certainly optimal. Note that $k = 0$ will not be reached in finite time.

Although in this particular case the parameters in the revenue function are small compared to those of he investment cost function, small investment expenditures will still be profitable if k is sufficiently large. This is true because marginal revenue increases linearly with k. Therefore, u will be positive for large values of k, but definitely zero for low values of the capital stock; see Figure 3.

5.3.1 Allowing for reversibility of investment

Since Figure 3 shows that $u = 0$ at a final time interval, it makes sense to consider the scenario here, where constraint (3) is replaced by the state constraint $k \geq 0$. Then the optimal trajectory in the phase diagram would converge to $k = 0$ within finite time. This is sketched in Figure 3a. Note that disinvestment does occurs here.

The proof is the same as in Example 8.8 on p.219 in Feichtinger and Hartl (1986).

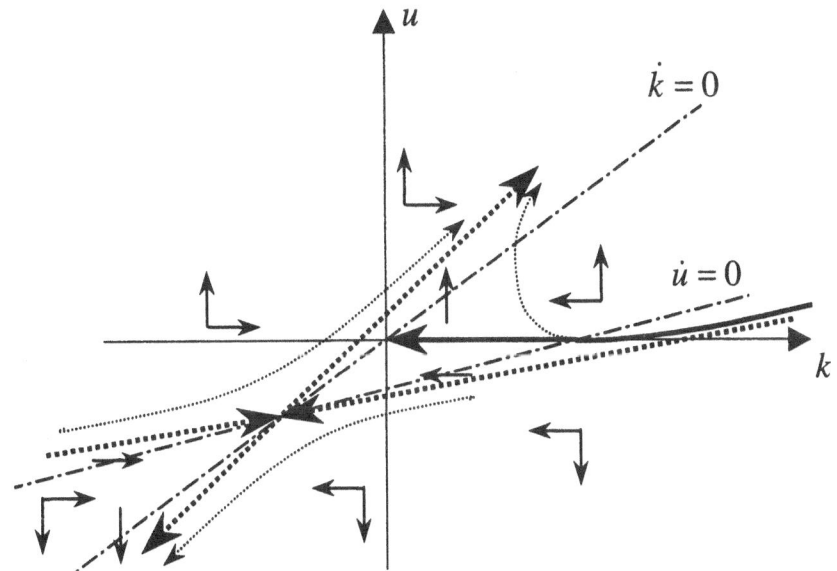

Fig. 3. The phase diagram in Case 3 where $c(\rho + \mu) - a > 0$ and $b - d\mu(\rho + \mu) < 0$.

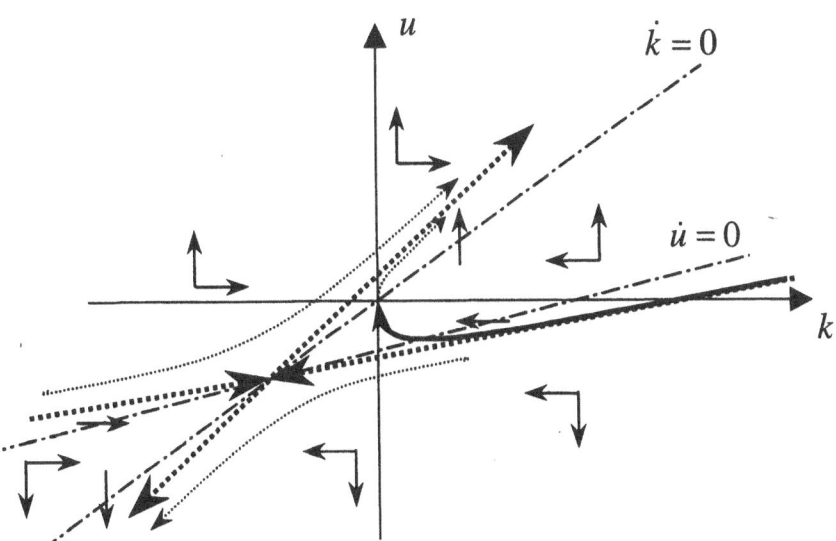

Fig. 3a. The phase diagram in Case 3 where investment is reversible.

236

5.4   Case 4: $c(\rho + \mu) - a > 0$ and $b - d\mu(\rho + \mu) > 0$

In this case the equilibrium is in the first quadrant and it is not a saddle point:

The equilibrium is an unstable focus. Except for the fact that the equilibrium now occurs for positive values of k and u this case is identical to Case 2.

No optimal solution exists, and the objective is unbounded. The calculations in Section 5.2 concerning constant and proportional investment, respectively, also apply to this case.

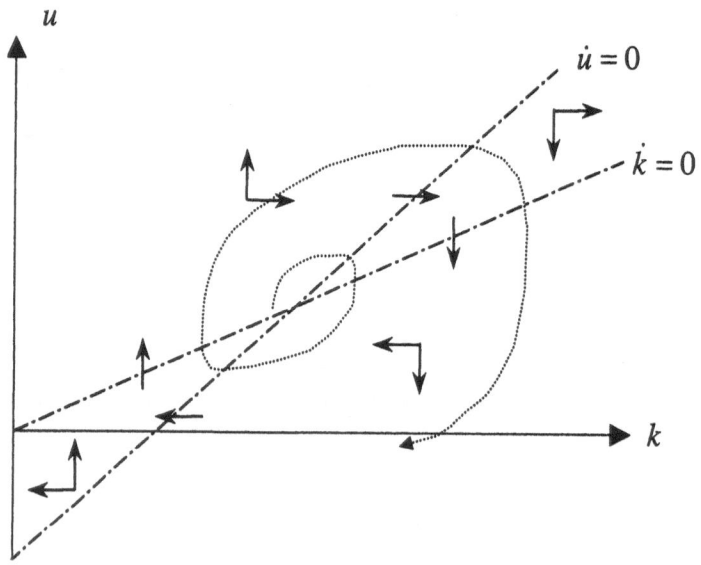

Fig. 4. The phase diagram in Case 4 where $c(\rho + \mu) - a > 0$ and $b - d\mu(\rho + \mu) > 0$.

6. Directions for future research

In this paper the standard capital accumulation model, but then with a strictly convex revenue function, was studied. Due to the quadratic specifications of the revenue and investment cost function, interesting results could be generated.

A straightforward extension is to make the revenue function a third order polynomial in which k^3 is multiplied with a negative parameter. In this way a convex-concave revenue function arises which makes it possible to redo the calculations of Dechert (1983) and Davidson and Harris (1981). As already mentioned in the Introduction, they arrived at solutions with multiple equilibria. They could identify levels of the capital stock, which we now call DNS (Dechert Nishimura Skiba)-points, where the firm is indifferent concerning to which equilibrium it should converge. Hopefully, it is possible to generate additional insights concerning these DNS-points in case we study the model with such a third order polynomial as revenue function.

A second interesting extension is to combine the just described convex-concave revenue function with introducing adjustment costs on changes in the investment rate (see Jorgensen and Kort (1983), Section 3.4.2). In such a model investments will be introduced as a second state variable, and the rate of change of investment is the control variable. Hence, the resulting model now contains two state variables and one control variable. It can be expected that also here multiple steady states exist. Depending on the location in the (k, u)-plane, it is optimal to converge to one of these steady states. It would be interesting to study whether so-called DNS-curves (which are DNS-points in a one-state-variable-model) exist, on which the firm is indifferent concerning to which steady state to converge to.

7. References

Feichtinger, G., Hartl, R.F.: Optimale Kontrolle ökonomischer Prozesse: Anwendungen des Maximumprinzips in den Wirtschaftswissenschaften, Berlin: de Gruyter 1986

Barucci, E.: Optimal investments with increasing returns to scale. International Economic Review 39, 789 - 808 (1998)

Davidson, R., Harris, R.: Non-convexities in continuous-time investment theory. Review of Economic Studies 48, 235-253 (1981)

Dechert, A.: Increasing returns to scale and the reverse flexible accelerator. Economic Letters 13, 69 - 75 (1983)

Dixit, A.K., Pindyck, R.S., Investment under Uncertainty, second printing, Princeton. Princeton University Press 1996

Eisner, R., Strotz, R.H., Determinants of Business Investments in Impacts of Monetary Policy, Englewood Cliffs, N.J.: Prentice Hall 1963

Hartl, R.F., Kort, P.M.: Capital accumulation of a firm facing environmental constraints. Optimal Control Applications & Methods 17, 253-266 (1996)

Jorgensen, S., Kort, P.M.: Optimal dynamic investment policies under concave-convex adjustment costs. Journal of Economic Dynamics and Control 17, 153-180 (1993)

Rothschild, M.: On the cost of adjustment, Quarterly Journal of Economics 85, 605-622 (1971)

Appendix 1

We now show that in Case 1 the profit $\Pi(k_0)$ is increasing in k_0.

We first show that that the coefficient of k_0^2 is positive:

$$\frac{b - d\beta^2}{\rho - 2\lambda_1} > 0 \quad \Leftrightarrow \quad \text{(note that } \rho - 2\lambda_1 > 0 \text{ because of (14))}$$

$$\frac{4b}{d} > (2\beta)^2 \qquad \Leftrightarrow \quad (\text{using } \beta = \mu + \lambda_1 \text{ and } (14))$$

$$\sqrt{\frac{4b}{d}} > \rho + 2\mu - \sqrt{(\rho + 2\mu)^2 - \frac{4b}{d}} \qquad \Leftrightarrow$$

$$\sqrt{\frac{4b}{d}} + \sqrt{(\rho + 2\mu)^2 - \frac{4b}{d}} > \rho + 2\mu \qquad \Leftrightarrow \text{. (since both sides are positive)}$$

$$\frac{4b}{d} + (\rho + 2\mu)^2 - \frac{4b}{d} + 2\sqrt{\frac{4b}{d}}\sqrt{(\rho + 2\mu)^2 - \frac{4b}{d}} > (\rho + 2\mu)^2 \qquad \Leftrightarrow$$

$$\sqrt{(\rho + 2\mu)^2 - \frac{4b}{d}} > 0 \qquad \text{which is clearly true.} \qquad \square$$

We now show that that the coefficient of k_0 is positive:

$$\frac{a + 2b\bar{k} - c\beta - 2d\beta\bar{u}}{\rho - \lambda_1} - 2\frac{b - d\beta^2}{\rho - 2\lambda_1}\bar{k} > 0 \qquad \Leftrightarrow$$

$$(a + 2b\bar{k} - c\beta - 2d\beta\bar{u})(\rho - 2\lambda_1) - 2(b - d\beta^2)(\rho - \lambda_1)\bar{k} =$$
$$(a - c\beta)(\rho - 2\lambda_1) + 2[(b - d\beta\mu)(\rho - 2\lambda_1) - (b - d\beta^2)(\rho - \lambda_1)]\bar{k} > 0$$

This is true because of:

$$(a - c\beta)(\rho - 2\lambda_1) > (a - c(\rho + \mu))(\rho - 2\lambda_1) > 0 \quad \text{in Case 1 (\textit{first term})}$$

and using $\beta = \mu + \lambda_1$ the bracket in the \textit{second term} can be written as:

$$(b - d\mu^2 - d\mu\lambda_1)(\rho - 2\lambda_1) - (b - d\mu^2 - d\lambda_1^2 - 2d\mu\lambda_1)(\rho - \lambda_1) =$$

$$(b\rho - d\rho\mu^2 - d\rho\mu\lambda_1) + (-2b\lambda_1 + 2d\mu^2\lambda_1 + 2d\mu\lambda_1^2)$$
$$+ (-b\rho + d\rho\mu^2 + d\rho\lambda_1^2 + 2d\rho\mu\lambda_1) + (b\lambda_1 - d\mu^2\lambda_1 - d\lambda_1^3 - 2d\mu\lambda_1^2) =$$

$$d\left[-\frac{b}{d} + \mu^2 + \rho\lambda_1 + \rho\mu - \lambda_1^2\right]\lambda_1 = 0$$

since from Section 4 we know that the eigenvalue satisfies

$$\lambda^2 - \rho\lambda - \mu(\rho + \mu) + \frac{b}{d} = 0$$

Thus, profit $\Pi(k_0)$ is higher, the larger the initial capital stock is.

A Note on Dynamic Transfer Price Bargaining

Steffen Jørgensen

University of Southern Denmark, Odense University, Campusvej 55, DK-5230 Odense M, Denmark.

Abstract. This note revisits the area of differential bargaining games, a field that attracted some interest in the late 1970s and early 1980s. This was also the period of time during which Gustav Feichtinger started his work on differential games and their applications. The bargaining model under consideration, as its predecessors, leads to "conservation of outcome". This means that negotiators are indifferent between settling at the outset of the game or waiting till offer equals demand. The phenomenon of instantaneous agreement has later been noted in discrete time, complete information bargaining games with alternating offers.

1. Introduction

The paper studies a bargaining problem. Basically, there are two approaches to such a problem. The first is the axiomatic one where a seminal result is the bargaining solution of Nash (1950). The axiomatic approach employs cooperative game theory and identifies principles (axioms) that intend to describe the *outcome* of the bargaining process. In games of complete information, the axiomatic approach has provided insights into important constructs such as bargaining power, equity, and efficiency. But since the bargaining process itself is not described (as the setup is static), no insight is provided into the process leading to the bargained outcome.

The strategic approach, on the other hand, is based on noncooperative game theory and employs an explicit model of the bargaining *process.* In this area, a seminal piece of work is Nash (1953). A disadvantage of the strategic approach is that the properties of the outcome are not necessarily revealed. An extensive form description of the bargaining process is used when the process evolves in discrete stages (typically with alternating offers). A notable example of the strategic approach is the alternating-offer bargaining model of Rubinstein (1982). Note that in the continuous-time setup of a differential game, the extensive form is suppressed and one avoids the need to specify the exact bargaining procedure. For a survey of bargaining games see, for instance, Chatterjee (1986).

This note employs the strategic approach and studies a noncooperative barganing game between a buyer and a seller. The problem is related to the differential game studies of labor-management bargaining of Leitmann and Liu (1974), Clemhout et al. (1975), Chen and Leitmann (1980). These works characterized bargaining policies (concessions) as Markovian strategies and the general assumption was that the players' concessions are made simultaneously, at any instant of time as long as the bargaining process goes on. A related, early contribution to concession bargaining is Rao and Shakun (1974) who used a discrete time setup to study a bargaining game with alternating concessions.

We proceed by presenting in Section 2 a simple two-player, noncooperative differential game of buyer-seller negotiations. Section 3 identifies a Markovian Nash equilibrium and Section 4 concludes.

2. The Model

A manufacturer and a retailer bargain over the transfer price of a shipment of goods, to be delivered from the former to the latter. (In practice, most purchase contracts among manufacturers and retailers are bargained. Actually, most purchases by institutions, government agencies, and commercial businesses are negotiated, cf. Angelmar and Stern (1978)). Our assumption is that the goods are ready to be delivered from the manufacturer's inventory and if an agreement on the transfer price is reached, they are delivered instantaneously to the retailer who immediately resells all the merchandise to her customers.

Let $y(t)$ denote the transfer price demanded by the manufacturer, and $x(t)$ the transfer price offered by the retailer at time t. These are the state variables of the game. Bargaining takes places over a (variable) time interval $[0,T]$ where T is the first instant of time at which $y(T) = x(T)$. If T is finite we say that the bargaining game terminates. The initial demand and offer are fixed and satisfy $y_0 > x_0 > 0$. Each player controls the rate at which she concedes on her demand/offer. Represent by $v(t)$ and $u(t)$ the concession rates of the manufacturer and the retailer, respectively. These rates are control variables of the players. Controls are defined on the interval $[0,T]$ and are constrained by $v(t), u(t) \geq 0$. (Balakrishnan and Eliashberg (1995) do not impose these conditions and, in some specific settings, they get negative concessions. Thus it can happen that the manufacturer increases her demand or the retailer reduces her offer during certain periods of time. We do not allow such behavior since it violates a basic principle of fairness in bargaining). The dynamics are given by the differential equations

$$\dot{y}(t) = -v(t), \; y(0) = y_0; \; \dot{x}(t) = u(t), \; x(0) = x_0. \tag{1}$$

Let $q = \text{const.} > 0$ denote the fixed size of the lot. If the game terminates, the retailer resells the merchandise at the consumer price p. We assume that she has committed to a fixed retail price (for instance, because she has already planned a promotion at that particular retail price or a fixed price is dictated by competition at the retail level). The manufacturer incurs a holding cost at the rate of $c = \text{const.} > 0$ per unit in stock. Disregarding the (sunk) cost of manufacturing the q units, the manufacturer's total cost over the time interval $[0,T]$ becomes $cq[1 - e^{-\rho_M T}]/\rho_M$ in which ρ_M is a positive and constant discount rate.

Each player wishes to choose her concession rate to maximize the present value of the profit stream over the interval $[0,T]$. Letting ρ_R denote the retailer's discount rate, the objective functionals of the two players can be stated in Mayer form

$$J_M = x(T)qe^{-\rho_M T} - cq[1 - e^{-\rho_M T}]/\rho_M \tag{2a}$$

$$J_R = [p - y(T)]qe^{-\rho_R T}. \tag{2b}$$

Note that the retailer's final offer and the holding cost determines the profit of the manufacturer while the profit of the retailer is determined by the manufacturer's final demand and the retail price. Due to discounting we expect that the game terminates: given that $x(T)$ and $y(T)$ are bounded, then for $T \to +\infty$ the present value of retailer profit tends to zero while that of the manufacturer tends to a negative number (viz., the present value of an infinite stream of holding costs).

Finally, it is convenient to use (1) to rephrase the problem (1) - (2) in Lagrange form

$$\max_{v \geq 0} \{J_M = x_0 + \int_0^T e^{-\rho_M t}[u(t) - \rho_M x(t) - c]dt\} \tag{3a}$$

$$\max_{u \geq 0} \{J_R = p - y_0 + \int_0^T e^{-\rho_R t}[\rho_R(y(t) - p) + v(t)]dt\} \tag{3b}$$

in which, without loss of generality, we have put $q = 1$.

3. A Markovian Nash Equilibrium

The players use Markovian strategies and since at any instant of time concessions are simultaneous, we look for a Nash equilibrium. Chen and Leitmann (1980) studies a Stackelberg game.

Strategies are given as $v(t) = \phi_M(x,y,t)$, $u(t) = \phi_R(x,y,t)$, that is, the choice of a player's concession rate at any instant of time is contingent upon the position of the game (represented by the current state pair and time).

Due to our assumptions, state constraints $y(t) \leq y_0$ and $x_0 \leq x(t)$ are automatically satisfied and (1) then shows the desired result: the manufacturer never increases her demand and the retailer never decreases her offer. We assume $p \geq y_0$, that is, the manufacturer is not allowed to make an extravagant initial demand, in excess of the retail price. Then it holds that $y(t) \leq p \; \forall t \in [0,T]$.

PROPOSITION. *The stationary concession strategies*

$$\phi_M^*(x,y) = \rho_R(p - y) \geq 0, \quad \phi_R^*(x,y) = \rho_M x + c > 0 \tag{4}$$

induce a Markovian Nash equilibrium of the bargaining game.

PROOF. By definition, a pair of admissible strategies is a Nash equilibrium iff it holds for all feasible initial states that

$$J_M(\phi_M^*, \phi_R^*) \geq J_M(\phi_M, \phi_R^*) \quad \text{for all } \phi_M \geq 0 \tag{5a}$$

$$J_R(\phi_M^*, \phi_R^*) \geq J_R(\phi_M^*, \phi_R) \quad \text{for all } \phi_R \geq 0 . \tag{5b}$$

Since $J_M(\phi_M, \phi_R^*) = x_0$ is constant, and hence independent of ϕ_M, (5a) holds with equality. Similarly, $J_R(\phi_M^*, \phi_R) = p - y_0$ is constant and (5b) is also satisfied with equality. $\qquad \square$

To determine the equilibrium state trajectory, insert from (4) into (1) to get the differential equations $\dot{y}(t) = -\rho_R[p - y(t)]$, $\dot{x}(t) = \rho_M x(t) + c$. Solving these equations, subject to the initial conditions, yields

$$y(t) = y_0 e^{\rho_R t} - p[e^{\rho_R t} - 1], \quad x(t) = x_0 e^{\rho_M t} + c[e^{\rho_M t} - 1]/\rho_M. \tag{6}$$

The control paths generated by the strategies (4) are found by differentiation with respect to time in (6), and using (1). The resulting derivatives show that manufacturer's demand $y(t)$ is a concave and decreasing function of time whereas the retailer's offer $x(t)$ is a convex and increasing function of time. Thus, throughout the game both players increase their concession rates. This behavior reflects their growing impatience by having not yet reached an agreement. The game terminates at the first instant of time, say T^*, at which it holds that

$$y(T^*) = x(T^*) \leftrightarrow F(T^*) = p + c/\rho_M \qquad (7)$$

in which $F(T^*) \triangleq (x_0 + c/\rho_M)\exp\{\rho_M T^*\} + (p - y_0)\exp\{\rho_R T^*\}$. Due to our assumption $y_0 > x_0$ it holds that $F(0) < p + c/\rho_M$ and since F is an increasing function, there exists a positive and finite T^* satisfying (7). (If the players have the same discount rate, an explicit expression for the termination date can be found).

Substituting $y(T)$ and $x(T)$ obtained from (6) into the payoffs (2) shows that in equilibrium the players receive the payoffs $J_M^* = x_0$, $J_R^* = p - y_0$. Equation (3) shows that these payoffs are the same as the ones that they would get by making an immediate agreement. Hence each player is indifferent between settling at $t = 0$, settling at $t = T^*$, or at any $t \in (0, T^*)$. The integrals in (3) are zero in equilbrium. The equilibrium strategies (5) lead to "conservation of outcome" (Leitmann and Liu (1974), Clemhout et al. (1975), Chen and Leitmann (1980)). If at time $t = 0$ the players agree to settle for the payoffs $J_M^* = x_0$, $J_R^* = p - y_0$, the manufacturer pays no holding costs but must accept the retailer's initial (and lowest) offer $x_0 < x(T^*)$. The retailer can resell at once at price p and does not have to wait until time T^* to collect her revenue. This comes at the cost that she must accept the manufacturer's initially demanded (and highest) transfer price $y_0 > y(T^*)$.

A similar result has been obtained in discrete-time, complete information bargaining games with alternating offers. For example, Rubinstein (1982) showed that there is a unique subgame perfect equilibrium in which acceptance of the first-moving player's offer is instantaneous. Recent work on wage bargaining, Coles and Smith (1998), reproduces this result: agreement on the wage is immediate and each player announces the reservation wage of the other player. Sákovics (1993), however, finds that when simultaneous offers are allowed in the discrete-time setup, there exist multiple subgame perfect equilibria that support delayed agreements.

4. Concluding remarks

This note has studied a differential game of concession bargaining. In a particular Markovian equilibrium, players are indifferent between settling at the initial instant of time or waiting till the time at which the demanded transfer price equals the transfer price offered. Thus the initial outcome, in terms of payoffs, is "conserved". In the game of this note there may be other equilibria than the one identified in the proposition. We make no claim that the equilibrium of the proposition is unique. Possible modifications of the game would be as follows:

(1) The initial values $y(0)$ and $x(0)$ were determined so as to satisfy the initial condition stated in (1). Alternatively, one could let these quantities be control parameters, to be chosen optimally at the initial instant of time. This formulation allows each player to make a strategic choice of her initial bargaining position.

(2) Let the retail price be time-dependent but (still) exogeneously given. Alternatively, let the retail price be a control of the retailer. Then one would need to introduce a demand function in the model.

(3) Play the game as a leader-follower game with, for instance, the manufacturer as the leader. In this case the retailer (the follower) conditions her concession rate upon state, time, and the current decision (concession) of the leader.

References

Angelmar, R. and Stern, L.W. (1978): Development of a Content Analytic System for Analysis of Bargaining Communication in Marketing. *Journal of Marketing Research* 15, 93-102.

Balakrishnan, P.V., Eliashberg, J. (1995): An Analytical Process Model of Two-Party Negotiations. *Management Science* 41, 2, 226-243.

Chatterjee, K. (1986): The Theory of Bargaining. In: Samuelson, L. (ed.), *Microeconomic Theory*, 159-195.

Chen, S.F.-H., Leitmann, G. (1980): Labour-Management Bargaining Modelled as a Dynamic Game. *Optimal Control Applications and Methods* 1, 11-25.

Clemhout, S., Leitmann, G., Wan Jr., H.Y. (1975): Bargaining under Strike: A Differential Game View. *Journal of Economic Theory* 11, 55-67.

Coles, M., Smith, E. (1998): Strategic Bargaining with Firm Inventories. *Journal of Economics Dynamics and Control* 23, 35-54.

Leitmann, G., Liu, P.T. (1974): A Differential Game Model of Labor-Management Negotiation During a Strike. *Journal of Optimization Theory and Applications* 13, 4, 427-444.

Nash, J. (1950): The Bargaining Problem. *Econometrica* 18, 155-162.

Nash, J. (1953): Two Person Cooperative Games. *Econometrica* 21, 128-140.

Rao, A.G. and Shakun, M.F. (1974): A Normative Model for Negotiations. *Management Science* 20, 10, 1364-1375.

Rubinstein, A. (1982): Perfect Equilibrium in a Bargaining Model. *Econometrica* 50, 97-110.

Optimal Production-Inventory Strategies for a Reverse Logistics System

Klaus-Peter Kistner, Imre Dobos

Lehrstuhl für BWL und UFO, Universität Bielefeld, Postfach 10 01 31, 33501 Bielefeld, Germany

Abstract. Reverse logistics is the term for logistic environments with reuse of product and materials. In these systems products can be manufactured and the returned units be remanufactured. These can be considered as new. The operation of such a system is described by the return process of used products and the disposal activity. The effectivity of a reverse logistics system can be measured by its costs. The costs are inventory holding costs, manufacturing and remanufacturing costs. We investigate two stores. The demand is satisfied from the first store, where the manufactured and remanufactured items are stored. The returned products are collected in the second store and then remanufactured. The costs of this system consist of the linear holding costs for these two stores and the convex non-decreasing manufacturing and remanufacturing costs. There is no delay between the using and return process.

Keywords: Remanufacturing; Reverse Logistics; Optimal Policies; Optimal Control

1. Introduction

Deterministic EOQ-type one-product two-store reverse logistics systems were studied extensively in the literature [1-8]. All these approaches assumed a predetermined control system and searched the optimal parameters for this systems. There are no results regarding the structure of optimal policies [9].

The aim of the paper is to find optimal inventory policies in these systems with special structure. The EOQ-type models are static in their structure. We make a dynamization of this models. It is assumed that the demand is a known continuous function in a given planning horizon and the return rate of used items is a given conststant. There is no delay between the using and return process. We investigate two stores. The demand is satisfied from the first store, where the manufactured and remanufactured items are stored. The returned products are collected in the second store and then remanufactured. The costs of this system consist of linear holding costs for these two stores and the convex non-decreasing manufacturing and remanufacturing costs.

The model can be represented by an optimal control problem with two state variables (inventory status in the first and second store) and with two control variables (rate of manufacturing and remanufacturing). The objective is to variables (rate of manufacturing and remanufacturing). The objective is to minimize the sum of holding costs in the stores and costs of manufacturing and remanufacturing. In this form, the model can be considered as a generalization of the well-known Arrow-Karlin model [10] with two warehouses.

2. The model

We will investigate a two-store reverse logistics model with continuous disposal. The following parameters are in our model:

- T	length of the planning horizon,
- $S(t)$	rate of demand, positive, continuous differentiable,
- r	return rate in the second store ($0 \leq r \leq 1$),
- h_1	inventory holding costs in the first store,
- h_2	inventory holding costs in the second store,
$-F_m(P_m(t))$	production costs for manufacturing; convex, monotonouosly non-decreasing, twice continuously differentiable,
$-F_r(P_r(t))$	production costs for remanufacturing, convex, monotonouosly non-decreasing, twice continuously differentiable.

Decision variables:

- $I_1(t)$	inventory level in the first store,
- $I_2(t)$	inventory level in the second store,
- $P_r(t)$	rate of the remanufacturing,
- $P_m(t)$	rate of the manufacturing (production).

The model can be written in the following form

$$\int_0^T \left[\{h_1 I_1(t) + F_m(P_m(t))\} + \{h_2 I_2(t) + F_r(P_r(t))\} \right] dt \to \min \tag{1}$$

such that

$$\dot{I}_1(t) = P_m(t) + P_r(t) - S(t), \quad I_1(0) = I_{10}$$
$$\dot{I}_2(t) = -P_r(t) + rS(t), \quad I_2(0) = I_{20} \tag{2}$$

$$I_1(t) \geq 0, \quad I_2(t) \geq 0$$
$$P_m(t) \geq 0, \quad P_r(t) \geq 0 \tag{3}$$

The material and cost flow of the model is depicted in Figure 1.

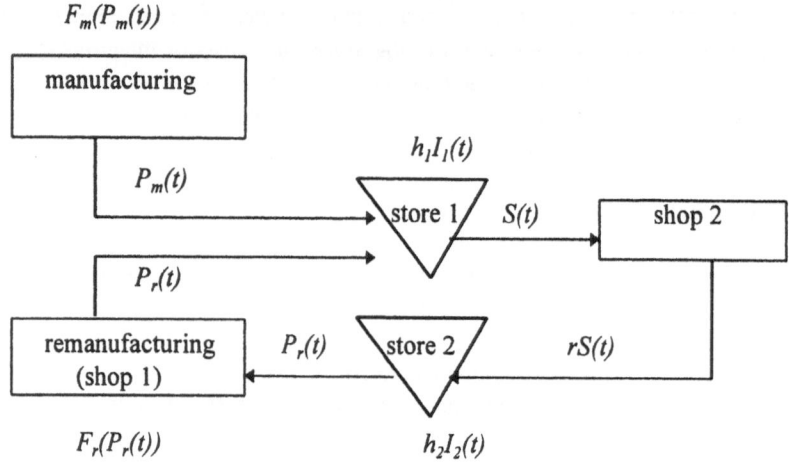

$$F_m(P_m(t))$$

Figure 1. The material and cost flow of the model

3. Some characteristics of the model

Let us first assume that we have solved the problem for $P_r(t)$ and $I_2(t)$. We look for the characteristics of the optimal manufacturing strategy. An auxiliary problem can be formulated in a simple form summing up the two differential equations for the known remanufacturing rate:

$$I_1(t) + I_2(t) = I_{10} + I_{20} + \int_0^t \left[P_m(\tau) - (1-r)S(\tau) \right] d\tau = \tilde{I}(t)$$

or $I_2(t) = \tilde{I}(t) - I_1(t) \geq 0$.

The problem can be reformulated in the following way:

$$\int_0^T \left[h_1 I_1(t) + F_m(P_m(t)) \right] dt + \int_0^T \left[h_2 I_2^0(t) + F_r(P_r^0(t)) \right] dt \to \min$$

such that

$$\dot{I}_1(t) = P_m(t) - \left(S(t) - P_r^0(t) \right) \quad I_1(0) = I_{10},$$

$$0 \leq I_1(t) \leq \tilde{I}(t), \quad P_m(t) \geq 0$$

In this form the problem is a classical Arrow-Karlin model with warehousing constraints [11-13], where the modified demand rate is equal to the difference of the demand rate and the rate of remanufacturing. This difference can be negative, but the modified foreward algorithm of Arrow and Karlin can be applied to this type of models.

Now we investigate the case, when manufacturing is not needed, i.e. the rate of manufacturing is equal to zero.

Lemma 1: If $I_{10} + I_{20} - (1-r)\int_0^T S(t)dt \geq 0$, then manufacturing is not needed to satisfy the demand: $P_m^0(t) = 0, \quad \forall t \in [0, T]$.

Proof. The solution of the relaxed problem for the rate of manufacturing is equal to zero, if

$$I_{10} - \int_0^t \left[S(\tau) - P_r^0(\tau) \right] d\tau \geq 0, \quad \forall t \in [0, T].$$

Our assumption is that we have solved the problem for the remanufacturing, so it is true

$$I_{20} - \int_0^t \left[P_r^0(\tau) - rS(\tau) \right] d\tau \geq 0, \quad \forall t \in [0, T].$$

If we summarize the two inequalities, we get the result of the lemma.

The result of this lemma can be interpreted, as manufacturing is not needed, if the sum of initial inventory level for end product and the initial inventory level for remanufactured items and the returned items is greater than the cumulated demand in the planning horizon, i.e. $I_{10} + I_{20} + r\int_0^T S(t)dt \geq \int_0^T S(t)dt$.

Next we investigate the optimal ending inventory levels.

Lemma 2: If $I_{10} + I_{20} - (1-r)\int_0^T S(t)dt < 0$, then in the optimal trajectory: $I_1^0(T) = 0$.

Proof. The proof is obvious. If the ending inventory level would be greater than zero, then the remanufacturing rate can be choosen zero at the end of the planning period, and we have achieved a better remanufacturing strategy with lower costs.

Now let us assume that the optimal manufacturing strategy $P_m(t)$ is known. Then we can formulate two auxiliary problems. These auxiliary problems can be reformulated with the introduced function $\tilde{I}(t)$:

if $h_1 \geq h_2$

$$\int_0^T \left[(h_1 - h_2) I_1(t) + F_r(P_r(t)) \right] dt + \int_0^T \left[h_2 \tilde{I}(t) + F_m(P_m^0(t)) \right] dt \rightarrow \min$$

such that

$$\dot{I}_1(t) = P_r(t) - \left(S(t) - P_m^0(t) \right) \quad I_1(0) = I_{10},$$
$$0 \leq I_1(t) \leq \tilde{I}(t), \quad P_r(t) \geq 0,$$

and if $h_1 < h_2$

$$\int_0^T \left[(h_2 - h_1) I_2(t) + F_r(P_r(t)) \right] dt + \int_0^T \left[h_1 \tilde{I}(t) + F_m(P_m^0(t)) \right] dt \rightarrow \min$$

such that

$$\dot{I}_2(t) = -P_r(t) + rS(t), \quad I_2(0) = I_{20},$$

$$0 \leq I_2(t) \leq \tilde{I}(t), \quad P_r(t) \geq 0.$$

With these formulations we can characterize the optimal remanufacturing strategy.

Lemma 3: If $h_1 \geq h_2$ and $I_{10} - \int_0^T S(t) dt \geq 0$, then manufacturing and remanufacturing are not needed to satisfy the demand: $P_m^0(t) = 0, P_r^0(t) = 0, \quad \forall t \in [0, T]$.

The proof is obvious. If the linear holding cost in the first shop is not lower than in the second shop, then the objective function has the lowest value, if we take the possible greatest inventory level. This possible greatest value is achieved, if the rate of remanufacturing is equal to zero:

$$I_2^0(t) = I_{20} + r \int_0^t S(\tau) d\tau, \quad \forall t \in [0, T].$$

If the rate of manufacturing is equal to zero, then from Lemma 1.

$$\tilde{I}^0(t) = I_{10} + I_{20} - (1 - r) \int_0^t S(\tau) d\tau \geq 0, \quad \forall t \in [0, T].$$

Using the inequalities for the inventory levels

$$\tilde{I}^0(t) = I_{10} + I_{20} - (1 - r) \int_0^t S(\tau) d\tau \geq I_{20} + r \int_0^t S(\tau) d\tau = I_2^0(t), \quad \forall t \in [0, T],$$

which is equivalent with our proposition. This lemma shows that a firm must remanufacture in a planning period, excepted, if the inventory holding costs in the first store are higher than those in the second store, and the initial inventory level in the first shop is enough to satisfy the demand in the planning period.

Last we investigate the optimal ending inventory level for store 2.

Lemma 4: If $h_2 > h_1$ and $I_{10} - \int_0^T S(t) dt < 0$, then in the optimal trajectory:

$$I_2^0(T) = 0.$$

This lemma is not proved here. If $h_1 \geq h_2$, then the ending inventory level in the second store can be strict greater than zero.

4. Solution of the model

To solve the problem, we apply the maximum principle of Pontryagin ([12], [14]). This problem is an optimal control problem with pure state variables constraints. The conditions of optimality is described in

Theorem 1: In order for $\{I_1^0(t), I_2^0(t), P_m^0(t), P_r^0(t)\}$ to be optimal the problem (1)-(3) it is necessary and sufficient that there exist continuous functions $\psi_1(t)$ and $\psi_1(t)$, where for all $0 \leq t \leq T$ we have $\begin{bmatrix} \psi_1(t) \\ \psi_2(t) \end{bmatrix} \neq \begin{bmatrix} 0 \\ 0 \end{bmatrix}$

(a)
$$\begin{bmatrix} \dot{\psi}_1(t) \\ \dot{\psi}_2(t) \end{bmatrix} = \begin{bmatrix} h_1 \\ h_2 \end{bmatrix} - \begin{bmatrix} \lambda_1(t) \\ \lambda_2(t) \end{bmatrix},$$

(b)
$$\max_{0 \le P_m(t)} \{\psi_1(t) P_m(t) - F_m(P_m(t))\} =$$
$$\psi_1(t) P_m^0(t) - F_m\left(P_m^0(t)\right)$$
$$\max_{0 \le P_r(t)} \{(\psi_1(t) - \psi_2(t)) P_r(t) - F_r(P_r(t))\} =$$
$$(\psi_1(t) - \psi_2(t)) P_r^0(t) - F_r\left(P_r^0(t)\right)$$

(c)
$$[\lambda_1(t) \quad \lambda_2(t)] \begin{bmatrix} I_1^0(t) \\ I_2^0(t) \end{bmatrix} = 0,$$
$$[\lambda_1(t) \quad \lambda_2(t)] \ge [0 \quad 0]$$

(d)
$$[\psi_1(T) \quad \psi_2(T)] \begin{bmatrix} I_1^0(T) \\ I_2^0(T) \end{bmatrix} = 0.$$
$$[\psi_1(T) \quad \psi_2(T)] \ge [0 \quad 0]$$

From conditions (b) the optimal manufacturing and remanufacturing rates can be simplified

$$P_m^0(t) = \begin{cases} 0 & \psi_1(t) \le F_m'(0) \\ [F_m']^{-1}(\psi_1(t)) & \psi_1(t) = F_m'\left(P_m^0(t)\right) > F_m'(0) \end{cases}$$

and

$$P_r^0(t) = \begin{cases} 0 & \psi_1(t) - \psi_2(t) \le F_r'(0) \\ [F_r']^{-1}(\psi_1(t) - \psi_2(t)) & \psi_1(t) - \psi_2(t) = F_r'\left(P_r^0(t)\right) > F_r'(0) \end{cases}$$

In the next section we characterize regions in a trajectory for which various sets of conditions hold.

5. Regional specialization of the optimal trajectory

Before constructing the optimal trajectory, we give the regional specialization of the model.

Case 1.: $I_1^0(t) > 0, \quad I_2^0(t) > 0, \quad t \in [t_1, t_2]$

In this case the variables $\lambda_1(t)$ and $\lambda_2(t)$ are equal to zero. The differential equations for the adjoint variables have the following form:

$$\begin{bmatrix} \dot\psi_1(t) \\ \dot\psi_2(t) \end{bmatrix} = \begin{bmatrix} h_1 \\ h_2 \end{bmatrix}.$$

Let us now assume that manufacturing and remanufacturing rate is greater than zero. Then the optimal rates can be computed, as

$$\begin{bmatrix} \psi_1(t) \\ \psi_1(t) - \psi_2(t) \end{bmatrix} = \begin{bmatrix} F_m'\left(P_m^0(t)\right) \\ F_r'\left(P_r^0(t)\right) \end{bmatrix}.$$

If we differentiate the last equation, we have two differential equations, which solve the problem for the rates

$$\begin{bmatrix} \dot\psi_1(t) \\ \dot\psi_1(t) - \dot\psi_2(t) \end{bmatrix} = \begin{bmatrix} h_1 \\ h_1 - h_2 \end{bmatrix} = \begin{bmatrix} F_m''\left(P_m^0(t)\right)\dot P_m^0(t) \\ F_r''\left(P_r^0(t)\right)\dot P_r^0(t) \end{bmatrix},$$

or

$$\dot P_m^0(t) = \frac{h_1}{F_m''\left(P_m^0(t)\right)}, \quad \dot P_r^0(t) = \frac{h_1 - h_2}{F_r''\left(P_r^0(t)\right)}. \tag{4}$$

If the initial values at time point t_1 are known, the differential equations can be solved.

Case 2.: $I_1^0(t) = 0, \quad I_2^0(t) > 0, \quad t \in [t_1, t_2]$

In this case the variable $\lambda_1(t)$ is not less than zero and $\lambda_2(t)$ is equal to zero. The differential equations for the adjoint variables have the following form:

$$\begin{bmatrix} \dot\psi_1(t) \\ \dot\psi_2(t) \end{bmatrix} = \begin{bmatrix} h_1 - \lambda_1(t) \\ h_2 \end{bmatrix}.$$

The sum of manufacturing and remanufacturing rates is equal to the demand rate. If we assume the nonnegativity manufacturing and remanufacturing rates, then the optimal rates can be computed, as

$$
\begin{bmatrix} \psi_1(t) \\ \psi_1(t) - \psi_2(t) \end{bmatrix} = \begin{bmatrix} F_m'\big(S(t) - P_r^0(t)\big) \\ F_r'\big(P_r^0(t)\big) \end{bmatrix},
$$

where $P_m^0(t) = S(t) - P_r^0(t)$.

If we differentiate the last equation, we have two differential equations, which solve the problem for the rates

$$
\begin{bmatrix} \dot{\psi}_1(t) \\ \dot{\psi}_1(t) - \dot{\psi}_2(t) \end{bmatrix} = \begin{bmatrix} h_1 - \lambda_1(t) \\ h_1 - h_2 - \lambda_1(t) \end{bmatrix} = \begin{bmatrix} F_m''\big(S(t) - P_r^0(t)\big)\big(\dot{S}(t) - \dot{P}_r^0(t)\big) \\ F_r''\big(P_r^0(t)\big)\dot{P}_r^0(t) \end{bmatrix},
$$

or

$$
\lambda_1(t) = h_1 - F_m''\big(S(t) - P_r^0(t)\big)\big(\dot{S}(t) - \dot{P}_r^0(t)\big) = h_1 - h_2 - F_r''\big(P_r^0(t)\big)\dot{P}_r^0(t) \ge 0.
$$

From this expression we get the next differential equation for the remanufacturing rate

$$
\dot{P}_r^0(t) = \frac{F_m''\big(S(t) - P_r^0(t)\big)\dot{S}(t) - h_2}{F_m''\big(S(t) - P_r^0(t)\big) + F_r''\big(P_r^0(t)\big)} \tag{5}
$$

If the initial value at time point t_1 is known, the differential equation can be solved.

Case 3.: $I_1^0(t) > 0, \quad I_2^0(t) = 0, \quad t \in [t_1, t_2]$

In this case the variable $\lambda_2(t)$ is not less than zero and $\lambda_1(t)$ is equal to zero. The differential equations for the adjoint variables have the following form:

$$
\begin{bmatrix} \dot{\psi}_1(t) \\ \dot{\psi}_2(t) \end{bmatrix} = \begin{bmatrix} h_1 \\ h_2 - \lambda_2(t) \end{bmatrix}.
$$

The remanufacturing rates is equal to the rate of return in the second store. If we assume the nonnegativity manufacturing and remanufacturing rates, then the optimal rates can be computed, as

$$
\begin{bmatrix} \psi_1(t) \\ \psi_1(t) - \psi_2(t) \end{bmatrix} = \begin{bmatrix} F_m'\big(P_r^0(t)\big) \\ F_r'\big(rS(t)\big) \end{bmatrix}.
$$

If we differentiate the last equations, we have two differential equations, which solve the problem for the rates

$$
\begin{bmatrix} \dot{\psi}_1(t) \\ \dot{\psi}_1(t) - \dot{\psi}_2(t) \end{bmatrix} = \begin{bmatrix} h_1 \\ h_1 - h_2 + \lambda_2(t) \end{bmatrix} = \begin{bmatrix} F_m''(P_m^0(t))\dot{P}_m^0(t) \\ F_r''(rS(t))r\dot{S}(t) \end{bmatrix},
$$

or

$$
\dot{P}_m^0(t) = \frac{h_1}{F_m''(P_m^0(t))},
$$

$$
\lambda_2(t) = F_r''(rS(t))r\dot{S}(t) - (h_1 - h_2) \geq 0,
$$

$$
\dot{S}(t) \geq \frac{h_1 - h_2}{rF_r''(rS(t))}
$$

If the initial value at time point t_1 are known, the differential equation for the manufacturing rate can be solved. The condition for the remanufacturing can be checked from the demand rate.

6. A forward algorithm to construct the optimal trajectory

Before applying the forward algorithm we must check the results of Lemmas 1. and 3. Let us assume that the conditions of the lemmas do not fulfill. Then we can investigate the general case. We can not distinguish between those cases in which store the holding cost is higher, but we check the terminal inventory level for store 2.

We give a modified forward Arrow-Karlin-type algorithm to construct the optimal trajectory. The goal of the algorithm is to minimize the inventory level under the assumption that the inventory levels in the stores are not negative. There are three regions for which sets of conditions hold: the inventory status is either positive or equal to zero in store 1. or equal to zero in store 2.. These three regions succeed each other.

In the first step we can assume that it is remanufactured. Then we solve the differential equations (4) with an initial value $P_m(0) = P_{m0}$, $P_r(0) = P_{r0}$.

So the inventory levels to these initial values are defined

$$
\left.
\begin{aligned}
I_1(t) &= I_{10} + \int_0^t P_m^0(\tau, 0, P_{m0})d\tau + \int_0^t P_r^0(\tau, 0, P_{r0})d\tau - \int_0^t S(\tau)d\tau \geq 0 \\
I_2(t) &= I_{20} - \int_0^t P_r^0(\tau, 0, P_{r0})d\tau + r\int_0^t S(\tau)d\tau \geq 0
\end{aligned}
\right\} \quad \forall t \in [0, T],
$$

where function $P_m^0(t,0,P_{m0})$ and $P_r^0(t,0,P_{r0})$ are the solution of the differential equations with initial values P_{m0} and P_{r0} at point of time 0. The inventory levels will be lower, if the initial value is decreased for the manufacturing and increased for the remanufacturing. Let us push the initial value P_{m0} downstairs for manufacturing and value P_{r0} upstairs for the remanufacturing, until there exists a point of time where one of the inventory levels is zero. Let us choose the greatest time point t_1 for which the one of the inventory levels is zero, i.e. either $I_1(t_1)=0$ or $I_2(t)=0$. If this greatest point of time is the end of the planning period T, then we have achieved the optimal solution. If not, then we follow the next step, and the optimal solution for the interval $[0,t_1]$ is

$$
\left.
\begin{aligned}
&P_r^0(t)= P_r^0(t,0,P_{r0}), \quad P_m^0(t)= P_m^0(t,0,P_{m0}) \\
&I_1^0(t) = I_{10} + \int_0^t P_m^0(\tau,0,P_{m0})d\tau + \int_0^t P_r^0(\tau,0,P_{r0})d\tau - \int_0^t S(\tau)d\tau \geq 0 \\
&I_2^0(t) = I_{20} - \int_0^t P_r^0(\tau,0,P_{r0})d\tau + r\int_0^t S(\tau)d\tau \geq 0
\end{aligned}
\right\} \quad \forall t \in [0,t_1].
$$

In the second step we choose the greatest time point t_2 on the interval $[t_1,T]$ so, that either

$$
I_1(t) = \int_{t_2}^t P_m^0(\tau,t_2,S(t_2)-P_r^0(t_2))d\tau + \int_{t_2}^t P_r^0(\tau,t_2,P_r^0(t_2))d\tau - \int_{t_2}^t S(\tau)d\tau \geq 0
$$

$$
I_2(t) = I_2^0(t_2) - \int_{t_2}^t P_r^0(\tau,t_2,P_r^0(t_2))d\tau + r\int_{t_2}^t S(\tau)d\tau \geq 0
$$

or

$$
I_1(t) = I_1^0(t_1) + \int_{t_2}^t P_m^0(\tau,t_1,P_m^0(t_1))d\tau + \int_{t_2}^t P_r^0(\tau,t_2,rS(t_2))d\tau - \int_{t_2}^t S(\tau)d\tau \geq 0
$$

$$
I_2(t) = -\int_{t_2}^t P_r^0(\tau,t_2,rS(t_2))d\tau + r\int_{t_2}^t S(\tau)d\tau \geq 0
$$

and on the interval $[t_1,t_2]$ the optimal solution is in the first case

$$P_r^0(t) = P_r^0\left(t, t_1, S(t_1) - P_r^0(t_1)\right) \quad P_m^0(t) = S(t) - P_r^0(t)$$

$$I_1^0(t) = 0, \quad I_2^0(t) = I_2^0(t_1) - \int_0^t P_r^0\left(\tau, t_1, S(t_1) - P_r^0(t_1)\right) d\tau + r \int_0^t S(\tau) d\tau \geq 0,$$

where function $P_r^0\left(t, t_1, S(t_1) - P_r^0(t_1)\right)$ is the solution of the equation (5) with initial value $S(t_1) - P_r^0(t_1)$ in point of time t_1, and in the second case

$$P_r^0(t) = rS(t), \quad P_m^0(t) = P_m^0(t, 0, P_{m0})$$

$$I_1^0(t) = I_1^0(t_1) + \int_{t_2}^t P_m^0(\tau, 0, P_{m0}) d\tau - (1 - r) \int_{t_2}^t S(\tau) d\tau, \quad I_2^0(t) = 0$$

As we see, in the second step we have checked whether the inventory level is equal to zero in either the first or the second store. If one of the terminal inventory levels is zero, then the optimum is achieved. If not, we choose the greatest point of time t_3 for which one of the inventory levels is zero, and repeat the second step again. The optimal solution is on interval $[t_2, t_3]$

$$P_r^0(t) = P_r^0\left(t, t_2, P_r^0(t_2)\right) \quad P_m^0(t) = P_m^0\left(t, t_2, P_m^0(t_2)\right)$$

$$I_1^0(t) = I_1^0(t_2) + \int_0^t P_m^0\left(\tau, t_2, P_m^0(t_2)\right) d\tau + \int_0^t P_r^0\left(\tau, t_2, P_r^0(t_2)\right) d\tau - \int_0^t S(\tau) d\tau \geq 0.$$

$$I_2^0(t) = I_2^0(t_2) - \int_0^t P_r^0\left(\tau, t_2, P_r^0(t_2)\right) d\tau + r \int_0^t S(\tau) d\tau \geq 0$$

Continuing this way the algorithm we can reach the optimum.

7. Conclusions

In this paper we have solved a reverse logistics problem. Because of the convexity of manufacturing and remanufacturing costs, the necessary conditions of optimality are sufficient, as well. After characterizing the model, it was given a forward algorithm to construct the optimal trajectory. This algorithm can be considered, as a generalization of forward algorithm of Arrow and Karlin. An other result is that the behavior of reverse logistics problems is affected by inventory holding costs for new and returned products. Remanufacturing controls the optimal path of the system, and the role of manufacturing is to broaden the production facilities, if the demand is not to satisfy from the remanufactured returned items.

References

1. Schrady, D.A. (1967): A deterministic inventory model for repairable items, Naval Research Logistics Quarterly, 14, 391-398
2. Richter, K. (1996): The EOQ repair and waste disposal model with variable setup numbers, European Journal of Operational Research, 96, 313-324
3. Richter, K. (1996): The extended EOQ repair and waste disposal model, International Journal of Production Economics 45, 443-447
4. Richter, K. (1997): Pure and mixed strategies for the EOQ repair and waste disposal problem, OR Spektrum 19, 123-129
5. Richter, K., Dobos, I. (1999): Analysis of the EOQ repair and waste disposal problem with integer setup numbers, International Journal of Production Economics, 59, 463-467
6. Teunter, R. (1998): Economic ordering quantities for remanufacturable item inventory systems, Working Paper; Preprint No. 31, Faculty of Economics and Management, Otto-von-Guericke University of Magdeburg.
7. Dobos, I., Richter, K. (1999): A remanufacturing model reconsidered: A technical note, Discussion paper 128, Viadrina European University, Faculty of Economics and Business Administration.
8. Dobos, I., Richter, K. (1999): The number of batch sizes in a remanufacturing model, Discussion paper 132, Viadrina European University, Faculty of Economics and Business Administration.
9. Fleischmann, M., Bloemhof-Ruwaard, J.M., Dekker, R., van der Laan, E.A., van Nunen, J.A.E.E., van Wassenhove, L.N. (1997): Quantitative models for reverse logistics: A review, European Journal of Operational Research 103, 1-17
10. Arrow, K.J., Karlin, S. (1958): Production over time with increasing marginal costs, In: K.J. Arrow, S. Karlin, H.Scarf (Eds.): Studies in the mathematical theory of inventory and production, Stanford Univ. Press, Stanford, 61-69
11. Stöppler, S. (1985): Der Einfluss der Lagerkosten auf die Produktionsanpassung bei zyklischem Absatz, OR-Spektrum 7, 129-142
12. Feichtinger, G., Hartl, R.F. (1986): Optimale Kontrolle ökonomischer Prozesse: Anwendungen des Maximumprinzips in den Wirtschaftswissenschaften, de Gruyter, Berlin
13. Dobos, I. (1990): The Modigliani-Hohn model with capacity and warehousing constraints, International Journal of Production Economics 24, 49-54
14. Mangasarian, O.L. (1966): Sufficient conditions for the optimal control of nonlinear systems, Journal SIAM Control 4, 139-153

Optimal Production Planning in Stochastic Jobshops with Long-Run Average Cost

Ernst Presman[1], Suresh P. Sethi[2], Hanqin Zhang[3]

[1] Central Economics and Mathematics Institute, The Russian Academy of Sciences, Moscow, Russia
[2] School of Management, The University of Texas at Dallas, TX 75083, USA
[3] Institute of Applied Mathematics, Academia Sinica, Beijing, 100080, China

Abstract. We consider a production planning problem for a general jobshop producing a number of products and subject to breakdown and repair of machines. The machine capacities are modeled as Markov chains. The objective is to choose the rates of production of the final products and intermediate parts on the various machines over time in order to meet the demand for the system's production at the minimum long-run average cost of production and surplus. The problem is formulated as a stochastic dynamic program. We prove a verification theorem and derive the optimal feedback control policy in terms of the directional derivatives of the so-called potential function. Finally, we construct a potential function in the special case of a jobshop producing only one final product.

1   Introduction

We consider the problem of a stochastic manufacturing system producing a variety of products in demand using machines in a general network configuration, which generalizes both the parallel and the tandem machine models. Each product follows a process plan, possibly out of a number of alternative process plans, that specifies the sequence of machines it must visit and the operations performed by them. A process plan may call for multiple visits to a given machine, as is the case in semiconductor manufacturing (Lou and Kager (1989), Srivatsan et al. (1994), Uzsoy et al. (1992, 1993)). The stochastic nature of the system is due to the machines that are failure-prone. The machine state capacities are assumed to be finite-state Markov chains. The problem we are interested in is that of obtaining the optimal feedback rates of production of intermediate parts and finished products in a manufacturing system consisting of a network of failure-prone machines. The objective is to meet demand for finished products at the minimum long-run average cost of production, inventories, and backlogs.

There has been a good deal of work related to the problems considered here. Sethi and Zhou (1994) and Presman et al. (1997) obtained the optimal control policy for such problems in the context of the discounted cost criterion.

For the long-run average cost criterion, Sethi *et al.* (1997) and Sethi *et al.* (1998) dealt with the case of parallel machine systems producing single or multiple products by using the vanishing discount approach. Extensions of these analyses to a flowshop consisting of two or more machines together must recognize the complications that arise due to the presence of internal buffers. The existence of internal buffers gives rise to state constraints and certain boundary conditions to be taken into account when using the vanishing discount approach. This was accomplished by Presman *et al.* (1999) in their analysis of the optimal control problem of an N-machine flowshop under the long-run average cost criterion.

Our purpose in this paper is to extend the results of Presman *et al.* (1999) to a stochastic jobshop problem. We use a vanishing discount approach to address the average-cost problem under consideration. Two major contributions are made in order to implement the vanishing discount approach. The first one constructs a control policy which takes any given system state to any other state in a time whose rth moment has a finite expectation in the case when the jobshop produces only one final product. The second one obtains a solution of the Hamilton-Jacobi-Bellman (HJB) equation for the problem in terms of directional derivatives by a limit procedure for the discounted cost problem as the discount rate tends to zero.

The plan of this paper is as follows. In the next section, we present a graph-theoretic representation of jobshops, analyze its structure, and give the system dynamics equations, and formulate four theorems representing the main results of the paper. In Section 3, we prove one of the theorems formulated in Section 2, which establishes the result needed for making use of the vanishing discount approach to analyze the average cost minimization problem under consideration. In Section 4, we conclude the paper.

2   Problem Formulation and Main Results

Sethi and Zhou (1994) defined a class of digraphs that are of interest, in the sense that any digraph in the class would correspond to the dynamics and the state constraints of a jobshop manufacturing system. The readers unfamiliar with the relevant graph theory concepts may consult Appendix I in Sethi and Zhang (1994).

Definition 2.1. A *manufacturing digraph* is a graph (V, A), where V is a set of $N + 2$, $N \geq 1$, *vertices* and A is a set of ordered pairs called arcs, satisfying the following properties:

(i) There is only one source, labeled 0, and only one sink, labeled $N + 1$, in the digraph.

(ii) No vertex in the graph is isolated.

(iii) The digraph does not contain any cycle.

Remark 2.1. Condition (ii) is not an essential restriction. Inclusion of isolated vertices is merely a nuisance. This is because an isolated vertex is like a warehouse that can only ship out parts of a particular type to meet their demand. Since no machine (or production) is involved, its inclusion or exclusion does not affect the optimization problem under consideration. Condition (iii) is imposed to rule out the following two trivial situations: (a) a part of type i in buffer i gets processed on a machine without any transformation and returns to buffer i, and (b) a part of type i is processed and converted back into a part of type j, $j \neq i$, and is then processed further on a number of machines to be converted back into a part of type i. Moreover, if we had included any cycle in our manufacturing system, the flow of parts that leave buffer i only to return to buffer i would be zero in any optimal solution. It is unnecessary, therefore, to complicate the problem by including cycles.

Definition 2.2. In a manufacturing digraph, the source is called the supply node, the sink represents the customers. Vertices immediately preceding the sink are called *external buffers*, and all others are called *internal buffers*.

In order to obtain the system dynamics from a given manufacturing digraph, a systematic procedure is required to label the state and control variables. For this purpose, let us suppose that our manufacturing digraph contains a total of $N + 2$ vertices including the source, the sink, m internal buffers, and $N - m$ external buffers for some integer m and N, $0 \leq m \leq N - 1$, $N \geq 1$.

Theorem 2.1. *We can label all the vertices from 0 to $N + 1$ in a way so that the label numbers of the vertices along every path are in a strictly increasing order, the source is labeled 0, the sink is labeled $N + 1$, and the external buffers are labeled $m + 1$, $m + 2$,...,N.*

Proof. The proof is similar to Theorem 2.2 in Sethi and Zhou (1994), and is omitted. $\square$

With the help of Theorem 2.1, one is able to formally write the dynamics and the state constraints associated with a given manufacturing digraph. But first let us give a few definitions.

Definition 2.3. For each arc (i, j), $j \neq N + 1$, in a manufacturing digraph, the rate at which parts in buffer i are converted to parts in buffer j is labeled as *control u_{ij}*. Moreover, the control u_{ij} associated with the arc (i, j) is called an *output* of i and an *input* to j. In particular, outputs of the source are called *primary controls* of the digraph. For each arc $(i, N + 1)$, $i = m + 1, ..., N$, the demand for products in buffer i is denoted by d_i.

In what follows, we shall also set

$$u_{i,N+1} = d_i, \ i = m + 1, ..., N,$$
$$u_{i,j} = 0, \text{ for } (i,j) \notin A, \ 0 \leq i \leq N, \ 1 \leq j \leq N + 1,$$

for a unified notation suggested in Presman *et al.* (1997). While d_i and $u_{i,j}$ for $(i, j) \notin A$ are not controls, we shall for convenience refer to $u_{i,j}$, $0 \leq i \leq N$, $0 \leq j \leq N + 1$, as controls. In this way, we can consider the controls as an

$(N + 1) \times (N + 1)$ matrix $\boldsymbol{u} = (u_{ij})$. The set of all such controls is written as $\mathcal{U}$, i.e.,

$$\mathcal{U} = \{\boldsymbol{u} = (u_{ij}) : 0 \leq i \leq N, 1 \leq j \leq N + 1, u_{ij} = 0 \text{ for } (i, j) \notin A\}.$$

Now we shall write the dynamics and the state constraints corresponding to a manufacturing digraph (V, A) containing $N + 2$ vertices consisting of a source, a sink, m internal buffers, and $N - m$ external buffers associated with the $N - m$ distinct final products to be manufactured. We label all the vertices according to Theorem 2.1. For simplicity in the sequel, we shall call the buffer whose label is i as buffer i, $i = 1, 2, ..., N$. We denote the surplus at time t in buffer i by $x_i(t)$, $i \in V \setminus \{0, N + 1\}$, and the control at time t associated with arc (i, j) by $u_{i,j}(t)$, $(i, j) \in A$. Note that if $x_i(t) > 0$, $i = 1, ..., N$, we have an inventory in buffer i, and if $x_i(t) < 0$, $i = m + 1, ..., N$, we have a shortage of finished product i. The dynamics of the system are therefore

$$\left\{ \begin{array}{ll} \dot{x}_i(t) = \sum_{\ell=0}^{i-1} u_{\ell i}(t) - \sum_{\ell=i+1}^{N} u_{i\ell}(t), & 1 \leq i \leq m, \\ \dot{x}_i(t) = \sum_{\ell=0}^{m} u_{\ell i}(t) - d_i = \sum_{\ell=0}^{m} u_{\ell i}(t) - u_{i,N+1}(t), & m + 1 \leq i \leq N \end{array} \right. \quad (1)$$

with $\boldsymbol{x}(0) := (x_1(0), ..., x_N(0)) = (x_1, ..., x_N) = \boldsymbol{x}_0$. The state constraints are

$$x_i(t) \geq 0, \ t \geq 0, \ i = 1, ..., m$$
$$-\infty < x_i(t) < +\infty, \ t \geq 0, \ i = m + 1, ..., N. \quad (2)$$

Let

$$\boldsymbol{u}_\ell(t) = (u_{\ell,\ell+1}(t), ..., u_{\ell,N}(t))', \ \ell = 0, ..., m,$$

and

$$\boldsymbol{u}_{m+1}(t) = (d_{m+1}, ..., d_N)'.$$

The relation (1) can be written in the following vector form:

$$\begin{array}{rl} \dot{\boldsymbol{x}}(t) & = (\dot{x}_1(t), ..., \dot{x}_N(t))' \\ & = B(\boldsymbol{u}_0(t), ..., \boldsymbol{u}_{m+1}(t))', \end{array}$$

where $B : R^J \to R^N$ is the corresponding linear operator with $J = (N - m) + \sum_{\ell=0}^{m}(N - \ell)$. Let $S = R_+^m \times R^{N-m}$. Furthermore, let S^b be the boundary of S, and $S^o = S \setminus S^b$. Now we introduce the control constraints. The control constraints depend on the placement of the machines, and the different placements on the same digraph will give rise to different jobshops. In other words, a jobshop corresponds to a unique digraph, whereas a digraph may correspond to many different jobshops. Therefore, to uniquely characterize a jobshop using graph theory, we need to introduce the concept of a placement of machines, or simply a placement. Let $m_c \leq r - N + m$, where r denotes the total number of arcs in A.

Definition 2.4. In a manufacturing digraph (V, A), a set $\mathcal{K} = \{K_1, K_2, .., K_{n_c}\}$ is called a *placement of machines* $1, 2, ..., n_c$, if $\mathcal{K}$ is a partition of $B = \{(i, j) \in A : j \neq N+1\}$, namely, $\emptyset \neq K_n \subset B$, $K_n \cap K_\ell = \emptyset$ for $n \neq \ell$, and $\cup_{k=1}^{n_c} K_k = B$.

A dynamic jobshop can be uniquely specified by a triple $(V, A, \mathcal{K})$, which denotes a manufacturing system that corresponds to a manufacturing digraph (V, A) along with a placement of machines $\mathcal{K} = (K_1, K_2, ..., K_{n_c})$.

Consider a jobshop $(V, A, \mathcal{K})$, where the dynamics of (V, A) are given in (1) and $\mathcal{K} = (K_1, K_2, ..., K_{n_c})$. Suppose we are given a stochastic process $m(t) = (m_1(t), ..., m_{n_c}(t))$, with $m_n(t)$ representing the capacity of the nth machine at time t, $n = 1, ..., n_c$. Therefore, the controls $u_{ij}(t)$ with $(i, j) \in K_n$, $n = 1, ..., n_c$, $t \geq 0$, in (1) should satisfy the following constraints:

$$0 \leq \sum_{(i,j) \in K_n} u_{ij}(t) \leq m_n(t) \text{ for all } t \geq 0, \ n = 1, ..., n_c, \tag{3}$$

where we have assumed that the required machine capacity p_{ij} (for unit production rate of type j from part type i) equals 1, for convenience in exposition. Moreover, the corresponding surplus process satisfies the constraints (2). The analysis in this paper can be readily extended to the case when the required machine capacity for the unit production rate of part j from part i is any given positive constants.

We are now in the position to formulate our stochastic optimal control problem for the jobshop defined by (1)-(3). For $m = (m_1, ..., m_{n_c})$, let

$$U(m) = \{u = (u_{ij}) : u \in \mathcal{U}, 0 \leq \sum_{(i,j) \in K_n} u_{ij} \leq m_n, \ 1 \leq n \leq n_c,$$

$$u_{i,N+1} - d_i, \ m + 1 \leq i \leq N\},$$

and for $x \in S$ and m,

$$U(x, m) = \left\{ u : u \in U(m) \text{ and } x_n = 0 \Rightarrow \sum_{i=0}^{n-1} u_{in} - \sum_{i=n+1}^{N} u_{ni} \geq 0, n = 1, ..., m \right\}.$$

Let the stochastic process $m(\cdot) = (m_1(\cdot), ..., m_{n_c}(\cdot))$ defined on a probability space $(\Omega, \mathcal{F}, P)$, denote the machine capacity-demand process.

Definition 2.5. We say that a control $u(\cdot) \in \mathcal{U}$ is admissible with respect to the initial state vector $x^0 = (x_1^0, \cdots, x_N^0) \in S$ and $m \in \mathcal{M}$ if

(i) $u(\cdot)$ is an $\mathcal{F}_t$-adapted measurable process with $\mathcal{F}_t = \sigma\{m(s) : 0 \leq s \leq t\}$;

(ii) $u(t) \in U(m(t))$ for all $t \geq 0$;

(iii) the corresponding state process $x(t) = (x_1(t), \cdots, x_N(t)) \in S$ for all $t \geq 0$.

The problem is to find an admissible control $u(\cdot)$ that minimize the long-ran average cost function

$$J(x^0, m^0, u(\cdot)) = \limsup_{T \to \infty} \frac{1}{T} E \int_0^T h(x(t), u(t)) dt, \tag{4}$$

where $h(\cdot,\cdot)$ defines the cost of surplus and production and m^0 is the initial value of $m(t)$. We impose the following assumptions on the process $m(t) = (m_1(t),\cdots,m_{n_c}(t))$ and the cost function $h(\cdot,\cdot)$ throughout this paper.

(A.1) Let $\mathcal{M} = \{m^1,\cdots,m^p\}$ for some integer $p \geq 1$, where $m^j = (m_1^j,\cdots,m_{n_c}^j)$, with m_n^j, $n = 1,\cdots,n_c$ denoting the capacity of the nth machine, $j = 1,\cdots,p$. The capacity process $m(t) \in \mathcal{M}$ is a finite state Markov chain with the following infinitesimal generator Q:

$$Qf(m) = \sum_{m' \neq m} q_{m'm}[f(m') - f(m)]$$

for some $q_{m'm} \geq 0$ and any function $f(\cdot)$ on $\mathcal{M}$. Moreover, the Markov process is strongly irreducible and has the stationary distribution p_{m^j}, $j = 1,\cdots,p$.

(A.2) Let $p_n = \sum_{j=1}^p m_n^j p_{m^j}$, $n = 1, ..., n_c$, and $n(i,j) = \arg\{(i,j) \in K_n\}$ for $(i,j) \in A$. Here p_n represents the average capacity of the machine n, and $n(i,j)$ is the machine number placed the arc (i,j). Assume that there exist $\{p_{ij} > 0 : (i,j) \in K_n\}(n = 1, ..., n_c)$ such that

$$\sum_{(i,j)\in K_n} p_{ij} \leq 1, \tag{5}$$

$$\sum_{\ell=0}^m p_{\ell i} p_{n(\ell,i)} > d_i, \quad i = m+1, ..., N, \tag{6}$$

$$\sum_{\ell=0}^{i-1} p_{\ell i} p_{n(\ell,i)} > \sum_{\ell=i+1}^N p_{i\ell} p_{n(i,\ell)}, \quad i = 1, ..., m. \tag{7}$$

(A.3) $h(\cdot,\cdot)$ is a non-negative, jointly convex function that is strictly convex in either x or u or both. There exist constants C_0 and $K_h \geq 1$ such that for all $x, x' \in S$ and $u, u' \in U(m^j)$, $j = 1, ..., p$,

$$|h(x,u) - h(x',u')| \leq C_0(1 + |x|^{K_h} + |x'|^{K_h})(|x - x'| + |u - u'|).$$

We use $\mathcal{A}(x^0,m^0)$ to denote the set of all admissible controls with respect to $x^0 \in S$ and $m(0) = m^0$. Let $\lambda(x^0,m^0)$ denote the minimal expected cost, i.e.,

$$\lambda(x^0,m^0) = \inf_{u(\cdot)\in\mathcal{A}(x^0,m^0)} J(x^0,m^0). \tag{8}$$

Consider now the following equation:

$$\lambda = \inf_{u\in U(x,m)} \{\partial_{Bu} f(x,m) + h(x,u)\} + Qf(x,\cdot)(m), \tag{9}$$

where λ is a constant, $f(\cdot,\cdot)$ is a convex function, and $\partial_r g(x)$ denotes the directional derivative of function $g(x)$ along the direction $r \in R^N$. We have the following verification theorem.

Theorem 2.2. *Assume* (i) $(\lambda, f(\cdot, \cdot))$ *with* $f(\cdot, \cdot)$ *being continuous convex satisfies* (9); (ii) *there exists* $u^*(x, m)$ *for which*

$$\inf_{u \in U(x,m)} \{\partial_{Au} f(x, m) + h(x, u)\} = \partial_{Au^*(x,m)} f(x, m) + h(x, u^*(x, m)),$$

and the equation $\dot{x}(t) = Au^*(x(t), m(t))$ *has for any initial condition* $(x^*(0), m(0)) = (x^0, m^0)$, *a solution* $x^*(t)$ *such that*

$$\lim_{T \to \infty} \frac{Ef(x^*(T), m(T))}{T} = 0.$$

Then $u^*(t) = u^*(x^*(t), m(t))$ *is an optimal control. Furthermore,* $\lambda(x^0, m^0)$ *does not depend on* x^0, m^0 *and coincides with* λ. *Moreover, for any* $T > 0$

$$
\begin{aligned}
f(x^0, m^0) &= \inf_{u(\cdot) \in \mathcal{A}(x^0, m^0)} E\left[\int_0^T (h(x(t), u(t)) - \lambda)\, dt + f(x(T), m(T))\right] \\
&= E\left[\int_0^T (h(x^*(t), u^*(t)) - \lambda)\, dt + f(x^*(T), m(T))\right].
\end{aligned}
$$

Note that Theorem 2.2 explains why equation (9) is the HJB equation for our problem and why the function $f(\cdot, \cdot)$ from Theorem 2.2 is called a potential function. The proof of Theorem 2.2 is standard, for a proof, see Presman *et al.* (1999).

Our goal is to construct a pair of $(\lambda, W(\cdot, \cdot))$ which satisfies (9). We are able to do this only in the case of one final product, i.e., when $m = N - 1$. To get the pair $(\lambda, W(., .))$, we use the vanishing discount approach. Consider a corresponding control problem with the cost discounted at the rate ρ. For $u(\cdot) \in \mathcal{A}(x^0, m^0)$, we define the expected discounted cost as

$$J^\rho(x^0, m^0, u(\cdot)) = E \int_0^\infty e^{-\rho t} h(x(t), u(t)) dt.$$

Define the value function of the discounted cost problem as

$$V^\rho(x^0, m^0) = \inf_{u(\cdot) \in \mathcal{A}(x^0, m^0)} J^\rho(x^0, m^0, u(\cdot)).$$

Theorem 2.3. *Assume that* $m = N - 1$. (i) *There exists a sequence* $\{\rho_k : k \geq 1\}$ *with* $\rho_k \to 0$ *as* $k \to \infty$ *such that for* $(x, m) \in \mathcal{S} \times \mathcal{M}$:

$$\lim_{k \to \infty} \rho_k V^{\rho_k}(x, m) = \lambda,$$

$$\lim_{k \to \infty} [V^{\rho_k}(x, m) - V^{\rho_k}(0, m^0)] = W(x, m),$$

where $W(x, m)$ *is convex in* x *for any given* m. (ii) *the pair* $(\lambda, W(\cdot, \cdot))$ *satisfies* (9) *in* $\mathcal{S}^0$. (iii) $\lambda(x^0, m^0)$ *does not depend on* (x^0, m^0).

Remark 2.2. Condition (A.2) is not needed in the discounted case. It is clear that this condition is necessary for the finiteness of the long-run average cost in the case when $h(\cdot, \cdot)$ tends to $+\infty$ as $x_N \to -\infty$. Theorem 2.3 states in particular that this condition is also sufficient.

To prove this theorem we need the following result, which is also of independent interest.

Theorem 2.4. *Assume that $m = N - 1$. For any $(x^0, m^0) \in \mathcal{S} \times \mathcal{M}$ and $(y, m') \in \mathcal{S} \times \mathcal{M}$, there exists a control policy $u(t), t \geq 0$ such that for any $r \geq 1$*

$$E\eta^r \leq C_1(r)\left(1 + \sum_{k=1}^{N} |x_k^0 - y_k|^r\right), \qquad (10)$$

where

$$\eta = \inf\{t \geq 0 : x(t) = y, \, m(t) = m'\},$$

and $x(t), t \geq 0$, is the surplus process corresponding to the control policy $u(t)$ and the initial condition $(x(0), m(0)) = (x^0, m^0)$.

Based on Theorem 2.4, the proof of (i) in Theorem 2.3 is analogous to the proof of the respective theorem, Theorem 2.2, in Presman *et al.* (1999) and will be omitted here. Note that the statement (iii) of Theorem 2.3 follows from Theorem 2.4. Statement (ii) in Theorem 2.3 follows from Theorem 2.4 and a simple limiting procedure, which is similar to Theorem 2.3 in Presman *et al.* (1999), here we omit the proof. Proof of Theorem 2.4 is given in the following section. In the next section we will see that the proof of the theorem is much more technique than the proof for the flowshop case.

3 Proof of Theorem 2.4

The proof of the theorem is divided into several steps, Here we give the outline of the proof. In order to find η, we alternate between these two policies described below. In the first policy, the production rate for each input is the maximum admissible capacity. In the second policy, we set $u_{0j}(t) = 0$, $\forall j$, and set the maximum possible production rate for all outputs that are not connected to the external buffers, under the restriction that the content of each buffer k, $1 \leq k \leq N - 1$, is not less than y_k. The first policy is used until such time when the content of the first buffer exceeds the value y_1 and the content of each buffer k, $2 \leq k \leq N$, exceeds the value $a + y_k$ for some $a > 0$. At that time we switch to the second policy. We use the second policy until such time when the content of the last buffer drops to the level y_N. After that we revert to the first policy, and so on. Using this alternating procedure, it is possible to specify η and provide an estimate for it.

To prove the theorem we need the following result concerning the difference between a Markov process and its mean.

Step 1. In this step we construct a family of auxiliary processes $x^0(t|s, x)$, $t \geq s \geq 0$, and $x \in S$. The function $u^0(x, m) = \{u^0_{i,j}(x, m) : (i, j) \in A\}$ is given by

$$u^0_{0\ell}(x, m) = p_{0\ell} m_{n(0,\ell)},$$

$$u^0_{ij}(x, m) = \begin{cases} p_{ij} m_{n(i,j)}, & \text{if } x_i > 0 \\ \dfrac{\left(\left(\sum_{\ell=0}^{i-1} u^0_{\ell i}(x, m)\right) \bigwedge \left(\sum_{\ell=i+1}^{N} p_{i\ell} m_{n(i,\ell)}\right)\right) p_{ij} m_{n(i,j)}}{\sum_{\ell=i+1}^{N} p_{i\ell} m_{n(i,\ell)}}, & \text{if } x_i = 0 \end{cases}$$

for $i = 1, ..., N - 1$. We define $x^0(t|s, x)$ as the process which satisfies the following equation (see (1)):

$$\dot{x}^0_i(t|s, x) = \sum_{\ell=0}^{i-1} u^0_{\ell i}(x^0(t|s, x), m(t)) - \sum_{\ell=i+1}^{N} u^0_{i\ell}(x^0(t|s, x), m(t)),$$
$$1 \leq i \leq N - 1,$$
$$\dot{x}^0_N(t) = \sum_{\ell=0}^{N-1} u^0_{\ell N}(x^0(t|s, x), m(t)) - d,$$

with $x^0(s|s, x) = x$. Clearly $x^0(t|s, x) \in S$ for all $t \geq s$. For fixed s, $x^0(t|s, x)$ is the state of the system with the production rate which is obtained by using the maximum admissible modified capacity at each machine. Define now the Markov time

$$\tau(s, x) = \inf\{t \geq s : x^0_1(t|s, x) \geq y_1, \ x^0_k(t|s, x) \geq a + y_k, \ k = 2, \cdots, N\}, \quad (11)$$

where $a > 0$ is a constant specified later. It follows from this definition that $\tau(s, x)$ is the first time when the state process $x^0(t|s, x)$ exceeds $(y_1, a + y_2, \cdots, a + y_N)$ under the production rate $u^0(x^0(t|s, x), m(t))$. From Assumption (A.2), the following lemma holds.

Lemma 3.1. *There exists a constant $C_2 = C_2(r)$ such that*

$$E\left(\tau(s, x) - s\right)^{2r} < C_2 \left(1 + \sum_{k=1}^{N} \left((y_k - x_k)^+\right)^r\right)^2.$$

We will prove this lemma later in Step 4.

Step 2. In this step we construct a family of auxiliary processes $x^1(t|s, x)$, $t \geq s \geq 0$ and $x \in S$. Consider the following function $u^1(x, m) = (u^1_{ij}(x, m)$, $(i, j) \in A\}$, which is defined only for x such that $x_i \geq y_i$ with $1 \leq i \leq N - 1$:

$$u^1_{0\ell}(x, m) = 0,$$

$$u^1_{ij}(x, m) = \begin{cases} p_{ij} m_{n(i,j)}, & x_i > y_i, \\ \dfrac{\left(\left(\sum_{\ell=0}^{i-1} u^1_{\ell i}(x, m)\right) \bigwedge \left(\sum_{\ell=i+1}^{N} p_{i\ell} m_{n(i,\ell)}\right)\right) p_{ij} m_{n(i,j)}}{\sum_{\ell=i+1}^{N} p_{i\ell} m_{n(i,\ell)}}, & x_i = y_i, \end{cases}$$

for $i = 1, ..., N-1$. We define $x^1(t|s, x)$ as a continuous process which coincides with $x^0(t|s, x)$ for $s \le t \le \tau(s, x)$, and which satisfies the following equation (see (1)):

$$\dot{x}_i^1(t|s, x) = \sum_{\ell=0}^{i-1} u_{\ell i}^1(x^1(t|s, x), m(t)) - \sum_{\ell=i+1}^{N} u_{i\ell}^1(x^1(t|s, x), m(t)),$$
$$1 \le i \le N - 1,$$
$$\dot{x}_N^1(t) = \sum_{\ell=0}^{N-1} u_{\ell N}^1(x^1(t|s, x), m(t)) - d,$$

when $t \ge \tau(s, x)$. Clearly $x^1(t|s, x) \in S$ for all $t \ge s$, and $x_i^1(t|s, x) \ge y_i$ $(1 \le i \le N - 1)$ for $t \ge \tau(s, x)$. This process corresponds to a policy in which after $\tau(s, x)$ we set $u_{0j}(t) = 0$ for any j, and set the maximum possible production rate for all other outputs that are not connected to the external buffers, under the restriction that the content of each buffer k, $1 \le k \le N - 1$, is not less then y_k.

We define now a Markov time

$$\theta(s, x) = \inf\{t \ge \tau(s, x) : x_N^1(t|s, x) = y_N\}. \tag{12}$$

Lemma 3.2. (i) *A constant a can be chosen in such a way that for all s, x*

$$P\left(x^1(\theta(s, x)|s, x) = y, \ m(\theta(s, x)) = m'\right) \ge 1 - q > 0.$$

(ii) *There exists a constant C_3 such that*

$$\frac{a}{d} \le \theta(s, x) - s \le \frac{1}{d}\left(\sum_{k=1}^{N} x_k - \sum_{k=1}^{N} y_k + C_3(\tau(s, x) - s)\right), \tag{13}$$

$$\sum_{k=1}^{N} x_k^1(\theta(s, x)|s, x) \le \sum_{k=1}^{N} x_k + C_3(\tau(s, x) - s). \tag{14}$$

This lemma will be proved in Step 5.

Step 3. Now we construct a process $x(t)$ ($t \ge 0$) and the corresponding control policy $u(t)$, which satisfies to the statement of Theorem 2.4.

Define a sequence of Markov times $(\theta_i)_{i=0}^{\infty}$ and the process $x(t)$ for $\theta_i \le t < \theta_{i+1}$ ($i = 1, 2, \cdots$) as follows: $\theta_0 = 0$, $\theta_1 = \theta(0, x^0)$ and $x(t) = x^1(t|0, x^0)$ with $0 \le t \le \theta_1$. If θ_i is defined for $i \ge 1$ and $x(t)$ is defined for $0 \le t \le \theta_i$, then we let $\tau_i = \tau(\theta_i, x(\theta_i))$, $\theta_{i+1} = \theta(\theta_i, x(\theta_i))$, and $x(t) = x^1(t|\theta_i, x(\theta_i))$ with $\theta_i \le t \le \theta_{i+1}$. According to the left inequality in (13), the process $x(t)$ is defined for all $t \ge 0$. The control policy corresponding to the process $x(t)$ is given by

$$u(t) = \begin{cases} u^0(x(t), m(t)), & \text{if } \theta_{i-1} \le t < \tau_i \\ u^1(x(t), m(t)), & \text{if } \tau_i \le t < \theta_i, \end{cases} \quad i = 1, 2 \cdots. \tag{15}$$

It is clear that this $u(t) \in \mathcal{A}(x^0, m^0)$.

For the process $x(t)$, a Markov time is defined as

$$\eta = \inf\{t \geq 0 : x(t) = y, \, m(t) = m'\}.$$

Let $\mathcal{S}_i = \{\omega : x(\theta_i) = y, m(\theta_i) = m'\}$. Using conditional probabilities, we have from Lemma 3.2 (i) that

$$P\left(\cap_{l=1}^{i} \mathcal{S}_l^c\right) \leq q^i, \quad i = 1, 2, \cdots. \tag{16}$$

Using (16) and the definition of $x(t)$ we get:

$$\eta^r = \sum_{i=1}^{\infty} \theta_i^r I_{\{\cap_{l=1}^{i-1} \mathcal{S}_l^c \cap \mathcal{S}_i\}}, \quad a.s., \tag{17}$$

where $\mathcal{S}_0^c = \Omega$. Using (13) and (14) we have for $n = 1, 2, \cdots$,

$$\theta_n - \theta_{n-1} \leq \frac{1}{d}\left(\sum_{k=1}^{N} x_k(\theta_{n-1}) - \sum_{k=1}^{N} y_k + C_3(\tau_n - \theta_{n-1})\right), \tag{18}$$

$$\sum_{k=1}^{N} x_k(\theta_n) \leq \sum_{k=1}^{N} x_k(\theta_{n-1}) + C_3(\tau_n - \theta_{n-1}). \tag{19}$$

Using (18) and (19) we have for $i = 1, 2, \cdots$,

$$\theta_i \leq \frac{1}{d}\left(\sum_{k=1}^{N}(x_k^0 - y_k)^+ + C_3\sum_{n=1}^{i}(\tau_n - \theta_{n-1})\right),$$

or

$$\theta_i^r \leq C_4 i^r \left(\sum_{k=1}^{N}\left((x_k^0 - y_k)^+\right)^r + \sum_{n=1}^{i}(\tau_n - \theta_{n-1})^r\right). \tag{20}$$

Note now that $x(\theta_n) \geq y$ for $n = 1, 2, \cdots$. Using the Schwarz inequality (Corollary 3 in page 106 of Chow and Teicher (1988)) we get from (16) and Lemma 3.1 that for $i \geq 1$

$$E\left(\tau_1^r I_{\{\cap_{l=1}^{i-1} \mathcal{S}_l^c \cap \mathcal{S}_i\}}\right) \leq q^{(i-1)/2}\left(E(\tau_1^{2r})\right)^{1/2}$$

$$\leq C_2(r)q^{(i-1)/2}\left(1 + \sum_{k=1}^{N}\left((y_k - x_k^0)^+\right)^r\right), \tag{21}$$

and for $2 \leq n \leq i = 2, 3, \ldots,$

$$E\left((\tau_n - \theta_{n-1})^r I_{\{\cap_{l=1}^{i-1} \mathcal{S}_l^c \cap \mathcal{S}_i\}}\right) \leq q^{(i-1)/2}\left(E((\tau_n - \theta_{n-1})^{2r})\right)^{1/2}$$

$$\leq C_2(r)q^{(i-1)/2}. \tag{22}$$

Substituting (20) into (17), taking expectation, and using (21) and (22) we get (10).

Step 4. Proof of Lemma 3.1. For the simplicity of exposition, we will write $\tau, \theta, x^r(t)$ and $u^r(t)(r = 0, 1)$ instead of $\tau(s, x), \theta(s, x), x^r(t|s, x)$ and $u^r(x^r(t|s, x), m(t))$ respectively, in the proofs of this lemma and Lemma 3.2.

First we sketch the idea of the proof. Lemma 3.1 follows from the fact that $P(\tau - s > t)$ decreases exponentially as t increases. To prove this fact, we divide the interval $(s, s + t)$ into two parts $(s, s + \varepsilon t)$ and $(s + \varepsilon t, s + t)$ in such a way that the behavior of all coordinates of $x^0(\cdot)$ is defined mainly by the behavior on the interval $(s + \varepsilon t, s + t)$. Using Lemma 3.3 in Presman *et al.* (1999), we show that all coordinates are positive on the interval $(s + \varepsilon t, s + t)$ with probability which exponentially tends to 1. If all coordinates are positive on the interval $(s + \varepsilon t, s + t)$, then $u^0(\cdot) = m(\cdot)$ on this interval, and using Lemma 3.3 in Presman *et al.* (1999) we can estimate a shift of process $x^0(\cdot)$.

Let $\overline{m}_k = \max_{1 \le j \le p} m_k^j$, $k = 1, \cdots, n_c$. Furthermore, let $\alpha_i = \sum_{\ell=0}^{i-1} p_{\ell i} p_{n(\ell, i)}$, $i = 1, ..., N$, $\beta_j = \sum_{\ell=j+1}^{N} p_{j\ell} p_{n(j,\ell)}$, $j = 1, ..., N - 1$, and $\beta_N = d$. We can choose $\varepsilon > 0$ such that

$$(\alpha_k - \beta_k)(1 - \varepsilon) - \varepsilon \sum_{\ell=k+1}^{N} p_{k\ell} \overline{m}_{n(k,\ell)} =: b_k > 0,$$

for all $1 \le k \le N - 1$, and $(\alpha_N - \beta_N)(1 - \varepsilon) - \varepsilon d =: b_N > 0$. Let $a_1 = 0$ and $a_k = a$ for $2 \le k \le N$. By the definition of τ,

$$P(\tau - s > t) \le \sum_{k=1}^{N} P(x_k^0(s + t) < a_k + y_k) \le P\left(\inf_{s+\varepsilon t \le v \le s+t} x_1^0(v) = 0\right)$$

$$+ \sum_{k=1}^{N-2} P\left(\cap_{i=1}^{k} \left\{\inf_{s+\varepsilon t \le v \le s+t} x_i^0(v) > 0\right\} \cap \left\{\inf_{s+\varepsilon t \le v \le s+t} x_{k+1}^0(v) = 0\right\}\right)$$

$$+ \sum_{k=1}^{N} P\left(\cap_{i=1}^{N-1} \left\{\inf_{s+\varepsilon t \le v \le s+t} x_i^0(v) > 0\right\} \cap \left\{x_k^0(s + t) < a_k + y_k\right\}\right). \quad (23)$$

First we estimate the first term on the right-hand side of (23). By $u_{1\ell}^0(v) \le p_{1\ell} \overline{m}_{n(1,\ell)}(\ell = 2, ..., N)$, it follows from Lemma 3.3 in Presman *et al.* (1999) that

$$P\left(\inf_{s+\varepsilon t \le v \le s+t} x_1^0(v) = 0\right)$$

$$\le P\left(\inf_{\varepsilon t \le v \le t} \int_s^{s+v} \left[p_{01} m_{n(0,1)}(r) - \sum_{\ell=2}^{N} p_{1\ell} m_{n(1,\ell)}(r)\right] dr \le 0\right)$$

$$\le P\left(\inf_{\varepsilon t \le v \le t} \int_s^{s+v} \left[p_{01} m_{n(0,1)}(r) - \sum_{\ell=2}^{N} p_{1\ell} m_{n(1,\ell)}(r) - (\alpha_1 - \beta_1)\right] dr\right.$$

$$\leq -\varepsilon \left[\alpha_1 - \beta_1\right] t)$$

$$\leq P\left(\sup_{\varepsilon t \leq v \leq t}\left|\int_s^{s+v}\left[p_{01} m_{n(0,1)}(r) - \sum_{\ell=2}^N p_{1\ell} m_{n(1,\ell)}(r) - (\alpha_1 - \beta_1)\right] dr\right|\right.$$

$$\left. \geq \varepsilon\left[\alpha_1 - \beta_1\right] t\right) \leq C_5 \exp\left\{-\varepsilon\left(\alpha_1 - \beta_1\right)\sqrt{t}\right\}. \tag{24}$$

If $\inf_{s+\varepsilon t \leq v \leq s+t} x_i(v) > 0$ for all $1 \leq i \leq k$ and $1 \leq k \leq N-2$, then

$$\sum_{\ell=0}^k u^0_{\ell,k+1}(v) = \sum_{\ell=0}^k p_{\ell,k+1} m_{n(\ell,k+1)}(v)$$

and

$$\sum_{\ell=k+2}^N u^0_{k+1,\ell}(v) \leq \sum_{\ell=k+2}^N p_{k+1\ell} m_{n(k+1,\ell)}(v)$$

for $v \in (s+\varepsilon t, s+t)$. So, just as in the proof of (24), we can show that for $k = 1, ..., N-2$,

$$P\left(\cap_{i=1}^k \left\{\inf_{s+\varepsilon t \leq v \leq s+t} x_i^0(v) > 0\right\} \cap \left\{\inf_{s+\varepsilon t \leq v \leq s+t} x_{k+1}^0(v) = 0\right\}\right)$$

$$\leq C_5 \exp\left\{-\varepsilon\left(\alpha_{k+1} - \beta_{k+1}\right)\sqrt{t}\right\}. \tag{25}$$

Now we consider the members of the last sum on the right-hand side of (23). According to the definition of $u^0_{ij}(\cdot)$, for $k = 1, ..., N$,

$$P\left(\cap_{i=1}^{N-1}\left\{\inf_{s+\varepsilon t \leq v \leq s+t} x_i^0(v) > 0\right\} \cap \left\{x_k^0(s+t) < a_k + y_k\right\}\right)$$

$$\leq P\left(\int_{s+\varepsilon t}^{s+t}\left[\sum_{\ell=0}^{k-1} p_{\ell k} m_{n(\ell,k)}(r) - \sum_{\ell=k+1}^N p_{k\ell} m_{n(k,\ell)}(r)\right] dr\right.$$

$$\left. + x_k - \varepsilon t \sum_{\ell=k+1}^N p_{k\ell}\overline{m}_{n(k,\ell)} < a_k + y_k\right)$$

$$\leq P\left(\int_{s+\varepsilon t}^{s+t}\left[\sum_{\ell=0}^{k-1} p_{\ell k} m_{n(\ell,k)}(r) - \sum_{\ell=k+1}^N p_{k\ell} m_{n(k,\ell)}(r)\right.\right. \tag{26}$$

$$\left.\left. - (\alpha_k - \beta_k)\right] dr < (a_k + y_k - x_k)^+ - b_k t\right),$$

with the notation convenience $\sum_{\ell=N+1}^N p_{N\ell} m_{n(N,\ell)}(r) = \sum_{\ell=N+1}^N p_{N\ell}\overline{m}_{n(N,\ell)} = d$. Applying Lemma 3.3 in Presman $et\ al.$ (1999) we have from (26), for $k = 1, ..., N$,

$$P\left(\cap_{k=1}^{N-1}\left\{\inf_{s+\varepsilon t \leq v \leq s+t} x_k^0(v) > 0\right\} \cap \left\{x_k^0(s+t) < a_k + y_k\right\}\right)$$

$$\leq \begin{cases} 1 & \text{for } 0 \leq t \leq (a_k + y_k - x_k)^+/b_k, \\ C_5 \exp\left\{-\frac{b_k t - (a_k + y_k - x_k)^+}{\sqrt{t}}\right\} & \text{for } t \geq (a_k + y_k - x_k)^+/b_k. \end{cases} \quad (27)$$

In the same way we can show that (27) is true for $k = N$. Note that $E(\tau - s)^{2r} = \int_0^\infty t^{2r-1} P(\tau - s > t)dt$. Substituting from (23), (24), (25), and (27) in this relation, we complete the proof of Lemma 3.1.

Step 5. Proof of Lemma 3.2. Taking the sum of all the equations in (1), we have $\sum_{k=1}^N x_k^1(t) = \sum_{k=1}^N x_k + \int_s^t (\sum_{\ell=1}^N u_{0\ell}^0(v) - d)dv$ for $s \leq t \leq \tau$. Consequently,

$$\sum_{k=1}^N x_k^1(\tau) \leq \sum_{k=1}^N x_k + \left(\sum_{\ell=1}^N p_{0\ell} \bar{m}_{n(0,\ell)} - d\right)(\tau - s). \quad (28)$$

Since $u_{0\ell}^1(t) = 0$ for $t > \tau$, we have as previously that $\sum_{k=1}^N x_k^1(\theta) = \sum_{k=1}^N x_k^1(\tau) - d(\theta - \tau)$. Since $\theta > \tau$ and $x_k^1(\theta) \geq y_k$, we have

$$\theta - \tau \leq \frac{1}{d}\left(\sum_{k=1}^N x_k^1(\tau) - \sum_{k=1}^N y_k\right), \quad \sum_{k=1}^N x_k^1(\theta) \leq \sum_{k=1}^N x_k^1(\tau). \quad (29)$$

From the definitions of θ and τ, we have that $y_N = x_N^1(\theta) = x_N^1(\tau) + \int_\tau^\theta (\sum_{\ell=0}^{N-1} u_{\ell N}^1(v) - d)ds \geq y_N + a - d(\theta - \tau)$, i.e., $\theta - \tau \geq a/d$. This relation together with (28) and (29) proves statement (ii) of Lemma 3.2.

To prove statement (i) we introduce the following notations.

$$\theta(k) = \inf\{t \geq \tau: x_k^1(t) = y_k\}, \quad k = 1, \cdots, N, \quad \tilde{S} = \{\omega: x^1(\theta) = y\},$$

$$S(k) = \{\omega: \inf_{0 \leq t < \infty} \int_\tau^{\tau+t} [\sum_{\ell=0}^{k-1} p_{\ell k} m_{n(\ell,k)}(v) - \sum_{\ell=k+1}^N p_{k\ell} m_{n(k,\ell)}(v)]dv > -\frac{a}{2}\},$$

$$k = 2, \cdots, N-1, S(N) = \{\omega: \inf_{0 \leq t < \infty} \int_\tau^{\tau+t} [\sum_{\ell=0}^{N-1} p_{\ell N} m_{n(\ell,N)}(v) - d]dv > -\frac{a}{2}\},$$

$$\tilde{S} = \cap_{k=2}^N S(k), \quad S(0) = \{\omega: m(\theta) = m'\}, \quad S = \tilde{S} \cap S(0).$$

Note that $\tilde{S} = \{\omega: \theta(N) \geq \max_{1 \leq k \leq N-1} \theta(k)\}$. From the definition of $u^1(x, m)$ and $x^1(t)$, it follows that if $\omega \in \tilde{S}$ then for $k = 2, ..., N-1$,

$$\sum_{\ell=k+1}^N u_{k\ell}^1(t) = \begin{cases} \sum_{\ell=k+1}^N p_{k\ell} m_{n(k,\ell)}(v) & \text{for } \tau < t \leq \theta(k), \\ 0 & \text{for } t > \theta(k), \end{cases}$$

and $\theta(k) - \theta(k-1) \geq (x_k^1(\theta(k-1)) - y_k)/(\sum_{\ell=k+1}^N p_{k\ell} \bar{m}_{n(k,\ell)}) \geq 0$, and $\theta(N) - \theta(N-1) \geq (x_N^1(\theta(N-1)) - y_N)/d \geq 0$. Therefore $\tilde{S} \subseteq \tilde{S}$ and

$$P[S^c] \leq \sum_{k=2}^N P(S^c(k)) + P(\tilde{S} \cap S^c(0)). \quad (30)$$

Note that if η_1 and η_2 are Markov times and $\eta_2 - \eta_1 > 1$, then there exists $q_1 < 1$ such that

$$\max_{1 \leq j \leq p} P(m(\eta_2) \neq m'|m(\eta_1) = m^j) < q_1 < 1. \tag{31}$$

By the definition of $\theta(N)$, we have $\theta(N) > a/d$. Thus, it follows from (31) that if $a > d$, then

$$P(\bar{S} \cap S^c(0)) < q_1 < 1. \tag{32}$$

Applying again Lemma 3.3 in Presman *et al.* (1999) we have

$$
\begin{aligned}
P(S^c(k)) \quad &\leq \textstyle\sum_{n=1}^{\infty} P\left(\int_{\tau}^{\tau+n}\left[\sum_{\ell=0}^{k-1} p_{\ell k} m_{n(\ell,k)}(v) - \sum_{\ell=k+1}^{N} p_{k\ell} m_{n(k,\ell)}(v)\right] dv \right. \\
&\qquad\qquad\qquad < \left. -a/2 + \textstyle\sum_{\ell=k+1}^{N} p_{k\ell} \bar{m}_{n(k,\ell)}\right) \\
&\leq \textstyle\sum_{n=1}^{\infty} P\left(\left|\int_{\tau}^{\tau+n}\left[\left(\sum_{\ell=0}^{k-1} p_{\ell k} m_{n(\ell,k)}(v) - \sum_{\ell=k+1}^{N} p_{k\ell} m_{n(k,\ell)}(v)\right)\right.\right.\right. \\
&\qquad\qquad \left.\left.\left. - (\alpha_k - \beta_k)\right] ds\right| > a/2 + n\,(\alpha_k - \beta_k) - \textstyle\sum_{\ell=k+1}^{N} p_{k\ell} \bar{m}_{n(k,\ell)}\right) \\
&\leq C_6 e^{-C_7 \sqrt{a}}.
\end{aligned}
\tag{33}
$$

It follows from (30), (32) and (33) that we can choose a and q such that $P(S^c) \leq q < 1$. This proves Lemma 3.2.

4 Concluding Remarks

In this paper, we have developed the theory of dynamic programming equation in terms of directional derivative for a jobshop with convex cost and the long-run average cost minimization criterion. Further research should focus on extending the analysis to this kind complex stochastic manufacturing systems with limited buffers. For two-machine flowshop with limited internal buffer, Presman *et al.* (2000).

Acknowledgment:

This research is supported in part by RFBR Grant 97-01-00684, NSERC Grant A4619, the Faculty Research Grant from the University of Texas at Dallas, a Distinguished Young Investigator Grant from NNSFC, and a Grant from Hundred Talents Program of the Chinese Academy of Sciences.

References

[1] Y. S. Chow and H. Teicher, *Probability Theory*, Springer Verlag, New York, 1988.

[2] S. Lou and P. W. Kager, A robust production control policy for VLSI wafer fabrication, *IEEE Trans. on Semiconductor Manufacturing*, Vol 2, 159-164, 1989.

[3] E. Presman, S. Sethi, H. Zhang, and A. Bisi, Average cost optimality for an unreliable two-machine flowshop with limited internal buffer, *Annals of Operations Research*, to appear, 2000.

[4] E. Presman, S. P. Sethi, H. Zhang and Q. Zhang, Optimal production planning in a stochastic N-machine flowshop with long-run average cost, *The Proceedings of the Indian National Science Academy*, December Issue, 1999.

[5] S. P. Sethi, W. Suo, M. I. Taksar and H. Yan, Optimal production planning in a multi-product stochastic manufacturing system with long-run average cost, *Discrete Event Dynamic Systems: Theory and Applications*, Vol 8, 37-54, 1998.

[6] S. P. Sethi, W. Suo, M. I. Taksar, and Q. Zhang, Optimal production planning in a stochastic manufacturing system with long-run average cost, *J. Optimization Theory and Applications*, Vol.92, 161-188, 1997.

[7] S. P. Sethi and Q. Zhang, *Hierarchical Decision Making in Stochastic Manufacturing Systems*, Birkhäuser Boston, Cambridge, MA, 1994.

[8] S. P. Sethi and X. Zhou, Stochastic dynamic jobshops and hierarchical production planning, *IEEE Trans. Auto. Contr.* Vol. 39, 2061-2076, 1994.

[9] N. Srivatsan, S. Bai, and S. Gershwin, Hierarchical real-time integrated scheduling of a semiconductor fabrication facility, in *Computer-Aided Manufacturing/Computer-Integrated Manufacturing*, Part 2 of Vol.61 in the series *Control and Dynamics Systems*, Leondes, C. (ed.), Academic Press, New York, 174-241, 1994.

[10] R. Uzsoy, Y. Lee, and L. Martin-Vega, A review of production planning and scheduling in the semiconductor industry, Part I: System characteristics, performance evaluation and production planning, *IIE Trans. on Scheduling and Logistics*, Vol. 24, 47-61, 1992.

[11] R. Uzsoy, Y. Lee, and L. Martin-Vega, A review of production planning and scheduling in the semiconductor industry, Part II: Shop floor control, *IIE Trans. on Scheduling and Logistics*, Vol. 25, 1993.

On an Extension of Classical Production and Cost Theory

Christoph Schneeweiss

Department of Operations Research, University of Mannheim, Germany

Abstract. The paper shows the insufficiency of traditional production and cost theory to capture production and cost phenomena. It is argued that already for the one-person situation production theory is not able to incorporate the organizational structure and, in addition, it is unable to deal with non-cost criteria, the cost value problem, and the aggregation problem. For the multi-person situation, transactional and transformational interactions cannot be described. As a consequence, the paper extends traditional production theory and shows that the theory of hierarchical planning provides a framework that is broad enough to amend the deficiencies mentioned. In particular, it is able to capture principal agent relationships and provides a general framework for modern cost and managerial accounting.

1. Introduction

For many years, production theory could be considered as a sound theoretical foundation of important parts of business administration. It was not only capable of describing central aspects of the production process but simultaneously it could be employed to construct appropriate cost systems like the well-known system of differential costing. Furthermore, many processes in other functional fields of business administration could be described using production theoretical notions, and the firm as a whole was characterized in terms of classical microeconomic theory.

Summarizing, classical production theory relies on the paradigm of a one person's description or of one decision maker. For modern analysis, however, this means a far too narrow perspective. Indeed, production theory should be extended to cope with a multi-person situation possessing asymmetric information, as might be met, e.g., in modern supply chain management. These persons may be described as a team or, at the other extreme, as opportunistically behaving decision makers. Clearly, this last case characterizes the principal agent setting (e.g, see [Spremann]) which, for more than two decades, has been developed in modern economic micro theory and which is more and more applied in today's theory of business administration (e.g., see [Schenk-Mathes]). Thus one may understand principal agent theory as a possible extension of the classical production and cost theory.

Replacing production and cost theory, however, by the broader framework of agency theory will not readily be accepted. First, most of the relations within a company may not be described as being antagonistic and second, it is formally not entirely obvious how the antagonistic (multi-person) setting is related to the one-person situation which, of course, should still be captured by a general theory.

In fact, principal agent theory can be interpreted as one particular specification within a broader conceptual framework. This framework is provided by a general theory of hierarchical planning which comprises team and non-team settings for symmetric and asymmetric information states.

Let us proceed as follows. First, let us characterize, from a decision theoretic point of view, the main features and drawbacks of classical production and cost theory. These deficiencies will lead us to the new concept of hierarchical planning which will briefly be explained in Section 3. In this section, we also show, in general terms, how the deficiencies of production and cost theory can be resolved. The remaining two sections will then be devoted to a more detailed discussion of the multi-person setting focusing on the team and non-team cases in Sections 4 and 5, respectively. Finally, Section 6 draws some general conclusions.

2. Classical Production and Cost Theory

In very general terms, production theory describes the transformation of input quantities (so-called 'production factors') into output quantities or 'products'. For technically efficient productions, this transformation is achieved through so-called production functions. Cost functions may then be interpreted as functions that transform costs of production factors into costs of output quantities. This transformation of cost values is closely related to the transformation of the adjoint physical quantities being described by production functions. One of the main problems of production and cost theory is to investigate different types of functions as to their ability to describe various production processes and cost situations.

Traditionally, production and cost theory may be considered as a mere phenomenological theory. A technologisation [Popper] of this theory, however, is straightforward. One simply defines input quantities as decision variables. Production functions are then no longer merely phenomenological statements (hypotheses) but rather they are describing and forecasting the consequences of a decision. Adopting this more general point of view, production functions may be interpreted as decision fields and cost functions as cost criteria of decision models. This broader view of production and cost theory will enable us to suggest a substantial extension to be presented in the sequel. It should be clear, however, that already the step from production and cost functions to fully described decision models is highly significant. This is due for at least two reasons:

1. Decision fields with their (in general) numerous constraints describe in a comparatively simple way highly complicated transformations that cannot be captured by analytic functions, particularly those investigated in traditional production theory. Decision fields with integer variables, or dynamic and stochastic relationships may only in the simplest cases be represented by traditional production functions. Such decision fields, however, are most common in production and operations management.

2. Traditionally, cost functions account for constraints in employing opportunity costs. Stating the decision field explicitly, this is no longer necessary. But what is more important, for more complicated decision fields (discrete, dynamic and/or stochastic), the simple concept of incorporating the decision field into a cost function via opportunity costs cannot be applied.

Embedding production and cost theory into the broader framework of a decision model being described by a quantitative decision field and a financial criterion constitutes an important step. However, this step is not comprehensive enough to capture problems of today's production planning and cost accounting. Already the simple task of determining a (linear) medium-term production plan causes distinct problems. To investigate the various difficulties, let us discuss first the one-person and then the multi-person situation.

2.1 One-Person or Team Situation

Even within the paradigm of one decision maker or of more than one decision maker forming a team, production theory meets with particular difficulties. This is mainly due to the (1) organizational structure of the system to be described, to the (2) non-cost nature of the criteria, and to the (3) 'cost value problem'.

(1) **Organizational structure**
Focusing only on two levels of an organization, production theory has serious difficulties to describe such a situation. Even if there is only one decision maker, one usually has the problem of information asymmetry. For instance, tactical decisions have to be made earlier than operational ones which implies that the information state at the tactical level is different from the information one is going to have at the time when the operational decision has actually to be made. Such a situation cannot be represented by a production function. Similarly, often the tactical level relies on aggregates of the operational level. As an example, a linear production program works with aggregate variables and coefficients. Hence, these coefficients are not original empirical values but have to be determined through an aggregation procedure. Thus, even if one restricts the description to the tactical level, the lower operational level cannot

be disregarded. This observation adds an additional aspect as to the nature of production theory. In most applications, this theory does not provide an empirical description as is usually claimed but contains normative elements as to the way how the necessary aggregation should be performed.

(2) Non-cost criteria

Usually a financial criterion is not the only one that should be taken into account. Often a cost function describes only a particular aspect of the entire problem. Consequently, one has a multi-criterion situation, and again cost theory tries to capture the additional aspects via opportunity costs. This is of course only one possibility to solve a multi-criterion decision problem.

Furthermore, cost functions (and a fortiori production functions) rely on a cardinal level of measurability. Hence one is not able to treat attributes like 'quality' or 'prestige', which may only be measurable at most at an ordinal level. Consequently, production and cost theory excludes many phenomena, particularly those that have to do with the leadership task.

(3) The cost value problem

In its simplest version, the value of cost parameters of a cost function are given by the costs of the input factors. As mentioned earlier, however, in many situations such a simplistic determination of cost parameters is not possible. Opportunity costs are a prominent example. Generally, to discuss the cost value problem in a rational way, it is necessary to assume the existence of a (meta-)criterion by which the cost parameters may be evaluated. No such construction can be found in traditional cost theory. Thus, again, even in the one-person setting, production and cost theory represent a very narrow concept which excludes problems that are of utmost importance for a theoretical understanding of business administration.

2.2 Multi-Person Non-Team Situation

The difficulties production and cost theory has to face in view of modern production planning and management accounting will increase for the non-team setting. Here, indeed, one is leaving the one-person paradigm for which traditional production theory had been developed. To capture the non-team case, it is necessary to describe the interaction of the persons involved. Let us focus on two types of interaction: the transactional and the transformational interaction. For the transformational interaction, the (cost) criterion is invariably changed, while in the transactional situation this is only the case as long as some incentives are effective.

(1) Transactional interactions

Principal agent relationships are of the transactional type. The agent influences (and hence 'determines') the financial criterion of the agent. The same is true for coordinating a multi-agent system in influencing the agents' financial criteria such that an honest exchange of information is induced.

(2) **Transformational interactions**

In transformational interactions, a long-term behavioral change of the decision criteria and, in particular, of the form of a cost function is achieved. Again, this type of influence cannot be described in production and cost theory, though it constitutes one of the central problems of modern management accounting. Hence, unlike the situation in traditional cost accounting systems, cost theory is not able to provide a production theoretic foundation for this important new area.

Summarizing, the difficulties of traditional production and cost theory are more than obvious. In particular, it cannot claim to provide a basis for production and operations management and modern cost accounting, respectively. Indeed, the traditional concept must be extended from a one-person to a multi-person description. This does not mean that the results of the traditional theory are obsolete. In treating more general problems, the simpler problems do still exist and are embedded in a more comprehensive setting. Clearly, this more comprehensive concept will be less specific than the questions being investigated in production theory. However, the concept will give an answer to the important questions raised in this section. In the following sections, we will now develop such a general conceptual framework.

3. A Brief Account of Hierarchical Planning

In the previous section, we already discussed a decision theoretic extension of traditional production and cost theory. Rather than talking of production and cost functions, we considered appropriate decision models. The advantage of such an extension can be seen in a richer and less condensed representation. This extension, however, did not suffice to treat an important number of unsolved problems. An even richer structure is necessary, i.e., a structure that allows the description of at least two decision makers. This structure is provided by the framework of distributed decision making and, in particular, by the theory of hierarchical planning (see [Schneeweiss 1999]).

In this theory, one considers two decision models, denoted as top- and base-model which interfere with each other as depicted in Fig. 1. Let us denote the top-model as $M^T(C^T, A^T)$ with top-criterion C^T and top-decision field A^T and analogously the base-model as $M^B(C^B, A^B)$ with criterion C^B and decision field A^B. The top-model may influence the base-model by an instruction IN while the bottom-up influence is exerted through an anticipation. In determining the instruction, the top-level observes the base-model and constructs an anticipated base-model $\hat{M}(\hat{C}^B, \hat{A}^B)$. Assuming a rational behavior of the base-level, the top-level now influences the anticipated base-model through the instruction IN and calculates the anticipated optimal reaction of the base-level. This reaction as a function of IN will be called anticipation function $AF = AF(IN)$. Having probed various instructions, the top-level finally ends up with an optimal instruction IN^* which is communicated to the base-model (s. Fig. 1).

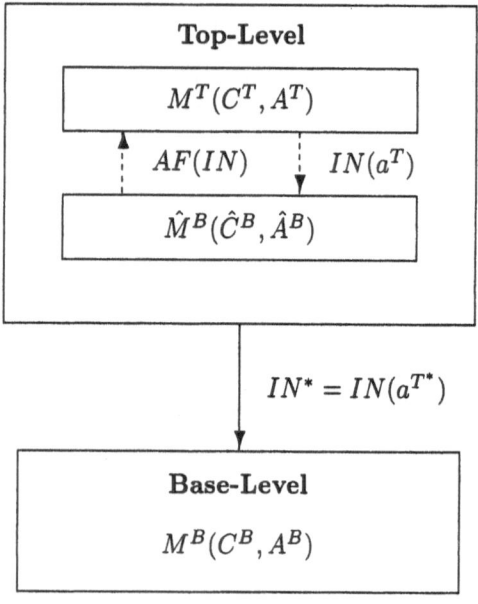

Fig. 1: Hierarchical Planning

This coupling of the two levels may readily be described more formally. In doing so, one does only need to introduce some additional structure. Indeed, it suffices to assume that C^T is a function of a private criterion C^{TT} and a top-down criterion C^{TB}. While the top-down criterion depends explicitly on the anticipation function AF, i.e., $C^{TB} = C^{TB}(AF)$, this is not the case for C^{TT}. The top-decision a^T is made in t_0 and the base-decision in t_1, later than t_0 ($t_1 \geq t_0$). Moreover, the top-information state $I_{t_0}^T$ comprises all stochastic information the top-level has (i.e., all random variables), and the same holds with $I_{t_1}^B$ for the base-level. Hence, a formal description of the two levels and their interference is given by the following pair of functional equations which will henceforth be called 'coupling equations'

$$a^{T^*} = \arg \underset{a^T \in A^T}{\text{opt}} \, E\left\{ C^T \left[C^{TT}(a^T), C^{TB}(AF(IN)) \right] \big| I_{t_0}^T \right\} \qquad (1a)$$

$$IN = IN(a^T)$$

$$AF(IN) = \arg \operatorname*{opt}_{\hat{a}^B \in \hat{A}^B_{IN}} E\left\{\hat{C}^B_{IN}(\hat{a}^B)|\hat{I}^B_{IN}\right\} \tag{1b}$$

$$a^{B*} = \arg \operatorname*{opt}_{_uB \in A^R_{IN*}} E\left\{C^B_{IN*}(a^B)|I^B_{IN*,t_1}\right\}. \tag{1c}$$

Note that not only the criterion and decision field of the base-level can be influenced but also the forecast base-information state $\hat{I}^B_{IN}$ or the actual information state I^B_{IN*,t_1} in t_1.

The reader will realize that the coupling equations describe a rich number of important settings. Fig. 2 provides an overview.

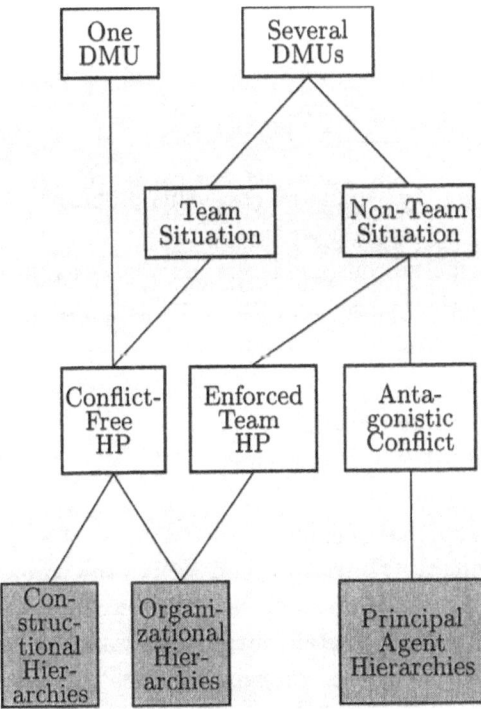

Fig. 2: Hierarchies in Distributed Decision Making

First the equations are able to describe both the one- and the multi-person case. For several decision making units (DMUs), they capture the team and the non-team-situation. The non-team case results in the principle agent type of models, while the team case ends up with a constructional or an organizational hierarchary, respectively. The constructional hierarchy is defined by a symmetric state of information, i.e., the two levels share their information states, whereas the organizational hierarchy describes a team which does not (or cannot) share its information, i.e., one has an asymmetric state of information.

Fig. 2 indicates that the coupling equations are general enough to cover the range from the one-person paradigm to the multi-person antagonistic case with asymmetric information. The following two sections will give evidence that the sketched framework of hierarchical planning will be capable of settling the open questions that were stated before as desiderata for a modern production and cost theory.

4. Hierarchical Planning as an Enhancement of Production and Cost Theory for the Team Case

In Section 2.1 we identified several drawbacks of the traditional theory which may be summarized in three main problems:

The inability of
1. properly treating the symmetric cost value problem,
2. of treating the aggregation problem, and
3. of describing simultaneously different organizational levels.

The first two issues refer to a symmetric information state, i.e., to a constructional hierarchy, while the third is concerned with asymmetric information.

4.1 Symmetric Information

(1) Cost value problem

Let us first consider the (symmetric) cost value problem. This is typically a hierarchical problem. The base-level describes the production model while the top-level performs the evaluation of the cost parameters of the base-level's cost function. Hence the general coupling equations can be specialized such that

$$p^* = \arg \operatorname*{opt}_{p \in P} E\{C^T(AF(p))|I^T\} \tag{2a}$$

and

$$AF(p) = \arg \operatorname*{opt}_{a^B \in A^B} E\{C^T(AF(p))|I^T\} \tag{2b}$$

with p denoting the cost parameter that has to be evaluated and P being the parameter set one is taking into account. Often the top-model describes an empirically given rather involved decision problem while the base-model represents a (mathematically) relaxed version of the top-model. This version is called decision generator [Schneeweiss 1987], and the cost evaluation turns out to be a parameter adaptation of the solution of the decision generator to the situation one is really interested in. Note that the criterion C^T of the more realistic situation can represent a multi-component criterion and not just a cost function as in the case for the (relaxed) decision generator. The optimal costs p^* are specific opportunity costs and are called steering costs ([Schneeweiss 1987, 1999]).

(2) Aggregation - Disaggregation

The aggregation-disaggregation problem can be represented by the coupling equations such that the top-equation describes the aggregated version of the base-equation. The aggregation is assumed to be achieved through the parameters $p_j, j = 1, \ldots, n$ of the base-level. The aggregated parameters $\bar{p}$ are constructed as a weighted sum $\bar{p} := \sum_{j=1}^{n} g_j p_j$ with weights depending on the solution $\hat{a}^B := \left(\hat{a}^B, \ldots, \hat{a}_n^B \right)$ of the anticipated base-model: $\bar{p} = \bar{p}(\hat{a}_1^B, \ldots, \hat{a}_n^B)$. Hence, in view of Eqs. (1),

$$a^{T^*} = \arg \operatorname*{opt}_{a^T \in A_{\bar{p}}^T} E\left\{ C^T \left(C^{TT}, C^{TB}(\bar{p}) \right) | I_t^T \right\} \tag{3a}$$

$$IN = IN(a^T)$$

$$\hat{a}^{B^0} = \arg \operatorname*{opt}_{\hat{a}^B \in \hat{A}_{IN}^B} E\left\{ \hat{C}_{IN}^B(\hat{a}^B) | \hat{I}_{IN}^B \right\} \tag{3b}$$

The anticipation equation (3b) is including the disaggregation device. Hence, for given aggregation and disaggregation devices the solution of the coupling equations results in an optimal aggregation-disaggregation procedure.

As an example, imagine the following situation. In hierarchical production planning [Hax/Meal], products are aggregated to product groups. This can be achieved in aggregating the products' cost parameters and capacity consumption rates. For holding cost parameters, e.g., the weights would depend on optimal inventories, and for the consumption rate one would take the weights to depend on production variables. These aggregated parameters result, at the aggregated production level, in capacities which are communicated to the (lower) production level as an instruction. The disaggregation may then be performed in terms of a capacity allocation for the single products. (For further analysis, see [Schneeweiss 1999].)

4.2 Asymmetric Information

Obviously, from its very construction hierarchical planning is capable of simultaneously describing different levels of an organization. Here again a parameter adaptation problem may occur. But now steering costs are used to coordinate decisions of two organizational levels. A particularly important type of costs are imputed costs, and, to be even more specific, so-called investment-oriented costs ([Küpper], [Schneeweiss 1999], [Eichin/Schneeweiss]). These costs adapt short-term production decisions to long-term investment decisions. Thus cost theory can be extended to capture simultaneously the long-term and the short-term level.

Also, the aggregation-disaggregation problem can be considered within the broader framework of asymmetric information. An extension of Eqs. (3) is straightforward. Moreover, the product aggregation and the 'time-aggregation' in determining imputed costs could be combined, which nicely shows that production theoretic aggregations and a cost theoretic determination of (imputed) opportunity costs may not be performed separately.

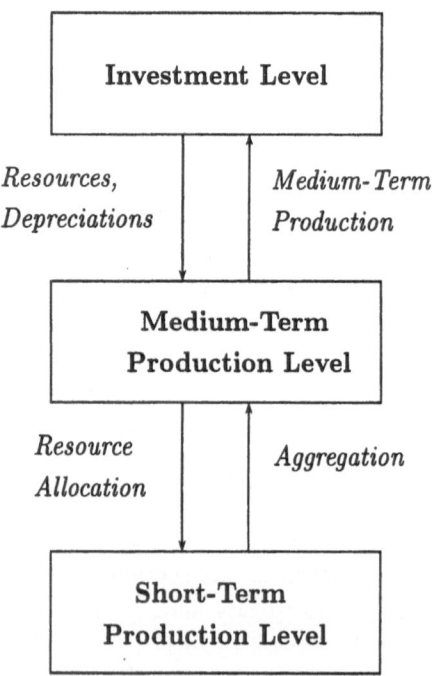

Fig. 3: Determination of Medium Term Parameters

As an important specific problem consider the determination of the parameters of a medium-term (linear) production plan. These parameters depend simultaneously on the investment-level and the short-term scheduling level (s. Fig. 3). This determination affords a three-step hierarchy with an aggregation-disaggregation between the short- and medium-term production levels and a determination of imputed costs between the tactical investment level and the medium-term production level. To describe the three-step hierarchy, the coupling equations can be extended in an obvious way (see the three-step hierarchy in [Schneeweiss 1999] Chapter 4). Again, traditional production and cost theory is not capable of tackling such a multi-level problem.

5. Hierarchical Planning as an Enhancement of Production and Cost Theory for the Non-Team Case

For the transactional influence, as mentioned in Section 2.2, the non-team asymmetric situation can be described by a principal agent theory. This theory turns out to be a particular case of the general coupling equations (1). For the transformational influence, hierarchical planning may be employed as well to provide at least a framework to discuss this rather involved type of problem.

5.1 Transactional influence

For the principal agent setting, the general coupling equations (1) may be specialized as follows

$$k^*(P) = \arg \opt_{k(P) \in A^T} E\{u^T(P - k(P)) \mid I^T\} \qquad (4a)$$

$$AF(k) = \arg \opt_{\hat{a}^B \in \hat{A}^B_{k(P)}} E\{\hat{u}^B(k(P) - \hat{V}(a^B)) \mid \hat{I}^B\} \qquad (4b)$$

$$k^*(P) = IN^*$$

$$a^{B^*}(k^*) = \arg \opt_{a^B \in A^B_{k*}} E\{u^B(k^* - V(a^B)) \mid I^B\}. \qquad (4c)$$

The top-equation (4a) (of the principal) determines incentives k as a function of profit $P : k = k(P)$ in optimizing expected utility $u^T(P - k(P))$ of its net profit. The (anticipated) profit is a function of the agent's optimal decision $\hat{a}^{B^o} \equiv AF(k), P = P(AF)$, and is determined by the anticipation equation (4b) with $\hat{V}(a^B)$ being the agent's disutility. The well-known participation relation [Spremann] can be captured by the agent's decision field $\hat{A}^B$. Extensions of this so-called standard principal agent setting are straightforward (e.g., see [Schneeweiss 1999]). Particularly, one may introduce further constraints describing incentives to induce a truthful communication of information.

5.2 Transformational influence

As mentioned in Section 2.3, the modern concept of behavioral costs may also be described within the framework of hierarchical planning. Again the general equations (1) can be specialized as

$$a^{T^*} = \arg \operatorname*{opt}_{a^T \in A^T} E\{C^T(C^{TT}(a^T), C^{TB}(AF(p))) \mid I^T\} \qquad (5a)$$

$$AF(p) = \arg \operatorname*{opt}_{\hat{a}^B \in \hat{A}^B} E\{\widehat{\tilde{C}}_p^{\,B}(\hat{a}^B) \mid \hat{I}^B\}, \qquad (5b)$$

with $\tilde{C}_p^B$ being the cost criterion of the DMU that is to be influenced transformationally.

6 Summary and Concluding Remarks

It has been shown that hierarchical planning provides a conceptual framework that, in contrast to traditional production and cost theory, is broad enough to comprise many important problems of production planning and cost accounting. In particular, the new concept is able to capture the following shortcomings of traditional production and cost theory:

(1) Hierarchical planning is not restricted to one level and to only one information state. In fact, it provides a rational way of investigating such problems as modern supply chain management which are of utmost significance for modern production planning.

(2) The cost value problem and, in particular, the problem of steering and opportunity costs can be investigated in a rational way. Moreover, the question can be settled which cost definition should be used and whether production functions can be considered as empirical or normative statements.

(3) Within the comprehensive concept of hierarchical planning, principal agent theory and production and cost theory can be related to each other.

(4) Hierarchical planning does not only accommodate transactional but also transformational influences. Hence, in contrast to production and cost theory, the broader concept of hierarchical planning allows to describe important aspects of the relationship between planning and leadership activities. This description is of significant importance for a science that claims to describe the management process.

Production and cost theory may be considered as a successful theory that allows the investigation of input-output processes. It may be used to identify production factors and to analyse particular combinations of these factors. The properties of production and cost theory are still existent in the wider theory of hierarchical planning which, however, in addition, is capable of analysing a large range of phenomena that cannot be treated within the scope of the traditional theory.

References

1. Eichin, R., Schneeweiss, Ch. (1999): Determining depreciations as a hierarchical problem. Discussion paper no. 62, Department of Operations Research, University of Mannheim.
2. Hax, A.D., Meal, D. (1975): Hierarchical integration of production planning and scheduling. In: Geisler, M.A. (ed.). Logistics, TIMS Studies in the Management Sciences. North Holland, Amsterdam.
3. Küpper, H.-U. (1991): Multi-Period Production Planning and Managerial Accounting. In: Fandel, G. and Zäpfel, G. (eds.). Modern Production Concepts, pp 46-62. Springer, Berlin, Heidelberg, New York.
4. Popper, K.R. (1992): The Logic of Scientific Discovery. Hutchinson, London.
5. Schenk-Mathes, H. (1999): Gestaltung von Lieferbeziehungen bei Informationsasymmetrie. Wiesbaden, Deutscher Universitätsverlag.
6. Schneeweiss, Ch. (1987): On a formalization of the process of quantitative model building. In: European Journal of Operational Research 29, pp 24-41.
7. Schneeweiss, Ch. (1999): Hierarchies in Distributed Decision Making. Springer, Berlin, Heidelberg, New York.
8. Spremann, K. (1987): Agent and Principal. In: Bamberg, G. and Spremann, K. (eds.). Agency Theory, Information, and Incentives, pp 3-37. Springer, Berlin, Heidelberg, New York.

Optimal Controls in Spatial Advertising Diffusion Models

Gernot Tragler

Institute of Econometrics, Operations Research and Systems Theory
Vienna University of Technology
Argentinierstr. 8/1192, A-1040 Wien, Austria
E-mail: tragler@e119ws1.tuwien.ac.at
Fax: +43 - 1 - 58801 - 11999
Tel.: +43 - 1 - 58801 - 11920

Summary. We start with a brief review of a descriptive multi population (multi market) advertising diffusion model, where word of mouth communication takes place both within single markets and between individuals from different markets. Assuming that a monopolist for one market who seeks to maximize profits has naive expectations about the states of the neighbouring markets, we derive an optimal advertising effort for this particular monopolist. Even if the according advertising strategy is not truly optimal in the sense that it does not take into account the variations in the neighbouring markets, in numerical analyses it has so far always proven to be better than constant or random advertising efforts. The technique proposed here may hence be seen as a fair trade-off between accuracy and practicability for multi state optimal control problems as the one presented in this paper.

1. Introduction

Recently, Tragler (1996) introduced a descriptive multi population (multi market) advertising model the starting point of which was an article by Yakowitz et al. (1990) about cellular automaton modeling of epidemics. Using their basic ideas gave a spatial advertising model on the micro-level which – applying the techniques proposed by Mazza and O'Connell (1992) – directly led to a generalization of the state dynamics of the well-known second advertising diffusion model by Gould (1970) including several subpopulations which interact through a neighbourhood structure to be specified.

While in Tragler (1996) emphasis is placed on the non-stationary behaviour of the uncontrolled system dynamics, the purpose of the current paper is to extend the model by deriving an *optimal advertising strategy* instead of assuming that the advertising effort is constant over all times. This is done by adopting an optimal control approach recently presented in a paper by Dawid and Feichtinger (1995), which itself is based upon the technique by Boldrin and Montrucchio (1986).

This paper is organized as follows. Section 2 reviews the model by Tragler (1996) and describes the optimal control model extension. In Section 3 the optimal solution under naive expectations will be derived and numerically tested. Section 4 concludes with a short discussion.

2. The Model

2.1 The Original Model

Assume that a given population V can be divided into L disjoint subpopulations $V_1, \ldots, V_L$ all of which have size g. In what follows, we will refer to these subpopulations as "markets". By $x_i(t)$ we denote the number of people in market V_i who are aware of a specific advertising campaign "AC" at time t.

The main idea behind the advertising diffusion model is that information made available by an advertiser spreads through the market by word of mouth rather than by an impersonal advertising medium (see, e.g., Dodson and Muller (1978)), where in Tragler (1996) the information exchange is assumed to take place both within single markets and between individuals from different markets, if these markets are "neighbours".

The full neighbourhood structure may be described with the help of a "weighted adjacency matrix" $[\rho_{ik}]_{i,k=1,\ldots,L}$ (all $\rho_{ik} \geq 0$, $\rho_{ik} = \rho_{ki}$) with

$$\rho_{ii} > 0 \quad \forall i = 1, \ldots, L$$

and

$$\rho_{ik} \left\{ \begin{matrix} > \\ = \end{matrix} \right\} 0 \text{ iff } V_i \text{ and } V_k \left\{ \begin{matrix} \text{are} \\ \text{are not} \end{matrix} \right\} \text{neighbours,}$$

where small/large values of ρ_{ik} indicate that there is rare/frequent contact between individuals from markets V_i and V_k, respectively.

The system dynamics describing the temporal evolution of the awareness of AC in each market V_i is then given by

$$x_i(t+1) = (1-r)x_i(t) + \left[s + p \left(\sum_{k=1}^{L} \rho_{ik} x_k(t) \right) \right] (g - x_i(t)) \quad i = 1, \ldots, L,$$

$$(2.1)$$

where the constant r expresses the fact that goodwill depreciates over time (i.e., people of course also forget about the advertising campaign), while s and p (both assumed to be constant in Tragler (1996)) are the so-called innovation and imitation coefficients, respectively.[1]

If we assume that there is only one market (i.e., $L = 1$), set $r = 0$, and normalize the subpopulation size to one (i.e., $g = 1$), then (2.1) reduces to the well-known advertising diffusion model by Bass (1969), the discrete-time version of which is given as (see, e.g., Feichtinger and Hartl (1986))

$$x(t+1) = x(t) + (\alpha + \beta x(t))(1 - x(t)),$$

---

[1] Note that the model is of discrete time; as Dawid and Feichtinger (1995) argue, *virtually all marketing models which have been empirically validated are based on a discrete-time scale; see Simon (1984), Lilien and Kotler (1983).*

where $x(t)$ is the portion of the population which is aware of AC at time t, α is the innovation coefficient, and β is the imitation coefficient.

This concludes the review of the original model by Tragler (1996). The following subsection describes the optimal control extension of this model.

2.2 The Optimal Control Extension

It is obvious that the size of the imitation coefficient p in the system dynamics (2.1) is a direct measure of the advertising efforts for the product under consideration: the higher the efforts to make the product known, the higher should be p, i.e. the higher is the "infection process" of being aware of AC. It is hence reasonable to assume that p is not constant over all times but rather changes according to the advertising efforts, i.e. $p = p(t)$. Even more realistic is the assumption that $p(t)$ changes according to some optimization background. We proceed with this idea which is based on the paper by Dawid and Feichtinger (1995).

The state dynamics of the optimal control advertising diffusion model by Dawid and Feichtinger (1995) is exactly (2.1) with $s = 0$, $\rho_{ii} = 1$, $\rho_{ik} = 0$ $(i \neq k)$ and $g = 1$. The technique they use is a variation of the technique by Boldrin and Montrucchio (1986), and their results may be applied directly to our model.

For that purpose, we consider the profit maximization problem for a monopolist M_i in market V_i, where the monopolist's control is the effort invested in advertising for V_i at t, $p_i(t)$. We assume that the monopolist is interested only in what happens in market V_i, and there is no other firm investing in advertisement in that particular market. As in Dawid and Feichtinger (1995), we keep $s = 0$, $\rho_{ii} = 1$ and $g = 1$, but we allow ρ_{ik} to be positive for $i \neq k$. Let now Π and Φ denote the monopolist's return and cost functions, respectively, then the optimization problem of the firm is formally stated as follows:

$$\max_{\{p_i(t)\}} \sum_{t=0}^{\infty} \delta^t \left(\Pi\left(x_i(t)\right) - \Phi\left(p_i(t)\right) \right) \tag{2.2}$$

subject to

$$x_i(t+1) = (1-r)x_i(t) + p_i(t)\left[\left(\sum_{k=1,\ldots,L,k\neq i} \rho_{ik}x_k(t)\right) + x_i(t)\right](1 - x_i(t)) \tag{2.3}$$

and

$$0 \leq p_i(t) \leq p_{max} \;\forall t,{}^2 \tag{2.4}$$

with the discount factor $\delta \in (0,1)$ and

---

[2] This condition guarantees that $x_i(t)$ never exceeds the market size 1; for details we refer to Tragler (1996).

$$0 < r < 1.$$

Instead of solving this full problem – which involves L states and is hence almost intractable – we assume that the monopolist maximizes the profit (2.2) under the assumption that the terms in the state dynamics depending on the states in the neighbouring markets are constant over all times; i.e., the firm has "naive expectations" about what is happening in the neighbourhood. These terms can thus be replaced by some constant C with

$$C = \sum_{k=1,\ldots,L,k\neq i} \rho_{ik} x_k$$

which will be assumed to be not greater than 1. The system dynamics (2.3) hence reduces to

$$x_i(l+1) = (1-r)x_i(t) + p_i(t)\,(C + x_i(t))\,(1 - x_i(t))\,. \qquad (2.5)$$

The following section describes the analysis of the optimization problem (2.2) under (2.5), (2.4), and the initial condition

$$x_i(0) = \bar{x}_i > 0.$$

3. Analysis

3.1 Analytical Results

Let the return and cost functions be given by

$$\Pi\,(x_i) = \beta^2 C^2 + \alpha x_i - \beta^2 (C + x_i)^2 (1 - x_i)^2 \quad \alpha, \beta > 0 \qquad (3.1)$$

and

$$\Phi(p_i) = p_i^2,$$

respectively.[3] We will assume that

$$\alpha \geq 2\beta^2 D$$

where

$$D = \max\left(C(1 - C), \frac{(1 + C)^3}{6\sqrt{3}}\right)$$

so that the return function Π is increasing on $(0, 1)$. The following theorem gives an optimal advertising effort $\hat{p}_i(t)$.

---

[3] The return function (3.1) has a concave-convex-concave shape and is depicted in Dawid and Feichtinger (1995), where the interested reader can also find a sophisticated interpretation for this shape.

Theorem 3.1. *Assume that $\alpha > 2\beta r$ and $\beta \leq \frac{4p_{max}}{(C+1)^2}$ holds. Then there exists a discount factor $\delta \in (0,1)$ such that*

$$\hat{p}_i(t) = \beta \left(C + x_i(t) \right) \left(1 - x_i(t) \right)$$

is the optimal advertising effort maximizing (2.2) under (2.5).

Proof. We will now show the optimality of

$$\hat{p}_i(t) = \beta \left(C + x_i(t) \right) \left(1 - x_i(t) \right)$$

in this model. For this purpose we will use the well-known Bellman equation as a sufficient condition of optimality.

To this end, we have to use the so-called reduced form of the model which we get by inserting the state equation into the cost function. Let

$$\Omega = \{(x_i, x_i') \,|\, (1-r)x_i \leq x_i' \leq (1-r)x_i + p_{max}\left(C + x_i\right)\left(1 - x_i\right), x_i \in (0,1)\}$$

denote the set of all pairs of states which can be realised in two consecutive periods and denote by U the reduced utility function which is given by

$$U\left(x_i, x_i'\right) = \Pi(x_i) - \Phi\left(\frac{x_i' - (1-r)x_i}{(C+x_i)(1-x_i)} \right).$$

Then the reduced form of the problem is:

$$\max_{\{x_i(t)\}} \sum_{t=0}^{\infty} \delta^t U\left(x_i(t), x_i(t+1)\right)$$

subject to

$$(x_i(t), x_i(t+1)) \in \Omega \; \forall t, \quad x_i(0) = \bar{x}_i > 0.$$

The reduced utility function $U\left(x_i(t), x_i(t+1)\right)$ denotes the profit the firm gets at time t, if its market share changes from $x_i(t)$ to $x_i(t+1)$ in the t'th period. A sequence $\{x_i(t)\}_0^{\infty}$ is called a program from $\bar{x}_i$ if $(x_i(t), x_i(t+1)) \in \Omega$ holds for all t and $x_i(0) = \bar{x}_i$. We will call a program $\{\hat{x}_i(t)\}_0^{\infty}$ from $\bar{x}_i$ an optimal program, if

$$\sum_{t=0}^{\infty} \delta^t U\left(\hat{x}_i(t), \hat{x}_i(t+1)\right) \geq \sum_{t=0}^{\infty} \delta^t U\left(x_i(t), x_i(t+1)\right)$$

holds for every program $\{x_i(t)\}_0^{\infty}$ with $x_i(0) = \bar{x}_i$. Further we define the value function by

$$\Psi\left(\bar{x}_i\right) = \sum_{t=0}^{\infty} \delta^t U\left(\hat{x}_i(t), \hat{x}_i(t+1)\right).$$

The value function will always be unique, but due to the fact that U will in general not be concave, we can not state a priori that the optimal program is unique. Therefore, the optimal policy function $\theta : (0,1) \to P\left((0,1)\right)$ will, in general, be a set-valued map, namely

$$\theta(x_i) = \{y \in (0,1) \,|\, y = \hat{x}_i(1) \text{ for an optimal program } \{\hat{x}_i\} \text{ from } x_i\}.$$

The Bellman equation is given by

$$G\left(x_i\right) = \max_{x_i' \in \Gamma(x_i)} \left[U\left(x_i, x_i'\right) + \delta G\left(x_i'\right)\right] \tag{3.2}$$

where

$$\Gamma\left(x_i\right) = \{y \in (0,1) \,|\, (x_i, y) \in \Omega\}$$

denotes the set of all states which are reachable from x_i within one period.

As U is continuous and bounded above, $(0,1)$ is convex and $\Gamma\left(x_i\right)$ is nonempty and compact for all $x_i \in (0,1)$, there exists a unique solution of this functional equation in continuous functions. This unique solution of (3.2) coincides with the value function Ψ (see Stokey and Lucas (1989)). This statement will be used in order to prove our theorem.

Let now $\delta = \frac{2\beta}{\alpha + 2(1-r)\beta}$. Due to the assumption that $\alpha > 2\beta r$, δ is in $(0,1)$. In order to show that

$$\Psi(x) = \frac{\beta^2 C^2}{1 - \delta} + \frac{2\beta}{\delta} x$$

is a solution of the Bellman equation (3.2) we define $W\left(x_i, x_i'\right) = U\left(x_i, x_i'\right) + \delta \Psi\left(x_i'\right)$. This yields (substituting $2\beta\left(\frac{1}{\delta} - 1 + r\right)$ for α)

$$W\left(x_i, x_i'\right) = 2\beta\left(\frac{1}{\delta} - 1 + r\right) x_i - \beta^2 (C + x_i)^2 (1 - x_i)^2 -$$

$$- \left[\frac{x_i' - (1-r)x_i}{(C + x_i)(1 - x_i)}\right]^2 + 2\beta x_i' + \frac{\beta^2 C^2}{1 - \delta}.$$

Setting $W_{x_i'}\left(x_i, x_i'\right)$ to zero leads to

$$x_i'^* = (1 - r)x_i + \beta(C + x_i)^2 (1 - x_i)^2.$$

Due to the assumption that $\beta \leq \frac{4p_{max}}{(C+1)^2}$ we have

$$0 < \beta(C + x_i)^2 (1 - x_i)^2 = [\beta\left(C + x_i\right)(1 - x_i)]\left(C + x_i\right)(1 - x_i) \leq$$

$$\leq p_{max}\left(C + x_i\right)(1 - x_i)$$

in $(0,1)$, and as $W_{x_i' x_i'}\left(x_i, x_i'\right)$ is negative on Ω, the maximum of $W\left(x_i, .\right)$ in $\Gamma(x_i)$ is attained at $x_i'^*$. Using this fact we have

$$\max_{x_i' \in \Gamma(x_i)} \left[U\left(x_i, x_i'\right) + \delta\Psi\left(x_i'\right)\right] = U\left(x_i, x_i'^*\right) + \delta\Psi\left(x_i'^*\right) = \frac{\beta^2 C^2}{1 - \delta} + \frac{2\beta}{\delta} x_i = \Psi\left(x_i\right).$$

This means that $\Psi(x_i)$ is a solution of (3.2). Therefore, $\Psi(x_i)$ is the unique value function, and

$$\theta(x_i) = (1-r)x_i + \beta(C+x_i)^2$$

is an optimal policy function of the given optimization problem implying that

$$\hat{p}_i(t) = \beta(C+x_i(t))(1-x_i(t))$$

is in fact optimal for our model. As $U(x_i, x_i') + \delta\Psi(x_i')$ is concave in x_i', the optimal policy is unique.

3.2 Numerical Results

In the preceding subsection we derived an optimal advertising effort $\hat{p}_i(t)$ for the monopolist M_i for market V_i under the assumption that M_i has naive expectations about the states in V_i's neighbourhood. $\hat{p}_i(t)$ is hence not the optimal advertisement spending for the "real" system, where the states change in all markets.

However, $\hat{p}_i(t)$ is the optimal advertising effort for the case that there is no (or constant) influence of other markets on V_i, and as long as the influence of neighbouring markets is not too big (i.e., ρ_{ik} small for $i \neq k$), this policy should perform well compared to most other strategies.

In an extensive numerical analysis $\hat{p}_i(t)$ has been tested against two other advertising strategies, which are easy to implement. The first alternative was to choose a constant advertising effort, i.e.

$$p_i(t) \equiv: p_c = \text{ constant } \forall\, t, \quad 0 \leq p_c \leq p_{max},$$

where different values of p_c were taken into consideration. In the second alternative, $p_i(t)$ was chosen randomly at each time step, i.e.

$$p_i(t) =: p_r(t) = p_{max} \cdot random(t) \quad \forall\, t,$$

where $random(t)$ denotes a random variable, which is assigned a new value every time step and satisfies the constraint $0 \leq random(t) \leq 1$.

In all cases investigated so far, the optimal advertising effort under naive expectations, $\hat{p}_i(t)$, has proven to be better (in terms of the profit (2.2)) than constant or randomly chosen advertising strategies.

Table 3.1. Specific choice of the parameter values for the numerical example presented in this subsection.

α	β	C	L	p_{max}	r
13	5	0.1	320	2.1	0.5

Fig. 3.1. Illustration of the arrangement of the L markets along a ring; the neighbours of each market are exactly the two adjacent markets and the market itself.

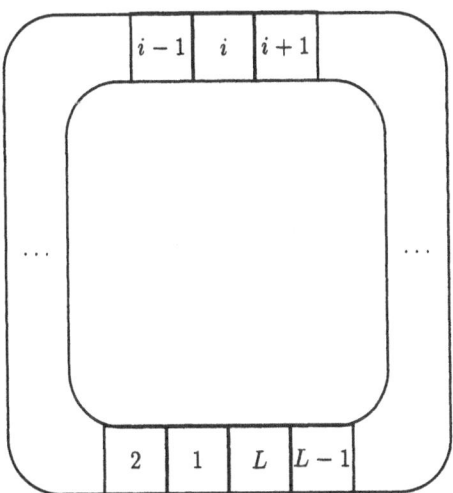

Let us have a brief look at a specific example with the parameter values chosen as given in Table 3.1. In this example, we assume the markets to be spaced along a ring (cf. the economic cellular automaton model by Keenan and O'Brien (1993)), where the neighbours of each market are exactly the two adjacent markets and the market itself (see Figure 3.1).

Table 3.2. M_i's profit for constant advertising effort p_c as percentage value of the profit under $\hat{p}_i(t)$; for each value of p_c there is an average value over ten runs as well as the best and worst result obtained within these ten runs.

p_c	Average [%]	Best [%]	Worst [%]
$0.25 \cdot p_{max}$	73.24	90.85	57.13
$0.5 \cdot p_{max}$	96.02	98.72	89.35
$0.75 \cdot p_{max}$	93.50	97.88	86.52

Table 3.3. M_i's profit for random advertising effort $p_r(t)$ as percentage value of the profit under $\hat{p}_i(t)$; there is an average value over thirty runs as well as the best and worst result obtained within these thirty runs.

Average [%]	Best [%]	Worst [%]
81.02	97.41	24.83

Tables 3.2 and 3.3 summarize the numerical results for this example. These tables give the monopolist's profit for constant and random advertising efforts, respectively, as percentage values of the profit under $\hat{p}_i(t)$. Due to the complexity of the model – recall that the total number of markets L is 320 – it is not surprising that the results depend pretty strongly on the initial states. For that reason, in the columns "Average" we give the average value over ten to thirty different simulation runs (each run with randomly chosen initial conditions), while in the columns "Best" and "Worst" we have the best and worst result obtained within these runs.

We see that not even the best result is ever as good as the strategy we derived in the previous subsection, so this strategy always "wins". However, if the constant advertising effort p_c is not too small, then this strategy is almost as good as $\hat{p}_i(t)$, even on the average, and the variance (i.e. the difference between best and worst case) is not too big. On the other hand, the random advertising effort $p_r(t)$ is without any doubt the big "loser" in this contest: the average is at 81%, the worst case at 25%, and the variance at more than 72%.

The results of this example reflect what has been found so far in all cases investigated: the best advertising strategy is $\hat{p}_i(t)$, followed by the constant effort p_c, while the random strategy definitely cannot be recommended.

4. Discussion

In this paper we derived an optimal advertising effort for a multi market advertising diffusion model, which has so far only been analysed in its descriptive form (Tragler (1996)). For that purpose, we made use of a recent approach by Dawid and Feichtinger (1995). It is worth to mention here that both Tragler (1996) and Dawid and Feichtinger (1995) placed emphasis on the chaotic behaviour of their descriptive and optimal control models, respectively. We can hence assume that also the model presented in this paper can exhibit chaotic behaviour, although this issue has not been investigated so far.

The optimal control approach presented in the previous section should be seen only as a first step in dealing with complex (i.e., multi state) dynamic optimization problems as the one analysed here. We conclude this paper by pointing to a few possible extensions / generalizations of this approach.

Concerning the particular advertising diffusion model by Tragler (1996), it is obvious that it is not necessary to restrict oneself to markets of only one size g, although some additional constraints have to be satisfied for the model parameters (like the constraint $p \leq p_{max}$). Apart from that, the neighbourhood structure does not need to be as symmetric as illustrated in Figure 3.1.

From a more general point of view, we can easily imagine how to apply the approach presented here to other multi state optimal control problems, if they have a structure comparable to that of our model. In particular, there are two mechanisms that make the optimal control under naive expectations, $\hat{p}_i(t)$, a good alternative for constant (p_c) or random $(p_r(t))$ strategies. On the one hand, $\hat{p}_i(t)$ is *optimal*, if there is *no* or *constant* influence from states other than $x_i(t)$, so that it remains a good strategy, as long as this influence is not too big. On the other hand, $\hat{p}_i(t)$ at least takes into account the variations in $x_i(t)$, which is not the case neither for constant or random strategies. Hence, the results of this paper suggest that a series of multi state optimal control problems (e.g., optimally controlled cellular automata) can approximatively be solved by assuming that most of the states are constant.

Acknowledgements

I wish to thank Herbert Dawid for many helpful comments on earlier versions of this paper.

References

Bass, F.M. (1969): A new product growth model for consumer durables. Management Science **15**, 215–227

Boldrin, M., Montrucchio, L. (1986): On the indeterminacy of capital accumulation paths. J. Economic Theory **40**, 26–39

Dawid, H., Feichtinger, G. (1995): Complex optimal policies in an advertising diffusion model. Chaos, Solitons & Fractals **5**(1), 45–53

Dodson, J.A., Muller, E. (1978): Models of new product diffusion through advertising and word-of-mouth. Management Science **24**(15), 1568–1578

Feichtinger, G., Hartl, R.F. (1986): Optimale Kontrolle ökonomischer Prozesse – Anwendungen des Maximumprinzips in den Wirtschaftswissenschaften. Walter de Gruyter, Berlin

Gould, J.P. (1970): Diffusion process and optimal advertising policy. In: Phelps, E.S. et al. (eds.): Microeconomic foundations of employment and inflation theory. Macmillan, London, 338–368

Keenan, D.C., O'Brien, M.J. (1993): Competition, collusion, and chaos. Journal of Economic Dynamics and Control **17**, 327–353

Lilien, G.L., Kotler, P. (1983): Marketing Decision Making. Harper, New York

Mazza, C., O'Connell, N. (1992): Microscopic and macroscopic aspects of epidemics. Appl. Math. Comput. **47**, 237–258

Simon, H. (1984): Goodwill und Marketingstrategie. Gabler, Wiesbaden

Stokey, N.L., Lucas, R.E. (1989): Recursive methods in economic dynamics. Harvard University Press, Cambridge

Tragler G. (1996): Chaotic behaviour in spatial advertising diffusion models. Institute of Econometrics, Operations Research and Systems Theory, Vienna University of Technology, Working Paper FB 205. [Forthcoming in: International Journal of Bifurcation and Chaos]

Yakowitz, S., Gani, J., Hayes, R. (1990): Cellular automaton modelling of epidemics. Appl. Math. Comput. **40**, 41–54

All about e_{60}

Charlotte Höhn[1]

[1] Federal Institute for Population Research, P.O.Box 5528, 65180 Wiesbaden, Germany

Abstract. This paper describes the development of life expectancy at birth and at age 60 in Austria, Germany, the Netherlands, Japan and Switzerland from 1955 to 1995. With the help of the programme LIFETIME it analyses contributions of mortality differences by age groups and causes of death to the change in life expectancy at birth in the four decades for Austria and by age groups in all the countries chosen.
The most important finding is that Austrian men aged 60 to 79 years have the most promising prospects of improvements in mortality reduction.

1. Dedication

Presumably, Gustav Feichtinger knows more about life tables, survivor functions and mortality differentials than many demographer colleagues including myself. Possibly, he would rather prefer to know all about Eve. And yet, I offer him a study on e_{60}, a few promising aspects of the future which, at least on average, is in store for him, simply statistically, of course. Whether these are more or less promising prospects lies in the eyes of the beholder. Personally, I wish Gustav Feichtinger a happy and rewarding time with friends and colleagues.

2. Trends in life expectancy at birth

As might have been expected I will not start with life expectancy at age 60 right away. Likewise, I will limit this study to a few countries only: first of all, of course, Austria, then her neighbouring countries Switzerland and Germany (separate for West and East Germany), The Netherlands (for reasons, I will reveal later) and the actual champion in life expectancy, Japan.

For this study WHO data have been used and analysed with the software package LIFETIME[1] developed by our common friend and colleague John H. Pollard[2]. The period covered is 1955-1995 with Switzerland up to 1994 and East Germany

[1] I would like to thank Karla Gärtner and Roland Greifelt, Federal Institute for Population Research, for having carried out the necessary calculations.

[2] For the first application of LIFETIME to a comparison of mortality patterns in East and West Germany with more information on this programme see Höhn, Charlotte and Pollard, John H., 1991, Mortality in the two Germanies in 1986 and trends 1976-1986, European Journal of Population, 7, 1-28

from 1969-1995. Annual, abridged life tables have been calculated so that not all values are comparable to results published by national statistical offices which usually calculate life tables for an average of more than one calendar year.

I start with the development of life expectancy for male new-borns (see fig. 1a). In 1955, life expectancy at birth is lowest for Japanese boys. This is hardly surprising since Japan is a latecomer in modernisation. The second lowest life expectancy have Austrian boys, followed by West German boys and Swiss boys. The highest life expectancy for male new-borns is found in the Netherlands in 1955.

The Netherlands in the following two decades exhibit a small decline in male life expectancy at birth followed by a moderate increase up to 1995 (and this is the reason why I included the Netherlands in this study) while all other countries show a more or less distinct increase. Starting in the mid-seventies one observes a convergence of the male e_0 for Austria, West Germany, the Netherlands and Switzerland. There is the same tendency for East Germany which is lagging behind to become the last as of 1984. From 1990 to 1991 (with Germany's reunification), there is even a decrease in male life expectancy at birth due to rapid motorisation on still old fashioned roads and highways with resulting car accidents mainly of young men.

Fig. 1a: Life expectancy at birth (male), 1955 - 1995

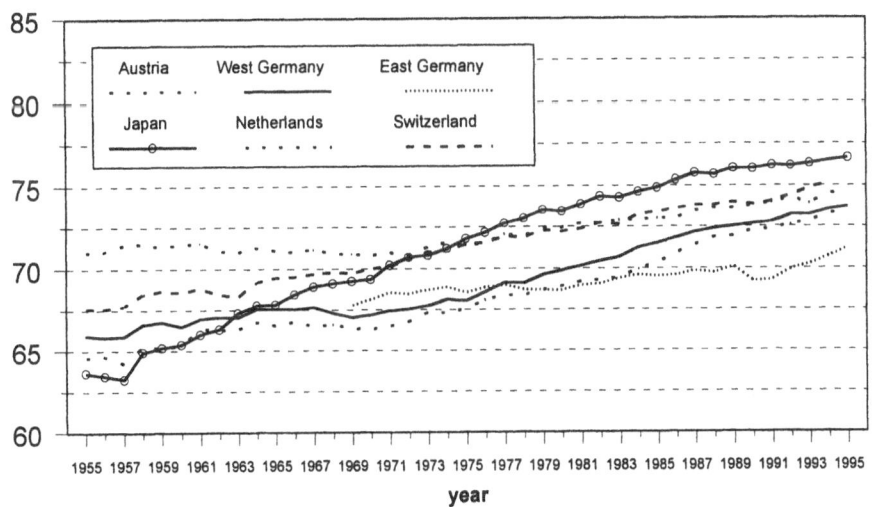

Source: WHO - data; own calculations with LIFETIME
BiB-H-FIG-1A

BiB

The strongest gains in life expectancy at birth enjoy Japanese boys bringing them to the top of the world in this respect. In our study, Japan so shoots from the last to the first rank. Switzerland is still second, the Netherlands fall back from the first to the third rank. Austria comes close to West Germany, and East Germany now holds the last rank with a tendency to improve in the future.

The trends of life expectancy at birth of girls show no decrease at no time studied (see fig. 1b). Needless to mention that the level is higher for girls than for boys. Concerning the ranking these are the same as for boys. With some interest we note that Dutch girls' life expectancy at birth stagnates since the late 80ies.

Fig. 1b: Life expectancy at birth (female), 1955 - 1995

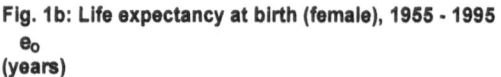

e_0
(years)

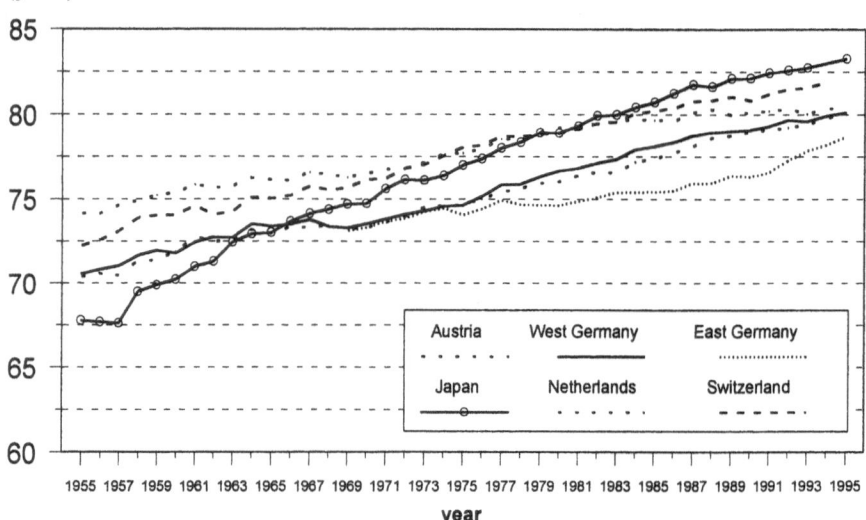

Source: WHO - data; own calculations with LIFETIME
BiB-H-FIG-1B
BiB

3. Trends in further life expectancy at age 60

Gains in further life expectancy at age 60 reveal a slightly different picture than what we saw for life expectancy at birth. Both men's and women's e_{60} stagnate in the fifties and sixties and diverge in the seventies through nineties.

Only in Japan, male e_{60} takes off already in the sixties to rush to the top (see fig. 2a). Systematic increases can also be found with Swiss men. Austria and West Germany display stagnation with backlashes until 1970 developing close to each other also in the increase up to 1995. The small decreases in the Netherlands continue until the mid-seventies followed by a very retarded and modest increase in the last decade studied. Let us note that life expectancy of East German men aged 60 stagnated until the 90ies leaving it on the last rank.

Fig. 2a: Further life expectancy at age 60, men, 1955 - 1995

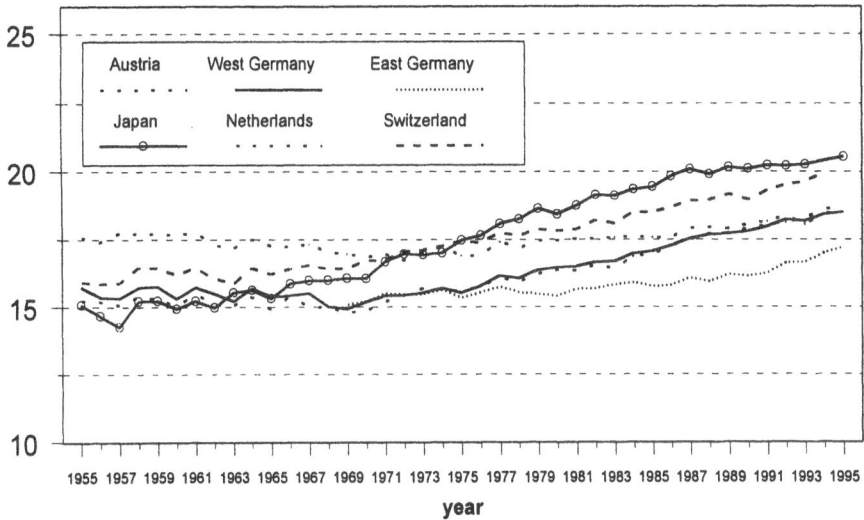

Source: WHO - data; own calculations with LIFETIME
BiB-H-FIG-2A

BiB

Again women aged 60 have positive trends of e_{60} throughout the decades studied (see fig. 2b). However, we also discern a divergence as with men. And the Dutch women have much less improvement in c_{60} in the eighties up to 1995 bringing them closer to Austrian and West German women.

While graphs provide a quick impression of levels and trends a table shows important details. Table 1 contains the differences in e_{60} between women and men reflecting excess male mortality. It is most interesting to note this plus in e_{60} in favour of women increased over the 4 decades studied. It, however, started to narrow in the Netherlands in the mid-eighties and it does not widen any longer in Austria in the nineties with a similar period between 1975 and 1980. In West Germany and Switzerland likewise this gap does not widen dramatically any longer since 1985 with periods of decrease. So, only in Japan the gap still widens and in addition the difference between men and women in e_{60} is highest. In Austria, the sex differential of e_{60} was highest in 1955 and is lowest in 1995. In conclusion it can be said that mortality of Austrian men, in relative terms to women's, decreased most rapidly.

302

Fig. 2b: Further life expectancy at age 60, women, 1955 - 1995

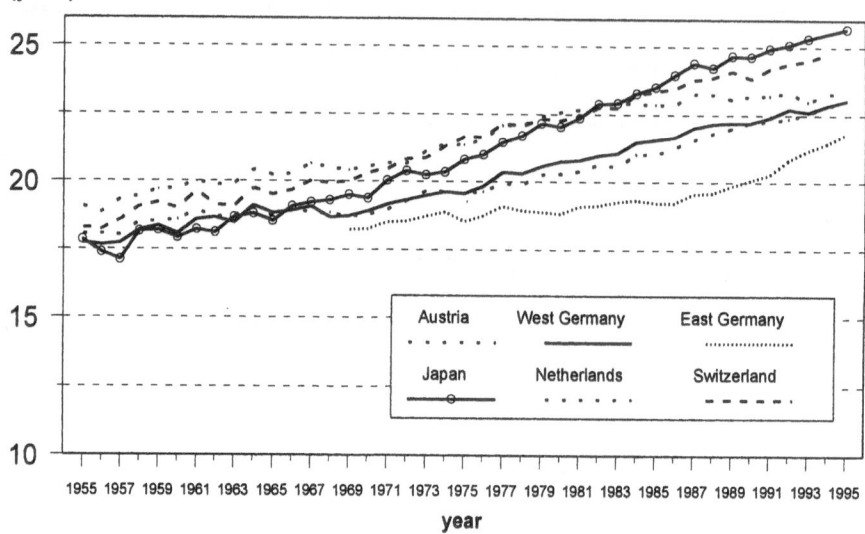

Source: WHO - data; own calculations with LIFETIME
BiB-H-FIG-2B

BiB

Tab. 1: Difference between women and men in further life expectancy at age 60

Year	Austria	Germany West	East	Netherlands	Japan	Switzerland
1955	2,80	2,00		1,49	2,75	2,36
1960	3,55	2,75		2,05	2,98	2,81
1965	3,79	3,43		2,96	3,22	3,30
1970	3,92	3,72	3,06	3,70	3,33	3,60
1975	3,95	4,03	3,20	4,46	3,33	4,20
1980	3,94	4,29	3,42	5,11	3,61	4,45
1985	4,06	4,50	3,47	5,30	4,06	4,85
1990	4,32	4,41	3,93	5,12	4,53	4,85
1995[1]	4,32	4,53	4,59	4,84	5,11	4,77

[1] Switzerland 1994; Source: WHO data, own calculations by LIFETIME

And yet, if we look at survivors at age 60 we clearly see that women fare so much better (see tab. 2). In 1995, at age 60 men just reach a percentage of survival that women reached already in 1960 with the exception of Japan. In Japan, men in 1995 survived to the same degree as Japanese women did in the mid-seventies. This is, of course, due to the late and fast improvement of life expectancy in Japan. Table 2 should be noted with care, particularly by policy-makers. 81 to 89 percent of males survive to the age of 60 and have a further life expectancy of 17 to 20 years. Females survive to age 60 even with a probability of 91 to 94 percent and then have a further life expectancy of nearly 22 to 26 years. The implications for old age security should be obvious.

Tab. 2: Survivors at age 60 in Austria, Germany, The Netherlands, Japan and Switzerland 1955 - 1995

| Year | Austria | Germany | | Netherlands | Japan | Switzerland |
		West	East			
			Men			
1955	73 000	75 184		82 793	71 124	77 527
1960	74 106	76 250		83 169	74 274	79 108
1965	76 301	77 731		82 411	78 396	80 729
1970	75 747	77 383	79 145	82 193	80 579	81 093
1975	77 075	78 298	79 699	83 437	83 789	82 976
1980	78 222	80 396	79 026	84 958	85 800	84 089
1985	80 374	82 838	80 220	86 171	86 888	85 631
1990	83 464	84 455	78 050	87 269	88 300	86 196
1995[1]	84 814	85 646	81 286	88 421	89 051	87 152
			Women			
1955	82 504	83 343		87 797	77 156	84 991
1960	84 457	85 037		89 075	81 497	87 713
1965	86 064	86 474		89 795	85 760	88 798
1970	86 506	86 563	86 953	89 722	87 902	89 561
1975	87 879	87 715	87 989	90 773	90 271	91 016
1980	89 186	89 593	88 360	91 686	92 113	91 591
1985	90 554	91 045	89 395	91 985	93 162	92 431
1990	91 640	91 898	89 569	92 380	93 969	92 851
1995[1]	92 567	92 450	91 656	92 577	94 368	93 228

[1] Switzerland 1994; Source: WHO data, own calculations by LIFETIME

4. Contributions of mortality differences by age and cause to the change in life expectancy

So far we have just described trends and levels of life expectancy in six countries. Now we should try to understand which causes of death in which age groups contributed to an increase or decrease of life expectancy. Since it would immensely blow up this article to present all findings that LIFETIME can provide[3] I limit myself to Austria, to the age groups 0, 1-19, 20-39, 40-59, 60-79 and 80+, to 6 large groups of causes of death plus the residual of causes of death and to the four decades studied. As to the presented four tables the following should be explained: The grand total is the increase of life expectancy at birth. It is given in hundredths of a year of life as are any other values. Negative values appear if a contribution by age and cause is decreasing life expectancy. The chosen causes of death are

- 1: Infectious and parasite diseases
- 2: Neoplasms
- 3: Circulatory system diseases
- 4: Digestive system diseases, excluding cirrhosis of liver
- 5: Cirrhosis of liver
- 6: External causes
- Residual: All other causes, including those related to infant mortality.

In the first decade studied (1955-1965), life expectancy at birth increased by 1.96 years for males and 2.6 years for females (see tab. 3a). The main contribution to this increase stems from the residual cause group, which contains all causes related to infant mortality (that is why I mentioned this in the above definition of the residual deliberately). Indeed, we find this contribution mainly in the column of age 0, and it is worth noting that male infants marked a higher contribution than females. The next important contribution comes for a reduction of infections mainly of the younger age groups. A reduction of infant mortality and of infectious diseases is quite typical for this stage of epidemiological transition.

Let us also note that for men we find more negative contributions than for women, most obvious for cirrhosis of liver where, however women too score a few negative contributions. Women fare much better in mortality reductions due to neoplasms and circulatory system diseases, while a reduction in external causes (traffic accidents) is higher for men aged 20-39.

In the decade 1965 to 1975, gains in life expectancy at birth are smaller: 1.02 years for males and 1.69 years for females (see tab. 3b). The gains in the residual causes are still the biggest but they no longer relate only to infant mortality, the bulk is now with persons over 40. There are remarkable negative contributions, that are increases of circulatory diseases, both for men and women in the age bracket 40+ years. For men these increases are much higher than for women at age 40 to 59 years while women score higher at age 80+ years. Negative contributions, mainly for men, also result from increases in cirrhosis of liver and external causes among the adult population.

The decade of 1975 to 1985 yields higher increases in life expectancy: 2.78 years for males and (slightly less) 2.72 years for women (see tab. 3c). Again there are the biggest gains in the residual, in infant mortality and in the age group 40+

---

[3] The mathematical basis for the decomposition of life expectancy can be found in Pollard, John H., 1988, On the decomposition of change in expectation of life and differentials in life expectancy. Demography, 25, 265-279

Tab. 3a: Contributions* of mortality differences by age and cause to the differential in complete expectation of life, Austria 1955 - 1965

Cause group	0		1 - 19		20 - 39		Age group 40 - 59		60 - 79		80+		All ages	
	M	F	M	F	M	F	M	F	M	F	M	F	M	F
1	6	6	11	10	10	10	9	7	5	7	0	1	41	40
2	-0	-0	0	-1	-4	3	8	4	-10	4	-0	-0	-8	10
3	1	1	5	5	2	4	-8	5	-9	19	4	8	-6	42
4	20	9	3	6	1	2	6	7	-0	1	0	0	30	27
5	-0	-0	0	-0	-1	-0	-7	-3	-7	-2	-0	-0	-16	-4
6	-2	-1	5	6	15	8	-1	-1	-2	0	-0	-0	17	9
Residual	121	85	10	9	1	13	10	15	-3	16	0	0	138	137
All	146	100	35	34	24	39	17	32	-26	48	2	8	196	260

Tab. 3b: Contributions* of mortality differences by age and cause to the differential in complete expectation of life, Austria 1965 - 1975

Cause group	0		1 - 19		20 - 39		Age group 40 - 59		60 - 79		80+		All ages	
	M	F	M	F	M	F	M	F	M	F	M	F	M	F
1	-0	1	1	2	2	2	10	4	9	4	0	0	25	14
2	0	-0	-1	0	2	2	0	8	14	9	-1	2	14	22
3	0	-0	-1	-0	-1	-3	-22	-6	-60	-63	-23	-45	-109	-117
4	2	3	4	1	-1	1	0	2	6	7	0	1	10	16
5	0	0	-0	-0	-3	-0	-12	-4	-2	-0	-0	0	-16	-5
6	-1	0	-10	-5	-10	-8	-3	2	-2	-0	0	-0	-26	-9
Residual	51	56	9	11	7	9	26	18	86	101	25	54	204	250
All	52	59	3	10	-2	3	-2	25	50	59	1	13	102	169

*Contributions are hundredths of a year of life
Source: WHO data, own calculations by LIFETIME

Tab. 3c: Contributions* of mortality differences by age and cause to the differential in complete expectation of life, Austria 1975 - 1985

Cause group	0		1 - 19		Age group 20 - 39		40 - 59		60 - 79		80+		All ages	
	M	F	M	F	M	F	M	F	M	F	M	F	M	F
1	-1	0	-0	1	0	0	2	-0	3	0	0	0	7	5
2	0	-0	-1	-0	-1	2	-5	5	8	1	-2	-1	0	5
3	0	0	1	-0	2	4	15	11	38	56	4	15	60	84
4	8	6	3	0	-3	-1	-25	-11	-12	1	-0	-0	-29	-5
5	0	0	0	0	6	2	33	11	21	9	1	1	62	23
6	3	2	19	11	15	5	7	1	10	5	0	3	56	26
Residual	60	56	11	10	6	7	13	17	26	30	7	13	122	133
All	71	65	32	25	27	17	39	35	97	101	11	31	278	272

Tab. 3d: Contributions* of mortality differences by age and cause to the differential in complete expectation of life, Austria 1985 - 1995

Cause group	0		1 - 19		Age group 20 - 39		40 - 59		60 - 79		80+		All ages	
	M	F	M	F	M	F	M	F	M	F	M	F	M	F
1	1	1	0	0	1	0	3	0	2	-0	0	0	7	2
2	0	1	3	1	5	3	11	6	8	13	2	2	29	26
3	0	0	-0	-0	4	2	30	10	69	73	23	46	126	132
4	2	1	-0	0	3	2	9	5	8	6	2	3	23	17
5	0	0	0	0	0	0	0	0	0	0	0	0	0	0
6	3	1	9	4	16	5	19	6	6	8	2	7	54	31
Residual	46	30	5	5	1	3	4	6	16	15	7	8	79	69
All	52	33	17	10	27	16	76	34	108	117	35	67	318	277

*Contributions are hundredths of a year of life
Source: WHO data, own calculations by LIFETIME

years. It is the first decade when increases in life expectancy at birth are no longer mainly due to decreases in infant mortality. Now the age group 60 to 79 years enjoys the biggest progress. We enter the mortality transition lead by declines in mortality of the senior population.

Opposite to the decade before, circulatory system diseases (more for women than for men), cirrhosis of liver and external causes (more for men than for women) now show mortality reductions. But there are also negative contributions with mortality increasing due to digestive system diseases (excluding cirrhosis of liver).

During the most recent decade studied (1985-1995), men again score a higher increase of life expectancy with 3.18 years (females 2.77 years) (see tab. 3d). Here we see virtually no negative contributions at all. Infant mortality is still declining. But the important gains are made by men and women aged 60-79 years and aged 80+ years where women, however, fare much better than men.

The biggest gains consequently do no longer emanate from the residual but from a mortality reduction in circulatory system diseases which is the leading cause of death in industrialised countries. For the first time we also note a reduction in mortality of neoplasms. In this decade there are also and again mortality reductions in digestive system diseases and external causes. But there is no change at all in the mortality level of cirrhosis of liver.

5. Contributions of mortality differences by age to life expectancy in Austria, Germany, The Netherlands, Japan and Switzerland

In this final part we compare Austrian gains in life expectancy at birth with those in West and East Germany, the Netherlands, Japan and Switzerland. From the tables similar to tables 3a-d for Austria we just consider the bottom lines which summarise the effects of decomposition by causes of death and age for the chosen age groups only, and we retain the four periods/decades (see tab. 4).

The increases in life expectancy over four decades show quite diverse developments in the six countries chosen. As in Austria also in West Germany the smallest gains were observed in the decade 1965/75. And while Austria has the biggest increases in the last decade West Germany has the highest one decade earlier: from 1975 to 1985.

In East Germany where we can only analyse the last two decades the improvement is biggest in the last decade as in Austria. But these increases are much smaller and, opposite to Austria, higher for women than for men.

Tab. 4: Contributions* of mortality differences by age to the differential in complete expectation of life in Austria, Germany, The Netherlands, Japan and Switzerland

Austria

Period	0		1 - 19		20 - 39		Age group 40 – 59		60 - 79		80+		All ages	
	M	F	M	F	M	F	M	F	M	F	M	F	M	F
1955/65	146	100	35	34	24	39	17	32	-26	48	2	8	196	260
1965/75	52	59	3	10	-2	3	-2	25	50	59	1	13	102	169
1975/85	71	65	32	25	27	17	39	35	97	101	11	31	278	272
1985/95	52	33	17	10	27	16	76	34	108	117	35	67	318	277

West Germany

Period	0		1 - 19		20 – 39		Age group 40 - 59		60 - 79		80+		All ages	
	M	F	M	F	M	F	M	F	M	F	M	F	M	F
1955/65	134	118	25	24	24	29	7	22	-35	67	8	24	163	284
1965/75	30	29	8	7	0	9	-3	17	11	55	0	9	47	125
1975/85	87	72	41	24	41	21	55	50	108	128	17	52	348	347
1985/95	30	26	13	11	8	8	51	23	95	80	27	55	224	203

East Germany

Period	0		1 - 19		20 – 39		Age group 40 - 59		60 - 79		80+		All ages	
	M	F	M	F	M	F	M	F	M	F	M	F	M	F
1975/85	51	40	16	11	2	5	-5	18	34	47	0	16	98	138
1985/95	34	26	13	12	-7	10	10	42	86	147	32	87	169	324

The Netherlands

Period	0		1 - 19		20 - 39		Age group 40 – 59		60 - 79		80+		All ages	
	M	F	M	F	M	F	M	F	M	F	M	F	M	F
1955/65	49	36	12	15	-3	16	-24	28	-37	75	11	27	9	198
1965/75	30	27	16	10	14	5	7	11	30	76	0	25	34	156
1975/85	22	16	27	18	10	8	47	21	51	81	7	58	163	200
1985/95	20	21	5	3	6	-0	43	8	76	28	1	13	151	72

Japan

Period	0		1 - 19		20 - 39		Age group 40 – 59		60 - 79		80+		All ages	
	M	F	M	F	M	F	M	F	M	F	M	F	M	F
1955/65	144	148	116	131	81	101	63	87	24	66	-7	-9	421	525
1965/75	69	59	32	27	47	38	73	74	150	156	26	44	397	397
1975/85	39	29	26	19	31	25	46	53	137	166	30	83	309	375
1985/95	8	10	10	5	12	9	47	27	62	103	35	98	176	254

Switzerland

Period	0		1 - 19		20 - 39		Age group 40 – 59		60 - 79		80+		All ages	
	M	F	M	F	M	F	M	F	M	F	M	F	M	F
1955/65	65	61	34	31	32	26	34	57	14	83	8	22	187	282
1965/75	54	53	9	7	3	7	33	39	83	142	20	51	201	299
1975/85	33	23	18	11	9	2	49	27	73	100	15	56	198	221
1985/94	15	13	18	10	-24	0	35	16	103	70	25	59	171	168

Source: WHO data, own calculations by LIFETIME

The Netherlands have the strangest pattern. Males improve their life expectancy up to 1985 with smaller but still sizeable gains in 1985/95. Dutch females have a remarkably low gain in this last period.

In Japan the gains are extremely high in the first period and become smaller in the following decades. Also Switzerland with the next highest level of life expectancy shows a smaller increase in the last decade. The path is nevertheless quite even over the decades.

So in the last decade studied Austria has the highest absolute gains, and, which is also special, more so for males than for females. The highest gains for females is to be found in East Germany.

As to the pattern of improvements we see the highest gains in infant mortality in the first decade 1955 to 1965. With time passing by the gains move to the middle aged adults and then to the seniors. In the age bracket 80+ years there are systematic improvements (more for women than for men) in Japan but also in Austria and East Germany.

Our main interest should focus on the age bracket 60 to 79 years. We note that in the decade 1955 to 1965 men scored a negative contribution in Austria, West Germany and the Netherlands. In Austria the situation of men aged 60 to 79 years improved from decade to decade as in East Germany, the Netherlands and in principle in Switzerland. The same trends hold true for Austrian and East German women, however on a higher level than men.

West German men and women and Dutch and Japanese women had the biggest improvements in the age bracket 60 to 79 years in the decade 1975 to 1985. Swiss women and Japanese men enjoyed the highest gains in the decade 1965/75. This means that the neat positive trend as in Austria cannot be found in West Germany and Japan, and with Dutch and Swiss women.

6. Personal conclusion

In a personal summary for Gustav Feichtinger let me say that he seems to live in the right country at the right age. Though predictions are always a risk, particularly when they concern the future, he grows into the healthiest environment for men surpassing the threshold of 60 years. In no other country so far men have higher gains in mortality reduction and the development in Japan for the oldest old (80+) leaves margin for improvements in the two decades ahead. Felix Austria has the soil, it seems, to develop the Swiss example where men aged 60 to 79 mark an even higher improvement than women. May Gustav Feichtinger enjoy the decades ahead in excellent health and spirit.

Quantifying Vicious Circle Dynamics: The PEDA Model for Population, Environment, Development and Agriculture in African Countries

Wolfgang Lutz[1], Sergei Scherbov[2]

[1] International Institute for Applied Systems Analysis, Schlossplatz 1, A-2361 Laxenburg, Austria
[2] University of Groningen, Population Research Centre, P.O. Box 800, NL-9700 AV Groningen, the Netherlands

Abstract. This paper develops a quantitative simulation model linking population parameters and education to land degradation, food production and distribution, and resulting in the proportion of the population which is food insecure. This model is inspired by the Vicious Circle Model of Dasgupta and others, but can be applied more generally to interactions between these variables. The model chooses a population-based approach which groups individuals into eight categories as defined by rural/urban place of residence, literacy status and food security status. Using the tools of multi-state population projections, each group is simulated by age and sex. The model links this population module to an agricultural production function and a food distribution function which considers the fact that not all people have equal access to the food produced. This model has been applied to several African countries. Here it is illustrated with an application to Burkina Faso.

1. Introduction: Population-Environment Models

Broadly speaking, in the field of population-environment models we can distinguish between two kinds of approaches: 1) comprehensive models that try to assess the full range of population-environment interactions for a specific region, and 2) other models that limit the focus to specific assumed chains of causation and therefore tend to be more focused and theory-driven. Both approaches can contribute to the better understanding of this complex field of studies. Both have their strengths and weaknesses. The more comprehensive (holistic) approach (1), which tries to evaluate all relevant factors, can help us to better understand the relative contribution of specific factors to the full picture. The series of PDE (Population-Development-Environment) case studies conducted by IIASA in different parts of the world are a good example for such comprehensive studies, which try to incorporate all relevant factors. The PEDA model presented in this paper follows the other, substantively more focused, strategy (2). It attempts to quantify one specific causal path which actually is an assumed loop or circle that follows a clearly-defined theoretical model. It is restricted to portraying factors that are relevant to that specific mechanism, leaving out all others. This approach is more in line with the tradition of economic modeling that tends to make *ceteris paribus* assumptions on all factors that are not directly relevant to the hypothesis studied, even though such factors may be very significant for the future of a country under a more comprehensive approach. For planning purposes and science-policy interactions, both the more comprehensive and the focused approaches have their virtues and shortcomings. In an ideal world a comprehensive super-model may incorporate several focused models, but this is difficult to achieve and may, indeed, suffer from some of the well-known shortages of mammoth models.

PEDA has been commissioned by the United Nations Economic Commission for Africa (UN-ECA) as an advocacy tool at the national level. In this it should compare with the widely used RAPID models but with a different focus and – this was the precondition of the authors' participation – a strong scientific foundation. It is meant to do advocacy not for any specific policy but for policies to take account of the strong existing nexus between population, environment and agricultural development. By the end of 1999 PEDA will have been initialized for six African countries (Burkina Faso, Cameroon, Madagascar, Mali, Uganda, and Zambia). It is planned to be applied to several more African countries in which food insecurity is a serious issue.

2. Focusing on the Vicious Circle Model (VCM)

In recent years a theoretical model, often labeled the "vicious circle model," has become a very influential paradigm for describing the interactions between population growth, low status of women and children, environmental degradation and food insecurity. It is essentially an extension of the basic neoclassical economic model which adds a poverty trap based not on the macroeconomic reasoning of Malthus, but on microeconomic effects at the household and community level (O'Neill et al. 1999). This model is based on the assumption that high fertility, poverty, low education and low status of women and children are bound up in a web of interactions with environmental degradation and declining food production, in such a way that stress from one of these sources can trap certain rural societies, especially those living in marginal lands, into a vicious circle of increasingly destructive responses.

Dasgupta (1993) presents this argument in a generalized form. The condition of poverty and illiteracy of the households concerned prevents substitution of alternative fuel sources or alternative livelihoods. A gender dimension is being added through the fact that the low status of women and girls also devalues the increasing amount of time and effort that they must devote to daily fuelwood gathering (Agarwal 1994; Sen 1994). The education of girls is blocked because girls are kept at home to help their mothers. The result is faster population growth, further degradation of the renewable resource base, increasing food insecurity, stagnating education levels, and yet a further erosion of women's status.

Leaving aside its actual empirical relevance, this vicious circle reasoning presents an important contribution to theory: it provides a unified framework for fertility, poverty, low female status and environmental degradation, which is also politically attractive because it explicitly addresses equity concerns. In this spirit, PEDA has been developed as a model that among other things can help quantify assumed vicious circle dynamics at the level of individual African countries. The PEDA model (as summarized in Figure 1) can be set up in a way that rapid population growth due to high fertility of the illiterate, food-insecure population in marginal rural areas[1] contributes to further degradation of the land, thus lowering agricultural production and further increasing the number of food-insecure. If not broken this vicious circle would lead to ever increasing land degradation and increases in the food-insecure population. The circle can be broken, however, through several possible interventions in the field of food production, food distribution, education, environmental protection and population dynamics. Such a quantitative model can help policy makers and other users to (a) view these interconnected aspects, and (b) think in terms of alternative outcomes of alternative policy scenarios.

Figure 1. Basic structure of the PEDA Model.

PEDA AFRICA
A Model Linking Population, Food Security and the Environment

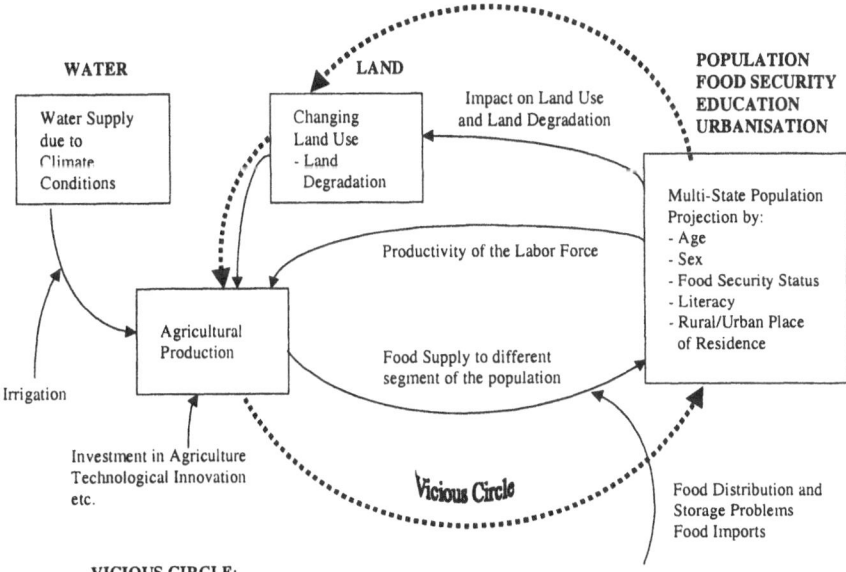

•••••• VICICOUS CIRCLE:
High population growth of the rural food insecure population will contribute to degrade the marginal lands. This decreases agricultural production which in turn still increases the number of food insecure persons.

---

[1] Unlike some of the theoretical models, PEDA does not assume that fertility increases due to land degradation. Fertility is an exogenous variable here for which scenarios can be set by the user for each of the eight sub-populations.

314

3. The Population-Based Approach to Population-Environment Modeling

The population-based approach views human beings with certain characteristics (such as age, sex, education, health, food security status, place of residence, etc.) as agents of social, economic, cultural and environmental change, but also as victims/beneficiaries who are at risk for suffering from repercussions of these changes, or benefiting from positive implications. In this sense the human population is seen as a driving force of these changes and is affected by the outcomes and consequences of these changes. The population-based approach does not assume that population growth or other demographic changes are necessarily the most important factors in shaping our future. It must not be misunderstood in the sense of a narrow view in which only demography matters. Instead the phenomena that we want to model are studied in terms of different characteristics that can be directly attached and (at least theoretically) measured with individual members of the population. Characteristics such as age, sex, literacy, place of residence and even nutritional status can be assessed at the individual level. The sum over these individual characteristics makes up the distribution in the total population. These individual characteristics are different from other frequently-used indicators such as the GNP per capita that cannot be directly measured with individuals. Although many of the powerful quantitative economic tools cannot be applied due to this choice of approach, other very powerful but less well known tools of multi-state demographic analysis and projection can be applied.

Figure 2. PEDA-population segment: A multi-state model by place of residence, education and food security status.

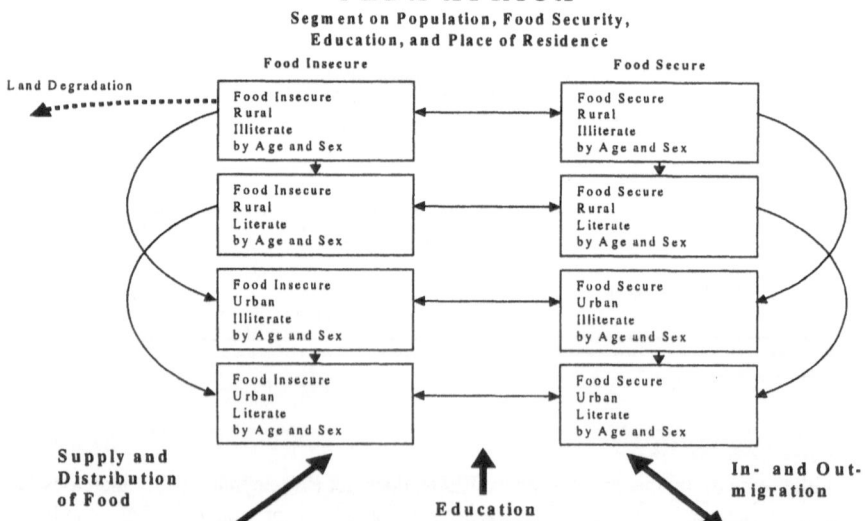

As shown in Figure 2, in PEDA the population of a country under consideration is broken down into eight sub-groups according to urban/rural place of residence, education and food security status. Place of residence and food security status are two dimensions which are core elements of the vicious circle reasoning as specified in this setting. Education, or more precisely literacy status, has been introduced into the model as one of the assumed key sources of population heterogeneity, which is related to both agricultural production and land degradation. Significant educational fertility differentials give the explicit consideration of education in the model a strong rationale. There is abundant literature on the significance of literacy in population-development-environment interactions (see, e.g., Lutz 1994). The potential of explicitly including education as a demographic dimension in multi-state population projection models has recently been evaluated (see Lutz et al. 1999) and is strongly recommended in the case of educational fertility (or other behavioral) differentials.

Each of these eight sub-groups further subdivides the population by age and sex, i.e., every one of the eight groups has its own age pyramid. During each one-year simulation step, a person will move up the age pyramid by one year within the same sub-group, or move to another sub-group while aging by one year. The movements between groups that are possible within each step are shown by arrows in Figure 2. For education and rural/urban migration, the model is hierarchical, i.e., people can only move in one direction, from lower to higher education and from rural areas to urban areas. Movement between food security states can happen in both directions, depending on the food conditions in the relevant year.

3.1. Fertility and mortality

In addition to the movements between sub-groups, each of the groups also experiences the vital events of births and deaths. Different sets of age-specific fertility rates are applied to the female populations in each of the sub-groups. The data for the starting year are based on empirical information about differential fertility rates by education and urban/rural place of residence. Fertility differentials by food security status typically need to be assumed because of the absence of empirical information. For the future years, fertility within each sub-group can either be held constant or be changed according to an assumed linear trend.

Age- and sex-specific mortality rates can also be set up independently for the different groups. This can have important implications for population dynamics since the food-insecure groups can be expected to have much higher mortality rates than the food-secure groups. Unfortunately, since almost no empirical information exists on these mortality differentials, most of the differentials have to be based on assumptions. Sensitivity analysis can then demonstrate the implications of different assumptions. Of course, special survey information on this could provide the necessary empirical data.

3.2. Education and rural/urban migration

As indicated above, it is assumed that persons can only move from the illiterate to the literate state, but if considered important, secondary illiteracy can be incorporated relatively easily. For simplicity, it is assumed that all education takes place in childhood, but again, adult literacy campaigns can be incorporated, if necessary.

In the model the transition to literacy can be defined in terms of the total educational transition rate, which defines the proportion of each cohort of girls or boys that will become literate.

In a similar fashion, the level of rural to urban migration can be defined through the total migration rate, giving the proportion of each rural cohort to move to urban areas. The only difference to education is that the migration is assumed to be less concentrated in a specific age group, but is spread over a broader age range according to typical age-specific migration patterns.

4. Land, Water and Agricultural Production

As indicated in Figure 1, the population module, i.e., the population by age and sex in the eight defined categories and for each year in time, affects the total agricultural production in two different ways. The productivity of the rural labor force as measured by the proportion literate of the rural population of productive age will directly enter the agricultural production function as discussed below. The other chain of causation is a direct reflection of the vicious circle reasoning: the factor land is degraded as a function of the increase in the number of people in the rural, food-insecure and illiterate category. In the current version of PEDA this impact is operationalized in the following manner: the total amount of high quality agricultural land enters the production function in an index form which is assumed to combine quantity and quality aspects. The higher the increase (as compared to the starting conditions) in this critical group of food-insecure, rural, illiterate population, the more this land factor will decline. The user can set a scenario variable, the land degradation impact factor, that determines to which degree a certain percentage increase in this critical population impacts on the land index.

Water is another important environmental factor for food security and agricultural production. No life can exist without water. And certainly no economic or human development is possible without water. But water only becomes a problem for food production if the demand (including household consumption and industrial consumption in addition to agricultural demand) exceeds the supply. Water supply, however, can vary greatly according to short- and long-term natural fluctuations in rainfall. It is greatly influenced by the surface structure, the pattern of river basins and groundwater systems, and it can be strongly influenced by human engineering (dams, irrigation, etc.). Because water is so important for sustainable development and follows complex non-linear dynamics, the PDE models developed by IIASA all have water modules that tend to be at least as complex as the population modules. For PEDA, however, water modeling is not a prime focus. In the vicious circle model, water only matters as one additional exogenous factor that can have impacts on the total food production, especially with respect to drought on the one hand, and irrigation efforts on the other. For this reason PEDA treats water as an index variable (with 1.0 reflecting the average conditions around the starting year) that can be enhanced through irrigation or natural changes and diminished through declines on water supply.

The total agricultural production in one year, measured in total calories produced, will then be a result of the input in terms of human labor force by different educational levels, water, land, and technological inputs such as fertilizers, mechanization, etc. Those additional inputs will also be treated in terms of externally-defined scenarios because these factors are not assumed to depend directly on other variables of the PEDA model. (If a user, however, wants to make, e.g., the rate of new agricultural investment dependent on population growth in either a positive or negative way, it is not difficult to do so and study the alternative results.) The specific agricultural production function used here is a Cobb-Douglas type production function estimated on time series data for a large number of developing countries, and has been derived from an internationally highly renown book in the field (Hayami and Ruttan 1971, p. 145, Q 19): .534 * Rural Labor Force (specified here as total rural adult population aged 15-60, calculated from combining the appropriate age groups in all four rural sub-groups); .088 * Total Agricultural Land (can be modified through land degradation or the clearing of new land as discussed); .162 * Fertilizer Use (will be treated as an exogenous scenario variable); .072 * Tractors Available (will be treated as an exogenous scenario variable called more broadly "mechanization"); .276 * Literacy (specified here as the proportion literate of the total rural population aged 10-45, calculated by combining both the food-secure and the food-insecure rural literate sub-populations); .158 * Technical Education (still treated here as an exogenous scenario variable, may later be related to the educational efforts parameter).

All these input variables to agricultural production are considered here on a relative scale, i.e., they are set to equal 1.0 in the starting year, and then change over time as it results from the other sectors of the model for the endogenous production factors or as defined in the scenario setting for the exogenous variables. (For example, an assumed increase in fertilizer input of 20 percent by 2003 would mean that the variable is set to gradually increase to 1.2 by that year.)

Unfortunately, in reality, not all the production will be consumed by individuals to satisfy their food needs. Some calories will be lost during the treatment of the food, others will be lost during transport and some will be lost due to inadequate storage. Of the food that will actually reach people for consumption, a certain fraction will go to urban areas and another to rural areas. All these factors can be assumed in PEDA as scenario variables specific for a country and can be changed over time, or alternative starting values can be assumed. More specifically, in the model there are three different scenario variables that the user can set: loss in transport and storage, food import/export, and an urban bias factor. The latter determines to what degree the total available food should be disproportionally distributed between urban and rural areas. If this factor is 1, then food will be distributed according to the population size of urban and rural areas. Within these areas, however not everybody will receive an equal amount of food. Reality shows that there are gross inequalities in access to food and therefore, PEDA has an explicit food distribution module.

5. Food Distribution

Even when the total amount of food reaching the (urban and rural total) population would be theoretically sufficient to provide the necessary minimum diet for everybody, in practice the distribution of food is unequal because some persons do have more purchasing power than others or have privileged access to food by other means. This will result in the fact that some people remain food-insecure even when the average total amount of food reaching the population is above the minimum.

There is abundant empirical evidence, backed up by theoretical considerations, clearly showing that the distribution of food is at least as important as the total production of food in explaining food insecurity. Especially the path-breaking work of Amartya Sen (1994) demonstrated that some of the worst famines occurred under conditions in which theoretically there would have been enough food for everybody if the distribution had been appropriate. For this reason it is evident that a model focusing on food security without paying attention to the distributional aspects would be incomplete, if not misleading. The main problem with considering such distributions, however, lies in the fact that hardly any empirical data exist on distributive mechanisms in the countries of Africa today, and that theoretical distributions are hardly appropriate because conditions tend to vary significantly from one country to another. As a solution to this problem, in PEDA we chose to approximate the food distribution function through an income distribution function, which exists for a number of African countries based on household income surveys. This allocation of food to urban and rural populations and the food distribution within these populations then determines the new sizes of the food-secure and food-insecure sub-populations in the following year.

Figure 3 shows such a food distribution function that is applied after allocating the total available food to urban and rural populations (according to an exogenously-defined "urban bias" variable). This figure shows a Lorenz curve with the cumulated proportion of the population on one axis and the cumulated calories available for distribution on the other. The available food is then distributed from left to right along the black curve. The given curve indicates that in this case, the first (most privileged) 10 percent of the population use 30 percent of the available food. Going further down the curve, about 23 percent of the population use half of the food, and half of the population uses 75 percent of the food. The borderline between the food-secure and the food-insecure population is then established by applying an externally-defined minimum calorie requirement per person. At the point where the remaining food supply falls below the minimum requirement times the remaining population, the border line for the population considered to be food-insecure is established. Over time the proportions food-insecure may change as a consequence of changes in the calories available for distribution or possible changes in the assumed food distribution function.

Figure 3. Food distribution function.

Food Distribution Function

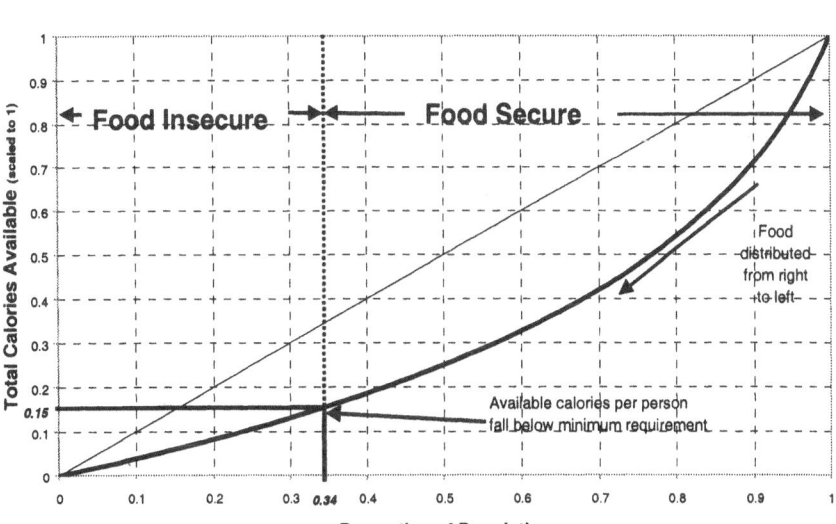

Proportion of Population

6. Operationalization and Sample Application to Burkina Faso

PEDA needs empirical data primarily for the starting conditions, for the transition rates between the different population groups, and for the factors associated with food production and food distribution. Since data sources for many African countries are limited, some of these data need to be based on estimations and assumptions. For the starting year one must have the distribution of the population by age: in 100 single-year age groups, sex (female/male), food security status (food-secure/food-insecure as defined below), literacy (literate/illiterate according to the UNESCO definition) and urban/rural place of residence (according to the national definition of a town). If the information is not readily available, one must estimate some of the distributions, as will be discussed below.

Before the model can be applied to a new country, it needs to be initialized. If all data would be readily available by single years of age, this would be a relatively straightforward process of entering data. Unfortunately, only a few countries in the world have this kind of data available, and none of them are in the African region. Hence, the process of initializing the model for African countries requires quite elaborate estimation techniques whose specific nature will vary from case to case, depending on the kind of empirical information available. In most cases, this requires special skills in mathematical demography and the programming of macros for spreadsheets. For this reason the process of initialization will not be described here. For the following description we assume that a country application has already been initialized.

The model is based on Excel and will run under Windows 95, 98, NT, 2000. It should run on any standard PC (given that all the necessary libraries have been installed) but the performance will depend on the speed of the hardware.

The model runs in single years of time. In the first year the people that are members of the eight different sub-groups will impact in two independent ways on the rest of the model. This is indicated in Figure 1 of the vicious circle by the two solid arrows leaving the population box on the right. Together with the other input factors these population factors will then result in a certain total calorie supply in the following year $(t+1)$. After adjustments for loss and import/exports in that year, the above-described food distribution function will then determine the proportions considered food secure and insecure in urban and rural areas. Together with the demographic transition rates (fertility, mortality, migration, education, etc.) applied between time t and $t+1$ this food distribution will then result in the new distribution of the total population by age and sex over the eight states at time $t+1$. This population will then serve as input to the agricultural production at time $t+2$, and so on.

We choose Burkina Faso as an example of a country that has already been initialized. It is a country with high food insecurity, high illiteracy, very high fertility and land degradation problems, which are rather typical for the whole Sahel region. For this application all data have been derived from internationally available sources. Missing data had to be estimated using various indirect procedures. It is, of course, desirable to ultimately run the model with as much empirical data as possible. This needs to be collected in collaboration with local scientists.

Current fertility levels in Burkina Faso are still very high. According to surveys, the rural illiterate population has a TFR of almost seven. The rural literate and the urban illiterate populations have rather similar fertility levels around five. Only the urban literate population has levels slightly below four. Because of a lack of empirical information about fertility by food security status, we assumed in these illustrative scenarios equal fertility levels for the food secure and insecure. These high fertility levels together with the fact that about three out of four women in Burkina Faso are rural and illiterate lets us expect very rapid population growth under the constant rates scenario.

The following four scenarios have been defined for illustration: *Constant Rates:* All parameters (including fertility and school enrolment) remain at their 1996 levels; *Increased Technological Inputs:* Fertilizer use, machinery and irrigation all increase by 3 percent per year (other parameters constant); *Strong Educational Efforts:* 80 percent of all girls and 90 percent of all boys learn to read and write (other parameters constant); *Combining Increased Technological Inputs, Strong Educational Efforts and Fertility Decline:* Levels decline to half of their 1996 values by 2030.

Figure 4. Scenario results for Burkina Faso for the total population (top lines) and for the food insecure population (bottom lines).

Scenario results for total and food insecure population

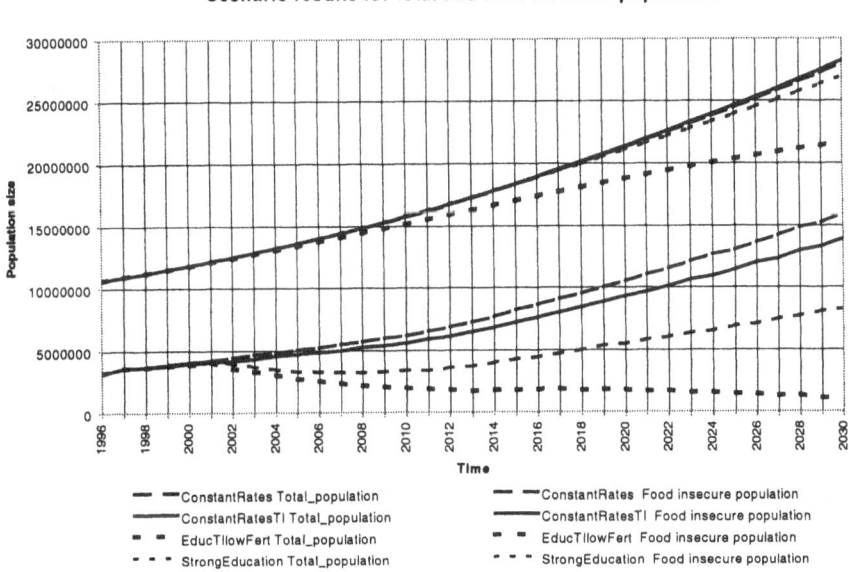

Figure 4 shows the results for the four scenarios for Burkina Faso to the year 2030 giving the total population (all eight sub-groups combined) and the food-insecure population (all four food-insecure sub-groups combined) in two separate sets of lines.

The constant rates scenario shows that over these 35 years the total population would increase by a factor of almost three. The food insecure population grows even more rapidly under this scenario with the proportion food insecure increasing from around 30 percent to more than 50 percent in 2030.

Adding the increased technological inputs to otherwise constant parameters does not affect the total population size but has a visible impact on reducing food insecurity which, under this scenario, would not reach 50 percent by 2030.

Adding the educational efforts to otherwise constant parameters (and no increased technological inputs) already has a visible impact on reducing total population size (through lower fertility of educated women) and also shows a very sizeable effect on reducing food insecurity (through the additional effect of higher productivity).

It is interesting to note that the effect of increased educational efforts only starts to become visible with a certain time lag, because education is concentrated in the younger ages that will only later enter the reproductive ages and the labor force. Hence in the very short run increased technological input has a great immediate effect, but in the longer run the educational effect is by far more important (at least under the specific assumptions made in this example). The decline in food insecurity brought along by education alone is not sustainable, however. After 2010-15 the proportion food-insecure starts to increase again because the still very high rates of population growth will outweigh the improvements in agricultural production.

Only the fourth scenario, combining the increased technological inputs with the education efforts and an assumed gradual halving of fertility rates by 2030, will result in a sustainable decline in the proportion food-insecure and in a path of total population growth that would "only" result in a doubling of the population by 2030.

As stated above, these four illustrative scenarios result from four specific rather *ad hoc* choices of model parameters. They provide some interesting stories and insights into the dynamics of the system. Other scenarios may tell other stories. The user of PEDA can calibrate innumerable different scenarios based on different parameter assumptions that he/she may find interesting or worth communicating to the policy makers.

It is our hope that this effort of trying to quantify vicious circle dynamics for specific African countries presents a useful step in further advancing work on the model itself, empirical applications to real world conditions, interdisciplinary dialogue on population-environment interactions, and finally, science-policy dialogue on the importance of dealing with these interconnected issues in an integrated manner.

References

1. Agarwal, B. (1994): The Gender and Environment Debate: Lessons from India. In: Arizpe, L., Stone, M.P., Major, D.C. (Eds.): Population and the Environment: Rethinking the Debate. Westview Press, Boulder, CO, 97-124.
2. Dasgupta, P.S. (1993): An Inquiry into Well-Being and Destitution. Oxford University Press, Oxford, U.K.
3. Hayami, Y., Ruttan, V. (1971): Agricultural Development: An International Perspective. Johns Hopkins University Press, Baltimore, MD.
4. Lutz, W. (Ed.) (1994): Population-Development-Environment: Understanding their Interactions in Mauritius. Springer Verlag, Berlin.
5. Lutz, W., Goujon, A., Doblhammer-Reiter, G. (1999): Demographic Dimensions in Forecasting: Adding Education to Age and Sex. In: Lutz, W., Vaupel, J.W., Ahlburg, D.A. (Eds.): Frontiers of Population Forecasting. A Supplement to Vol. 24, 1998, Population and Development Review. Population Council, New York, 42-58.
6. O'Neill, B.C., MacKellar, F.L., Lutz, W. (1999): Population and Climate Change. To be published by Cambridge University Press, forthcoming.
7. Sen, A. (1994): Women, Poverty, and Population: Issues for the Concerned Environmentalist. In: Arizpe, L., Stone, M.P., Major, D.C. (Eds.): Population and the Environment: Rethinking the Debate. Westview Press, Boulder, CO, 67-86.

The Complexity of the Malthusian Trap and Potential Routes of Escape[*]

Alexia Prskawetz[1] and Alessandra Gragnani[2]

[1] Max Planck Institute for Demographic Research, Rostock, Germany
[2] Department of Electronics and Computing, Politecnico of Milan, Italy

1. Introduction

One of the most challenging subjects in economic growth theory is the study of economic development ranging from Malthusian stagnation to the modern growth regime within a unified model. Recent work in this area includes Becker et al. (1999), Galor and Weil (1998), Goodfriend and McDermott (1995), Jones (1999) and Kremer (1993). While the Malthusian regime is characterised by low and non-growing per capita output and high mortality and fertility, constant growth rates of per capita output and low levels of fertility and mortality will prevail in the modern growth regime. Moreover, the relation between population growth and per capita income will be positive in the Malthusian regime and turn negative in the modern growth regime. "Demographic transition" is the theoretical concept that explains the change in the relation between per capita income and population growth.

In order to understand the transition, one has to understand how society moves from the low-level Malthusian equilibrium to the high level equilibrium of the modern growth regime. Common to most of the models mentioned above is the assumption that population growth initiates specialisation, human capital accumulation, and technological progress so that decreasing returns to scale due to a fixed factor in production (as prevalent in the Malthusian regime) will be overcome in the long run. As per capita income continuously increases, the demographic transition will be initiated by a decline in mortality, followed by a decline in fertility. During the transition period high levels of population growth and per capita income will coexist. Ultimately the prevalence of increasing returns to human capital accumulation will lead people to substitute quality for quantity of children and the resulting decline in fertility will reduce population growth. This characterises the final stage of the transition, where increasing levels of per capita income are associated with declining rates of population growth and the economy converges towards a balanced growth regime.

[*] This paper was written while the second author was visiting the Max Planck Institute for Demographic Research in May and July 1999. The views expressed here are the authors' own views and do not necessarily represent those of the Max Planck Institute for Demographic Research.

But understanding the mechanisms of how society can move from one equilibrium to another does not yet explain why some countries have been stuck in the Malthusian trap while others managed to escape. In this paper we refer to a simple model of economic development and the demographic transition as developed in Prskawetz et al. (1999). Within this framework we focus our attention on the configuration of Malthusian equilibria and the escape therefrom. We assume that both are dependent on the parameters of technological progress and population development. We show that the Malthusian trap need not be unique and that it can be a limit cycle. We then demonstrate two different routes of escape from the Malthusian trap. The first route of escape is the traditional path of 'history dependence', i.e., initial conditions will determine the long-term dynamics. We refer to the second route of escape if the economy is able to escape the Malthusian trap independent of initial conditions.

2. The model

We assume one sector of production which depends on the stock of human capital H_t and technology W_t

$$Y_t = H_t^\alpha W_t^\gamma \tag{1}$$

where $0 < \alpha \le 1$, and $\gamma > 0$ denote the corresponding production elasticities. Total output Y_t can be used on a one-to-one basis for consumption and investment.

Human capital depends on the number of people L_t and the level of education E_t in the economy:

$$H_t = L_t^\epsilon E_t^{1-\epsilon} \tag{2}$$

with $0 \le \epsilon \le 1$. In the literature, human capital and education are either equated or education denotes the flow of human capital. Such a notion of human capital excludes the demographic component. Equation (2) emphasises the components of human capital: population and embodied education. The production of human capital is assumed to exhibit constant returns to scale with respect to both inputs but decreasing marginal productivity of each input separately.

The dynamics of labor, education, and technology are given by[1]

$$\dot{L} = [b(y) - d(y)]L \tag{3}$$
$$\dot{E} = s(y)Y - d(y)E \tag{4}$$
$$\dot{W} = w(h)W \tag{5}$$

---

[1] Time arguments are omitted in the following.

where $b(y)$, $d(y)$, and $s(y)$ are the endogenous birth rate, death rate, and investment rate into education as a function of per capita income $y = \frac{Y}{L}$, and $w(h)$ is the growth rate of technological progress as a function of per capita human capital $h = \frac{H}{L}$.

We define the birth rate, death rate, and investment rate into education as in Strulik (1999):

$$b(y) = b_{nat} + \frac{b_{max}}{1 + exp\left[bs(y - yb)\right]} \tag{6}$$

$$d(y) = d_{nat} + \frac{d_{max}}{1 + exp\left[ds(y - yd)\right]} \tag{7}$$

$$s(y) = \frac{s_{max}}{1 + exp\left[-ss(y - ys)\right]}. \tag{8}$$

The constants b_{nat} and d_{nat} are the natural birth and death rates, and s_{max} is the maximum investment rate. These values are reached as per capita output becomes large. For low levels of per capita income, $b_{nat} + b_{max}$ and $d_{nat} + d_{max}$ are the maximum birth and death rates. The constants yb, yd, and ys determine the turning point of the logistic functions, which are also called half saturation constants since they represent the level of per capita income at which the logistic functions obtain half their maximum value. The constants bs, ds, and ss determine the slope of the logistic functions which are highest at the value of half the saturation constants. During the demographic transition, first mortality rates decline with increasing levels of per capita income, as modeled by the assumption $yd < yb$. Moreover, mortality rates are more sensitive than birth rates to changes in per capita income, hence $ds > bs$.

To generate economic growth we postulate a spillover of per capita human capital onto total productivity.[2] Technology is exogenously given and does not use up human capital. But the ability to utilise technology depends on the individual's skill level in the developing economy, as represented by the endogenous growth rate $w(h)$. We assume that the rate of technological progress initially rises at an increasing rate with increasing levels of average human capital. Eventually the increase in the rate of technological progress starts to decline and converges to its long-term maximum.[3] This specification of the rate of technological progress turns out to be capable of generating multiple Malthusian traps, in addition to being capable of generating the low and high per capita income equilibria as outlined in, e.g., Becker et al. (1990) and Strulik (1999). More specifically, we assume the following endogenous rate of technological progress:

---

[2] A similar spillover effect is modeled by Lucas (1988), who defines the production function $AK^{\beta}[uhN]^{1-\beta}h_a(t)^{\gamma}$, with N being the population size, K physical capital, u time spent in the labour force, hN effective labour, and h_a the average level of human capital.

[3] An analogous functional form is used in Becker et al. (1990) with respect to the rate of return of human capital investment.

$$w(h) = \frac{w_{max}}{1 + exp\left[-ws(h - hw)\right]}. \tag{9}$$

The constant w_{max} denotes the maximum rate of technological progress, hw is the half saturation constant and ws determines the slope of the logistic function.

Logarithmically differentiating total output (1) and human capital (2) and substituting in the dynamics of labor (3), education (4), and technological progress (5) results in a two-dimensional system of differential equations in the variables per capita output $y = \frac{Y}{L}$ and per capita education $e = \frac{E}{L}$

$$\hat{y} = \frac{\dot{y}}{y} = (\alpha - 1)[b(y) - d(y)] + \alpha(1 - \epsilon)\hat{e} + \gamma w(h) \tag{10}$$

$$\hat{e} = \frac{\dot{e}}{e} = s(y)\frac{y}{e} - b(y) \tag{11}$$

with per capita human capital $h = e^{1-\epsilon}$.

Applying a local stability analysis (see Prskawetz et al. (1999)) it can be shown that the determinant and the trace of the Jacobian, evaluated at the equilibria, are given by

$$detJ = yb(1 - \alpha)n_y - Ay\gamma w_e$$
$$trJ = y[(\alpha - 1)n_y + \alpha(1 - \epsilon)A/e] - b$$

where $A = s[\eta_s + 1 - \eta_b]$ with η_s and η_b denoting the income elasticity of the savings rate and the birth rate, i.e. $\eta_s = \frac{ds}{dy}\frac{y}{s}$ and $\eta_b = \frac{db}{dy}\frac{y}{b}$. A is always positive since the income elasticity of the savings rate is positive and the income elasticity of the birth rate is negative. The constant w_e denotes the derivative of the growth rate of technology with respect to per capita education and is positive as well. The difference between birth and death rates is equal to the growth rate of population $n(y)$, and its derative n_y can be either positive or negative. Whenever $n_y < 0$, the determinant is negative and the equilibrium will be a saddle point. Only in the case of $n_y > 0$ can the determinant of the Jacobian be positive. In this case, the sign of the trace of the Jacobian determines the stability of the system. While the latter equilibrium, $n_y > 0$, is termed a *Malthusian low-level equilibrium*, an equilibrium with $n_y < 0$ is called a *high level equilibrium*. This notation relates to the observation that population growth increases with per capita income, i.e. $n_y > 0$, in the Malthusian regime, which characterises rather primitive societies with low and non-growing income per capita. During the course of the demographic transition, which is initiated by the increase in per capita income, the relation between population growth and per capita income turns negative, i.e. $n_y < 0$.

The preliminary numerical results in Prskawetz et al. (1999) have shown that the Malthusian equilibrium must not be unique and that it can loose its stability. While the Malthusian trap can obviously become very complex, the high-level equilibrium will always be unique and be characterised by saddle point stability, as known from previous results in Prskawetz et al. (1999). History (in terms of the initial conditions of per capita income and per capita education) will determine whether an economy can 'escape' the Malthusian trap or whether it will be pushed back to the low-level equilibrium. The stable manifold of the high-level equilibrium (the saddle point) delineates these two regions.

In the next sections we shall present a more in-depth analysis of the complexity of the Malthusian trap as depending on the parameters of the rate of technological progress and demographic development and show distinct routes to escape from it. The results are obtained through a detailed numerical local and global bifurcation analysis (Champneys and Kuznetsov (1994) and Strogatz (1995)) of the system (10)-(11).[4]

3. A numerical example

In Prskawetz et al. (1999) it was shown that the Malthusian trap might not be unique, and that it could consist of two stable equilibria separated by a saddle point. As Figure 1 illustrates, the Malthusian trap might not only be non-unique but it could also be of higher dimension, namely a limit cycle. Within the convex region bordered by the stable manifold M_s of the high-level equilibrium S, initial conditions will determine whether the economy ends up in the low-level equilibrium E or whether it will be trapped by the stable limit cycle C_s, where per capita income and per capita education will follow persistent cycles over time. An unstable limit cycle C_u separates the regions of attraction of the two stable Malthusian traps E and C_s. For example, a trajectory starting from point 1 ends up in the periodic regime C_s, while a trajectory starting inside the unstable limit cycle C_u will converge to the equilibrium E. To escape the Malthusian traps, the economy has to start off outside the convex region bordered by the stable manifold M_s – at point 2, for example. In other words, history determines the long-term dynamics of the system.

A central parameter determining the configuration of the long-term dynamics is the sensitivity of savings as given by ss. An increase in ss will foster the accumulation of education for low values of per capita income. Consequently, per capita human capital and the ability to use technological progress will increase. Plotting per capita output at an equilibrium or on a cycle as a function of the parameter ss illustrates the complexity of the Malthusian trap (Figure 2).

For values of ss below point ss_F, the Malthusian trap consists of a unique stable equilibrium E which is almost independent of the specific values of ss.

---

[4] The analysis is performed with LOCBIF, a professional software package for dynamic systems based on a continuation technique (Khibnik et al. (1993)).

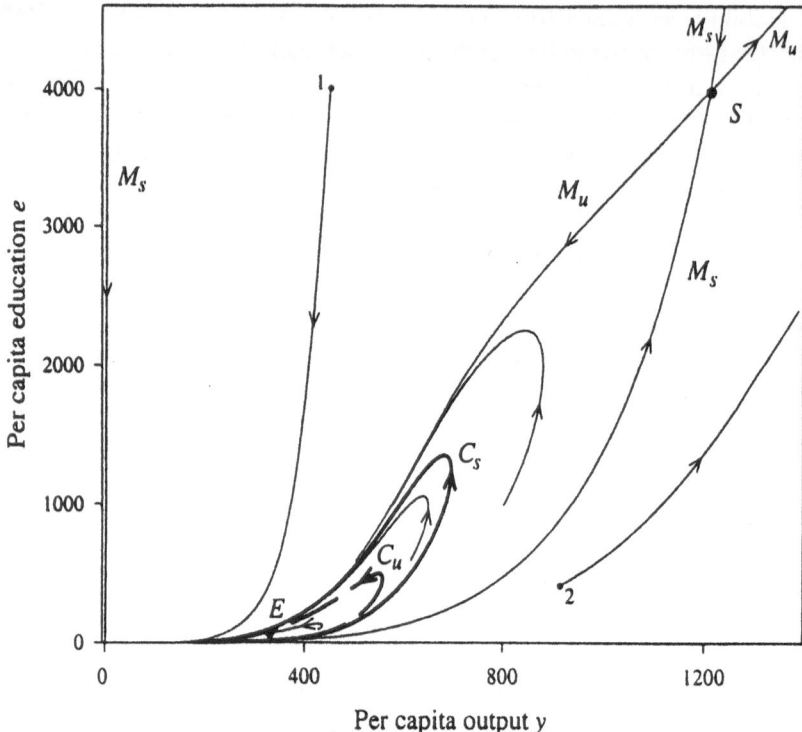

Fig. 1. Phase-space diagram for the parameter set $b(y)$: $b_{nat} = 0.015, b_{max} = 0.065, bs = 0.003, yb = 1500, d(y) : d_{nat} = 0.01, d_{max} = 0.1, ds = 0.005, yd = 300,$ $s(y)$: $s_{max} = 0.2, ss = 0.0129, ys = 600, w(h) : w_{max} = 0.05, ws = 0.1, hw = 10$ and production elasticities set equal to $\alpha = \gamma = \epsilon = 0.5$

Through a fold bifurcation at ss_F a stable limit cycle, C_s, and an unstable one, C_u, appear (the limit cycles are represented by their minimum and maximum values of y) while the former stable equilibrium E persists. As the parameter ss further increases, the amplitude of the stable limit cycle grows and that of the unstable limit cycle shrinks. Ultimately, the unstable limit cycle and the stable equilibrium collide through a subcritical Hopf bifurcation at point ss_{H+}. Thus, the system has two alternative stable modes of behavior between the points ss_F and ss_{H+}, namely, the equilibrium E and the limit cycle C_s. The basin of attraction of the equilibrium E is bounded by the unstable limit cycle C_u (see also Figure 1), and it shrinks as ss increases. As ss further increases, the stable limit cycle C_s moves closer and closer to a saddle point S which exists for all values of ss. The cycle touches the saddle and becomes a homoclinic orbit at the supercritical homoclinic bifurcation point ss_{P-}. Hence, between ss_{H+} and ss_{P-} the system has only one stable Malthusian trap which is a limit cycle. Once the saddle connection breaks (ss greater than ss_{P-}) the Malthusian trap is unique and unstable, so that independently of the initial levels of per capita income and per capita education, the economy will enjoy sustained economic growth in the long run.

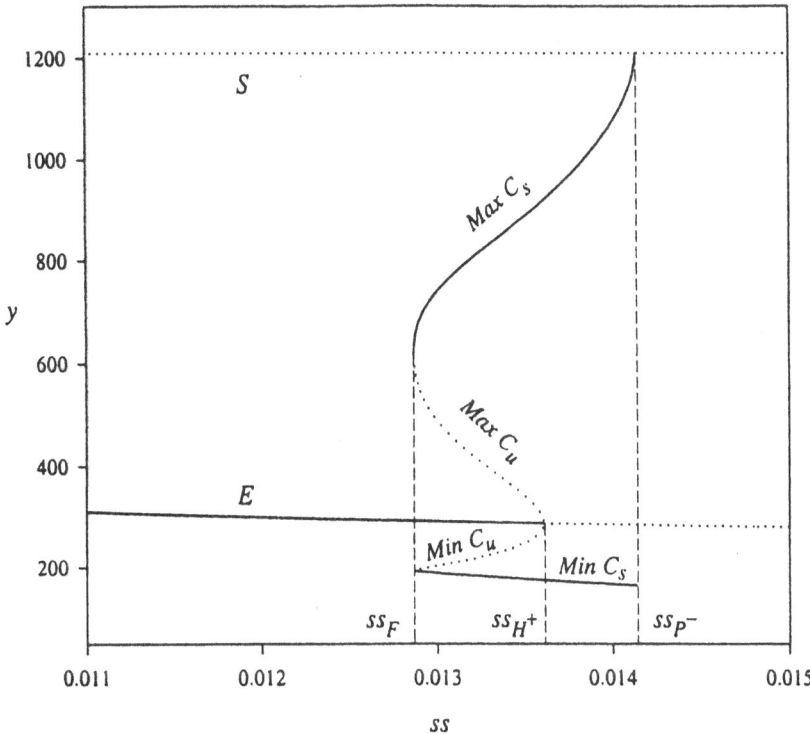

Fig. 2. One-dimensional bifurcation diagram with respect to sensitivity of savings ss. All other parameters are set as in Figure 1. Solid (dashed) lines refer to stable (unstable) equilibria or cycles

Figure 2 illustrates not only the complexity of the Malthusian trap, but it also shows that there exist two qualitatively different routes for escaping the Malthusian trap. Either the economy starts outside the convex region bordered by the stable manifold of the saddle equilibrium (*route 1* as illustrated in Figure 1) or ss increases to such an extent that the homoclinic bifurcation point is overcome and there are no more stable Malthusian traps, which means that the economy can always escape (*route 2*).

4. Bifurcation analysis

To obtain a more detailed picture of the possible configuration of Malthusian traps which depend on technological progress, we vary the half saturation constant of technological progress hw together with the parameter ss. By increasing the value of hw, higher levels of average human capital are required to utilise technology.

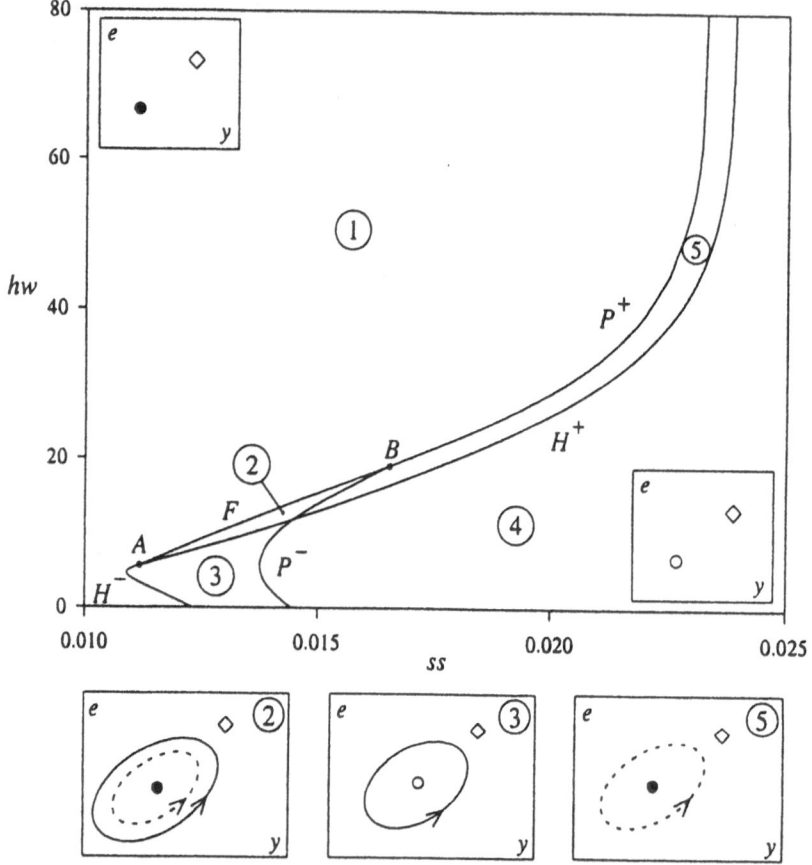

Fig. 3. Two-dimensional bifurcation diagram with respect to sensitivity of savings ss and half saturation constant of the rate of technological progress hw. All other parameters are set as in Figure 1. In the five sketches, solid (dashed) lines refer to stable (unstable) cycles, full (empty) dots refer to stable (unstable) equilibria and diamonds refer to saddles

Figure 3 shows bifurcation curves of model (10)-(11) in the two-dimensional parameter space (ss, hw). The organising centre is the codimension-2 bifurcation point A, at which three bifurcation curves merge: a supercritical (H^-) and subcritical (H^+) Hopf and a fold of cycles (F). There is another codimension-2 bifurcation point B, where a supercritical (P^-) and subcritical (P^+) homoclinic bifurcation curve merge with the fold F. As a result, there are five possible regions, characterised by different dynamic behaviours identified as $1, 2, ..., 5$ and described with a sketch of the corresponding stable and unstable equilibria and cycles. In regions 1 and 4 the Malthusian trap is unique and either a stable (region 1) or an unstable equilibrium (region 4). In regions 2, 3 and 5 there are multiple Malthusian traps. Region 2 is characterised by a stable equilibrium and a stable and unstable limit cycle (compare Figure 1), while in region 3 (region 5) an unstable (stable) equilibrium is surrounded by a stable (unstable) limit cycle. Note that the high-level equilibrium exists in every region; it is unique and it is always a saddle.

These regions differ not only with respect to the complexity of the Malthusian trap, but also with respect to the specific route by which the economy can escape the trap. In regions 1, 2, and 3 the economy can only escape the Malthusian trap if it starts outside the convex region bordered by the stable manifold of the high-level equilibrium (compare Figure 1). The second route for escaping the Malthusian trap (compare Figure 2) can be observed in region 4. Since there are no stable Malthusian traps, in the long run continued economic growth is guaranteed, independent of initial conditions. Note that region 4 can either be reached via a supercritical homoclinic bifurcation (P^-) or via a subcritical Hopf bifurcation (H^+). The first possibility was discussed in section 3. In the second case (crossing curve H^+ from region 5 to 4) the stable equilibrium collides with a co-existing unstable limit cycle and becomes unstable thereafter. Most interestingly, in region 5 there exists another route for escaping the Malthusian trap, which is qualitatively similar to the first route, where initial conditions determine whether the economy can escape. But in contrast to regions 1, 2 and 3, where the economy has to start off outside a convex region bordered by the stable manifold of the saddle equilibrium, it is sufficient if the economy starts outside the unstable limit cycle.

The results of the analysis (as it concerns the stable Malthusian traps) are summarised in Figure 4, where the grey region is characterised by a stable equilibrium, the striped region by a stable cycle, and the white region has no stable equilibrium. As a result, the economy can escape via route 1 in the grey and striped regions and via route 2 in the white one. As Figure 4 illustrates, whether or not the economy can escape the Malthusian trap independently of the initial conditions (route 2) depends mainly on the value of the parameter ss. This follows from the shape of the white region in Figure 4, which extends over the whole range of values of hw if ss increases sufficiently.

Note that the region where the Malthusian trap can become a limit cycle is restricted to a small area only, situated in the lower left-hand corner of the bifurcation diagram where both parameters are relatively small. This area might also be very sensitive to further parameter changes, such as in the rate of technological progress ws, for example. As illustrated in Figure 5, decreasing the value ws increases the range of values ss for which a limit cycle exists, and the period of the limit cycle decreases at the same time. Hence, the probability that a limit cycle will be observed in reality will increase with lower levels of ws.

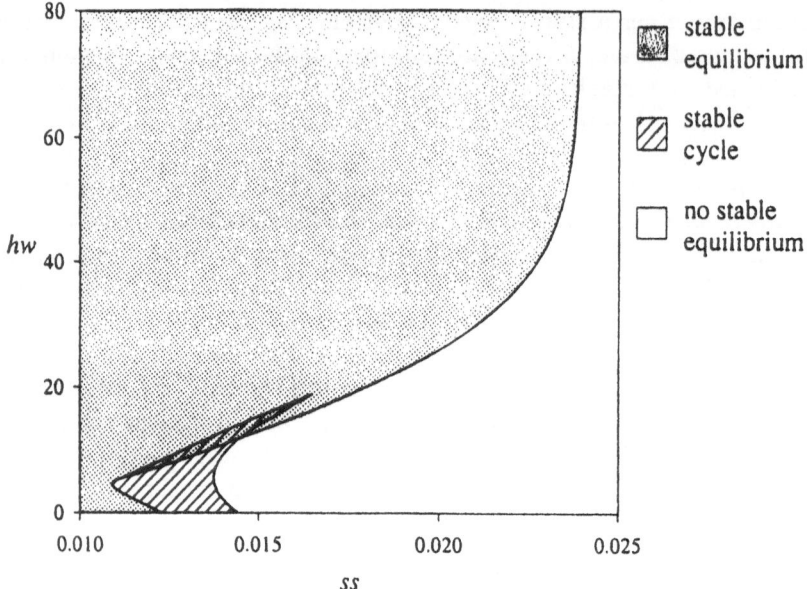

Fig. 4. Regions obtained from Figure 3 and characterised by different stable Malthusian traps

As shown in Strulik (1999) population policy consisting of an increase in income dependency of fertility for low values of per capita output (i.e. by reducing the threshold level yb) might be a more 'efficient' way to initiate an escape from the Malthusian trap as compared to technological parameters. This proposition is confirmed by the diagram shown in Figure 6. In fact, a decrease in the value of the parameter yb facilitates an escape from the Malthusian trap via the second route already for small values of the parameter ss. Moreover, a new configuration of the Malthusian traps as represented by two stable equilibria (dotted region) might appear.

5. Discussion

We have analysed in this paper a unified model that is capable of explaining the Malthusian low-level trap together with the regime of sustained economic growth. We have shown that the model allows for complex Malthusian traps, where multiple equilibria and cycles can coexist. As a result, the growth rate of per capita output must not necessarily equal zero in a Malthusian trap – it can fluctuate over time, as happens along the path of a Malthusian limit cycle. The potential success of an escape from a low-level Malthusian trap depends on the pace of technological progress as determined by the average level of human capital in the economy.

We have shown that there exist two qualitatively different ways in which the transition from the Malthusian low-level trap to the regime of sustained

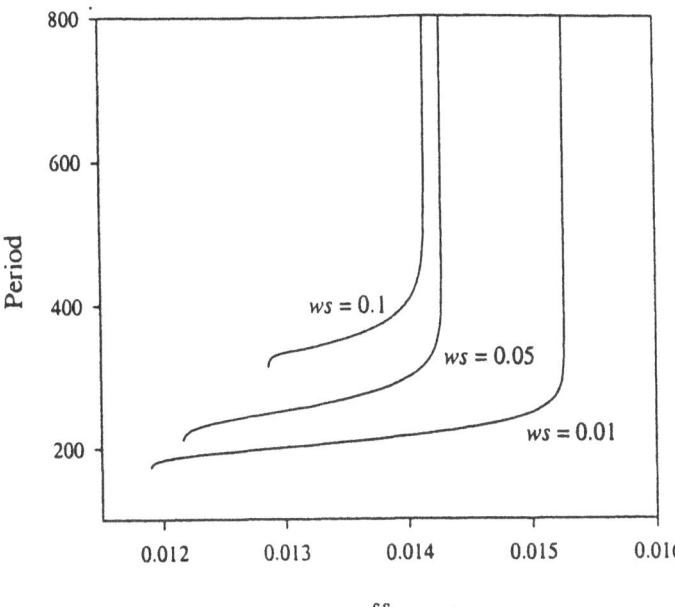

Fig. 5. Period of limit cycles (corresponding to different values of parameter ws) with respect to parameter ss. All other parameters are set as in Figure 1

economic growth can be operationalised. Either the economy is pushed into the growth regime by a change in initial conditions of per capita income and per capita education (route 1), as could be induced by, e.g., development aid. The second route of escape takes advantage of the economy's own ability to escape the Malthusian trap. Route 2 might be initiated by government policies that raise the sensitivity of savings ss or reduce the constant per capita income levels (hw and yb) at which the rate of technological progress and the birth rate are most sensitive to income changes. It is definitely more favourable to escape from the Malthusian trap via route 2 than via route 1, the reason being that an economy will be more robust to exogenous shocks since it can always recover through its own forces. In the case of route 1, an exogenous shock could push the economy so far backwards that it will end up in the Malthusian trap again.

Figure 7 summarises the transitional dynamics of the growth rate of per capita output $\hat{y}$, the growth rate of the population n, and the rate of technological progress w for three economies that start out from the same initial conditions and differ only in the sensitivity of savings ss.[5] Although all three economies start out at the maximum rate of technological progress w, per capita output growth will initially be negative. This can be explained by the low initial level of per capita income y, which implies high levels of popula-

---

[5] We therefore present different transitional dynamic modes along the bifurcation diagram in Figure 2.

334

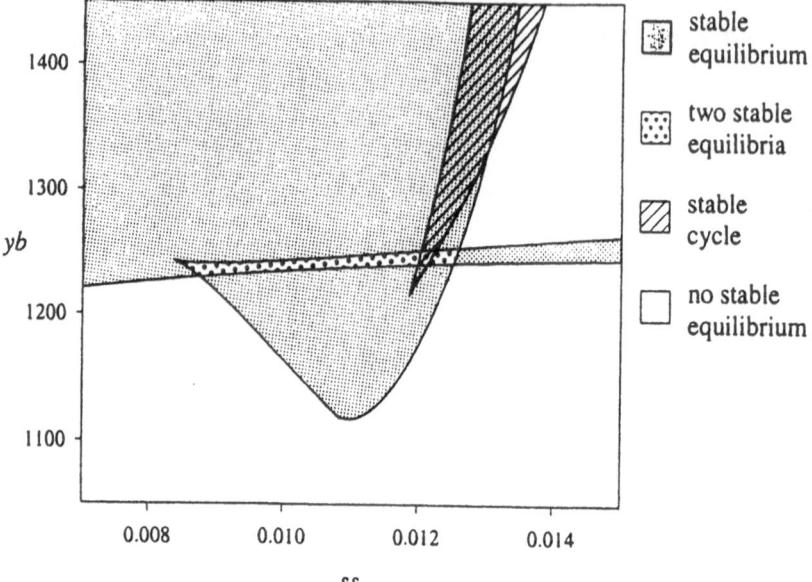

Fig. 6. Regions characterised by different stable Malthusian traps obtained from a two-dimensional bifurcation diagram with respect to sensitivity of savings *ss* and half saturation constant of birth rate *yb*. All other parameters are set as in Figure 1

tion growth that put downward pressure on per capita output. As per capita output decreases, the rate of investment into education falls and the average human capital declines. As a result, the rate of technological progress drops. After about 100 – 120 periods, all three economies have recovered to near zero economic growth $\hat{y}$ and low rates of population growth n and technological progress w. From here on, the rate of investment into education, as determined by the parameter *ss*, will determine the long-term dynamics. For low values of *ss* the economy will either be trapped in a Malthusian equilibrium (*a*) or in a Malthusian limit cycle (*b*) while higher values of *ss* facilitate an escape from the Malthusian regime (*c*). The population dynamics are 'Malthusian' in the first two economies, i.e., an increase in per capita output y (positive values of $\hat{y}$) will lead to higher population growth, which in turn puts pressure on per capita income ($\hat{y}$ becomes negative) so that population growth will decline. Since education does not keep pace with population growth during the regime of falling per capita levels, the average level of human capital declines and the rate of technological progress w will decrease. Consequently, economic growth will either stagnate (*a*) or it will fluctuate over time (*b*). Only in (*c*) will the economy escape from the Malthusian regime.[6] Note that the modern growth regime is entered at period 221. From

---

[6] Note that the value of the parameter *ss* in the escape scenario in Figure 7 (*c*) belongs to the range in the bifurcation diagram (shown in Figure 2), where one stable Malthusian limit cycle exists. Hence, all illustrations in Figure 7 refer to

this point onward population growth decreases though per capita income is still growing. As a result, the average level of human capital is high and the maximum rate of technological progress can be sustained. The economy will converge towards a balanced growth regime with population growth tending towards its natural level of $b_{nat} - d_{nat}$ and per capita output growth equal to $\frac{\gamma w_{max} - (1-\alpha)(b_{nat} - d_{nat})}{1-\alpha(1-\epsilon)}$.

Summing up our results, an escape from the Malthusian trap can be facilitated either through development aid (a change in initial conditions) or by development policies (a change in specific parameter values). The extent of development policies, i.e., the absolute change in the values of parameters such as ss, hw or yb, will determine whether these policies are sufficient to facilitate an escape from the low-level equilibrium independent of initial conditions (compare the white regions in the bifurcation diagrams of Figure 4 and Figure 6). Development policies characterised by values of parameters near the bifurcation curves will be very unstable, in the sense that small changes in the parameters can lead to different long-term dynamics.

References

Becker, G.S., Glaeser, E.L. and Murphy, K.M. (1999): Population and Economic Growth. American Economic Review **89(2)**, 145-149

Becker, G.S., Murphy, K. and Tamura, R. (1990): Human capital, fertility, and economic growth. Journal of Political Economy **98**, 12-37

Champneys, A.R. and Kuznetsov, Yu.A. (1994): Numerical detection and continuation of codimension-two homoclinic bifurcations. International Journal of Bifurcation and Chaos 4, 785-822

Galor, O. and Weil, D.N. (1998): Population, technology and growth: from the Malthusian regime to the demographic transition. CEPR Discussion Paper Series No. 1981, London.

Goodfriend, M. and McDermott, J. (1995): Early Development. American Economic Review **85(1)**, 116-133

Jones, C.I. (1999): Was an industrial revolution inevitable? Economic growth over the very long run. Working paper, Stanford University.

Khibnik, A., Kuznetsov, Yu., Levitin, V. and Nikolaev, E. (1993): Continuation techniques and interactive software for bifurcation analysis of ODEs and iterated maps. Physica D **62**, 360-371

Kremer, M. (1993): Population growth and technological change: one million B.C. to 1990. Quarterly Journal of Economics **108**, 681-716

Lucas, R. (1988): On the mechanics of economic development. Journal of Monetary Economics **22**, 3-42

---

economies that will escape the Malthusian trap via the first route, i.e., initial conditions will determine the long-term dynamics. In fact, as the parameter ss increases the convex region bordered by the stable manifold of the saddle, S shrinks and moves further to the right of the e-axes. Consequently the region of initial conditions for which an escape becomes feasible is widened.

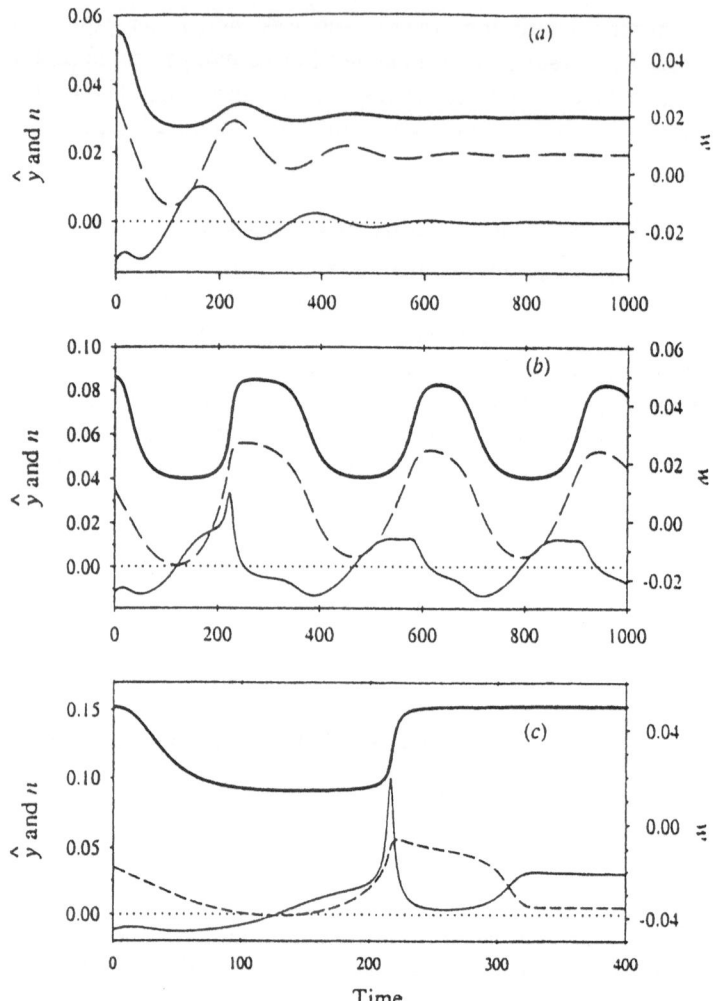

Fig. 7. Time series of growth rates of per capita output $\hat{y}$ (solid line), population n (dashed line), and technological progress w (bold line) for the same initial condition $y = 450, e = 4000$ and for different values of the parameter ss: (a) $ss = 0.01$, (b) $ss = 0.0129$, (c) $ss = 0.014$. All other parameters are set as in Figure 1

Prskawetz, A., Steinmann, G. and Feichtinger, G. (1999): Human capital, techno-
   logical progress and the demographic transition. Forthcoming in Mathematical
   Population Studies.
Strogatz, S.H. (1995): Nonlinear Dynamics and Chaos. Addison-Wesley Publishing
   Company, Reading, Massachusetts
Strulik, H. (1999): Demographic transition, stagnation, and demoeconomic cycles
   in a model for the less developed economy. Journal of Macroeconomics **21(2)**,
   397-413

Natural Resources, Standards of Living, and the Demographic Transition

Gunter Steinmann[1] and Alexia Prskawetz[2] *

[1] Department of Economics, Martin-Luther University Halle-Wittenberg, Halle, Germany
[2] Max Planck Institute for Demographic Research, Rostock, Germany

1. Introduction

In the very long run both the development of population size and the rise of income per capita have been characterised by stagnation or, at most, by very modest growth with cyclical ups and downs. It was not until two centuries ago that the age of industrialisation changed this trend and initiated an unprecedented increase in the standard of living (income per capita) as well as the population size. Industrialisation occurred first in Great Britain, and later in other western European countries, North America, Australia, and Japan – and eventually, most recently, in many Asian and Latin American countries. The coincidence of rapid economic and demographic growth has been a common feature in the initial stage of the modern process of development in these various countries.

The aim of our paper is to present a unified model of economic and demographic development[1] that matches the stylised historical experiences and presumed future of:

1. a very long period of slow economic and demographic growth or stagnation (*pre-transition age*),
2. a relatively short period of rapid economic and demographic growth (*age of demographic transition*), and eventually,
3. a period of continuing economic growth with zero population growth (*post-transition age*).

Our model is based on a production function with the four inputs labor quantity, labor quality (education), physical capital and natural resources. We assume constant returns to scale with respect to education, physical capital, and natural resources. This allows the possibility of economies of scale in the case of population growth unless the production capacity is limited by increasing scarcity of natural resources. Population growth does not necessarily lead to greater scarcity of natural resources because resource investment can augment the stock of potential resources (carrying capacity).

---

* The views expressed here are the authors' own views and do not necessarily represent those of the Max Planck Institute for Demographic Research.

[1] A similar but stochastic framework for modelling long-term economic and demographic development is presented in Komlos and Artzrouni (1990).

Further, we assume that constant population size represents the rule and that positive population growth is the exception in the demographic process. Positive population growth occurs only within a relatively short historical period of time, when an improvement in the standard of living leads first to a reduction of mortality and - due to a lagged adjustment - only later to a corresponding reduction in fertility. Therefore, positive population growth takes place during the period of demographic transition but is absent in all other periods before or after this transition.

The long-term pattern of economic growth shows some similarities to the demographic process. In the pre-transition age, per capita income gradually grows with the expansion of physical capital, education and carrying capacity. When per capita income has passed a certain threshold value it initiates a decline in mortality (hence population growth), which in turn accelerates the process of economic growth due to economies of scale. The higher per capita income allows more investment in resources, education, and physical capital and, as consequence, increases the carrying capacity and decreases the scarcity of natural resources. Therefore, the growth rate of per capita income remains relatively high even in the post-transition age, when the population growth comes to an halt.

But economic development is not necessarily a panacea for continued economic growth as was recently suggested by Chu (1998), chapter 14. Increasing specialisation and scale of economic activities as evidenced in advanced (developed) societies might also imply a substantial dependency of economic development on the stage of the natural environment. We include these considerations into our model by distinguishing between the likelihood and the scale of resource catastrophes. We assume that, whereas the likelihood of resource shocks declines with higher income, the extent of resource shocks increases owing to more specialisation and dependence on certain resources with increasing development.

Another crucial assumption of our model is the distinction between potentially-usable resources (carrying capacity) and actually-used resources. The actually-used resources are a fraction of the potentially-usable resources and will coincide in the long run with the latter unless the carrying capacity grows. While the growth of actually-used resources is defined by biological growth, the potentially-usable resources can be increased by investment, and it will depreciate as the consequence of an environmental shock.

2. The model

As the framework of our analysis we choose a descriptive, discrete-time dynamic model with one sector of production[2]:

---

[2] A similar model, but without natural resources, is presented in Jäger and Steinmann (1997) and Steinmann et al. (1998).

$$Y_t = E_t^\alpha K_t^\beta R_t^{1-\alpha-\beta} L_t^\gamma \tag{1}$$

with $0 < \alpha, \beta < 1, \alpha + \beta < 1$ and $\gamma > 0$. The inputs into the production of Y_t are education E_t, physical capital K_t, natural resources R_t, and labour L_t. We assume a Cobb Douglas production function with constant returns to scale with respect to inputs E_t, K_t, and R_t. Population L_t provides for increasing returns to scale.[3]

The dynamics of education, physical capital, and population are given by

$$
\begin{align}
E_{t+1} &= E_t + s_E Y_t - d(y_t) E_t \tag{2} \\
K_{t+1} &= K_t + s_K Y_t - \delta K_t \tag{3} \\
L_{t+1} &= L_t + [b(y_t) - d(y_t)] L_t \tag{4}
\end{align}
$$

where $y_t = Y_t/L_t$ denotes per capita income and $b(y_t), d(y_t)$ are the endogenous birth and death rates. The investment rates into education s_E and physical capital s_K are assumed to be exogenously determined. While we assume a constant depreciation of physical capital equal to δ, education is assumed to depreciate through the death of people.

For the endogenous death rate $d(y_t)$ we postulate the flexible logistic form as given in the Appendix. Below a minimum threshold y^{min}, the death rate is at its maximum value d^{max}, while increasing per capita income leads to a decrease in the death rate until the lowest possible death rate d^{min} is attained. In accordance with the concept of the demographic transition, we postulate that the birth rate follows the death rate with a lag equal to τ time units, i.e. $b(y_t) = d(y_t - \tau)$.

The dynamics of the renewable resources R_t (the actually-used resources) are described by the standard model of mathematical bioeconomics (Clark (1985))

$$R_{t+1} = R_t + a R_t (1 - R_t/Q_t) \tag{5}$$

where the variable Q_t determines the carrying capacity (saturation level) of the resource stock (i.e., Q_t is the stationary solution of R_t) and the parameter a determines the speed at which the resource regenerates.

The carrying capacity itself is dynamic and can be augmented as well as destroyed by human activities:

$$Q_{t+1} = Q_t + s_Q Y_t - p(Q_t/L_t) g(Q_t/L_t) Q_t. \tag{6}$$

As motivated in the introduction, we assume that economic development increases the carrying capacity through increased investment $s_Q Y_t$ (where

---

[3] Recent work on combining endogenous economic growth and endogenous population growth (e.g. Kremer (1993)) emphasises the importance of population dynamics to determine the long-term dynamics of economic systems. While population growth is mainly regarded as hindering economic growth, these authors even demonstrate the necessity of population growth to initiate economic growth.

s_Q denotes the exogenously given rate of investment) and through a decline in the likelihood $p(.)$ of a resource shock. As the level of per capita carrying capacity $q_t = Q_t/L_t$ is itself a measure of economic development, we assume that the likelihood of an environmental shock depends negatively on the level of q_t and is bounded between one and zero. But economic development also increases the vulnerability of the environment once a resource shock takes place. We model this latter impact by assuming that the extent $g(.)$ of an environmental shock increases with the level of per capita carrying capacity from some minimum level g^{min} to a maximum level g^{max}. The functional forms of the probability and extent of an environmental shock are summarised in the Appendix.

Recalling equation (1) and omitting time arguments in the following, the growth rate of total output $\hat{Y}$ equals

$$\hat{Y} = \alpha\hat{E} + \beta\hat{K} + (1 - \alpha - \beta)\hat{R} + \gamma\hat{L}, \tag{7}$$

with $\hat{E} = s_E Y/E - d(y)$, $\hat{K} = s_K Y/K - \delta$, $\hat{R} = a(1 - R/Q)$ and $\hat{L} = b(y) - d(y)$. Constant growth rates of education and physical capital require $\hat{E} = \hat{K} = \hat{Y}$. Furthermore, population growth will only be a transient phenomena during the demographic transition, such that $\hat{L} = 0$ in the long run. As a result, equation (7) reduces to

$$\hat{Y} = \hat{R}. \tag{8}$$

Long-term economic growth will therefore be determined by the growth rate of natural resources, which in turn equals the growth rate of the carrying capacity $\hat{Q}$ in the long run. Consequently, the investment rate into Q, together with the probability and strength of an environmental shock, will determine whether the economy grows, stagnates, or shrinks over time.

3. Numerical simulations

While we have briefly sketched the possible long-term dynamics in the previous section, we have to rely on numerical simulations to understand transitional dynamics.[4] We start out from a regime of low population and economic growth and investigate the dynamics of natural resources and population growth that can initiate a demographic transition and continuous economic growth.

Figure 1 represents the baseline scenario that replicates the stylised historical experience as summarised in the introduction. After a long period of slow economic and demographic growth, a short period of simultaneous rapid economic and demographic growth sets in. After the demographic transition has been completed, population growth settles down to its pre-transitional

---

[4] A Pascal program can be obtained from the authors on request.

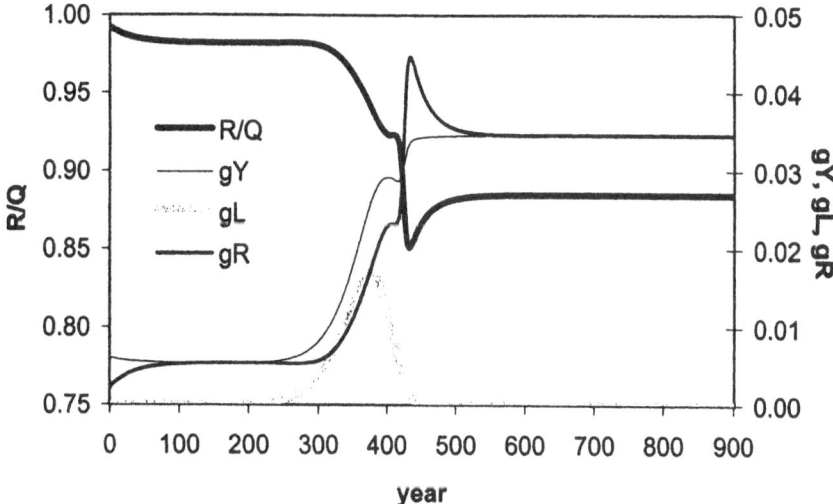

Fig. 1. Time series of the ratio of actually-used to potentially-useable resources R/Q and growth rates of output $\hat{Y} = gY$, population $\hat{L} = gL$ and natural resources $\hat{R} = gR$ for the parameter set $\delta = 0.1, s_K = 0.06, s_E = 0.06, s_Q = 0.03, a = 0.3, \tau = 50$ and production elasticities set equal to $\alpha = 0.5, \beta = 0.3, \gamma = 0.1$. The parameters for the endogenous functions of $d(y_t), p(q_t), g(q_t)$ are set as stated in the Appendix

level while economic growth continues to accelerate until it converges to its long-term maximum value. Note, that the growth rate of actually-used resources is highest at the end of the demographic transition and that it falls thereafter.

If the potential resource stock is already jeopardised at low levels of economic development, the demographic transition may not set in and output will stagnate. Such a scenario is represented in Figure 2, where we have reduced the levels of $q2^{min}$ and GSS, which corresponds to a shift of the curve $g(q_t)$ (representing the extent of an environmental shock) to the left. As population growth is zero and the ratio of actual to potential resources converges towards one, the production function (1) exhibits decreasing returns to scale in the long run.

Alternatively, the increasing returns to scale effect of population growth in the production of output Y is intensified if population growth increases and extends over a longer time period (Figure 3). This effect corresponds to an increase in the lag with which the birth rate follows a decline in the death rate. Consequently, the economic growth rate will converge towards a higher level than in Figure 1.

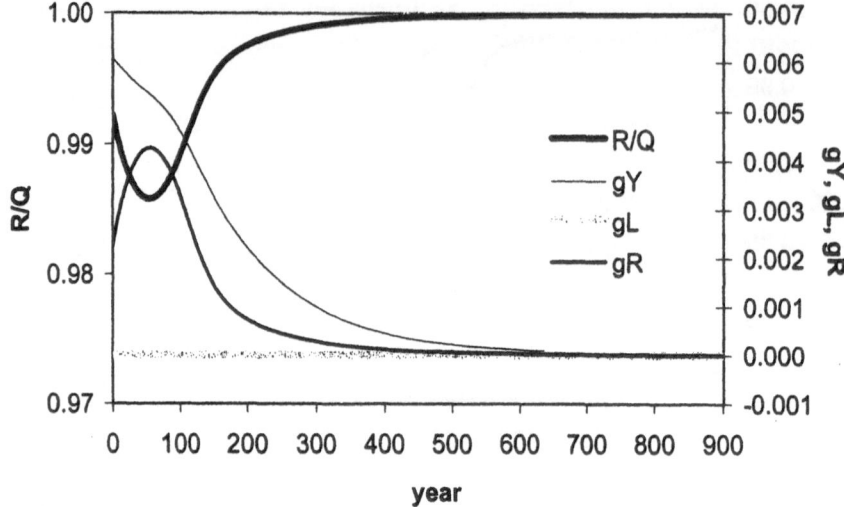

Fig. 2. Time series of the ratio of actually-used to potentially-useable resources R/Q and growth rates of output $\hat{Y} = gY$, population $\hat{L} = gL$, and natural resources $\hat{R} = gR$. All parameters are set as in Figure 1 except for the endogenous function of $g(q_t)$, where we set $q2^{min} = q_0$, $GSS = 2q_0$

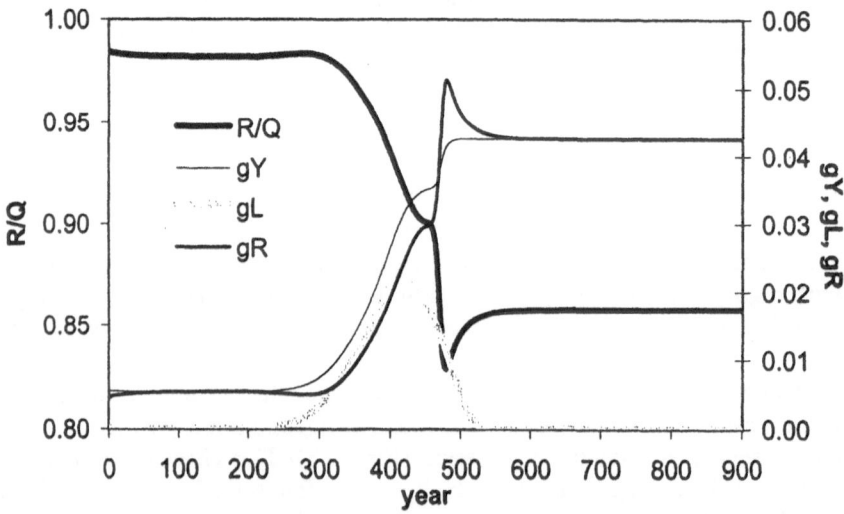

Fig. 3. Time series of the ratio of actually-used to potentially-useable resources R/Q and growth rates of output $\hat{Y} = gY$, population $\hat{L} = gL$, and natural resources $\hat{R} = gR$. All parameters are set as in Figure 1 except for the endogenous function of $b(y_t)$, where we set $\tau = 100$

4. Conclusions

The aim of our paper is to emphasise the *revisionist view* of a long-term future characterised by a large but stagnant population, resource creation, and increasing income rather than the *alarmist view* of population growth, resource depletion, and economic misery. It is obvious that the world's population cannot grow forever because the land surface of the earth is physically limited. But there is a great amount of evidence that population growth will come to an end long before the last square-metre land has been populated. The epoch of rapid population growth is a relatively short period in the history of mankind, and it takes place transitionally when mortality decline is not yet met by the corresponding decline in fertility. The adjustment of fertility to lower mortality is under way in all countries with persistently low mortality (Theory of Demographic Transition). It is only the length of the lag that varies between the countries.

Our paper analyses the effects both of decreasing mortality at the end of the pre-transition age and of positive and – in many cases – rapid population growth during the transition age on physical capital, education, resources, and per capita income. Population growth from lower mortality is certainly benign to the growth of the stock of education capital because the improved life expectancy allows longer use of education (a lower rate of depreciation). Population growth can stipulate economic growth and lead to an escape from economic misery due to economies of scale (increasing returns of scale) unless dilution of capital, education, and resources - or, even worse, resource depletion – counteract the positive effects and set limits to growth.

We are convinced that the long-term history of mankind shows more evidence for decreasing resource scarcity and improving standards of living than for increasing resource scarcity and lasting economic misery. Our paper deals with the two contradictory forces of resource creation and resource depletion induced by growing population. Simulations 1 and 3 demonstrate that population growth and high rates of investment in research and development of new resources create favourable conditions for rapid, permanent economic growth. But high risks of resource shocks and the serious damage of resource destruction can undermine the take-off and constitute insurmountable barriers for an escape from economic misery. Simulation 2 represents an example of persistent demographic and economic stagnation and misery.

References

Clark C. W. (1985): Bioeconomic Modelling and Fisheries Management. John Wiley, New York

Chu C. Y. C. (1998): Population Dynamics: A New Economic Approach. Oxford University Press, New York

Jäger, M. and G. Steinmann (1997): Mortality and economic growth. Volkswirtschaftliche Diskussionsbeiträge, Discussion Paper 4, Martin-Luther Universität Halle-Wittenberg

Komlos J. and Artzrouni M. (1990): Mathematical investigations of the escape from the Malthusian trap. Mathematical Population Studies 2(4), 269-287

Kremer, M. (1993): Population growth and technological change: one million B.C. to 1990. Quarterly Journal of Economics 108, 681-716

Steinmann, G. (1974): Bevölkerungswachstum und Wirtschaftsentwicklung. Duncker & Humblot, Berlin

Steinmann, G., Prskawetz, A. and Feichtinger, G. (1998): A model on the escape from the Malthusian trap. Journal of Population Economics 11, 535-550

Appendix

For the endogenous death rate, as well as for the endogenous probability and extent of an environmental shock, we postulate the flexible functional form introduced in Steinmann (1974):

$$d(y_t) = \begin{cases} d^{max} & \text{if } y_t \leq y^{min} \\ d^{max} - \frac{d^{max}-d^{min}}{1+MD}(\tanh\left[\frac{M}{MSS}LD(y_t)\right] + MD) & \text{if } y_t > y^{min} \end{cases}$$

$$p(q_t) = \begin{cases} 1 & \text{if } q_t \leq q1^{min} \\ 1 - \frac{1}{1+PD}(\tanh\left[\frac{P}{PSS}OD(q_t)\right] + PD) & \text{if } q_t > q1^{min} \end{cases}$$

$$g(q_t) = \begin{cases} g^{min} & \text{if } q_t \leq q2^{min} \\ g^{min} + \frac{g^{max}-g^{min}}{1+GD}(\tanh\left[\frac{G}{GSS}FD(q_t)\right] + GD) & \text{if } q_t > q2^{min} \end{cases}$$

where $LD(y_t) = (y_t - y^{min} - \frac{MSS}{2})$, $MD = -\tanh(-\frac{M}{2})$, $OD(q_t) = (q_t - q1^{min} - \frac{PSS}{2})$, $PD = -\tanh(-\frac{P}{2})$, $FD(q_t) = (q_t - q2^{min} - \frac{GSS}{2})$ and $GD = -\tanh(-\frac{G}{2})$ with the exogenous fixed parameters M, P, G and the parameters $MSS/2 + y^{min}$, $PSS/2 + q1^{min}$ and $GSS/2 + q2^{min}$ determining the points of inflection of the death rate, the probability, and extent of an environmental shock.

For the numerical simulations presented in section 3 we set $M = P = G = 5$, $g^{max} = 0.08$, $g^{min} = 0.02$, $d^{max} = 0.04$ and $d^{min} = 0.01$. The points of inflections are defined in relation to the initial level of per capita output y_0 and per capita carrying capacity q_0 as follows: $y^{min} = 3y_0$, $MSS = 6y_0$, $q1^{min} = q2^{min} = 3q_0$, $PSS = GSS = 6q_0$.

How Mortality Improvement Increases Population Growth

J.W. Vaupel[1] and V. Canudas Romo[2]

[1] Founding Director of the Max Planck Institute for Demographic Research, Rostock, Germany
[2] Sergio Camposortega Cruz Ph.D Fellow at the Max Planck Institute for Demographic Research

1. Introduction

Pseudo-stable and quasi-stable population models were developed by Feichtinger (1979), Coale (1972), Dinkel (1989) and others. Building on an article by Bennett and Horiuchi (1981), Preston and Coale (1982), Arthur and Vaupel (1984), and Kim (1986) further extended these models to develop general relationships for arbitrary population surfaces over age and time. In the spirit of Feichtinger and these other demographers, we present and prove in this chapter some formulas that capture the impact of mortality improvement on population growth. Goldstein and Schlag (forthcoming) also investigate this question but from a different perspective.

It is sometimes forgotten that in addition to babies and immigrants, people whose lives are saved also augment populations. Over the past century, most of the momentous increase in the world's population has been fueled by increased survival. Birth rates have tended to fall, often sharply, but death rates have decreased even faster. As a result, the population of the world has multiplied. In the future, global birth rates may fall below two children per woman, as they have already done in Europe and parts of the Far East. Population declines caused by such low fertility will be somewhat offset by mortality improvements, to the extent such improvements continue and are substantial. Even without net in-migration, a country's "replacement level of fertility" may be less than 2.0 if lifespans continue to lengthen.

Consider a population closed to migration with a continuous population surface $N(x, y)$ over age x and time y. See Keiding (1990) and Arthur and Vaupel (1984) for a discussion of this fundamental but elusive quantity, which is often referred to as age-specific population size. Let the intensity of population growth be denoted by

$$\rho(x, y) = \frac{dN(x, y)/dy}{N(x, y)} = \acute{N}(x, y), \tag{1}$$

where the acute accent, here and elsewhere in this chapter, denotes the relative derivative with respect to time. Let the intensity of population growth for the population as a whole be denoted by

$$\bar{p}(y) = \frac{\int_0^\omega p(x,y)N(x,y)dx}{\int_0^\omega N(x,y)dx}, \tag{2}$$

where ω is the highest age attained. Letting

$$T(y) = \int_0^\omega .N(x,y)dx, \tag{3}$$

be the total population, note that

$$\bar{p}(y) = \acute{T}(y) = \frac{dT(y)/dy}{T(y)}. \tag{4}$$

The quantity $\bar{p}(y)$, which is often called the population growth rate, is of prime interest to us.

Let the intensity of mortality (also known as the hazard of death or force of mortality) be given by

$$\mu(x,y) = -\frac{dN(x+a,y+a)/da}{N(x,y)}. \tag{5}$$

and let the intensity of improvement in mortality be denoted by

$$\acute{\mu}(x,y) = -\frac{d\mu(x,y)/dy}{\mu(x,y)}. \tag{6}$$

Our focus is on how $\bar{\mu}(x,y)$ affects $\bar{p}(y)$.

As shown by Preston and Coale (1982),

$$N(x,y) = B(y)s(x,y)R(x,y), \tag{7}$$

where $B(y)$ denotes the number of births at time y, $s(x,y)$ is the period survival function,

$$s(x,y) = e^{-\int_0^x \mu(a,y)da} \tag{8}$$

and

$$R(x,y) = e^{-\int_0^x p(a,y)da}. \tag{9}$$

As emphasised by Arthur and Vaupel (1984),

$$N(x,y) = B(y-x)s_c(x,y), \tag{10}$$

where $B(y-x)$ denotes the number of births at time $y-x$ and $s_c(x,y)$ is the cohort survival function

$$s_c(x,y) = e^{-\int_0^x \mu(a,y-x+a)da}. \tag{11}$$

It follows from (7) and (10) that

$$R(x,y) = \frac{B(y-x)}{B(y)} \frac{s_c(x,y)}{s(x,y)}, \qquad (12)$$

so $R(x,y)$ can be considered to be a cohort-period adjustment that captures the dissimilarity between actual cohort and synthetic period values.

To understand how population growth is related to changes in the number of births, to improvements in mortality, and to the cohort-period adjustment, it makes sense to start with a general result that is useful for a variety of different kinds of decomposition. Consider some total or sum $V(y)$:

$$V(y) = \int v(x,y)dx \qquad (13)$$

or

$$V(y) = \sum_x v(x,y). \qquad (14)$$

Suppose

$$v(x,y) = u_1(x,y)u_2(x,y)...u_n(x,y). \qquad (15)$$

Then

$$\acute{V}(y) = \sum_{i=1}^{n} E\left[\acute{u}_i(x,y)\right], \qquad (16)$$

where

$$E\left[\acute{u}_i(x,y)\right] = \frac{\int \acute{u}_i(x,y)v(x,y)dx}{\int v(x,y)dx} \qquad (17)$$

or

$$E\left[\acute{u}_i(x,y)\right] = \frac{\sum_x \acute{u}_i(x,y)v(x,y)}{\sum_x v(x,y)}. \qquad (18)$$

To prove this result, note that

$$\acute{v}(x,y) = \acute{u}_1(x,y) + \acute{u}_2(x,y) + ... + \acute{u}_n(x,y) \qquad (19)$$

and

$$\acute{V}(y) = \frac{\int \acute{v}(x,y)v(x,y)dx}{\int v(x,y)dx}. \qquad (20)$$

The general result of formula (16) can be used in equation (3) after substituting (7) to decompose population growth into three components:

$$\bar{p}(y) = \acute{B}(y) + \bar{\acute{s}}(y) + \overline{\acute{R}}(y), \qquad (21)$$

where $\acute{B}(y)$ is the intensity of change in births, $\bar{\acute{s}}(y)$ is the average intensity of change in survival,

$$\bar{\dot{s}}(y) = \frac{\int_0^\omega \frac{ds(x,y)/dy}{s(x,y)} N(x,y)dx}{\int_0^\omega N(x,y)dx} \tag{22}$$

and $\bar{R}(y)$ is the average intensity of change in the cohort-period adjustment

$$\bar{\dot{R}}(y) = \frac{\int_0^\omega \frac{dR(x,y)/dy}{R(x,y)} N(x,y)dx}{\int_0^\omega N(x,y)dx}. \tag{23}$$

It should be noted, however, that the second term on the right-hand side of (21) reflects both current changes in mortality and historical factors that have determined the current population structure. To eliminate the influence of past changes in fertility and mortality (and perhaps migration), the following result, due to Vaupel (1992), is useful:

$$E_{w_2}(v) - E_{w_1}(v) = \frac{Cov_{w_1}(v,\varphi)}{E_{w_1}(\varphi)}, \tag{24}$$

where

$$\varphi \equiv \varphi(x) = \frac{w_2(x)}{w_1(x)}. \tag{25}$$

Note that expected values are given by

$$E_{w_i}(v) = \frac{\int_0^\omega v(x)w_i(x)dx}{\int_0^\omega w_i(x)dx} \tag{26}$$

and the covariance is given by

$$Cov_{w_1}(v,\varphi) = \frac{\int_0^\omega v(x)\varphi(x)w_1(x)dx}{\int_0^\omega w_1(x)dx} - \frac{\int_0^\omega v(x)w_1(x)dx}{\int_0^\omega w_1(x)dx} \frac{\int_0^\omega \varphi(x)w_1(x)dx}{\int_0^\omega w_1(x)dx}. \tag{27}$$

Formula (24) is readily proved by substitution and simplification.
  Results (21) and (24) imply

$$\bar{p}(y) = \dot{B}(y) + \dot{e}_o(y) + R^*(y), \tag{28}$$

where

$$R^*(y) = \bar{\dot{R}}(y) - Cov_N(\dot{s}(x,y), R^{-1})\frac{T(y)}{B(y)e_o(y)}. \tag{29}$$

To prove this, let $w_1(x) = N(x,y)$ and $w_2(x) = B(y)s(x,y)$ and note that

$$E_{s(x,y)}(\dot{s}(x,y)) = \dot{e}_o(y), \tag{30}$$

because

$$e_o(y) = \int_0^\omega s(x,y)dx. \tag{31}$$

Formula (28) permits decomposition of the current population growth rate into

(1) the current intensity of change in births,

(2) the current intensity of change in period life expectancy (which captures the impact of current mortality change), and

(3) a residual term that reflects the influence of historical fluctuations that have resulted in a population size and structure that is different from the stationary population size and structure implied by current mortality and births.

Note that (7) implies that

$$R(x,y) = \frac{N(x,y)}{B(y)s(x,y)}. \tag{32}$$

Regardless of whether a population is open or closed to migration, the value of $R(x,y)$ can thus be interpreted as the ratio of the actual population to the stationary life-table population. Hence, if the actual population structure is the same as the life-table structure, i.e.,

$$N(x,y) = B(y)s(x,y), \tag{33}$$

then

$$R(x,y) = 1, \text{ for all } x \tag{34}$$

and

$$R^*(y) = 0. \tag{35}$$

Table 1 provides some illustrative examples of the decomposition in (28). The population growth rate was estimated by

$$\bar{\rho}(y) \approx \frac{\ln\left[\frac{T(y+5)}{T(y-5)}\right]}{10}. \tag{36}$$

The intensity of change in births was estimated by

$$\dot{B}(y) \approx \frac{\ln\left[\frac{B(y+5)}{B(y-5)}\right]}{10}. \tag{37}$$

Similarly, the intensity of change in period life expectancy was estimated by

$$\dot{e}_o \approx \frac{\ln\left[\frac{e_o(y+5)}{e_o(y-5)}\right]}{10}. \tag{38}$$

Table 1. Decomposition of Population Growth Rate

	Year	$\bar{p}(y)\%$	$\dot{B}(y)\%$	$\dot{e}_o(y)\%$	$R^*(y)\%$
World	1960	1.92	1.40	1.21	−0.69
	1990	1.60	−0.11	0.39	1.32
	2040	0.55	0.07	0.20	0.28
Austria	1990	0.60	0.14	0.39	0.07
Belgium	1990	0.27	0.13	0.30	−0.16
Finland	1990	0.41	0.04	0.28	0.09
Germany	1990	0.48	−0.62	0.30	0.80
Italy	1990	0.12	−1.02	0.30	0.84
United Kingdom	1990	0.33	−0.25	0.26	0.33
USA	1990	0.99	−0.49	0.24	1.24

Source: Data from Eurostat (1998); World and USA from United Nations (1996).

Finally, $R^*(y)$ was simply estimated as the residual

$$R^*(y) = \bar{p}(y) - \dot{B}(y) - \dot{e}_o(y). \tag{39}$$

In Keyfitz (1977) and Vaupel (1986) the impact on life expectancy of changes in age-specific death rates is analyzed. Their results, when combined with (28), shed light on how mortality change affects population growth. If the rate of mortality improvement is constant over age,

$$\dot{\mu}(x,y) = \dot{\mu}(y), \text{ for all x}, \tag{40}$$

then Keyfitz (1977) shows that

$$\dot{e}_o(y) = \dot{\mu}(y)H(y), \tag{41}$$

where $H(y)$ is given by

$$H(y) = -\frac{\int_0^\omega s(x,y)\ln[s(x,y)]dx}{\int_0^\omega s(x,y)dx} \tag{42}$$

and can be interpreted as the entropy of the survival function. If mortality follows a Gompertz trajectory,

$$\mu(x,y) = a_y e^{bx}, \tag{43}$$

then Vaupel (1986) indicates that

$$\dot{e}_o(y) \approx \frac{\dot{\mu}(y)/b}{e_o(y)} \tag{44}$$

and

$$e_o(y+1) - e_o(y) \approx \frac{\dot{\mu}(y+0.5)}{b}. \tag{45}$$

Note that (45) implies that a constant rate of mortality improvement will continue to increase life expectancy by about the same absolute amount. On the other hand, (44) implies that as life expectancy increases, the relative rate (i.e., intensity) of improvement will fall.

Because Vaupel's derivation of these approximations is in an unpublished working paper (Vaupel (1985)), we provide a derivation here. Approximation (45) follows from (44) via the approximation

$$\dot{e}_o(y) \approx \frac{e_o(y+1) - e_o(y)}{e_o(y)}. \tag{46}$$

Approximation (44) can be derived from (41) by showing that

$$e_o(y)H(y) \approx \frac{1}{b}. \tag{47}$$

Substituting (43) in (8) and then in (42) and then substituting the left hand side of (43), (8) and (31) yields

$$e_o(y)H(y) = \frac{1}{b}\left[\int_0^\omega \mu(x,y)s(x,y)dx - a_y\int_0^\omega s(x,y)dx\right] \tag{48}$$

$$= \frac{1}{b}[1 - a_y e_o(y)]. \tag{49}$$

If $a_y \ll e_o$, as it generally is in low-mortality populations, then the approximation follows. The approximation gets better as mortality improvements are made, because (44) implies that a_y declines faster than e_o rises.

Consider now the relationship between the total fertility rate (TFR) and the intensity of change in the number of births. Let A be the average age of childbearing. Then the net reproduction rate (NRR) is approximately given by

$$NRR \approx \pi \, s(A) \, TFR, \tag{50}$$

where π is the proportion of female births and $s(A)$ is a girl's chance of surviving to the average age of childbearing (Coale (1972)). Under current conditions in developed countries, $NRR \approx 0.48 \, TFR$, so that an NRR of about one will be produced by a TFR of about 2.1. In a stable population (i.e., with fixed age-specific fertility and mortality rates), a well-known result from Lotka (1934) implies

$$\dot{B} \approx \frac{NRR - 1}{A}. \tag{51}$$

Assume $R^* = 0$. Let the TFR be 1.99, let life expectancy at birth be 80, and let the pace of mortality improvement be 1.5% per year (which is close to the average current level in some developed countries). Assume mortality increases exponentially at a rate of 0.1. Then

$$\bar{p}(y) = \dot{B}(y) + \dot{e}_o(y) = -0.0015 + 0.0019 = 0.0004. \qquad (52)$$

Although stylised, this result shows that population growth can be positive even if the TFR is below the so-called replacement level.

References

Arthur, W.B. and Vaupel, J.W. (1984): Some general relationships in population dynamics. Population Index **50(2)**, 214-26

Bennett, N. and Horiuchi, S. (1981): Estimating the completeness of death registration in a closed population. Population Index **47(2)**, 207-21

Coale, A.J. (1972): *The Growth and Structure of Human Populations: A Mathematical Investigation*. New York: Princeton University Press

Dinkel, R.H. (1989): *Demographie. Band 1: Bevölkerungsdynamik*. Munich: Vahlen

Eurostat (1998), NewCronos CD 98: Different Statistics from European Countries. Luxemburg.

Feichtinger, G. (1979): *Demographische Analyse und populations-dynamische Modelle*. Wien New York: Springer-Verlag

Goldstein and Schlag (forthcoming): Longer life and population growth. Population and Development Review

Keyfitz, N. (1977): *Applied Mathematical Demography*. New York: Wiley

Keiding, N. (1990): Statistical inference in the Lexis Diagram. Philosophical Transactions of the Royal Society of London, 487-509

Kim, Y.J. (1986): Examination of the generalized age distribution. Demography **23**: 451-61

Lotka, A.J. (1934): *Théorie Analytique des associations biologiques*. France: Hermann éditeurs

Preston, S.H. and Coale, A.J. (1982): Age structure, growth, attrition and accession: a new synthesis. Population Index **48(2)**, 217-59

United Nations Population Division Databases, Population Prospects (1996). United Nations, New York.

Vaupel, J.W. (1992): *Analysis of Population Changes and Differences*. Unpublished book

Vaupel, J.W. (1986): How change in age-specific mortality affects life expectancy. Population Studies **40**, 147-157

Vaupel, J.W. (1985): How change in age-specific mortality affects life expectancy.WP-**85-17**. (Laxenburg, Austria: International Institute for Applied Systems Analysis, 1985)

The Evolution of Drug Initiation:
From Social Networks to Public Markets

Jonathan P. Caulkins

Carnegie Mellon University, H. John Heinz III School of Public Policy and
    Management, Pittsburgh, PA 15213-3890 USA
RAND, Drug Policy Research Center, Santa Monica, CA 90407 USA

Abstract. This paper seeks to integrate two competing notions of what drives initiation into illicit drug use, the so-called "snowball model" of sellers recruiting new customers and the "social contagion" model of current users recruiting friends into drug use. The model hypothesizes that both occur, but in distinct phases of the drug epidemic. The data are insufficient to validate the model, but inasmuch as it is valid the principal policy conclusion is that interventions that reduce the "snowball effect" in Phase I may substantially reduce consumption in Phase II and overall.

1. Introduction

Illicit drugs impose large costs on society, so there is naturally interest in understanding better the process of initiation into drug use. One aspect of that question is whether new initiates are recruited by drug dealers (for a profit motive) or by other users (as a social phenomenon), or both. This paper introduces a new model of this recruitment process and explores some of its policy implications.

The belief that dealers recruit new users has a colorful history. The notion of dealers "pushing" their wares on innocent children is reflected in the term "drug pusher", and it is embodied in the lyrics of a 1950s comedy song by Tom Lehrer in which "The Old Dope Peddler" "gives the kids free samples because he knows full well that today's young innocent faces are tomorrow's clientele." This perspective also finds adherents among those who favor drug legalization. If initiation into drug use is caused by drug dealers trying to expand their profitable business, then eliminating the profits by legalizing drugs, will eliminate the impetus for initiation into drugs.

This view has been attacked on several fronts. Kaplan (1983) argues that drug dealers' careers are short and brand loyalty weak so dealers can not expect to monopolize new customers' supply for long. Furthermore, it takes more than a few doses to become "hooked," and empirically most users report being introduced to drug use by a friend or family member, not a dealer.

In economic terms, recruiting new users may be riskier than selling to current users, and recouping the cost of this initial "investment" is difficult in a competitive market in which users can find alternative dealers. It is hard to charge the new recruits higher prices. If the dealer tried, the newly recruited user could try to switch to an alternative supplier.

Finally, there is a mathematical argument against the dealers-recruit-users theory. Since some new users escalate to heavy use, and heavy users often try to finance their habit by selling drugs, if sellers recruited new users there could be a positive feedback loop. Sellers would recruit new users, who would become heavy users and then sellers, who would in turn recruit still more new users, etc. This "snowball effect" could make the system grow uncontrollably, but real epidemics do not grow without bound.

These arguments have led analysts to focus on models in which initiation is driven by the number of users (Behrens et al., 1999; Tragler et al., forthcoming). Indeed, in many respects these models have relegated dealers to a passive, or reactive role. Consistent with a "risks and prices" perspective (Reuter and Kleiman, 1986), dealers are assumed to flow freely in and out of the market in a manner that equates the expected return from dealing with the expected return from alternative activities, whether the alternative is legitimate work or another form of crime (cf. Caulkins, 1993).

In this paper we raise the possibility that perhaps both models (dealers recruit users and users recruit users) are correct – but at different points in a drug epidemic. Drug epidemics evolve continuously, but we will simplify by envisioning an epidemic with just two phases. In the first phase the drug is relatively uncommon. It is used and spreads within social networks because the critical mass of users and sellers necessary to support a "public" or visible market (e.g., a street market) does not exist. In this phase, transactions occur primarily between people who associate for reasons other than conducting drug transactions (i.e., in the course of "routine activities"). The second phase is the more familiar one today. In it, distribution in social networks is supplemented and, in some cases, even supplanted by "mature" markets in which users can locate sellers fairly easily, and anonymous or "arms's length" transactions exist.

We hypothesize that drug sellers have an incentive to recruit new users in phase I but not phase II because in phase I sellers can charge users "above market" prices. "Above market" is in quotes because in phase I there is no single market price. Rather, prices are negotiated between individual sellers and users. The exact price any given user pays will depend on a variety of factors (e.g., the buyer's negotiating skill and "reservation price" relative to that of the seller), but the prices will generally be higher than in phase II for at least two reasons. First, poor information flows and inability to negotiate with multiple sellers mean that sellers are better able to extort some degree of monopoly rents in phase I. Second, the market volume is larger in phase II, so enforcement swamping will tend to dilute enforcement intensity and the associated "tax" on prices (Kleiman, 1993).

The concepts of "enforcement swamping" and an "enforcement tax" are simple, but may appear mysterious to those not yet familiar with them. The idea behind the "tax" is that drug sellers demand monetary compensation for the risks of enforcement. The greater the enforcement risk they bear, the greater the markup they will demand and, hence, the higher the price (Reuter and Kleiman, 1986). The "swamping" idea is that for any given level of enforcement, the risk incurred by selling a unit of drugs is inversely related to the size of the market. The larger the market, the more thinly the enforcement is spread, and the lower the risk per unit.

This does not necessarily mean that enforcement risk approaches infinity when the market is very small. Rather, when the market is small, enforcement risk is driven by the statutory penalties. As the market grows, initially the total amount of punishment grows in proportion, keeping the enforcement intensity (i.e., the enforcement level divided by market size) roughly constant. But at some point, the criminal justice system cannot keep up with the expanding market, e.g., because prison over-crowding forces authorities to release offenders before they have served their entire sentence, and the amount of punishment grows less than proportionately with the market. Then enforcement intensity declines, and competition will tend to bring prices down.

We also hypothesize that other types of initiation will be different in phase I vs. phase II in three respects. First, in phase I there is no public market, so availability is restricted for people who do not have friends who are users. This might reduce initiation by "innovators," people who decide to start using on their own, not at the urging of a user or seller.

Second, users in phase I will recruit new users more slowly than they do in phase II. In part that may be due to the generally higher prices in phase I making it difficult to convince people to try the drug (and perhaps making current users reluctant to share). Even more importantly, in phase II when enforcement is stretched thin, authorities will tend to triage enforcement needs, and focus scarce resources on the more serious offenders, namely the sellers. In phase II, arrested users will often not be prosecuted if they do not have a history of prior arrests and have committed no other crime (e.g., a violent crime or possession of large quantities suggesting that they are working as a courier). And, if arrested users are prosecuted and convicted, they may be sentenced to nothing more serious than probation. Hence, the risks to users of exposing their drug use status to others in phase II are lower than in phase I. Finally, the more drug users there are, the lower the social approbation associated with being a user. This could make users in phase II more willing to talk to non-user about becoming users. For all these reasons, we hypothesize that the rate at which users expose potential new users to the possibility of using drugs is lower in phase I than it is in phase II.

Third, in phase I initiation will not be moderated as substantially by a negative reputation for the drug. Whether a contact turns into an initiation depends in part on how appealing drug use would seem to the potential new user. When the drug has acquired a reputation for being dangerous, potential users who are offered the opportunity to use may decline (Musto, 1987). Behrens et al. (1999) hypothesize that this can be modeled in terms of a "reputation effect" which restrains the rate of recruitment and which is governed by the relative number of heavy and light users. Heavy users are more likely to manifest the ill-effects of excessive use, so the more heavy users there are relative to light users, the worse the reputation of the drug and the fewer potential initiations become actual initiations.

We accept this reputation model for phase II but suggest that it may not apply in phase I. The reputation of a drug to a potential initiate is not actually driven by the number of light and heavy users, but by the number of light and heavy users the individual in question knows about. In phase II, when public markets exist, the drug is likely to have attracted media attention, so most potential users will be aware of the existence of heavy users and associated risks of excessive drug use. In phase I, however, the heavy users are hidden (since there are no public markets), and the media are less likely to focus on a drug that is not obviously a significant public problem. E.g., through the 1970s, the US media was much less negative about cocaine than they were in the 1980s, particularly the late 1980s after crack spawned the creation of flagrant public cocaine markets in many US cities. Hence, we hypothesize that this moderating influence of a drug's negative reputation can be ignored in phase I. This means that the actual rate of initiation per current user could be higher or lower in phase II than it was in phase I. In phase I it will have one constant value. In phase II, the constant will be larger but it will be modulated by the reputation effect. When the reputation is relatively benign, then initiation may be higher than in phase I. When the reputation is particularly malevolent, it could be lower.

2. Mathematical Model

2.1 Equations Governing Dynamics

We are now in a position to generalize the Behrens et al. (1999) model to incorporate this image of two distinct phases of an epidemic. We retain Behrens et al.'s model without change for phase II. In particular, in phase II we imagine that the dynamics of the drug epidemic are governed by:

$$\dot{L} = I(L,H) - (a+b)L$$
$$\dot{H} = bL - gH \quad\quad\quad (1)$$
$$I(L,H) = \tau + sLe^{-qH/L},$$

where

$L(t)$ = number of light users at time t,
$H(t)$ = number of heavy users at time t,
$I(L,H)$ = initiation of new users,
a = constant rate of desistance from light use,
b = constant rate of escalation from light to heavy use,
g = constant rate of desistance from heavy use,
τ = rate of spontaneous initiation by "innovators",
s = rate at which light users would recruit new users if the drug's reputation were benign, and
$e^{-qH/L}$ = adverse reputation of the drug that moderates initiation.

In phase I we modify the initiation function by (i) reducing the rate of spontaneous initiation from τ to τ_I, (ii) reducing the parameter governing the rate at which light users contact potential initiates from s to s_I, (iii) eliminating the modulating effect of negative reputation, and (iv) adding a term reflecting a tendency for some dealers (denoted by D) to recruit new users at some constant rate (k). I.e., the equation governing initiation becomes:

$$I(L,H) = \tau_I + s_I L + kD. \quad\quad\quad (2)$$

This raises the question of how to model D, the number of dealers who will actively recruit new users. One approach is to hypothesize that the number of dealers is proportional to the volume of drug sales, which in turn is proportional to a weighted average of the number of light and heavy users, weighted by their respective consumption rates. Absorbing the proportionality constants into the parameter k, this suggests that Equation (2) can be rewritten as

$$I(L,H) = \tau_I + s_I L + k (L + f H), \qquad (2')$$

where f is the ratio of the average consumption rate for heavy users relative to light users.

2.2. Solution

Assuming the number of dealers is proportional to $L + f H$ means the dynamics in phase I form a simple system of two linear differential equations that can be solved explicitly. If $s_I + k + k f b / g \neq a + b$ then the equilibrium is

$$\hat{L} = \tau_I / (a + b - s_I - k - k f b / g) \text{ and} \qquad (3)$$
$$\hat{H} = b \, \hat{L} / g.$$

The roots of the characteristic equation are

$$\lambda_1, \lambda_2 \;=\; \frac{v - g +/- \sqrt{\left(v - g\right)^2 + 4\left(b k f + g v\right)}}{2},$$

where $v = s_I + k - a - b$. Since b, k, and f are positive constants, rewriting this as

$$\lambda_1, \lambda_2 \;=\; \frac{v - g +/- \sqrt{\left(v + g\right)^2 + 4 b k f}}{2}$$

shows that the eigenvalues are real and distinct. There are two types of solutions.

Case I: If $v + b k f / g < 0$ then the absolute value of the square root term is smaller than the absolute value of $v - g$ and $v \quad g < 0$. So there are two negative eigenvalues, and the equilibrium, which is in the first quadrant ($\hat{L} > 0$, $\hat{H} > 0$), is a stable node.

Case II: If $v + k f b / g > 0$ then the absolute value of the square root term is larger than the absolute value of $v - g$, so there is one positive and one negative eigenvalue, and the saddle point equilibrium is in the third quadrant. Furthermore, for all initial conditions in the first quadrant, the trajectories will eventually grow without bound.

If $v + k f b / g = 0$ then the $\dot{L} = 0$ and $\dot{H} = 0$ isoclines are parallel. Again for any initial conditions $L(0) \geq 0$ and $H(0) \geq 0$, the trajectories will eventually grow without bound so this situation is similar to Case II.

The Phase II model has been solved by Behrens et al. (1999). In short, for the parameter values used below, there is a stable focus in the first quadrant, with trajectories spiraling in counter-clockwise.

2.3 Transition from Phase I to Phase II

We next need to define the transition from phase I to phase II. To the best of our knowledge this issue has not been studied empirically. We know that markets for some drugs in some cities are public (e.g., heroin in New York City, Amsterdam, and Frankfurt and cocaine in many US cities), and for other drugs in other cities they are not (heroin in many smaller US cities; cocaine in many European cities;

XTC in most places). But we really do not know what determines which condition pertains when and where.

We hypothesize, though, that enforcement swamping is a key issue. If enforcement resources are adequate, then people trying to sell publicly will be arrested. Hence, no one tries to sell publicly and distribution is confined to markets embedded in social networks. If, however, enforcement pressure is so diluted that it becomes profitable for a seller to step forward and sell publicly, a seller will do so. When this happens, it becomes easier for other sellers to do so, setting off a chain reaction. This suggests there may be a very rapid transition from phase I to phase II when the market gets large enough relative to the enforcement resources. The transition would of course literally be continuous, especially if one considers an aggregate of geographically dispersed markets. If the change is rapid, however, the discrete two-stage approximation may not be too bad.[1]

Let f denote the average heavy user's consumption rate relative to that of a light user, so the size of the market is proportional to $L + fH$. For any given level of enforcement, the critical ratio of enforcement to market size will be reached when the quantity $L + fH$ grows to some constant.

We assume that once the market transitions into Phase II it never reverts back to Phase I. Theoretically, if the trajectories in Phase II were such that $L + fH$ fell below the critical constant, enforcement pressure might drive drug dealing from public markets back into social networks. We ignore this possibility, however, because it simplifies the analysis and because we can think of no such example in recent history.

The last thing we need is a performance measure. We do not have an objective function per se because we have no controls. But we do want to explore how the system responds to changing the values of various parameters (e.g., the rate at which dealers recruit new users in Phase I). This could be done by looking at how entire trajectories change, but it is useful to have a scalar summary measure of the severity of the drug epidemic.

For this purpose we follow the lead of Rydell et al. (1996) by focusing on the discounted quantity consumed. Since we do not model prices, this is equivalent to tracking the discounted weighted number of light and heavy users, weighting heavy users f times as heavily as light users.

Discounted consumption in Phase I can be calculated explicitly. The solutions to the phase I problem are

$$L(t) = c_1 e^{\lambda_1 t} + c_2 e^{\lambda_2 t} - \frac{g\tau}{vg + bkf} \quad \text{and}$$

$$H(t) = \frac{\lambda_1 - v}{kf} c_1 e^{\lambda_1 t} + \frac{\lambda_2 - v}{kf} c_2 e^{\lambda_2 t} - \frac{b\tau}{vg + bkf},$$

where c_1 and c_2 are constants determined by the initial conditions. Thus

$$L(t) + fH(t) = \alpha e^{\lambda_1 t} + b e^{\lambda_2 t} + \gamma,$$

---

[1] Note, enforcement swamping is probably more of a necessary than a sufficient condition given that marijuana is still sold substantially within social networks.

where
$$\alpha = (1 + f(\lambda_1 - v)/kf) c_1,$$
$$\beta = (1 + f(\lambda_2 - v)/kf) c_2, \text{ and}$$
$$\gamma = -(fb + g) \tau/(vg + bkf).$$

If the transition from Phase I to Phase II occurs at time T, the net present value of consumption in Phase I:

$$= \int_0^T e^{-rt} \left(\alpha e^{\lambda_1 t} + \beta e^{\lambda_2 t} + \gamma \right) dt = \frac{\alpha \left(e^{(\lambda_1 - r)T} - 1 \right)}{\lambda_1 - r} + \frac{\beta \left(e^{(\lambda_2 - r)T} - 1 \right)}{\lambda_2 - r} - \frac{\gamma \left(e^{-rT} - 1 \right)}{r}.$$

The present value of consumption in Phase II discounted back to time T can be computed numerically as a function of the Phase II initial conditions $(L(T), H(T))$, as in Behrens et al. (1999). Multiplying this by e^{-rT} to discount back to time 0 and adding it to the discounted consumption from Phase I gives the overall total discounted consumption, measured as a multiple of the amount a light user consumes in a year.[2]

2.4 Parameters

We parameterize the model for the recent US cocaine epidemic. The phase II parameters are taken from Behrens et al. (1999), specifically $a = 0.163$, $b = 0.024$, $g = 0.062$, $s = 0.61$, $\tau = 50{,}000$, and $q = 7.0$. To be consistent with prior analyses (e.g., Behrens et al., 1999, Rydell et al., 1996) we discount at $r = 4\%$ per annum, assume initial values of $L(1962) = 330{,}000$ and $H(1962) = 0$, and assume the light and heavy users consume at average rates of 16.42 and 118.93 grams per year, respectively. Thus, $f = 7.25$ to three significant digits.

As mentioned, the character of a drug epidemic varies continuously; it does not display a discrete transition from one phase to another. Youth started selling cocaine in street markets in New York City around 1973.[3] Since drug trends in New York City lead those in the rest of the US, we assume that the cocaine market as a whole was still in phase I as late as 1973. At the other extreme, many major cities in the US had flagrant crack markets by 1988, so we take that as the latest possible time for the transition from Phase I to Phase II. We do sensitivity analysis with respect to the transition point in five-year increments, specifically considering transitions at the market sizes observed in 1973, 1978, 1983, and 1988. Everingham and Rydell (1994) find that $L + 7.25H = 3.575$, 9.180, 16.283, and 17.265 (in millions), in these four years respectively. So below we compute results for transitions characterized by these four market sizes.

There is little empirical basis for estimating the other three parameters (τ_1, s_1, and k). We choose them to be consistent with what scant data exist on initiation and numbers of light and heavy users in the early years of the cocaine epidemic, in a manner similar to that done by Caulkins et al. (1999). Everingham and Rydell

---

[2] For any initial conditions of interest, the Phase II model comes very close to the steady state within the first 50 years. So, for convenience, we only compute the NPV of consumption for the first 50 years of Phase II. This quantity differs from the infinite horizon NPV by approximately $e^{-r50} * (\hat{L} + f\hat{H})/r$.

[3] Bruce Johnson, personal communication, March 16, 1999.

(1994) provide the only data on numbers of light and heavy users in the early phase of the cocaine epidemic, with figures going back to 1962.

There are two sources of initiation data, Everingham and Rydell (1994) and Johnson et al. (1996), but both are problematic. The Everingham and Rydell initiation estimates are the average of two sets of estimates, those produced by the "difference" and "retrospective" methods. For the "difference" estimate they assumed that the number of users in 1972 (the first year prevalence was measured directly) reflected constant incidence between 1962 and 1972. That there would be no increase in initiation when the epidemic was growing rapidly is not truly plausible, but Everingham and Rydell had little incentive for developing more realistic time trends. "Neither the year of the start of the epidemic nor the shape of the incidence curve before the first survey year is critical" to their analysis (p.32), but it clearly is for present purposes.

Johnson et al.'s (1996) initiation data show a smooth, steady increase from 1962 to 1972, but the overall level of initiation is implausibly low. In particular, the total of all initiation reported by Johnson et al. between 1962 and 1971 is less than the increase in the number of users reported by Everingham and Rydell over that time period. The actual rates of initiation must, of course, be greater than the increase in the number of users because of quitting. If one tries to reproduce the past epidemic using parameters values estimated based only on the Johnson et al. data, the size of the resulting modeled epidemic is well below what was observed. This may be at least in part because the Johnson et al. figures are based only on the household population, whereas Everingham and Rydell augment those data with estimates for the homeless and incarcerated who are excluded from the household surveys.

In an effort to capture the complementary strengths of both data sources, we choose parameter values that fit the shape of the Johnson et al. initiation data, but then scale them up so they can reproduce the level of the epidemic as described by Everingham and Rydell. Table 1 shows the values of the three Phase I initiation parameters which give the least squares fit to the Johnson et al. initiation data for blocks of ten years, starting with the earliest year-range for which data are available. The clear pattern in the data is that for the early years of the epidemic, the best fit is obtained when initiation is driven by the "snowball" model of dealers recruiting new users. Then, there is a sharp transition. Starting with the year range 1972-1981, the best fit is obtained by having that "snowball" parameter set equal to zero. (Note: the values of τ and s_I are not meaningful as estimates of Phase II parameters because, by assumption, reputation effects should be included in Phase II.)

In light of Table 1, we set the base case values of $\tau = s_I = 0$ and choose k to be the value that gives the least-squares fit to the Everingham and Rydell epidemic data through 1978.[4] The resulting value ($k = 0.237$) can be interpreted as

---

[4] In minimizing the sum of the squared differences we weight the squared differences of light and heavy users by the inverse of the square of the average number of light and heavy users from 1962 to 1978. I.e., we weight squared differences in heavy users 122.7 times as heavily as squared differences in the number of light users. Because there were so many

suggesting that Phase I sellers expanded their sales by 23.7% per year by recruiting new users.

Table 1: Phase I Initiation Parameters that Give the Least Squares Fit to Johnson et al.'s (1996) Initiation Data, for Various Ten-Year Blocks of Data

Year Range	τ_I	s_I	k
1962-1971	0	0	0.09
1963-1972	0	0	0.10
1964-1973	0	0	0.13
1965-1974	0	0	0.14
1966-1975	0	0	0.14
1967-1976	0	0	0.15
1968-1977	0	0	0.15
1969-1978	0	0	0.15
1970-1979	0.08	0	0.13
1971-1980	0.17	0.03	0.09
1972-1981	0.32	0.16	0
1973-1982	0.41	0.14	0
1974-1983	0.45	0.13	0
1975-1984	0.50	0.12	0
1976-1985	0.69	0.10	0
1977-1986	0.72	0.09	0
1978-1987	0.67	0.10	0
1979-1988	0	0.18	0
1980-1989	0	0.17	0
1981-1990	0	0.18	0
1982-1991	0	0.17	0

We are skeptical that all Phase I initiation actually occurred through the "snowball" mechanism for both theoretical reasons (surely some friends introduced other friends to cocaine in Phase I) and empirical reasons (allowing s_I to be positive actually yields better fits to the Everingham and Rydell data and, at any rate, the above estimation rests on slender data). Hence, we also found combinations of the Phase I initiation parameters fit the data roughly as well and represented a sort of "average" of the extreme snowball model (only $k > 0$) and an image that Phase I initiation is dominated by τ and s_I, as discussed below.

3. Results

3.1 Numerical Results

Figure 1 plots the Phase I trajectory obtained with $\tau = s_I = 0$, $k = 0.237$ through 1978 alongside the corresponding data from Everingham and Rydell (1994).

---

more light than heavy users in these years, failing to weight in this way would result in effectively only fitting to the number of light users.

Table 2 summarizes the key associated quantities for various phase transition points including the net present value (NPV) of consumption.

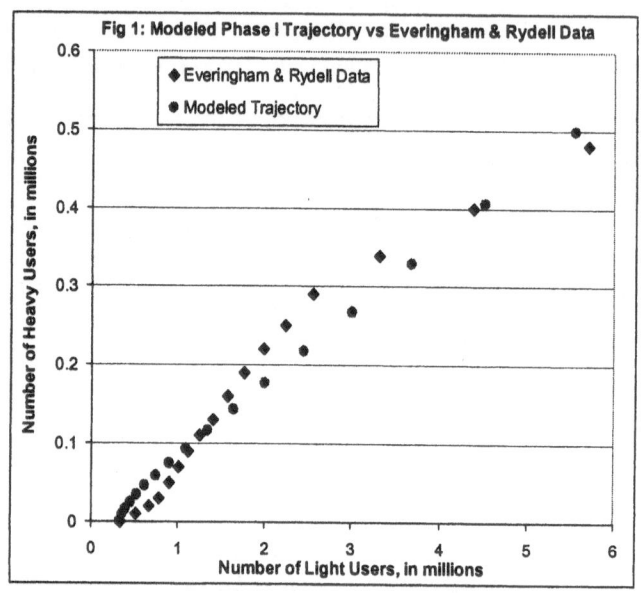

Fig 1: Modeled Phase I Trajectory vs Everingham & Rydell Data

Table 2: Base case results with $\tau = s_I = 0$, $k = 0.237$

	Transition from Phase I to Phase II Defined by Weighted # of Users in Year:			
	1973	1978	1983	1988
NPV Consumption in Phase I[a]	12	27	44	47
NPV Consumption in Phase II[a]	135	227	338	352
NPV Consumption in Total[a]	147	254	382	399
T (Transition time, years after 1962)	11	16	19	19
$L(T)$ in millions	2.2	5.6	9.9	10.5
$H(T)$ in millions	0.2	0.5	0.9	0.9
Reputation at time $T = e^{-q\,H(T)/L(T)}$	0.54	0.53	0.53	0.53

[a] Consumption given in units of millions of light-user year equivalents of use

Several things are apparent from the table. First, the longer an epidemic is allowed to "snowball" before a negative reputation begins to suppress initiation and dealers no longer have an incentive to recruit users, the more severe the epidemic will be in terms of total quantity of drugs consumed.[5] Second, the later the transition to Phase II the greater the proportion of total consumption that occurs in Phase I. Third, the ratio of heavy to light users at the transition time T

---

[5] This statement only necessarily holds for the parameter values investigated here. It may not be true generally. It is easy to see that the instantaneous rate of consumption is not monotonic in the transition time for all future times, so it may be that there are sets of parameter values, including the discount rate, such that this would not be true.

and, hence, the reputation at time T, is very nearly the same for all four transitions thresholds, a point to which we will return shortly.

As mentioned, we do not have a great deal of confidence in the finding that τ and $s_I = 0$ and only $k > 0$ in Phase I. Hence, the quantities in Table 2 were also evaluated for the sets of Phase I initiation parameters obtained as follows. The parameters τ and s_I were set equal to a given proportion of their Phase II values, and k was selected to give the best fit to the Everingham and Rydell epidemic data given those values of τ and s_I. This was done for proportions ranging from 0 (which gives the results reported in Table 2) to 0.525 (for which the corresponding value of k is almost 0). That is, the proportion was varied to consider the range of possibilities between all and none of the Phase I initiation being driven by the "snowball effect" of dealers recruiting new users. All of the quantities in Table 2 vary monotonically in this proportion, and describing the relationship as a linear function of the proportion is not a bad approximation.[6] Hence, in Table 3 we just show the percent change in the Table 2 quantities that is generated by varying this proportion from 0 to 0.525.

Table 3: Change in results as move from $\tau = s_I = 0$, $k = 0.237$ to $\tau = 26.25$ $s_I = 0.32025$, $k = 0.004286$

	Transition from Phase I to Phase II Defined by Weighted # of Users in Year:			
	1973	1978	1983	1988
NPV Consumption in Phase I	+8%	+21%	+25%	+26%
NPV Consumption in Phase II	-15%	-35%	-41%	-42%
NPV Consumption in Total	-13%	-29%	-34%	-34%
T (Transition time, years after 1962)	-7%	+4%	+9%	+9%
$L(T)$ in millions	-4%	-8%	-9%	-9%
$H(T)$ in millions	+6%	+12%	+14%	+14%
Reputation at time $T = e^{-q\,H(T)/L(T)}$	-6%	-13%	-15%	-15%

Two themes emerge from Table 3. First, except in the first column, the less that initiation in Phase I is driven by the "snowball effect," the longer it takes to transition to Phase II. This leads to greater consumption in Phase I, but allows time for more Phase I users to escalate to heavy use. This implies that the drug's reputation at the beginning of Phase II is worse, which reduces Phase II consumption by enough to lower consumption overall.

Second, shifting from an all "snowball effect" model of Phase I to a no "snowball effect" model does not change any of the projected quantities by more than 50%, so the overall qualitative results are relatively robust with respect to that parameter uncertainty.

---

[6] The maximum absolute percent deviation between the actual value and a straight line between the values obtained when the proportion was 0 and 0.525 is 2.5%. If one focuses on outcomes other than consumption in just one phase or the other (i.e., ignores the first two rows), the maximum absolute percent error in the linear approximation is 1.4%.

From a policy perspective, the most interesting question is how much these quantities change if the snowball effect parameter (k) is reduced. Presumably the greater the enforcement risk, the lower k will be. Undercover "buy-bust" enforcement in particular should make dealers wary of aggressively recruiting new customers (Moore, 1973). Figure 2 shows how reducing k affects several key quantities when the transition between phases is defined by the level of use observed in 1978 (second data column in tables above). Results for transitions at other levels of use are similar. The plot assumes baseline initiation parameter values of $\tau = 12{,}500$, $s_I = 0.1525$, and $k = 0.125$. I.e., the figure shows the effect of reducing the snowball effect parameter k by a given proportion from an initial value that is only half as great as blind adherence to the parameter estimation above would suggest. Or, in a rough sense, the plot shows the effect of reducing k from a baseline in which the snowball effect is responsible for about half of Phase I initiation. Focusing on this figure rather than the corresponding one for which $\tau = s_I = 0$ and $k = 0.237$ is conservative in the sense that it reduces the apparent importance of the parameter k.[7]

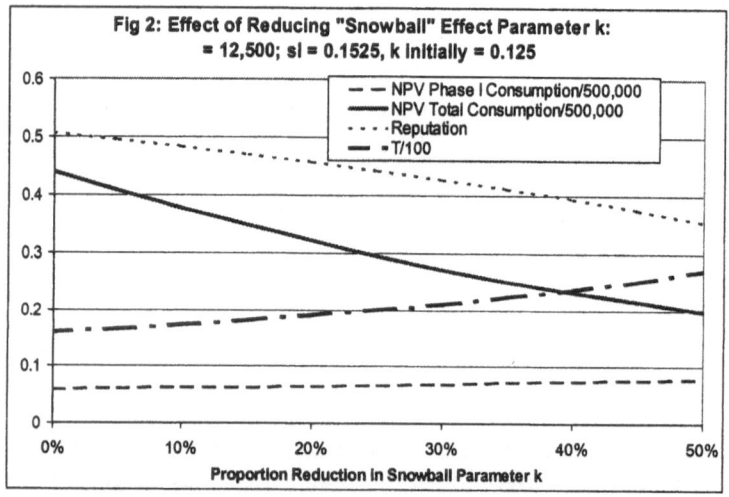

Fig 2: Effect of Reducing "Snowball" Effect Parameter k: = 12,500; sl = 0.1525, k initially = 0.125

Despite this conservatism, Figure 2 clearly suggests that reducing the rate of initiation through the snowball effect is consequential. Cutting k by 10% would delay the transition to Phase II by a year and a half, thereby reducing the net present value of combined drug consumption in Phases I and II by 14%. The "arc elasticity" of the effect on consumption (% change in NPV total consumption

---

[7] The slopes and curvature of the lines in the graph with $\tau = s_I = 0$ and $k = 0.237$ are the same, and the change in Phase I consumption is very similar, but as one moves from left to right across the graph the transition time T increases from 16 to 41, not just 16 to 27; the reputation at time T decreases from 0.53 to 0.3, not from 0.51 to 0.36; and total consumption is reduced by 74% not 56%.

divided by the % change in k) decreases as the change in k gets larger, but remains above one (in absolute value) for reductions in k as great as 60%.

Whether buy-bust enforcement or other interventions designed to reduce the snowball effect parameter k are cost-effective depends on how difficult it is to reduce k and on the nature and extent of collateral effects. But to the extent that this model captures important dynamics of the spread of a drug epidemic, it suggests that slowing the rate at which dealers recruit new users in the early stages of an epidemic, when distribution occurs primarily within social networks, may have important beneficial effects on the long-term evolution of the drug epidemic.

3.2 Approximate Analytical Explanation of Results

Some observations emerged in the course of this analysis. First, total consumption is dominated by consumption during Phase II. Second, at least for the four transition points considered here, the net present value of Phase II consumption is monotonically decreasing in the ratio of heavy to light users at the transition time T. Third, for any given ratio of heavy to light users at the transition, the net present value of future consumption never varied by more than about one-third whether the transition occurred at market sizes observed historically in 1973, 1978, 1983, or 1988. That is, consumption in Phase II is driven more by the reputation of the drug when it emerges in public markets than by the size of the market at that time. Finally, for all sets of parameter values considered here, the Phase I model trajectories plotted in H-L space become very nearly straight lines before the transition from Phase I to Phase II occurs.

Thus the overall magnitude of the epidemic is dominated by Phase II consumption, and Phase II consumption is dominated by what occurs in Phase I (specifically the ratio of heavy to light users toward the end of Phase I), not by the timing of the transition from Phase I to Phase II. That is fortuitous because the timing of the transition is not well understood (and, indeed, is not even well defined to the extent that the transition is continuous not discrete). It also implies that it is important to understand what drives the ratio of heavy to light users in Phase I. The answer is quite simple.

The ratio of heavy to light users quickly approaches a constant because the negative eigenvalue (λ_2) is small, both in absolute terms (typically around -0.2) and relative to the positive eigenvalue (λ_1, which is usually at least 0.15 or so). Since the transition time T is generally at least 16 or so, by time T the $e^{\lambda_2 t}$ terms in the expressions for the numbers of light and heavy users are dominated by the $e^{\lambda_1 t}$ terms. Thus around time T,

$$\frac{H(T)}{L(T)} \approx \frac{-v-g+/-\sqrt{(v+g)^2+4bkf}}{2kf}.$$

Numerically this expression is quite sensitive to k, but not very sensitive to s_l, for the baseline values of a, b, g, and f. Again this suggests that the leverage point in Phase I for reducing the severity of the overall epidemic is the "snowball parameter" k.

4. Conclusion

The goal of this paper was to propose a plausible model that integrates two heretofore competing notions of what drives initiation into drug use, sellers recruiting new customers (the "snowball model") and existing users introducing friends to use (a "social contagion model"). The proposed model allows both to occur, but at different phases or stages of the drug epidemic. In particular, dealers are hypothesized to only recruit new users in the first phase, when distribution is embedded in social networks and users have difficulty locating alternate suppliers, thereby giving sellers the chance to reap personally the benefits of successfully recruiting a new user.

The model seems consistent with available historical data, but the data are so thin that this in no way constitutes a true validation. The model is best construed as a hypothesis communicated in a stylized mathematical model.

To the extent that the model has any validity, it suggests that slowing the rate at which dealers recruit new users in Phase I (slowing the "snowball effect") can substantially reduce the total quantity consumed throughout the epidemic, not just in Phase I. The mechanism is that slowing the rate of growth of the epidemic gives some users time to escalate their use to a destructive level. Then, when selling shifts from social networks to more public markets, the initial reputation of the drug will be less positive, tempering the spread of the drug in Phase II, when initiation is dominated by the social contagion model. Thus the encouraging message of this model for enforcement in the early stages of the drug epidemic is that merely slowing or delaying the spread of the drug might substantially reduce the severity of the overall drug epidemic. It may not be necessary to prevent completely the eventual emergence of a public market, let alone to eradicate the market altogether in order to produce substantial benefits.

These policy conclusions are conjectural, but what is clear is that epidemic models of drug use display "interesting" behavior in the sense that small changes in certain parameters early in an epidemic can produce large changes in the overall course of the modeled epidemic. This has also been observed by Behrens et al. (1999) and suggests the need for further research and modeling of the determinants of initiation in particular and the evolution of drug use in the early stages of an epidemic more generally.

Acknowledgements

The author thanks Doris Behrens for doing the numerical Phase II calculations and Peter Reuter and Gernot Tragler for many helpful suggestions. This material is based upon work supported in part by RAND's Drug Policy Research Center and by the National Science Foundation under Grant No. SBR-9357936. Any opinions, findings, and conclusions or recommendations expressed in this material are those of the authors and do not necessarily reflect the views of the National Science Foundation.

References

Behrens, D.A., Caulkins, J.P., Tragler, G., Haunschmied, J.L., and Feichtinger, G. (1999): A Dynamical Model of Drug Initiation: Implications for Treatment and Drug Control. *Mathematical BioSciences* 159, 1-20.

Caulkins, J.P. (1993): Local Drug Markets' Response to Focused Police Enforcement. *Operations Research* 41(5), 848-863.

Caulkins, J.P., Rydell, C.P., Everingham, S.S., Chiesa, J., and Bushway S. (1999): *An Ounce of Prevention, a Pound of Uncertainty: The Cost-Effectiveness of School-Based Drug Prevention Program.* RAND, Santa Monica, CA.

Everingham, S.S. and Rydell, C.P. (1994): *Modeling the Demand for Cocaine.* RAND, Santa Monica, CA.

Johnson, R.A., Gerstein, D.R., Ghadialy, R., Choy, W., and Gfroerer, J. (1996): *Trends in the Incidence of Drug Use in the United States, 1919-1992.* US Department of Health and Human Services, Washington, DC.

Kaplan, J. (1983): *The Hardest Drug: Heroin and Public Policy.* The University of Chicago Press, Chicago.

Musto, D. (1987): *The American Disease.* Yale University Press, New Haven, CT.

Reuter, P. and Kleiman, M.A.R. (1986): Risks and Prices: An Economic Analysis of Drug Enforcement. In: M. Tonry and N. Morris (Eds.): *Crime and Justice: An Annual Review of Research, Vol. 7.* University of Chicago Press, Chicago IL.

Rydell, C.P., Caulkins, J.P., and Everingham, S.S. (1996): Enforcement or Treatment: Modeling the Relative Efficacy of Alternatives for Controlling Cocaine. *Operations Research* 44(6), 687-695.

Tragler, G., Caulkins, J.P., and Feichtinger G. (Forthcoming): Optimal Dynamic Allocation of Treatment and Enforcement in Illicit Drug Control. *Operations Research.*

On the Structure of Cointegration

Manfred Deistler[1] and Martin Wagner[2]

[1] Institut für Ökonometrie, Operations Research und Systemtheorie, Technische Universität Wien, Argentinierstraße 8, A-1040 Wien.
email: Manfred.Deistler@tuwien.ac.at

[2] Abteilung für Ökonomie und Finanzwirtschaft, Institut für Höhere Studien, Stumpergasse 56, A-1060 Wien.
email: mwagner@ihs.ac.at

Abstract In this paper we deal with structural properties of cointegrated systems in ARMA form, where the integration and cointegration orders are arbitrary. In particular we analyze the relation between the structural properties and the Smith-McMillan form.

1. Introduction

Cointegration analysis definitely has become one of the most popular fields of modern econometrics for almost two decades now. The two major reasons for this are, first, that many economic time series show apparent nonstationarities in form of trends in means and variances and second that in economic theory long run static equilibrium relations between the variables are of primary interest.

It should be noted that in the past for econometric model building often differenced data, in order to achieve stationarity, have been used. In this case important information at frequency zero is neglected. In a certain sense cointegration is a counterrevolution against such differencing. In addition the modelling of cointegrated processes is an important contribution to the identification of linear unstable systems. However, it should be kept in mind that nonstationarities coming from unit root models are very special. The main justification for considering such a highly "non-generic" case is that genuine features of many economic time series can be explained this way.

During the last years an august statistical theory for cointegration has been developed (see e.g. Johansen 1991, Phillips 1991). Here we consider problems, which are prior to statistical analysis in the narrow sense, which we call structural problems (Hannan and Deistler 1988), where the analysis commences from the population second moments of the observations or from the transfer function, rather than from data.

The results of such an analysis nevertheless turn out to be important for the statistical analysis. In this contribution we discuss the relation between the structure of cointegration for general integration and cointegration orders and the Smith-McMillan form of the transfer function.

2. Basic Definitions

Let $\mathbb{N}$ denote the natural numbers, $\mathbb{Z}^+$ the non-negative integers and let $(\Omega, \mathcal{A}, \mathcal{P})$ be the underlying probability space. We consider stochastic processes $(x_t) = (x_t | t \in \mathbb{N})$, where $x_t : \Omega \to \mathbb{R}^n$. By z we denote the backward-shift on $\mathbb{N}$, i.e. $z(x_t | t \in \mathbb{N}) = (0, x_1, x_2, ...)$ as well as a complex variable $z \in \mathbb{C}$. The concept of cointegration has been introduced in Granger (1981) and Engle and Granger (1987).
In this contribution we restrict ourselves to ARMA processes (x_t), i.e. to the solution of ARMA systems

$$a(z)x_t = b(z)\varepsilon_t \tag{2.1}$$

where (ε_t) is white noise with covariance matrix $\Sigma = \mathbb{E}\varepsilon_t \varepsilon_t'$ and where

$$a(z) = \sum_{j=0}^{p} A_j z^j, \qquad b(z) = \sum_{j=0}^{q} B_j z^j; \qquad A_j, B_j \in \mathbb{R}^{n \times n}$$

Note that here for the sake of simplicity we only consider solutions on $\mathbb{N}$ for zero initial values. Throughout we impose the following assumptions:

The roots of $\det a(z)$ are restricted to be outside the unit circle or at $z = 1$.

$$\det\ a(z) \neq 0 \qquad |z| < 1 \quad \text{or for } |z| = 1 \quad \text{and } z \neq 1 \tag{2.2}$$

The strict miniphase assumption:

$$\det\ b(z) \neq 0 \qquad |z| \leq 1 \tag{2.3}$$

The matrices (a,b) are left coprime, i.e.

$$(a(z), b(z)) \quad \text{has rank } n \quad \forall z \in \mathbb{C} \tag{2.4}$$

The covariance matrix Σ of ε_t is non singular:

$$\Sigma > 0 \tag{2.5}$$

The polynomial matrix $a(z)$ as well as an ARMA process (x_t) are called *stable* if $\det a(z) \neq 0$ for all $|z| \leq 1$. Note that a stable process is asymptotically stationary. The transfer function is then given by

$$k(z) = a^{-1}(z)b(z) = \sum_{j=0}^{\infty} K_j z^j, \quad |z| < 1, \quad K_j \in \mathbb{R}^{n \times n} \tag{2.6}$$

This defines the solution of (2.1) corresponding to zero initial values

$$x_t = \sum_{j=0}^{t-1} K_j \epsilon_{t-j} \tag{2.7}$$

which is the only solution considered in this paper. A rational matrix $k(z)$ satisfying (2.2) and (2.4) ((2.3) and (2.4)) is called stable (strictly miniphase). We call a process (x_t) *integrated of order* $d \in \mathbb{Z}^+((x_t) \in I(d))$ if $(1 - z)^d(x_t)$ is stable, but $(1 - z)^{d-1}(x_t)$ is not stable.

A stochastic process $(x_t) \in I(d)$ is called *cointegrated*, if for some $\alpha \in \mathbb{R}^n, \alpha \neq 0$, the process $(\alpha' x_t)$ is integrated of order lower than d, i.e. $(\alpha' x_t) \in I(d - s_\alpha)$, $s_\alpha > 0$. The integer s_α is called the cointegration order corresponding to α. The largest s_α over all $\alpha \in \mathbb{R}^n$, $\alpha \neq 0$ is called the cointegration order s of (x_t) and we use the symbol $(x_t) \in CI(d, s)$.

3. The Structure of General Cointegrated Processes

In this section we relate some properties of the Smith-McMillan (SM) form of the transfer function k(z) to some structural properties relevant for cointegration. The main conclusion is that the diagonal matrix of the SM form, in general, does not contain the complete information about the cointegration structure of the process.

The SM form (see e.g. Kailath 1980, Hannan and Deistler 1988) is a rather straightforward generalization of the famous Smith form for the special case of polynomial matrices to rational matrices. As a consequence the rational transfer function $k(z)$ in (2.6) can be written as

$$k(z) = u(z)\Lambda(z)v(z) \tag{3.1}$$

where u(z) and v(z) are unimodular polynomial matrices and $\Lambda(z)$ is diagonal of the form

$$\Lambda(z) = \begin{pmatrix} \frac{\epsilon_1(z)}{\psi_1(z)} & & & \\ & \ddots & & 0 \\ & & \ddots & \\ 0 & & & \frac{\epsilon_n(z)}{\psi_n(z)} \end{pmatrix}$$

where $\epsilon_i(z), \psi_i(z)$ are relatively prime monic (i.e. the leading coefficient is equal to 1) polynomials, ϵ_i divides ϵ_{i+1}, $i = 1, 2, \ldots, n - 1$ and ψ_{i+1} divides ψ_i, $i = 1, 2, \ldots, n - 1$.

Here $\Lambda(z)$ is unique for given $k(z)$, whereas $u(z)$ and $v(z)$ are not unique in general. Note that k(z) is nonsingular (as a matrix over the field of rational functions) and thus all diagonal entries of $\Lambda(z)$ are non-zero. Thus, from (2.2) and (2.3), we have

$$k(z) = u(z) \underbrace{\begin{pmatrix} (1-z)^{-n_1} & & & 0 \\ & (1-z)^{-n_2} & & \\ & & \ddots & \\ 0 & & & (1-z)^{-n_n} \end{pmatrix}}_{\Lambda_1(z)} \times \qquad (3.2)$$

$$\times \underbrace{\begin{pmatrix} \frac{\epsilon_1(z)}{\tilde{\psi}_1(z)} & & 0 \\ & \ddots & \\ 0 & & \frac{\epsilon_n(z)}{\tilde{\psi}_n(z)} \end{pmatrix}}_{\Lambda_2(z)} v(z)$$

where n_i is the multiplicity of the zero of $\psi_i(z)$ at $z = 1$ and $\tilde{\psi}_i(z) = \psi_i(z)(1-z)^{-n_i}$ where the rational matrix $\Lambda_2(z)$ is stable and strictly miniphase.

As directly can be seen from (3.2), $l(z) = (1-z)^{n_1} k(z)$ is stable and miniphase (the latter means that $\det l(z) \neq 0$ for $|z| < 1$), but for $n_n < n_1$ not strictly miniphase.

Since $l(z)$ is a rational stable transfer function we may expand it into a power series in a certain neighborhood around $z = 1$:

$$l(z) = \sum_{j=0}^{\infty} L_j (1-z)^j \qquad (3.3)$$

here

$$l(1) = L_0 = u(1) \begin{pmatrix} I_{n-d} & 0 \\ 0 & 0 \end{pmatrix} \Lambda(1) v(1) \neq 0 \qquad (3.4)$$

where d is the number of elements $\psi_j(z)$ in (3.1) where the multiplicity of the zero at $z = 1$ is less than n_1, and thus $l(1)$ has rank $n - d$.

Thus we have the following representation for $k(z)$:

$$k(z) = L_0(1-z)^{-n_1} + \ldots + L_{n_1-1}(1-z)^{-1} + \underbrace{\sum_{j=n_1}^{\infty} L_j(1-z)^{j-n_1}}_{k^*(z)} \qquad (3.5)$$

Since $l(z)$ is analytic in a region containing the closed unit circle we see that $k^*(z)$ is stable.

From (3.5) we obtain

$$x_t = L_0(1-z)^{-n_1}\varepsilon_t + \ldots + L_{n_1-1}(1-z)^{-1}\varepsilon_t + k^*(z)\varepsilon_t$$

Now it is straightforward to see that n_1 is the order of integration of (x_t) and that for every $\alpha \neq 0$ such that $\alpha' L_0 = 0$ holds, $(\alpha' x_t)$ is integrated of order less than n_1.

Now by the *cointegration structure* of the transfer function we denote the vector of all feasible pairs $((m_1, d_1), \ldots, (m_s, d_s))$ where m_i are the different cointegration orders, ordered according to increasing m_i and d_i are the dimensions of the corresponding cointegrating spaces.

Theorem 3.1. *The cointegration structure as well as the corresponding cointegrating spaces of the transfer function $k(z)$ are not changed by postmultiplying $k(z)$ with a rational, stable and strictly miniphase square matrix $h(z)$, i.e. a matrix which is rational and has no poles and no determinantal zeros inside or on the unit circle.*

Proof:
Write $h(z) = \sum_{j=0}^{\infty} H_j (1 - z)^j$, then

$$
\begin{aligned}
k(z)h(z) \;=\;& L_0 H_0 (1-z)^{-n_1} + (L_0 H_1 + L_1 H_0)(1-z)^{-n_1+1} + \ldots \\
& + (L_0 H_{n_1-1} + \ldots + L_{n_1-1} H_0)(1-z)^{-1} + g(z)
\end{aligned}
$$

where $g(z)$ is stable, since $(1 - z)^{n_1} k(z)h(z)$ is stable. Since $h(1) = H_0$ is nonsingular $\tilde{x}_t = k(z)h(z)\varepsilon_t$ is integrated of order n_1. Let $lker(A)$ denote the left kernel of A. Now if $\alpha \neq 0$ is a cointegrating vector of cointegration order m_j for $x_t = k(z)\varepsilon_t$, then from (3.5) we have that this is true if and only if $\alpha \in lker L_0 \cap \ldots \cap lker L_{m_j-1}$ and $\alpha \notin lker L_{m_j}$ for $(\alpha' x_t)$ nonstationary or $\alpha \in lker L_0 \cap \ldots \cap lker L_{m_j-1}$ for $(\alpha' x_t)$ stationary. But this is equivalent to $\alpha \in lker L_0 H_0 \cap \ldots \cap lker(L_0 H_{m_j-1} + \ldots + L_{m_j-1} H_0)$ and $\alpha \notin lker(L_0 H_{m_j} + \ldots + L_{m_j} H_0)$ or $\alpha \in lker L_0 H_0 \cap \ldots \cap lker(L_0 H_{m_j-1} + \ldots + L_{m_j-1} H_0)$ for $(\alpha' \tilde{x}_t)$ stationary.

Pre-multiplication of the transfer function with a unimodular matrix does in general change the cointegration structure. This can be seen from the following example: Let

$$
\begin{pmatrix} (1-z)^2 x_{1t} \\ x_{2t} \end{pmatrix} = \begin{pmatrix} \varepsilon_{1t} \\ \varepsilon_{2t} \end{pmatrix}
$$

For this system, which is integrated and cointegrated of order 2, the expansion (3.5) of the transfer function $k(z)$ is given by

$$
k(z) = \begin{pmatrix} 1 & 0 \\ 0 & 0 \end{pmatrix}(1-z)^{-2} + \begin{pmatrix} 0 & 0 \\ 0 & 0 \end{pmatrix}(1-z)^{-1} + \begin{pmatrix} 0 & 0 \\ 0 & 1 \end{pmatrix}
$$

Thus the left kernel of L_0 is spanned by $(0, \alpha_2)$, with $\alpha_2 \neq 0$, which is also contained in the left kernel of $L_1 = 0$. So this system has a cointegration structure $(m_1, d_1) = (2, 1)$. Now, if we premultiply the transfer function with the unimodular matrix

$$
u(z) = \begin{pmatrix} 1 + (1-z)^2, & 1-z \\ 1-z, & 1 \end{pmatrix}
$$

the transfer function $k^u(z)$ of the resulting process (x_t^u) is given by

$$k^u(z) = \begin{pmatrix} (1-z)^{-2} + 1, & 1-z \\ (1-z)^{-1}, & 1 \end{pmatrix}$$

The expansion of this transfer function at $z = 1$ is given by

$$k^u(z) = \begin{pmatrix} 1 & 0 \\ 0 & 0 \end{pmatrix}(1-z)^{-2} + \begin{pmatrix} 0 & 0 \\ 1 & 0 \end{pmatrix}(1-z)^{-1}$$

$$+ \begin{pmatrix} 0 & 0 \\ 0 & 0 \end{pmatrix} + \begin{pmatrix} 0 & 1 \\ 0 & 0 \end{pmatrix}(1-z)$$

From the expansion of $k^u(z)$ it is directly seen that the intersection of the left kernels of the matrices corresponding to the terms $(1-z)^{-2}$ and $(1-z)^{-1}$ is 0. So this system is integrated of order 2 and cointegrated of order 1, the indices (m_1, d_1) are given by $(1,1)$.

The non-invariance of the cointegration structure with respect to pre-multiplication of the transfer function with unimodular matrices stems from the fact that with unimodular transformations it is possible to include differences of components of higher integration orders in the subsequent components with originally lower integration orders. These additional components of higher integration orders then cannot necessarily be wiped out by static linear combinations. In the above example this effect is seen directly by looking at

$$x_t^u = \begin{pmatrix} (1-z)^{-2} + 1, & 1-z \\ (1-z)^{-1}, & 1 \end{pmatrix}\begin{pmatrix} \varepsilon_{1t} \\ \varepsilon_{2t} \end{pmatrix}$$

from which it becomes clear that the second component is now integrated of order 1. Thus the vector $(0, \alpha_2)$, $\alpha_2 \neq 0$, is now a cointegration vector of order 1.

4. Cointegration and Factor Models

Let $(x_t) \in CI(1,1)$. Then $(1-z)x_t = l(z)\varepsilon_t = L_0\varepsilon_t + (1-z)l^*(z)\varepsilon_t$ with $l^*(z)$ a stable and miniphase transfer function. $L_0 = l(1)$ is rank deficient. Thus (x_t) can be written as

$$x_t = \frac{1}{1-z}L_0\varepsilon_t + l^*(z)\varepsilon_t \tag{4.1}$$

$$= L_0\sum_{t=1}^{T}\varepsilon_t + l^*(z)\varepsilon_t \tag{4.2}$$

(4.1) can be interpreted as a special factor model as follows: Let r denote the dimension of the cointegrating space, it is the dimension of the left kernel of

L_0. Then L_0 can be written as $\phi\eta'$, where $\phi, \eta \in \mathbb{R}^{n \times r}$ and both matrices have full rank r.

Now $\eta' \sum_{t=1}^{T} \varepsilon_t$ can be interpreted as an r dimensional factor process, usually called common trends. And $l^*(z)\varepsilon_t$ is a stationary noise process.

Note that this is a special factor model where in general noise and factor are correlated and where the factors are nonstationary whereas the noise is stationary.

Now let us exemplify the general idea with $I(2)$ models. Here we can write

$$(1 - z)^2 x_t = l(z)\varepsilon_t = L_0\varepsilon_t + (1 - z)L_1\varepsilon_t + (1 - z)^2 l^*(z)\varepsilon_t$$

where the L_j correspond to the expansion of $l(z)$ at $z = 1$, i.e. $l(z) = \sum_{j=0}^{\infty} L_j(1 - z)^j$.
And thus

$$
\begin{aligned}
x_t &= \frac{1}{(1 - z)^2} L_0\varepsilon_t + \frac{1}{1 - z} L_1\varepsilon_t + l^*(z)\varepsilon_t \\
&= L_0 \sum_{u=1}^{T} \sum_{t=1}^{u} \varepsilon_t + L_1 \sum_{t=1}^{T} \varepsilon_t + l^*(z)\varepsilon_t \\
&= \phi_0\eta_0' \sum_{u=1}^{T} \sum_{t=1}^{u} \varepsilon_t + \phi_1\eta_1' \sum_{t=1}^{T} \varepsilon_t + l^*(z)\varepsilon_t
\end{aligned}
$$

therefore in this case in general two kinds of common trends, $\eta_0' \sum_{u=1}^{T} \sum_{t=1}^{u} \varepsilon_t$ and $\eta_1' \sum_{t=1}^{T} \varepsilon_t$ are present. This can be extended in a straightforward manner to the general case.

References

1. Engle R.F. & C.W.J. Granger, 1987, Co-Integration and Error Correction: Representation, Estimation and Testing, Econometrica, 55, 251 – 276.
2. Granger C.W.J., 1981, Some properties of Time Series Data and their use in Econometric Model Specification, Journal of Econometrics, 16, 121 – 130.
3. Hannan E.J. & M. Deistler, 1988, The Statistical Theory of Linear Systems, New York: Wiley.
4. Johansen S., 1991, Estimation and Hypothesis Testing of Cointegration Vectors in Gaussian Vector Autoregressive Models, Econometrica, 59, 1551 – 1580.
5. Kailath Th., 1980, Linear Systems, Englewood Cliffs, N.J.: Prentice-Hall.
6. Phillips P.C.B., 1991, Optimal Inference in Cointegrated Systems, Econometrica, 59, 283 – 306.

Model and Reality -
The Principle of Simplicity within the Empirical Sciences

Wolfgang Eichhorn[1] and Ulrike Leopold-Wildburger[2]

[1] Department of Economic Theory and Operations Research,
University of Karlsruhe, D-76128 Karlsruhe
[2] Department of Statistics and Operations Research,
Karl Franzens-University of Graz, A-8010 Graz

1. Introduction

Albert Einstein's request to those who create models in physics or even more general in the empirical sciences was: " Make things as simple as possible, but not simpler! " The opinion of further famous physicists including Werner Heisenberg was that connections in physics are based on simple principles. These ideas lead us to the following questions, which we want to ask in connection with the natural sciences as well as with other empirical sciences, including economics:

(a) How simple should a model (of a piece) of reality be?

"Reality" is here the field in which we try to find connections empirically between phenomena and quantities and to measure them.

(b) What are the reasons for simplicity?

Are the connections from a section of reality described by models frequently simple because reality is (structured) simply or because the models are (constructed) simply? With other words: Is reality a priori based on simple principles or is the "principle of simplicity" only valid for the construction method of lots of models of reality?

We will try to find answers to question (a) within section 2 and answers to question (b) will be given within section 3. Within section 4 we concentrate on the sense of using *ideal models* and idealized connections. Within such models, if-then statements are always if–then statements of logic. *Real models* are to be understood as models with if-then statements: "If A, then with high probability also B."

Further, we will try to enlarge our answers to questions (a) and (b) with a couple of examples.

In general our answer to question (a) is the following: A model should be as simple as possible, such that it can fulfil the *purpose* for which it was created quite well.

Our opinion in answering question (b) is the following: reality is *not* structured simply, it is *not* always based on simple principles; the principle of simplicity can only be a guideline for *model building*. It frequently happens that we can navigate perfectly in complex reality with the help of a simple model, at least in respect to certain aims.

In section 4 we demonstrate the sense or nonsense of *ideal models*, dependent on their purpose. Further we emphasize that in *real models* one has to take account of a *"law of diminishing probabilities of events or results"* being valid for chains of implications of the following kind: Some are not logical implications but are only valid with certain high probabilities.

2. How simple should a model of reality be?

We argue in favour of the fact that a model (even of a complex detail) of reality ought to be as simple as possible for the aim it has been constructed for, such that it is a sensible aid for the purposes connected with this goal.

2.1 Having in mind a model of Austria with the purpose of getting information on how to drive between Austrian's cities (Graz, Innsbruck, Linz, Salzburg and Vienna) using highways as much as possible and minimizing distances on small roads; a detailed all-round model of "Austria's reality", as well as a map of Austria in scale 1:100.000 including roads for bikes and hiking would be much less useful than a rather simple map. Similar examples can be found in the work of the great economist Joan Robinson [1903 – 1983].

2.2 A further example: One of Galilei's descendants wonders about the distance covered by a stone dropped from a height of 125m, after one, two, three, four or five seconds, more generally after $t \leq 5$ seconds. She tries to use a formula of a model. She wants to know the length of the distances dependent on t with a precision of only +/- 1 %.

Under the assumption that an exact connection between time and (covered) distance exists at all, the aim of a precision of, let us say, 10^{-4} % is only to be reached within a rather complicated model. The model should include the resistance of the air and its dependence on the velocity of the falling stone, its form, its volume and its specific weight, the state and condition of the geoid in relation to the place where it has been thrown from and a lot of further *facts*.

Instead of all these considerations we could easily start with the hypothesis that the velocity of the thrown stone, that is

$$\frac{ds(t)}{dt}$$ (s (t) distance after t sec),

is proportional to the used time:

$$\frac{ds(t)}{dt} = gt$$ (g is a positive constant) (1)

such that for the acceleration of the falling stone the following equation is valid:

$$\frac{d^2 s(t)}{dt^2} = g = const.$$ (2)

By integration of (1) we obtain:

$$s(t) = \frac{g}{2} t^2 + c$$

with a constant c which is equal to zero because of the assumption s(0) = 0. Hence the distance after t seconds is

$$s(t) = \frac{g}{2} t^2.$$ (3)

The conclusions (2) und (3) from hypothesis (1) and the assumption that s(0) = 0 are purely logical assumptions. They are therefore valid under the presuppositions
s(0) = 0 and (1).

Is, however, equation (1) valid? We have to check this by measurements.

Obviously we know that within experiments the values of the measurements always vary. This fact can be caused by the errors of measurement, which appear obligatory, but can also be caused by additional influences. Further, it might happen that a precise, unique connection between the quantities (in our case time t and distance s) does *not* exist at all.

We measure the two, more or less dependent, quantities and illustrate the values within the system of coordinates. By doing so, clusters of points are always created. In our example Figure 1 and Figure 2 illustrate the values of the velocity ds/dt and the distance s dependent of time t. (For the purpose of demonstration we have enlarged the variances).

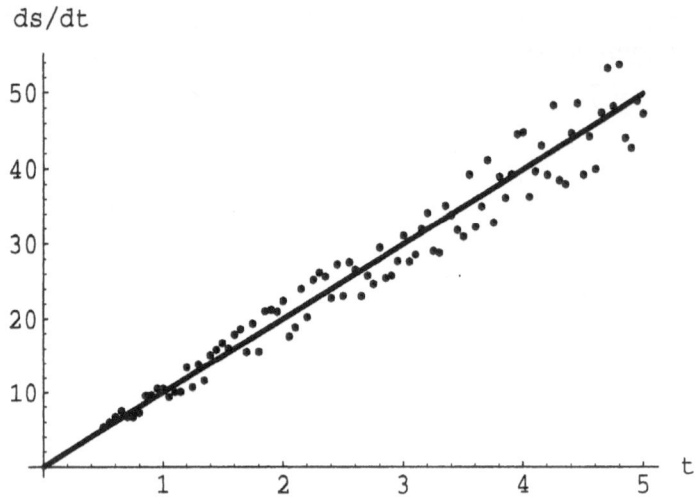

Figure 1: Part of a straight line starting from the origin and aimed at showing the trend of a cluster of points

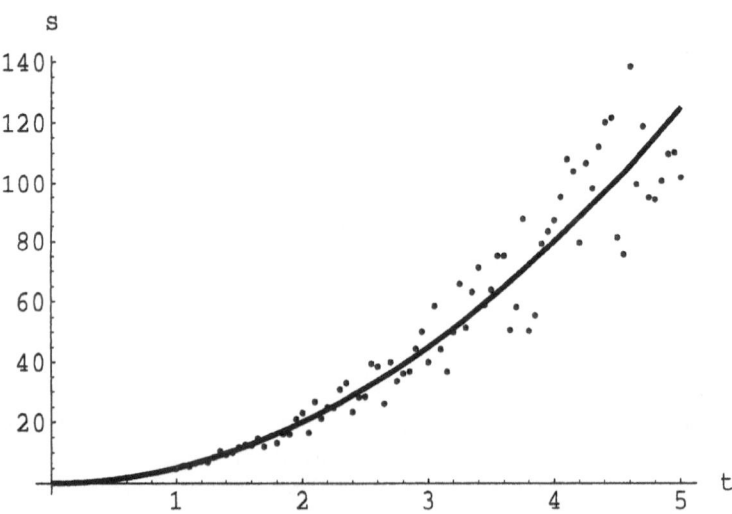

Figure 2: Part of a parabolic curve starting from the origin and aimed at showing the trend of a cluster of points

If we want to represent the cluster of points in Figure 1 by a curve in the best possible way, the obvious thing to do would be to use a straight line starting from the origin of the system of coordinates.

For Figure 2 we obviously could choose the piece of a parabolic curve as shown in the graphic. The hypothesis (1), the logical consequence (3) deduced from (1)

and the equation s(0) = 0 thus would be confirmed rather satisfactorily to a certain extent. As we know, we get numerical (mean) values for the constant g in (1) and (3) from the (Gauß)condition that is by determining the straight line in Figure 1 and the parabolic curve of Figure 2 in the following way: Minimize the sums of the squares of the distances between the points of the clusters and the points of the curves (Carl Friedrich Gauß [1777 - 1855]).

It is nice when the values of g determined by this method always match satisfactorily with reality. If they do, we are allowed to call the constant g, as is usual, (compare equation (2)) the earth's acceleration due to gravity constant.

If, however, they do not, we would try to make errors of measurement responsible *or* we had to doubt the simple hypothesis (1) and with it equation (3). Perhaps these equations were too simple to present reality with the required precision. It might therefore be necessary to deduce complex formulae within a complicated model to achieve the preciseness we are looking for.

But if we wish less details, for instance only what we sketched at the beginning of this section, we can uphold the principle of simplicity of a model, even if the complexity of the real connections is high.

3. Is the principle of simplicity valid for models **and** reality?

We have already raised this question in section 2.2 and we want to continue our arguments in terms of further examples.

3.1   As a first example we take the so-called psychophysical law, also called Fechner-Weber law (Ernst Heinrich Weber [1795-1878] and Gustav Fechner [1801-1888], see Weber (1851), Fechner (1860)). This law postulates relationships between psychological and physical facts; that is, relationships between (the intensities of) stimuli and (the intensities of) sensations or responses (as reactions to the stimuli).

Tests with experimental subjects generate
- upon stimulation of sensory organs
- in certain intervals $A_1 \leq x \leq A_2$ of bearable intensities x of stimuli
- for the intensities y of the sensations
a cluster of points as shown in Figure 3.

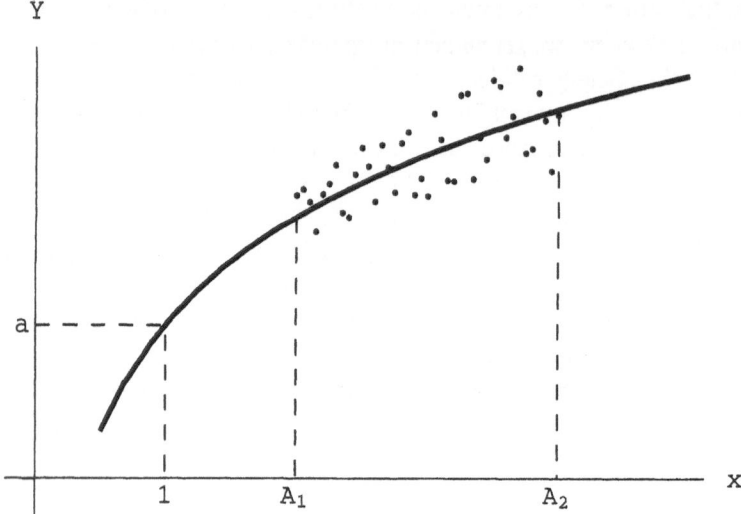

Figure 3: Graph of a function of the form (5), which runs smoothly in a balanced way through the cluster of points above the interval from A_1 to A_2.

In the relevant technical literature the psychophysical law is often identified with one of the following formulae:

$$y = c \ln x \qquad (c > 0, \text{constant}), \qquad (4)$$

$$y = a + b \ln x \qquad (a \geq 0, b > 0, \text{both constant}). \qquad (5)$$

Actually the graphs of the functions (4) and (5) respectively can run quite well through the clusters of points as Figure 3 shows. (The constants a, b and c can be determined for instance by the method of least squares).

Both formula (4) and (5) are bound to give the impression that the connection between the intensities x of a stimulus and the intensities y of a sensation
(i)   is unique for each subject and each sensory organ, and
(ii)   can best be represented with the help of the natural logarithm function ln x.

There is, however, no presumption that this is the correct interpretation.
(i')   It could also be the case, that a unique (deterministic) connection does <u>not</u> exist; the connection could be by stochastic, namely in *both directions:* the inexactness of the respective measurements, *as well as* the inexactness of the phenomena at hand.

(ii') It is much more convenient to draw a (from below) strictly concave function through the cluster of points of Figure 3 instead of drawing a straight line or a (from below) strictly convex function. It might, however, under certain circumstances be equally good or even better to take completely different classes of functions, e.g. the class of polynomials with a fixed degree n ≥ 3.

These and similar considerations can be avoided if our aims are less ambitious. For instance, our aims of modelling the possible connection between the intensities x of a stimulus and the intensity y of the sensation might only be with the aim to get answers to the following questions:

(α) Do the clusters of points above the intervals of tolerance (as shown in Figure 3) always have the same shapes? Can their trends always be represented by (from below) strictly concave functions much better than by straight lines or by (from below) convex functions?

(β) Supposing that the measurements are as precise as possible and independent from each other, do the measured values (x_i, y_i) always lie within relatively small intervals?

$$y_i \in \left\{ c * \ln x_i + d \middle| \begin{array}{l} -\varepsilon \le d \le \varepsilon \quad (\varepsilon > 0, \text{fixed}, \text{"small"}), \\ c* > 0, \text{suitably determined}, 0 < A_1 \le x_i \le A_2 \end{array} \right\}$$

respectively

$$y_i \in \left\{ a* - b * \ln x_i + d* \middle| \begin{array}{l} -\varepsilon \le d* \le \varepsilon*, \quad a* \ge 0, (\varepsilon* > 0, \text{fixed}, \text{"small"}), \\ b* > 0, \text{suitably determined}, 0 < A_1 \le x_i \le A_2 \end{array} \right\}$$

In other words: this example also shows, that for certain problems or aims there *are reasons to create principally simple models*. Obviously this does not imply that reality follows simple rules. Presumably reality is rather complex within the context of psychophysics. This is indicated by the variances of measured values, among other things. For a number of questions, however, it is not relevant to detect or even to model this complexity.

3.2   In our next example we shall refer to a well-known economist, sociologist and statistician: Vilfredo Pareto [1848 - 1923]. Like many others, Pareto (1897) was interested in the inequality of income and wealth in several countries, regions and towns. It has been pointed out by various authors that the Pareto distribution

$$P(x) = Ax^{-\alpha} \qquad \text{(A, \alpha positive constants)} \qquad (6)$$

fits the data fairly well towards the higher levels of income or wealth.

In (6), P (x) denotes the number of households with income (wealth) $\geq$ x. The constants A and α have been determined for several countries (regions, towns).

It is worth noting that the values for α fall in a very narrow range. In the literature published on this topic, α ranges from 1.6 to 2.4 for income data and from 1.3 to 2.0 for wealth data (depending on the year and the country, region or town under consideration). The results seem to indicate a common underlying mechanism. Because of this reason we talk of Pareto's law.

The numerical values for α and A are found with the help of the following method. Let x_0 be a relatively high level of income or wealth in a country (region, town). Determine, in a certain year, for the levels x_0, x_1, x_2, ..., x_n ($x_0 < x_1 < x_2 < ... < x_n$) the numbers y_0, y_1, y_2, ..., y_n of households with income (wealth) $\geq x_0$, x_1, x_2, ..., x_n, respectively. As a matter of fact, $y_0 \geq y_1 \geq y_2 \geq ... \geq y_n$. We then have a situation like the one in Figure 4.

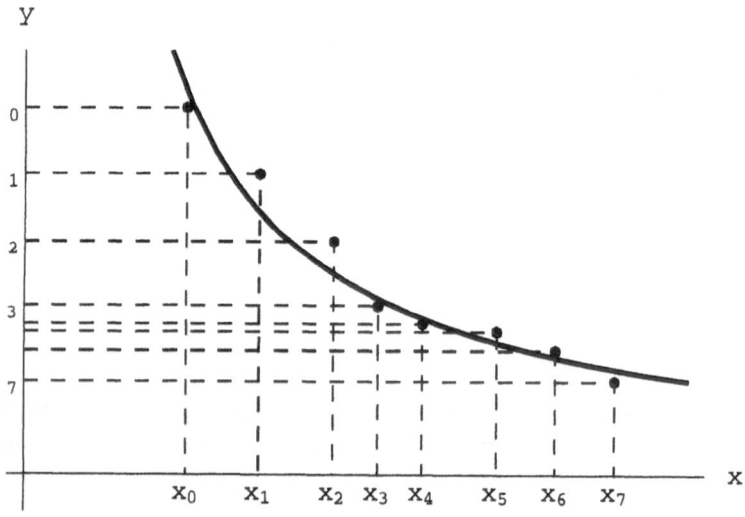

Figure 4: Cloud of points around the graph given by the function $y=A*x^{-\alpha}$

Now determine the numerical values α^* and A^*, say, for α and A by solving the problem of minimizing

$$\left(y_0 - Ax_0^{-\alpha}\right)^2 + \left(y_1 - Ax_1^{-\alpha}\right)^2 + \left(y_2 - Ax_2^{-\alpha}\right)^2 + \cdots + \left(y_n - Ax_n^{-\alpha}\right)^2$$

(Method of least squares, Gauß). We point out here *that* in our case *the inequality of income (wealth)* ≥ 0 *is not given by* $P(x) = A * x^{-\alpha^*}$, $x \geq x_0$.

All we can say is that $P(x) = A * x^{-\alpha^*}$ *fits the data fairly well towards* $x \geq x_0$. In other words, the "underlying mechanism" mentioned above does *not* imply that the points (x_0, y_0), (x_1, y_1), (x_2, y_2), ...,(x_n, y_n) belong to the graph of a *unique* function given by (6). The mechanism may be *very complex*, but Pareto's law is *very simple*. We have again a simple model that is sufficient to answer certain questions related to a reality that is probably very complex.

3.3   Are there examples for situations in which both the model of (a part of) reality and (this part of) reality are simple? According to our definition of "reality" in our introduction such an example would exist if there were a model which assumes only a connection $y = f(x)$ between two quantities x and y and if every empirically determined point (x_j, y_j) would exactly belong to the graph of the function f.

Yet, already because of errors of observation this is not to be expected. That is, the question concerning the simplicity of at least some aspects of reality cannot be decided in this way. We are firmly convinced that this question, although theoretically interesting, is not important with regard to applications. In respect to applications, what matters is that a simple model does a good job concerning a particular purpose.

In this connection it should be noted that even Newton's law of gravitation (Isaac Newton [1643 - 1727]), that is a simple model about the force attracting two particles in the universe

(*)   does not necessarily prove that the corresponding reality is relatively simple,

(**)   is not necessarily useful for the solution of *every* practical problem in astronomy.

Newton's law of gravity states that any particle of matter in the universe attracts any other particle with a force P as in formula (6), where x is the distance between the two particles, $\alpha = 2$ (the simple natural number 2) and $A = G\, m_1\, m_2$ (m_1, m_2 are the masses of the particles, G is a constant called the "universal gravity constant").

Is this simple formula only a satisfactory approximation to reality or does it always exactly conform to reality? An unambiguous *"yes"* to the second question cannot be given because of the following reasons:

(α)   If one determines the forces y_1, y_2, ..., y_n empirically depending on the distances x_1, x_2, ..., x_n, respectively, not all of the points (x_1, y_1), (x_2, y_2), ..., (x_n, y_n) will exactly lie on the graph of $y = Ax^{-2}$. It is impossible to prove that only errors of observation are responsible for that.

(β)   If we determine *that* α^* for which $Ax^{-\alpha}$ best fits the data (x_1,y_1), (x_2,y_2), ...,(x_n, y_n) according to the method of least squares, then in most cases we will get $\alpha^* \neq 2$.

It is not sure that the arithmetic means of two, three, ... α^*'s (independently determined for other distances and masses) will exactly tend to 2.

(γ)   Let a particle of mass m have the potential Gm/x. The derivative with respect to x, that is the force with which the particle attracts another particle of mass 1 and distance x, is $-Gm/x^2$. Our question whether the "simple" exponent 2 is the right one is equivalent to the question whether the potential of a particle of mass m is Gm/x. There is no proof that *this* simple formula is more than an (often useful) approximation to reality.

We want to quote, in this connection, Felix Klein [1849 - 1925] (1928, p.19, 20): "Die genaue Formulierung der Naturgesetze durch einfache Formeln beruht auf dem Wunsche, die äußere Erscheinung durch möglichst einfache Hilfsmittel zu beherrschen." Klein presents examples from the literature on Newton's law. These quotations perfectly confirm our assertion (α). Next he writes: "Die Genauigkeit, mit der die allgemein geltenden Naturgesetze zwingend durch das Experiment bewiesen sind, ist selbst im Falle des Newtonschen Anziehungsgesetzes nur sehr beschränkt." (We would prefer to write "acknowledged" instead of "geltend" (being valid)).

Klein also mentions the American astronomer Hall who changed Newton's law from $y = Ax^{-2}$ to

$$y = Ax^{-2-0.1574\bullet10^{-7}}$$

in order to take into account certain irregularities in the motion of Mercury; see, in this connection, our assertion (**).

Instead of "repairing" Newton's formula $y = Ax^{-2}$ only a little bit, Hall may have got even better results with the aid of a less simple formula, for instance

$$y = \frac{A}{bx + cx^d} \qquad \text{(A as above)},$$

where the parameters b, c and d had to be determined from the data.

4. Idealized simplified connections versus realized complicated corrections

Let us idealize the simple (or simplified?) connections dealt with in sections 2 and 3:

$$y = gx^2/2 \qquad \text{(law of gravity)}, \qquad (3)$$
$$y = a* + b* \ln x \qquad \text{(Fechner-Weber law)}, \qquad (7)$$
$$y = A * x^{-\overset{*}{\alpha}} \qquad \text{(Pareto's law)}, \qquad (8)$$
$$y = Ax^{-2} \qquad \text{(Newton's law)}. \qquad (9)$$

As we have shown that these connections are idealized in the following sense: If we put $x = \overline{x}$ we *determine (calculate)* $y = \overline{y}$ by inserting $x = \overline{x}$ on the right hand side of the formulae (3), (7), (8), (9) and regard the $\overline{y}$'s as belonging to (connected with) the corresponding $\overline{x}$'s. This is a mathematical idealization since empirical measurements assign to the value $\overline{x}$ values $\overset{=}{y}$, which are unequal to $\overline{y}$ in general.

In mathematical theories containing (3), (7), (8) or (9) we use logical if-then statements of the following kind:

If $x = \overline{x}$ then $g\overline{x}^2/2 =: \overline{y}$,   (3')

if $x = \overline{x}$ then $a* + b* \ln \overline{x} =: \overline{y}$,   (7')

if $x = \overline{x}$ then $A * \overline{x}^{-\overset{*}{\alpha}} =: \overline{y}$ (8')

if $x = \overline{x}$ then $A\overline{x}^{-2} =: \overline{y}$ (9')

or shorter:

$x = \overline{x}$ $\Rightarrow$ $g\overline{x}^2/2 =: \overline{y}$,

$x = \overline{x}$ $\Rightarrow$ $a* + b* \ln \overline{x} =: \overline{y}$,

$x = \overline{x}$ $\Rightarrow$ $A * \overline{x}^{-\overset{*}{\alpha}} =: \overline{y}$

$x = \overline{x}$ $\Rightarrow$ $A\overline{x}^{-2} =: \overline{y}$.

In reality things are not that simple: Instead of the logical implications (3'), (7'), (8'), (9') one can at most apply "if-then" statements of the kind

$$x = \overline{x} \xrightarrow[p(\varepsilon)]{} y = \overline{y} .$$

(10)

The meaning of this is the following: If $x = \overline{x}$ then the probability that a value $\overline{\overline{y}}$ as defined above satisfies

$$(1-\varepsilon)\overline{y} \le \overline{\overline{y}} \le (1+\varepsilon)\overline{y}$$

is p (ε) ($0 \le \varepsilon \le 0.3$, say, $\overline{y}$ as defined in (3'), (7'), (8') or (9')).

To give an example, let us start with purely mathematical and logical conclusions based on the assumption

A_1: Pareto's distribution (6) is the right formula to calculate the number $P(x)$ of households with wealth $\ge x_0$ (x_0 sufficiently large) in a given country.
A (logical) conclusion from A_1 is the assertion

B_1: If the constants A and α in (6) are numerically determined for a country then we have: If $x = \overline{x}$ is numerically given then the number $\overline{y}$ of the households with wealth $\ge \overline{x}$ can be determined for this country: $\overline{y} = A\overline{x}^{-\alpha}$.

Now we inform:

A_2: For the distribution of wealth in Italy, for example, the numerical values of A, α are A^*, α^*, respectively.
From B_1, A_2 we (logically) conclude:

B_2: In Italy the number of households with wealth $\ge \overline{x}$ is $y^* = A^* \overline{x}^{-\alpha^*}$.

Let this information, combined with an assumption A_3, (logically) yield B_3, and so on.

As can be seen, by this or a similar procedure we get "chains of logical conclusions" of the following kind:

$$A_1 \Rightarrow \left. \begin{array}{c} B_1 \\ A_2 \end{array} \right\} \Rightarrow \left. \begin{array}{c} B_2 \\ A_3 \end{array} \right\} \Rightarrow \cdots \Rightarrow \left. \begin{array}{c} B_{k-1} \\ A_k \end{array} \right\} \Rightarrow B_k .$$

(11)

Note that from A_1 follows B_1, from both B_1 and (an additional) A_2 follows B_2, and so on. In such an ideal or mathematical model or theory it is clear that if the assumptions A_1, A_2, ..., A_k are true and the k logical implications "$\Rightarrow$" are right, then B_k is true.

If we want to apply such a model or theory to a real-world problem we have to check whether the k implications "$\Rightarrow$" are indeed logical implications or only if - then statements of the kind (10) or the kind

$$A \xrightarrow{p} B \tag{12}$$

meaning: if A is true then there is a probability of p that B is true.

We are firmly convinced that a model or theory that is based on empirically observable or measurable connections can *not* contain only implications "$\Rightarrow$". In such a *real* or *realistic* model or theory there will be if-then statements of the kind "$\xrightarrow{p(\varepsilon)}$" (see (10)) and/or "$\xrightarrow{p}$" (see (12)).

This means for practical applications, i.e. applications to real-world problems: We cannot build on "chains of logical conclusions" as in (11), but only on "chains consisting of both logical conclusions *and* implications that are valid only with probabilities" such as

$$A_1 \xrightarrow{p_1} \left.\begin{array}{c} B_1 \\ A_2 \end{array}\right\} \xrightarrow{p_2} \left.\begin{array}{c} B_2 \\ A_3 \end{array}\right\} \xrightarrow{p_3} \cdots \xrightarrow{p_{k-1}} \left.\begin{array}{c} B_{k-1} \\ A_k \end{array}\right\} \xrightarrow{p_k} B_k . \tag{13}$$

Here the p_j (j=1, 2,..., k) are probabilities, that is, $0 \le p_j \le 1$.

As in (10), p_j may depend on ε. If $p_j = 1$ for j=1,2,...,k then we have a situation similar to that in (11), i.e., we can say: If A_1 then B_k with probability 1.

But notice: the probability that B_k "follows" from A_1 and so on can be very small, even if the p_j's in (12) all are close to 1. If, for instance, $p_1 = p_2 = \ldots = p_k = 0.9$ and suitable independence assumptions are made then B_k "follows" from A_1 (and so on) with certain (high) probability

0.59 (k=5),   0.35 (k=10),   0.21 (k=15),   0.12 (k=20),
0.072 (k=25),   0.042 (k=30),   0.025 (k=35),   0.015 (k=40),
0.009 (k=45),   0.005 (k=50).

Obviously there is a *"law of diminishing probabilities of events or results"* in the context of "chains of implications, where some are not logical ones but are valid only with certain (high) probabilities".

This example shows: The principle of simplicity in the empirical sciences suggests looking for ideal (often idealized) simple (often simplified) connections. Working with such connections can take us far away from reality. As soon as this is realized, corrections should be made. This is often complicated. The reason for all this: reality and also many parts of reality are *not* simple.

References

1. Fechner, G.T. (1860), Elemente der Psychophysik, Breitkopf Härtel, Leipzig.
2. Klein, F. (1928), Elementarmathematik vom höheren Standpunkt aus, Vol. 3, third Edition, Springer-Verlag, Berlin.
3. Pareto, V. (1897), Cours d'économie politique, vol.2, Librairie de l'Université, Lausanne.
4. Robinson, J. (1979), The Generalization of the General Theory and Other Essays, 2.ed., Basingstoke, Macmillan.
5. Weber, E.H. (1851), Annotationes anatomicae et physiologicae, Leipzig.

A Bayesian Semiparametric Analysis of ARCH Models

Hideo Kozumi[1] and Wolfgang Polasek[2]

[1] Faculty of Economics, Hokkaido University, Kita-ku, Kita 9 Nishi 7, Sapporo 060-0809, Japan
[2] Institute of Statistics and Econometrics, University of Basel, Holbeinstrasse 12, 4051 Basel, Switzerland

Summary. This paper provides a Bayesian analysis of a semiparametric autoregressive conditional heteroscedasticity (ARCH) model. We propose a semiparametric ARCH model using a Dirichlet process prior and show a Markov chain Monte Carlo method for the posterior inference. The model is estimated with a data set of monthly exchange rate for the Deutsche Mark to the U. S. Dollar.

1. Introduction

In financial time series analyses, the autoregressive conditional heteroscedasticity (ARCH) model has played a central role to explore the phenomenon of volatility clustering. Since its introduction by Engle (1982), the ARCH model has been extended to various directions and applied intensively to exchange rates, interest rates and stock prices. There are now more than several hundred articles discussing theoretical properties of the ARCH model and its variants as well as empirical applications (see Bollerslev *et al.*, 1992, for a good review).

Although most extension of the ARCH model has followed a parametric framework, several authors have proposed nonparametric techniques for ARCH modeling. Following Robinson (1987), Pagan and Ullah (1988) and Robinson (1988) advocate the kernel method to estimate the functional form of the conditional variance. Gallant and Tauchen (1989, 1992) also propose an alternative semiparametric approach and prove that the maximum likelihood estimator is consistent under mild regularity conditions. Their basic idea is to correct the Gaussian ARCH model with a modified Hermite series expansion. A model derived by such a correction is rich enough to approximate densities with skewness or fat tails.

In Bayesian nonparametric analyses, it is necessary to specify a prior distribution for an unknown distribution which is an infinite dimensional parameter and a Dirichlet process is a common choice in practice. The Dirichlet process, as introduced by Ferguson (1973), is a probability measure with large support on the set of all probability measure on the sample space and indexed two parameters. An advantage of the Dirichlet process is its easy interpretation of its parameters. It also has the desirable property that it can be updated easily by data in a Bayesian framework. However, the Dirichlet process almost surely selects a discrete distribution function (see Blackwell and

McQueen, 1973). To solve the problem, Antoniak (1973) formalize the basic idea of a Dirichlet process mixing. While the use of the Dirichlet process prior was recognized very useful, its application was restricted only to simple cases because of its computational difficulty. Therefore, Bayesian nonparametric analyses were silent for a long time.

Since Gelfand and Smith (1990) introduce the Gibbs sampling, Markov chain Monte Carlo (MCMC) methods have become a powerful tool for Bayesian inference. Some of the earlier references on MCMC methods include Geman and Geman (1984), Hasting (1970), Metropolis, *et al.* (1953), and Tanner and Wong (1987). See also Besag and Green (1993) or Smith and Roberts (1993) for a review of MCMC methods. Furthermore, recent development of MCMC method opened the way to Bayesian nonparametric analyses. Escobar (1994) and Escobar and West (1995) analyze the hierarchical Bayesian model with the Dirichlet process prior in the context of density estimation and demonstrate that the Gibbs sampling can overcome the computational difficulty. Since then, the use of the Dirichlet process prior has become increasingly popular for Bayesian modeling (see, for example, Doss, 1994; Kuo and Mallick, 1995; Müller, Erkanli and West, 1996; West, Müller and Escobar, 1994).

The objective of this paper is to develop a Bayesian semiparametric ARCH model with the Dirichlet process prior. The ARCH model has usually conditional variances which are expressed as a linear function of past squared values of the observations. We consider an alternative specification of the conditional variances as in Geweke (1986) and Pantula (1986) and rewrite the ARCH model in an autoregressive (AR) form. Then, to construct a Bayesian semiparametric model, it is assumed that the parameters in the model are a sample from an unknown distribution, and we employ the Dirichlet process prior. Consequently, we obtain a semiparametric class of ARCH models.

As noted in Gallant, Rossi and Tauchen (1993), inference on the probability distribution of a time series process is a fundamental statistical object. Therefore, we are also interested to estimate a predictive density which summarizes information about a future observation within a Bayesian framework.

The plan of this paper is as follows. In Section 2, we provide a brief review of Dirichlet processes and develop a Bayesian semiparametric ARCH model using the Dirichlet process prior. Section 3 describes the Gibbs sampling algorithm suggested by MacEachern and Müller (1994). In Section 4 we apply our model to monthly exchange rates of the Deutsche Mark to the U. S. Dollar. Finally, conclusions are given in Section 5.

2. Model description

The ARCH model proposed by Engle (1982) is written as

$$z_t = \sqrt{h_t}\epsilon_t, \quad \epsilon_t \sim N(0,1), \tag{2.1}$$

$$h_t = \beta_0 + \sum_{i=1}^{p} \beta_i z_{t-i}^2, \tag{2.2}$$

where z_t is the time series observation at time t $(t = 1, \ldots, T)$, ϵ_t is an error term with mean zero and unit variance. To ensure the conditional variances h_t to be positive, the parameters must satisfy

$$\beta_0 > 0, \quad \beta_i \geq 0 \ (i = 1, \ldots, p).$$

Since the restrictions are cumbersome for the later analysis, we consider an alternative specification of h_t given by

$$\log h_t = \beta_0 + \sum_{i=1}^{p} \beta_i \log z_{t-i}^2. \tag{2.3}$$

This logarithmic parameterization of h_t is discussed by Geweke (1990) and Pantula (1986). Therefore, the ARCH model can be expressed as

$$\log z_t^2 = \beta_0 + \sum_{i=1}^{p} \beta_i \log z_{t-i}^2 + \log \epsilon_t^2, \tag{2.4}$$

that is, $\log z_t^2$ is modeled as the AR(p) model. It is well known that the first second moments of $\log \epsilon_t^2$ are given as $E(\log \epsilon_t^2) = -1.27$ and $Var(\log \epsilon_t^2) = \pi^2/2$. However, $\log \epsilon_t^2$ is no longer normally distributed and its distribution has a complicated form. One approach to handle the situation is to treat $\log \epsilon_t^2$ as if it were normally distributed, but this might result in inefficient estimation results.

Instead of assuming a certain distribution for the error terms, we employ a semiparametric approach in the Bayesian framework and rewrite the model (2.4) compactly as

$$y_t \sim N(x_t'\beta_t, \sigma_t^2), \tag{2.5}$$

where $y_t = \log z_t^2$, $x_t = (1, \log z_{t-1}^2, \ldots, \log z_{t-p}^2)'$, $\beta_t = (\beta_{t0}, \beta_{t1}, \ldots, \beta_{tp})'$. Here we assume that $\theta_t = (\beta_t, \sigma_t^2)$ is a sample from an unknown distribution G. Thus, integrating out θ_t yields the density

$$f(y_t|x_t, G) = \int f(y_t|x_t, \theta_t)dG(\theta_t). \tag{2.6}$$

The probability distribution of y_t is no longer normal but depends on the past observations x_t, and we obtain a semiparametric class of models. Since the equation (2.5) does not describe the first moment of z_t, it is assumed that $E(z_t) = 0$. Now the conditional variances of z_t are given by

$$E[\exp(y_t)] = E(z_t^2) = h_t = \exp(x_t'\beta_t + \sigma_t^2/2). \tag{2.7}$$

Since Bayesian analyses require a prior distribution for G, we assume that G has a Dirichlet process prior introduced by Ferguson (1973, 1974). Here we provides a briefly definition of the Dirichlet process and review its feature. Full mathematical details of the Dirichlet process can be found in Ferguson (1973) and Antoniak (1974). Let $(\mathcal{X}, \mathcal{B})$ be an arbitrary measurable space and G_0 be a probability distribution defined on $(\mathcal{X}, \mathcal{B})$. Then, a random probability measure G is said to follow a Dirichlet process, i.e., $G \sim DP(\alpha G_0)$, if for any partition $B_1, \ldots, B_m$ of $\mathcal{X}$, the random vector $(G(B_1), \ldots, G(B_m))$ has a Dirichlet distribution with parameter vector $(\alpha G_0(B_1), \ldots, \alpha G_0(B_m))$. A key feature of the Dirichlet process is that it gives mass one to the set of discrete distributions, that is, any realizations $\theta = (\theta_1, \ldots, \theta_T)$ are almost surely discrete. Thus, the distribution of $\log \epsilon_t^2$ is approximated as a mixture of normal distributions in our model. A similar approximation is also employed in Shephard (1994).

It is also worth mentioning that our model has two distinct features compared with the original ARCH model. First, the distribution of z_t is not necessarily normal and can be nonnormal. The second is that the model allows the parameterization of $\log h_t$ to be nonlinear. Therefore, our model is considered to be a nonnormal and nonlinear extension of the ARCH model.

To complete the model description, we specify prior distributions for the rest of the parameters. As for the base distribution G_0 we choose normal and inverted gamma distributions, that is,

$$dG_0(\theta_t) = N(\mu, \Sigma)IG(n_0/2, \tau/2),$$

since the model (2.5) has a regression form. Moreover, we use prior distributions for μ, Σ and τ for hierarchical modeling as follows:

$$
\begin{aligned}
\mu &\sim N(\mu_0, V_0), \\
\Sigma^{-1} &\sim W(\nu_0, \Sigma_0), \\
\tau &\sim Ga(m_0/2, \tau_0/2).
\end{aligned}
$$

Following Escobar and West (1995), we assume that α follows a gamma distribution

$$\alpha \sim Ga(a_0, b_0).$$

3. Markov chain Monte Carlo

Our Bayesian semiparametric class of ARCH models is summarized as

$$
\begin{aligned}
y_t &\sim N(x_t'\beta_t, \sigma_t^2), \\
\theta_t &\sim G, \\
G &\sim DP(\alpha G_0), \\
dG_0 &= N(\mu, \Sigma)IG(n_0/2, \tau/2), \\
\mu &\sim N(\mu_0, V_0),
\end{aligned}
$$

$$\Sigma^{-1} \sim W(\nu_0, \Sigma_0),$$
$$\tau \sim Ga(m_0/2, \tau_0/2),$$
$$\alpha \sim Ga(a_0, b_0).$$

Collecting all observations $Y = (y_1, \ldots, y_T)'$, the model in (2.5) is written as

$$f(Y|G) = \prod_{t=1}^{T} \int f(y_t|x_t, \theta_t) dG(\theta_t). \tag{3.1}$$

Since it is difficult to sample G, we rewrite the model with the latent variables θ_t but marginalize over G, obtaining

$$\prod_{t=1}^{T} f(y_t|x_t, \theta_t) \cdot \pi(\theta). \tag{3.2}$$

As mentioned in the previous section, the Dirichlet process has the discreteness property that any realizations of T parameters θ_t lie in a set of $k \leq T$ distinct values, denoted by $\theta^* = (\theta_1^*, \ldots, \theta_k^*)'$. It should be noted that θ^* is a sample from G_0 (see Antoniak, 1974). Let $S = (S_1, \ldots, S_T)$ be the configuration vector, i.e., $S_t = j$ if and only if $\theta_t = \theta_j^*$. The configurations S_t map θ^* into θ by $\theta_t = \theta_i^*$ if $S_t = i$. We can also define the vector $n = (n_1, \ldots, n_k)'$ with $n_i = \#\{t : S_t = i\}$. Then, the conditional distribution of θ_t is derived as

$$\theta_t|\theta^{(t)} \sim \frac{\alpha}{\alpha + T - 1} G_0 + \frac{1}{\alpha + T - 1} \sum_{i=1}^{k^{(t)}} n_i^{(t)} \delta(\theta_i^{*(t)}), \tag{3.3}$$

where $\theta^{(t)} = (\theta_1, \ldots, \theta_{t-1}, \theta_{t+1}, \ldots, \theta_T)$. Also, $k^{(t)}$ denotes the number of distinct values in $\theta^{(t)}$ with $n_i^{(t)}$ common values $\theta_i^{*(t)}$, and $\delta(x)$ stands for the distribution giving mass one to the point x.

Using the fact that the knowledge of θ is equivalent to that of θ^*, S and k, West, Müller and Escobar (1994) provide an efficient algorithm to sample θ^*, S and k. It is necessary in their algorithm to evaluate the integral $\int f(y_t|x_t, \theta_t) dG_0(\theta_t)$, which can be obtained analytically only when G_0 has a conjugate form. However our prior specification is nonconjugate and it is difficult to evaluate the integral. Instead, we use another algorithm developed by MacEachern and Müller (1994) to handle nonconjugate case. They suggest to expand the parameter vector $(\theta_1^*, \ldots, \theta_k^*)$ to include $(\theta_{k+1}^*, \ldots, \theta_T^*)$. Note that the corresponding n_j $(j > k)$ are zero. MacEachern and Müller (1994) introduce the prior distribution on $(\theta_1^*, \ldots, \theta_T^*, S)$ which induces the same prior distribution as the original. Consequently sampling S_t, θ^* and k can be done in the following way:

(a) Generate S_t as follows:

$$S_t = \begin{cases} i, & i = 1, \ldots, k^{(t)}, & \text{with probability } q_i, \\ k^{(t)} + 1, & & \text{with probability } q_{k^{(t)}+1}, \end{cases}$$

where

$$q_i = \begin{cases} n_i^{(t)} f(y_t | \theta_t = \theta_i^*) & i = 1, \ldots, k^{(t)}, \\ \frac{\alpha}{k^{(t)}+1} f(y_t | \theta_t = \theta_{k^{(t)}+1}^*) & i = k^{(t)} + 1. \end{cases}$$

If $n_{S_t}^{(t)} = 0$, then with probability $1/k^{(t)}$ resample S_t as above, otherwise leave S_t unchanged.

(b) Generate $\theta_i^* = (\beta_i^*, \sigma_i^{2*})$ from the complete conditional distributions

$$\begin{aligned} \beta_i^* &\sim N(\mu_{\beta_i^*}, \Sigma_{\beta_i^*}), \\ \sigma_i^{2*} &\sim IG(n_{i*}/2, s_{i*}/2), \end{aligned}$$

where

$$\begin{aligned} \Sigma_{\beta_i^*}^{-1} &= \Sigma_{t \in I_i} \frac{x_t x_t'}{\sigma_i^{2*}} + \Sigma^{-1}, \\ \mu_{\beta_i^*} &= \Sigma_{\beta_i^*} \left(\sum_{t \in I_i} \frac{x_t y_t}{\sigma_i^{2*}} + \Sigma^{-1} \mu \right), \\ n_{i*} &= n_0 + n_i, \\ s_{i*} &= \tau + \sum_{t \in I_i} (y_t - x_t' \beta_i^*)^2, \end{aligned}$$

and $I_i = \{t : S_t = i\}$.

It should be noted that sampling k is implicitly accomplished in (a).

The complete conditional distributions of μ, Σ^{-1} and τ are easily obtained and given as

$$\begin{aligned} \mu &\sim N(\mu_*, V_*), \\ \Sigma^{-1} &\sim W\left(\nu_0 + k, \Sigma_0 + \sum_{i=1}^{k}(\beta_i^* - \mu)(\beta_i^* - \mu)'\right), \\ \tau &\sim Ga\left((m_0 + n_0 k)/2, (\tau_0 + \sum_{i=1}^{k} 1/\sigma_i^{2*})/2\right), \end{aligned}$$

respectively, where

$$\begin{aligned} V_*^{-1} &= k\Sigma^{-1} + V_0^{-1}, \\ \mu_* &= V_*\left(\Sigma^{-1} \sum_{i=1}^{k} \beta_i^* + V_0^{-1} \mu_0\right). \end{aligned}$$

Escobar and West (1995) note that if we augment the model with a variable η whose complete conditional distribution is $Be(\alpha + 1, T)$, then the complete conditional distribution of α is given by

$$\alpha \sim wGa(a_0 + k, b_0 - \log(\eta)) \tag{3.4}$$
$$+(1 - w)Ga(a_0 + k - 1, b_0 - \log(\eta)), \tag{3.5}$$

where

$$\frac{w}{1 - w} = \frac{a_0 + k - 1}{T(b_0 - \log(\eta))}.$$

The predictive density for the next observation is given by

$$f(y_{T+1}|Y) = \int \int f(y_{T+1}|Y, \theta_{T+1})p(\theta_{T+1}|Y)d\theta_{T+1}. \tag{3.6}$$

In order to estimate the predictive density, it is necessary to sample θ_{T+1}, and this is an easy routine in Gibbs sampling. After delivering a simulated sample of ϕ_{T+1}^m and τ^m ($m = 1, \ldots, M$), the predictive density can be approximated by

$$f(y_{T+1}|Y) \approx \frac{1}{M} \sum_{m=1}^{M} f(y_{T+1}|Y, \theta_{T+1}^m). \tag{3.7}$$

4. Illustrative example

We illustrate our Bayesian semiparametric ARCH models with a data set of monthly exchange rate for the Deutsche Mark against the U. S. Dollar. The data was transformed in one hundred times log differences, giving 267 observations of monthly growth rates. The data covers the period December 1972 to February 1995. The plot of the data is shown in Figure 4.1.

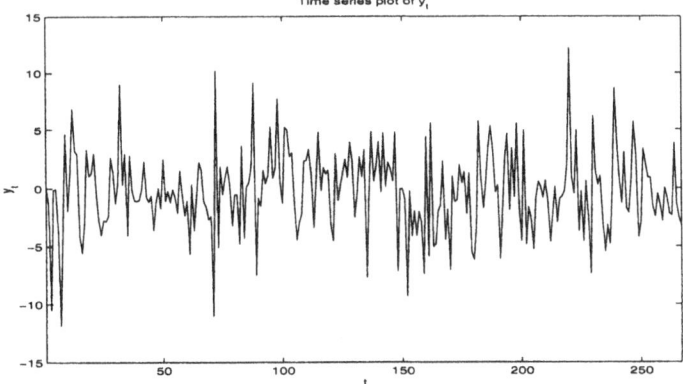

Fig. 4.1. The time series plot of the data.

Table 4.1. Posterior estimates of the ARCH models.

| | Parametric ARCH | Semiparametric ARCH | |
		Case I	Case II
β_0	2.547		
	(0.104)		
β_1	0.035	0.063	
	(0.034)	(0.056)	
β_2	-0.086	-0.079	
	(0.035)	(0.058)	
k		71.62	62.75
		(16.03)	(15.21)

(Standard deviations are in parentheses.)

We estimated the ARCH model given in (2.1) and (2.3) as well as the semiparametric ARCH models for a comparative analysis. As for estimation of the semiparametric models, we considered two cases (Case I and Case II): Case I assumes that only β_{0t} is associated with the Dirichlet process prior and the other β's are assigned normal distribution as a prior. In Case II the Dirichlet process prior is applied to all β's as in Section 2.

Since most of empirical results show that the ARCH model with small order is enough to explain monthly data, we set $p = 2$ in all the cases. The results reported below were based on the following specification of hyper–parameters: We set $\mu_0 = 0$ and $\nu_0 = n_0 = m_0 = \tau_0 = 5$. V_0 and Σ_0 were diagonal matrices whose elements were 1000 and 0.01, respectively. Also $a_0 = 5$ and $b_0 = 0.1$ were chosen. As for Gibbs sampling, we ran a single chain of 30000 iterations of the Gibbs sampling. The burn–in period is 15000, giving 15000 simulation samples.

The results in Table 4.1 suggest that similar posterior estimates were drawn from both of the parametric and semiparametric models. The posterior means of β's are similar and their standard deviations for the semiparametric model are slightly larger than those for the parametric model. In the results for the semiparametric models, the posterior means of k are quite large.

Figure 4.2 plots the posterior means of β_t for Case II and reveals their sharp fluctuations. Interestingly the movement of β_{0t} is highly related to that of the data shown in Figure 4.1. Recent work reports that the stochastic volatility (SV) model is preferable to the ARCH model to explain the volatility movement (see, *e.g*, Jacquier, *et al.*, 1994). One of the distinct differences between those models is that conditional variances of the SV model are random variable given the past information, but those of ARCH model are not. Our findings, large posterior means of k and the sharp fluctuations of β_{0t}, might be an evidence to support the SV model. Thus, the ARCH model does not explain the volatility movement well, and it is necessary to take account of other sources for modeling the volatility. For further examination, three–dimensional plots of the estimates of $\log h_t$ are shown in Figure 4.3. The plot of $\log y_t^2$ is also shown in Figure 4.3 since it can be considered as

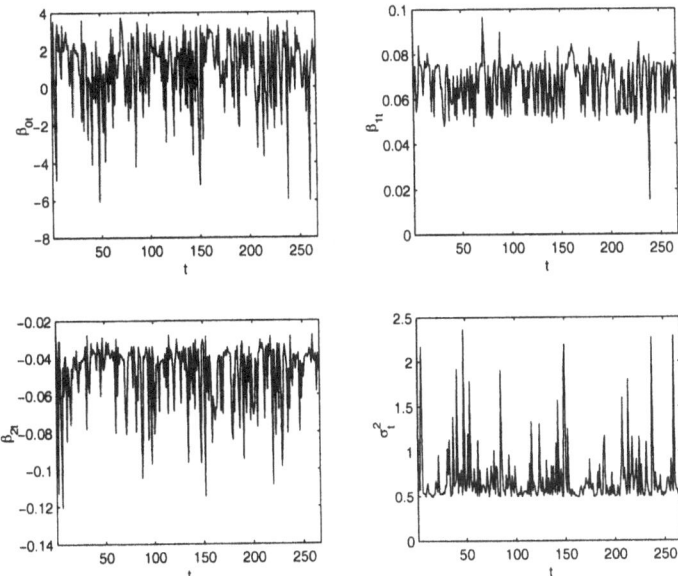

Fig. 4.2. The time series plot of θ_t.

an approximation of the logarithm of conditional variances. Compared with the ARCH model, the results for the semiparametric model are very similar to those for the data. These results also confirm the previous discussion.

As described in Section 1, a predictive density summarizes information about the probability distribution of data. We estimated the predictive densities of the model based on the method given in (3.7). The result is shown in Figure 4.4 with a histogram of y_t. From Figure 4.4, it can be seen that the estimate of a predictive density shows definite skewness and reflects the non-normality of $\log \epsilon_t^2$. Therefore, it should be avoided to treat the distribution of $\log \epsilon_t^2$ as if it were normal.

5. Conclusion

We have developed a Bayesian semiparametric ARCH model using the Dirichlet process prior. The model proposed in this paper constructs a semiparametric class of ARCH models. We have also showed the Gibbs sampling scheme to make posterior inference possible.

The proposed model was illustrated with monthly exchange rate of Deutsche Mark to U. S. Dollar. From the empirical analyses, we found that our semiparametric ARCH model can explain the data quite well,

Our methodology is the first step to a Bayesian semiparametric approach to ARCH modeling. Most of recent work is conducted with daily data and more frequently observed data. To handle such data sets, it is known that the generalized ARCH (GARCH) model and the SV model are more appropriate

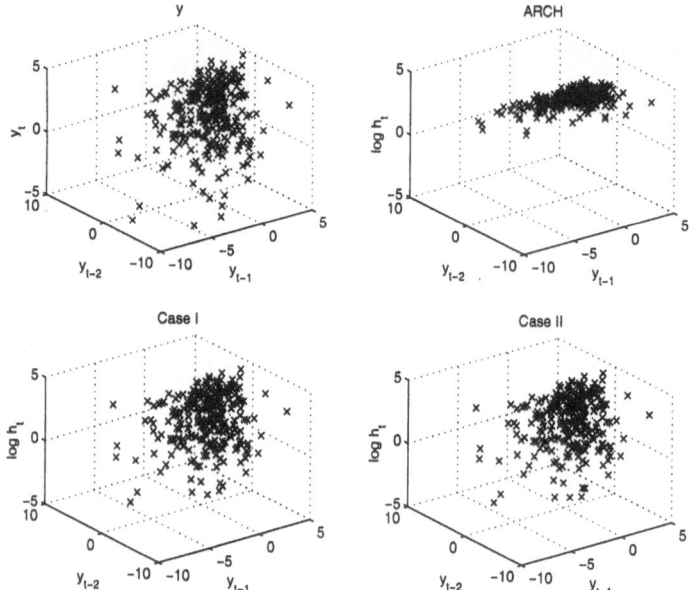

Fig. 4.3. The three–dimensional plots of $\log h_t$.

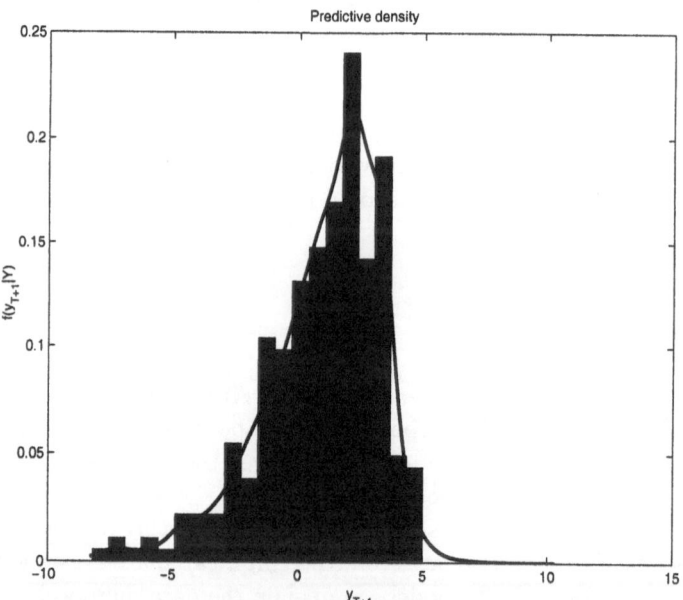

Fig. 4.4. The estimate of the predictive density.

than the ARCH model since the ARCH model requires a long lag length. Therefore, it is necessary to apply our approach to those models with some modifications, and this topic is left for the future research.

Acknowledgement. This paper was written while the first author was visiting at Institute of Statistics and Econometrics, University of Basel. The authors are grateful to Jeffrey Pai and Michael Escobar for their comments on the earlier version of this paper.

References

Antoniak, C.E. (1974): Mixtures of Dirichlet processes with applications to nonparametric problems. *The Annals of Statistics* **2**, 1152–1174

Besag, J. and Green, P.J. (1993): Spatial statistics and Bayesian computation. *Journal of the Royal Statistical Society Ser.* **B 55**, 25–37

Blackwell, D. and MacQueen, J.B. (1973): Ferguson distributions via Pólya urn schemes. *The Annals of Statistics* **1**, 353–355

Bollerslev, T., Chou, R.Y., and Kroner, K.F. (1992): ARCH modeling in finance. *Journal of Econometrics* **52**, 5–59

Doss, H. (1994): Bayesian nonparametric estimation for incomplete data via successive substitution sampling. *The Annals of Statistics* **22**, 1763–1786

Engle, R.F. (1982): Autoregressive conditional heteroskedasticity with estimates of the variance of U.K. inflation. *Econometrica* **50**, 987–1008

Escobar, M.D. (1994): Estimating normal means with a Dirichlet process prior. *Journal of the American Statistical Association* **89**, 268–277

Escobar, M.D. and West, M. (1995): Bayesian density estimation and inference using mixtures. *Journal of the American Statistical Association* **90**, 577–587

Ferguson, T.S. (1973): A Bayesian analysis of some nonparametric problems. *The Annals of Statistics* **1**, 209–230

Gallant, A.R. and Tauchen, G. (1989): Seminonparametric estimation of conditionally constrained heterogeneous processes: Asset pricing applications. *Econometrica* **57**, 1091–1120

Gallant, A.R. and Tauchen, G. (1992): A nonparametric approach to nonlinear time series analysis: Estimation and simulation. In *New Directions in Time Series Analysis*, eds. D. Brillinger, P. Caines, J. Geweke, E. Parzen, M. Rosenblatt and M. Taqqu, 71–92. New York: Springer–Verlag

Gelfand, A.E. and Smith, A.F.M. (1990): Sampling based approaches to calculate marginal densities. *Journal of the American Statistical Association* **85**, 398–409

Geman, S. and Geman, D. (1984): Stochastic relaxation, Gibbs distributions and the Bayesian restoration of images. *IEEE Transactions on Pattern Analysis and Machine Intelligence* **6**, 721–741

Geweke, J. (1986): Modeling the persistence of conditional variances: A comment. *Econometric Reviews* **5**, 57–61

Hasting, W.K. (1970): Monte Carlo sampling methods using Markov chains and their applications. *Biometrika* **57**, 97–109

Jacquier, E., Polson, N.G., and Rossi, P.E. (1994): Bayesian analysis of stochastic volatility models (with discussion). *Journal of Business and Economic Statistics* **12**, 371–417

Kuo, L. and Mallick, B.K. (1997): Bayesian semiparametric inference for the accelerated failure time model. *Canadian Journal of Statistics in press.*

MacEachern, S.N. and Müller, P. (1994): Estimating mixture of Dirichlet process models. Tech.Rep.94–11, ISDS, Duke University.

Metropolis, N., Rosenbluth, A.W., Rosenbluth, M.N., Teller, A.H., and Teller, E. (1953): Equations of state calculations by fast computing machines. *Journal of Chemical Physics* **21**, 1087–1091

Müller, P., Erkanli, A. and West, M. (1996): Bayesian curve fitting using multivariate normal mixtures. *Biometrika* **83**, 67–79

Pagan, A.R. and Ullah, A. (1988): The econometric analysis of models with risk terms. *Journal of Applied Econometrics* **3**, 87–105

Pantula, S.G. (1986): Modeling the persistence of conditional variances: A comment. *Econometric Reviews* **5**, 71–74

Robinson, P.M. (1987): Asymptotically efficient estimation in the presence of heteroskedasticity of unknown form. *Econometrica* **55**, 875–891

Robinson, P.M. (1988): Semiparametric econometrics: A survey. *Journal of Applied Econometrics* **4**, 35–51

Shephard, N. (1994): Partial non–Gaussian state space. *Biometrika* **81**, 115–131

Smith, A.F.M. and Roberts, G.O. (1993): Bayesian computations via the Gibbs sampler and related Markov chain Monte Carlo methods. *Journal of the Royal Statistical Society Ser.* **B 55**, 3–23

Tanner, M.A. and Wong, W.H. (1987): The calculation of posterior distribution by data augmentation (with discussion). *Journal of the American Statistical Association* **82**, 528–550

West, M., Müller, P. and Escobar, M.D. (1994): Hierarchical priors and mixture models, with application in regression and density estimation. In *Aspects of Uncertainty: A Tribute to D.V. Lindley*, eds. A.F.M. Smith and P. Freeman, 363–385. New York: Wiley

Screening Policies for Control of an Infectious Disease

George Leitmann
University of California Berkeley, CA 94720, USA
gleit@coe.berkeley.edu

Abstract

Utilizing Lyapunov stability theory *constructively* we deduce constant screening policies for use by a public health authority in the control of an infectious disease such as gonorrhea.

1. Introduction

As pointed out by Cromer [1], gonorrhea is the most reported comunicable disease in the USA; e.g., see [2-9]. Earlier researchers have been quite successful in modelling seasonal fluctuations in the infected population. As for control strategies, Cromer [1] examined the effect of seasonal screening and noted a resultant decrease in the infected population.

Here we consider a two-population model similar to that of Cromer's single-population one. Moreover, we allow for uncertainty in the system parameters. We seek control strategies, i.e., screening rates, which are constrained *and* independent of the current state of the system, i.e., the fractions of the populations which are infected at any given instant of time, and which assure exponential convergence to zero or at least to computable levels of infected.

2. Problem Formulation

Let $I_i(t)$ and $S_i(t)$, respectively, denote the fractions of population $i = 1, 2$, which are infected and susceptible at time t. Since gonorrhea does not confer immunity, nor is there an inoculation against infection, immediately upon recovery a cured individual becomes susceptible again. On the other hand, since the incubation period of the disease as well as the time period of cure are relatively very brief, we assume that the total population i consists only infected and susceptible individuals. Thus, $I_i(t) + S_i(t) = 1$. Furthermore, as in Cromer's model, we assume that the fraction of infected increases at a rate proportional to the product of the fractions of susceptibles and of infected in the other population (contact effect), and decreases at a rate proportional to the fraction of infected which seek treatment in the absence of intervention (screening) by a public health authority.

Thus,

$$
\begin{aligned}
\dot{I}_i(t) &= a_1(t)I_2(t)[1 - I_1(t)] - d_1(t)I_1(t) \\
\dot{I}_2(t) &= a_2(t)I_1(t)[1 - I_2(t)] - d_2(t)I_2(t),
\end{aligned}
\tag{1}
$$

where the contact rates $a_i(t)$ and the cure rates $d_i(t)$ are uncertain; namely,

$$
0 \le a_i(t) \le \bar{a}_i ,
$$
$$
\quad i = 1, 2 \tag{2}
$$
$$
0 < \underline{d}_i \le d_i(t) ,
$$

and the bounds $\bar{a}_i$ and $\underline{d}_i$ are assumed known (i.e., for which reliable estimates can be obtained).

Next we consider intervention by a public health authority in the form of screening, i.e., identifying and curing infected individuals. Let $s_i(t)$ denote the fractional rate of screening the population i at time t, and let $I_i'(t)$ denote the fraction of the screened which is infected. To allow for $I_i'(t) \ne I_i(t)$, we introduce the uncertain coefficient of screening efficiency

$$
b_i(t) := I_i'(t)/I_i(t) , \quad i = 1, 2 \tag{3}
$$

where we assume

$$
0 < \underline{b}_i \le b_i(t) \tag{4}
$$

Again assuming that the time period of curing is negligibly short so that infected individuals identified via screening are cured immediately, intervention by screening results in the modified model

$$
\begin{aligned}
\dot{I}_1(t) &= a_1(t)I_2(t)[1 - I_1(t)] - d_1(t)I_1(t) - s_1(t)b_1(t)I_1(t) \\
\dot{I}_2(t) &= a_2(t)I_1(t)[1 - I_2(t)] - d_2(t)I_2(t) - s_2(t)b_2(t)I_2(t).
\end{aligned}
\tag{5}
$$

For reasons such as shortages of facilities and personnel, the fractional rate of screening is limited, that is

$$
0 \le s_i(t) \le \bar{s}_i , \quad i = 1, 2 \tag{6}
$$

where $\bar{s}_i$ is prescribed.

The problem then is to determine screening rates $s_i(t)$ which assure exponential convergence of the infected fractions $I_i(t)$ to zero or, if not possible due to screening rate constraint (6), to a computable neighborhood of zero.

3. Screening Policies

Let us consider *constant* screening rates

$$s_i(t) \equiv s_i = \text{constant}, \ i = 1, 2 \tag{7}$$

with

$$0 \leq s_i \leq \bar{s}_i \ . \tag{8}$$

The closed loop system (5) then becomes

$$\begin{aligned} \dot{I}_1(t) &= a_1(t)I_2(t)[1 - I_1(t)] - c_1(t)I_1(t) \\ \dot{I}_2(t) &= a_2(t)I_1(t)[1 - I_2(t)] - c_2(t)I_2(t) \ , \end{aligned} \tag{9}$$

where

$$c_i(t) := d_i(t) + b_i(t)s_i \ , \quad i = 1, 2 \tag{10}$$

so that

$$0 < \underline{c}_i \leq c_i(t) \tag{11}$$

and

$$\underline{c}_i := \underline{d}_i + \underline{b}_i s_i \ . \tag{12}$$

Let $\mathcal{T}$ be the region of *viable states*, that is

$$\mathcal{T} := \{(I_1, I_2) : 0 \leq I_i \leq 1 \ , \ i = 1, 2\} \tag{13}$$

Then we have the obvious *Invariance Result*. Region $\mathcal{T}$ is invariant for the uncertain system (9).

Next introduce

$$\beta := (\underline{c}_1 \underline{c}_2 / \bar{a}_1 \bar{a}_2)^{1/2} \ . \tag{14}$$

Then, as shown in [10] for the single population case and in [11, 12] for the two-population case, using the Lyapunov function

$$V(I_1, I_2) = I_1^2 + \mu I_2^2 \ , \tag{15}$$

where $\mu := \bar{a}_1 / \bar{a}_2$, leads to the

Stability Result 1. If $\beta > 1$, then every viable solution of (9) decays exponentially to zero with rate of convergence.

$$\alpha = \frac{\underline{c}_1 + \underline{c}_2 - \sqrt{(\underline{c}_1 - \underline{c}_2)^2 + 4\bar{a}_1\bar{a}_2}}{2} \ . \tag{16}$$

Before considering $\beta \leq 1$, define the *region of convergence*

$$\mathcal{R}(\eta) := \{(I_1, I_2) : (I_1, I_2) \ \epsilon \ \mathcal{T} \text{ and } I_1^2 + \mu I_2^2 \leq \eta = \text{ constant } \geq 0\} \tag{17}$$

We note that

$$(a) \quad \mathcal{R}(\eta) \subset \mathcal{T} \text{ if } < 1 + \mu \ .$$
$$(b) \quad \mathcal{R}(\eta) = \mathcal{T} \text{ if } \geq 1 + \mu$$
$$(c) \quad \mathcal{R}(\eta) \to \{0, 0\} \text{ as } \eta \to 0.$$

Now consider any non-negative $\alpha < \min \{\underline{c}_1, \underline{c}_2\}$ and let

$$\beta_\alpha := \left[\frac{(\underline{c}_1 - \alpha)(\underline{c}_2 - \alpha)}{\bar{a}_1 \bar{a}_2}\right]^{1/2} \ , \tag{18}$$

and, without loss of generality, assume that $\bar{a}_1 \leq \bar{a}_2$. Then, as shown in [11, 12], we have the

Stability Result 2. Every viable solution of (9) converges to region $\mathcal{R}(\eta_\alpha)$ with rate α, where

$$\eta_\alpha = \begin{cases} 0 & \text{if } \beta_\alpha \geq 1 \\ 4(1 - \beta_\alpha)^2 & \text{if } 1 \geq \beta_\alpha \geq 1/2 \\ 1 + \mu(1 - 2\beta_\alpha)^2 & \text{if } 1/2 \geq \beta_\alpha > 0 \ . \end{cases}$$

Furthermore, $\mathcal{R}(\eta_\alpha)$ is invariant.
Detailed derivations as well as other results can be found in [11, 12].

4. Simulation Results
Here we present some simulation results; considerably more extensive ones can be found in [11, 12].

Figures 1-4 show time histories of the fractions of infected, $I_i(t), i = 1, 2$, for two parameter realizations, and without and with screening intervention. For constant parameters, e.g.,

$$a_1(t) \equiv 4 , \quad a_2(t) \equiv 6$$
$$d_1(t) \equiv 2 , \quad d_2(t) \equiv 1$$
$$b_1(t) \equiv 1 , \quad b_2(t) \equiv 1$$

and
$\beta = 0.2887 < 1$, the system evolution in the absence of screening $(s_i(t) \equiv 0)$ is illustrated in Figure 1. In the presence of sufficient screening, e.g., $s_1 = s_2 = 4.4$, so that $\beta = 1.2 > 1$, the system evolution is illustrated in Figure 2.
For time-variable parameter realizations, e.g.,

$$a_1(t) = 4(0.6 + 0.4\sin 5t)$$
$$a_2(t) = 6(0.6 + 0.4\sin 5t)$$
$$d_1(t) \equiv 2 , \quad d_2(t) \equiv 1$$
$$b_1(t) \equiv 1 , \quad b_2(t) \equiv 1$$

and again $\beta = 0.2887 < 1$, the system evolution in the absence of screening $(s_i(t) \equiv 0)$ is illustrated in Figure 3. Again, in the presence of sufficient screening, e.g., $s_1 = s_2 = 4.4$, so that $\beta = 1.2 < 1$, the system evolution is illustrated in Figure 4.

5. **Conclusions** We analyze the use of screening to control the levels of infection of an endemic disease such as gonorrhea, utilizing a two-population model with uncertain, time-varying parameters. We find that constant and bounded screening rates assure that the levels of infection go to zero exponentially at a computable rate, or else, due to the impressed upper bound on the screening rates, converge exponentially to a computable neighborhood of zero at a given rate.

In particular, it is found that *constant* screening rates assure the desired convergence of the levels of infection. This conclusion is vital since the screening rates do not depend on the knowledge of the current levels of infection as is usually the case with robust feedback control; e.g. [13].

References

1. Cromer, T.L. (1988): Seasonal control for an endemic disease with seasonal fluctuations. Theoretical Population Biology **33**, 115-125.

2. Hethcote, H.W., (1973): Asymptotic behavior in a deterministic epidemic model. Bulletin of Mathematical Biology **35**, 607-614.

3. Lajmanovich, A., and Yorke, J.A., (1976): A deterministic model for gonorrhea in a nonhomogeneous population. Mathematical Biosciences **28**, 221-236.

4. Hethcote, H.W. and Yorke, J.A., (1985): Gonorrhea Transmission Dynamics and Control. Springer Verlag, New York.

5. Cooke, K.L., and Kaplan, J.L., (1976): A periodicity threshold theorem for epidemics and population growth. Mathematical Biosciences, **31**, 87-104.

6. Smith, H.L., (1977): Periodic solutions of a delay integral equation modeling epidemics. Journal of Mathematical Biology **4**, 69-80.

7. Nussbaum, R.D., (1978): A periodicity threshold theorem for some nonlinear integral equations. SIAM Journal of Mathematical Analysis **4**, 356-376.

8. Cromer, T.L., (1985): Asymptotically periodic solutions to Volterra integral equations in epidemic models. Journal of Mathematical Analysis and Applications **100**, No. 2, 483-494.

9. Yorke, J.A., Hethcote, H.W., and Nold, A., (1978): Dynamics and control of the transmission of gonorrhea. Sex Transm. Dis. **5**, No. 2, 51-56.

10. Leitmann, G., (1988): New screening policies for the control of a communicable disease. 8th Japanese-German seminar on Nonlinear Problems in Dynamical Systems, University of Kobe, Japan, October.

11. Leitmann, G., Lee, C.S., and Coreless, M., (1998): One approach to robust control and its application to screening strategies for some communicable diseases, Robustness in Identification and Control. University of Siena, Italy, July 1998.

12. Corless, M., and Leitmann, G., (1998): Analysis and Control of Communicable Disease. Nonlinear Analysis, to appear.

13. Leitmann, G., (1998): The use of screening for the control an an endemic disease. International Series of Numerical Mathematics, **124**, Birkhaüser Verlag, Basel, 291-300.

Der Herr des Chaos

Alexander Mehlmann

Institut für Ökonometrie, Operations Research und Systemtheorie, Technische Universität Wien, Argentinierstraße 8, A-1040 Wien.

(für den Institutsvorstand und Weltreisenden
anläßlich der Rückkehr aus der Südsee verfaßt
und unter beträchtlichen Mühen anno 2000
rekonstruiert, verdichtet und erweitert)

Der Südsee Strände sind nun leer.
Aloa! Abschied fällt so schwer.
Nur Wien erstrahlt in stillem Glück:
Der Herr des Chaos kehrt zurück.

Die Mitarbeiter sind erwacht
Aus langem Schlaf, aus dunkler Nacht
Und sind verzweifelt und betrübt,
Weil's gar nichts zu berichten gibt.

Rein gar nix? Läßt der Herr von Zych
Uns beim Kongreß denn nicht im Stich?
Und ist der Fonds nicht aufgeschreckt,
Da weit im Soll steckt das Projekt?

Auch Deistlers Leute merken's schon;
Verschleudern uns're Dotation,
Verletzten ständig und zur Gänze
Die heilige Abteilungsgrenze.

Dies alles mag wohl grad so sein,
Doch sonst nur eitel Sonnenschein;
Und trotz der störenden Studenten
Träumt jedermann stets von der Rente(n).

Herr Hartl überspringt die Hürden
Zu fernen Professorenwürden
Und da zur Zeit bloß Assistent,
Scheint's, daß er selbst sich nicht mehr kennt.

Luptacik nimmt sich ständig frei
Und missioniert die Slovakei,
Wo Tausende auf allen Vieren
Bereits nach seinem Output gieren.

Frau Toda hält die Zügel fest
Und uns die Knute spüren läßt.
Selbst die Quästur zeigt nun Respekt;
Hält sich im Untergrund versteckt.

Als Resultat all dieser Pein
Stellt man bereits das Forschen ein.
Und alle Institutsberichte
Enthalten längst nur noch Gedichte.

Gustav Feichtinger
Curriculum Vitae

Personal Data

Date of Birth:	July 14, 1940
Place of Birth:	Wiener Neustadt, Austria
Marital Status:	Married to Ingrid
Children:	two daughters Susanne, Evelyn

Education and Professional Experience

University of Vienna (Mathematics and Physics) 1958 – 1963

Ph.D. in Mathematics (Algebra) 1963

IBM Research Laboratory Vienna 1964 - 1965

University of Bonn (Statistics and Operations Research, Assistant and Associate Professor) 1965 – 1972

Habilitation in Statistics 1969/70

University of Technology Vienna (Full Professor for Operations Research) since 1972

Director of the Department for Theoretical Demography at the Institute of Demography of the Austrian Academy of Science since 1977

Part-time position at the International Institute for Applied Systems Analysis (IIASA) since 1992

Member of the Advisory Board of the Max Planck Institute for Demographic Research, Rostock

Associate Editorships

Central European Journal of OR

Dynamics and Control

International Abstracts in Operations Research

International Game Theory Review

International Transactions in Operational Research

Journal of Evolutionary Economics

Journal of Population Economics

Mathematical Population Studies

Springer Lecture Notes in Economics and Mathematical Systems

Publications

Books and Edited Volumes

B1. *Lernprozesse in stochastischen Automaten.* 'Lecture Notes in Operations Research and Mathematical Systems', Vol. **24**, Springer, Berlin, 1970.

B2. *Stochastische Modelle demographischer Prozesse.* 'Lecture Notes in Operations Research and Mathematical Systems', Vol. **44**, Springer, Berlin, 1971.

B3. *Bevölkerungsstatistik.* de Gruyter, Berlin, 1973.

B4. *Stationäre und schrumpfende Bevölkerungen. Demographisches Null- und Negativwachstum in Österreich.* 'Lecture Notes in Economics and Mathematical Systems', Vol. **149**, Springer, Berlin (Ed.), 1977.

B5. *Demographische Analyse und populationsdynamische Modelle: Grundzüge der Bevölkerungsmathematik.* Springer, Wien, 1979.

B6. *Optimal Control Theory and Economic Analysis.* North-Holland, Amsterdam (Ed.), 1982.

B7. *Operations Research in Progress.* Reidel, Dordrecht (with P. Kall, Eds.), 1982.

B8. *Optimal Control Theory and Economic Analysis 2.* North-Holland, Amsterdam (Ed.), 1985.

B9. *Optimale Kontrolle ökonomischer Prozesse: Anwendungen des Maximumprinzips in den Wirtschaftswissenschaften.* de Gruyter, Berlin (with R.F. Hartl), 1986.

B10. *Optimal Control Theory and Economic Analysis 3.* North-Holland, Amsterdam (Ed.), 1988.

B11. *Methods of Operations Research: OR 1990,* Anton Hain, Frankfurt (with R.F. Hartl, W. Janko, W.E. Katzenberger, and A. Stepan, Eds.), 1991.

B12. *Dynamic Economic Models and Optimal Control.* North-Holland, Amsterdam (Ed.), 1992.

B13. *Operations Research Proceedings: Papers of the International Conference on Operations Research 1990.* Springer, Berlin (with W. Bühler, R.F. Hartl, F.J. Radermacher, and P. Stähly, Eds), 1992.

B14. *Nonlinear Methods in Economic Dynamics and Optimal Control.* Annals of Operations Research **37**, Baltzer, Basel (with R.F. Hartl, Eds.), 1992.

B15. *Optimal Control and Differential Games.* Annals of Operations Research **88** (with H. Dawid and R.F. Hartl, Eds), 1999.

B16. *Nonlinear Dynamical Systems and Adaptive Methods.* Annals of Operations Research **89** (with H. Dawid and R.F. Hartl, Eds.), 1999.

B17. *Modelling and Decisions in Economics. Essays in Honor of Franz Ferschl.* Springer, Heidelberg, (with U. Leopold-Wildburger and K.-P. Kistner, Eds.), 1999.

Selected Scientific Papers

1. Über die Anzahl aller nichtisomorphen transitiven Automaten von Primzahlordnung. *Elektronische Informationsverarbeitung und Kybernetik* **3**, 275-282, 1967.

2. Beiträge zur Theorie abstrakter Automaten. *Elektronische Datenverarbeitung* **8**, 361-366, 1967.

3. Über isomorphe Unterautomaten in abstrakten Automaten. *Elektronische Datenverarbeitung* **9**, 411-415, 1967.

4. Der Quotientenautomat nach einem direkten Faktor einer Automorphismengruppe. *Computing* **3**, 1-8, 1968.

5. Zur Theorie abstrakter stochastischer Automaten. *Zeitschrift für Wahrscheinlichkeitstheorie und verwandte Gebiete* **9**, 341-356, 1968.

6. Automatentheorie und dynamische Programmierung. *Elektronische Informationsverarbeitung und Kybernetik* **4**, 347-352, 1968.

7. Eine automatentheoretische Deutung des einelementigen Lernmodells der Stimulus Sampling Theorie. *Grundlagenstudien aus Kybernetik und Geisteswissenschaft* **9**, 3-19, 1968.

8. Ein automatentheoretischer Zugang zu Lernprozessen. *Kybernetik* **5**, 85-88, 1968.

9. Zur mathematischen Lerntheorie. *Grundlagenstudien aus Kybernetik und Geisteswissenschaft* **9**, 39-47, 1968.

10. Stochastische Automaten als Grundlage linearer Lernmodelle. *Statistische Hefte* **10**, 5-21, 1969.

11. Optimale Darbietungen (Lernstrategien) zweier Items in der statistischen Lerntheorie. *Elektronische Informationsverarbeitung und Kybernetik* **5**, 109-125, 1969.

12. Ein Markoffsches Lernmodell für Zwei-Personen-Spiele. *Elektronische Datenverarbeitung* **7**, 322-325, 1969.

13. Anwendungen der Automatentheorie auf Lernprozesse. 2. Internationaler Kongreß *Datenverarbeitung im europäischen Raum*, 17-21, 1969.

14. Endliche stochastische Automaten als Modelle in den Verhaltenswissenschaften. *Seminarbericht* **17** *des Instituts für Theorie der Automaten und Schaltnetzwerke*, GMD Bonn, 1969.

15. Einige Resultate aus der Bevölkerungsmathematik. In: R. Henn (Ed.), *'Operations Research Verfahren VIII'*, 89-106, Hain, Meisenheim, 1970.

16. Über ein Entscheidungsproblem in der Erziehungsplanung. *Unternehmensforschung* **14**, 140-151, 1970.

17. Kybernetische Begriffe und Arbeitsweisen in der Lerntheorie. *Zeitschrift für experimentelle und angewandte Psychologie* **17**, 37-51, 1970.

18. Zur Erlernung optimalen Verhaltens in Entscheidungssituationen. *Statistische Hefte* **12**, 14-21, 1970.

19. BAYES-Automaten. *Elektronische Informationsverarbeitung und Kybernetik* **6**, 371-379, 1970.

20. Über Automatenmodelle in der statistischen Lerntheorie. *Kybernetik* **6**, 237-242, 1970.

21. Gekoppelte stochastische Automaten und sequentielle Zwei-Personen-Spiele. *Unternehmensforschung* **14**, 249-258, 1970.

22. "Wahrscheinlichkeitslernen" in der statistischen Lerntheorie. *Metrika* **18**, 35-55, 1971.

23. Zur Erlernung optimalen Verhaltens in Entscheidungssituationen bei Ungewißheit. *Statistische Hefte* **12**, 14-21, 1971.

24. Stochastische Automaten mit stetigem Zeitparameter. *Angewandte Informatik* **4**, 156-164, 1971.

25. Über Lernprozesse mit variabler Transitionsstruktur in Entscheidungssituationen bei Ungewißheit. *Jahrbücher für Nationalökonomie und Statistik* **186**, 14-28, 1971.

26. Absorbing Markov chains in population mathematics. *Proceedings of the 38th Session of the International Statistical Institute*, Washington, D.C, 1971.

27. Markoff-Modelle zur sozialen Mobilität. *Jahrbuch für Sozialwissenschaft* **22**, 300-316, 1971.

28. Automatentheoretische Lernmodelle. *Umschau in Wissenschaft und Technik* **71**, 166, 1971.

29. Bevölkerungswissenschaft vor neuen Problemen. *Umschau in Wissenschaft und Technik* **71**, 596-597, 1971.

30. Über Sinn und Ziel der Kybernetik. In: J. Meurers (Hrsg.), *'Philosophia naturalis 13'*, 185-190, Hain, Meisenheim, 1972.

31. Über die Kovarianzen von Maximum Likelihood-Schätzungen gewisser Parameter bei multiplen Dekrementtafeln. *Mathematische Operationsforschung und Statistik* **3**, 217-226, 1972.

32. Über ein N-elementiges Modell der Reiz-Stichproben-Theorie mit stetigem Zeitparameter und zwei Antwortklassen. *Metrika* **18**, 94-109, 1972.

33. Stochastische Dekrementmodelle der Bevölkerungsstatistik. *Biometrische Zeitschrift* **14**, 106-125, 1972.

34. Zur Bayes-Analyse statistischer Entscheidungsprobleme. *Zeitschrift für Betriebswirtschaft* **7**, 42. Jg, 1972.

35. Neue Entwicklungen der formalen Demographie I. *Mitteilungsblatt der Österreichischen Gesellschaft für Statistik und Informatik* **5**, 23-29, 1972.

36. Neue Entwicklungen der formalen Demographie II. *Mitteilungsblatt der Österreichischen Gesellschaft für Statistik und Informatik* **6**, 53-64, 1972.

37. Stochastische Prozesse im Lichte ihrer Anwendungen. *Angewandte Informatik* **3**, 104-114, 1972.

38. Markovian models for some demographic processes. *Statistische Hefte* **4**, 311-334, 1973.

39. Bemerkungen über lineare Modelle der Populationsdynamik. In: R. Henn (Ed.), *'Operations Research-Verfahren XV'*, 40-62, Hain, Meisenheim (with M. Deistler), 1973.

40. On the linear stochastic model in population dynamics. *Proceedings of the 39th Session of the International Statistical Institute* **4**, 301-306 (with M. Deistler), 1973.

41. OR-Modelle sozio-demographischer Prozesse. In: P. Gessner et al, *'Proceedings in Operations Research 3'*, 195-229, Physica, Würzburg, 1974.

42. Bemerkungen über stochastische Modelle der Straffälligkeit. *Allgemeines Statistisches Archiv* **58**, 198-221, 1974.

43. Einblicke in bevölkerungsmathematische Modelle und deren Anwendungsmöglichkeiten. *Jahrbuch für Sozialwissenschaft* **25**, 86-105, 1974.

44. Stochastische Modelle des Manpower Planning. *Antrittsvorlesung an der TH Wien*, Nr. **42**, TH Wien, 1974.

45. Von der Bevölkerungsstatistik zur demographischen Forschung. *Mitteilungsblatt der Österreichischen Gesellschaft für Statistik und Informatik* **13**, 15-36, 1974.

46. Nullwachstum der Bevölkerung frühestens in 100 Jahre. *Umschau in Wissenschaft und Technik* **74**, 551-552, 1974.

47. The linear model formulation of a multitype branching process applied to population dynamics. *Journal of the American Statistical Association* **69**, 662-664 (with M. Deistler), 1974.

48. Bemerkungen über Wahrscheinlichkeitsmodelle für rekurrente demographische Prozesse. *Zeitschrift für Bevölkerungswissenschaft* **3/4**, 57-72, 1975.

49. Weltbevölkerungsjahr, -Konferenz und - Aktionsplan 1974. *Jahrbuch für Sozialwissenschaft* **26**, 123-137, 1975.

50. Die demographische Revolution. *Österreichische Gesellschaft für Familienplanung*, Jahrestagung in Igls, 1975.

51. Nullwachstum der Bevölkerung im linearen populationsdynamischen Modell. *Zeitschrift für Bevölkerungswissenschaft* **2**, 39-53 (with M. Deistler), 1975.

52. Are economically dependent groups likely to become a significant larger proportion of the population as a whole? In: *'Proceedings of the Colloque on the Changing Population Structures in Europe and Rising Social Costs'*. Council of Europe, Strasbourg, 1976.

53. Some economic consequences of declining fertility in the Federal Republic of Germany. In: *'Les Methodes d'Analyse en Demographie Economique, Dossiers et recherches 1'*. Institut National d'Etudes Demographiques, Paris, 1976.

54. Stabile Populationsdynamik. In: R. Henn (Ed.), *'Operations Research Verfahren XXV'*, 434-442, Hain, Meisenheim, 1976.

55. On the generalization of stable age distributions to Gani-type manpower models. *Advances in Applied Probability* **8**, 39-53, 1976.

56. The recruitment trajectory corresponding to particular stock sequences in Markovian person-flow models. *Mathematics of Operations Research* **1**, 175-184 (with A. Mehlmann), 1976.

57. Ursachen und Konsequenzen des Geburtenrückganges. In: *'Soziale Probleme der modernen Industriegesellschaft'*. Arbeitstagung des Vereins für Socialpolitik in Augsburg 1976, 393-434, Duncker & Humblot, Berlin, 1977.

58. Methodische Probleme der Familienlebenszyklus-Statistik. In: H. Albach, Helmstädter and R. Henn (Eds.), *'Quantitative Wirtschaftsforschung'*. Wilhelm Krelle zum 60. Geburtstag, Mohr, Tübingen, 1977.

59. Bevölkerung. In: W. Albers et al. (Ed.) *'Handwörterbuch der Wirtschaftswissenschaft (HdWW)'*, 610-631, G. Fischer, Stuttgart, 1977.

60. Bevölkerungsmodelle. *Allgemeines Statistisches Archiv* 4, 325-348, 1977.

61. Aussichten für das Erwerbspotential und den Anteil wirtschaftlich abhängiger Personen in europäischen Bevölkerungen. *Schriftenreihe des Instituts für Demographie der Österreichischen Akademie der Wissenschaften* 2, 1977.

62. The impact of mortality on the life cycle of the family in Austria. *Zeitschrift für Bevölkerungswissenschaft* 4, 51-79 (with H. Hansluwka), 1977.

63. Bevölkerungsschrumpfung in Österreich. Konsequenzen rückläufiger Fruchtbarkeit für die demographische Entwicklung. *Schriftenreihe des Instituts für Demographie der Österreichischen Akademie der Wissenschaften* 3 (with J. Muzicant), 1977.

64. Optimales Wachstum stabiler Bevölkerungen in einem neoklassischen Modell. *Zeitschrift für Bevölkerungswissenschaft* 1, 63-73 (with M. Deistler, M. Luptacik and A. Wörgötter), 1978.

65. Analyse der Fertilitätsentwicklung in Österreich nach Heiratsjahrgängen. *Schriftenreihe des Instituts für Demographie der Österreichischen Akademie der Wissenschaften* 5 (with A. Haslinger), 1978.

66. Pseudostabile Bevölkerungen: Populationsdynamik bei gleichmäßig sinkender Fertilität. *Schriftenreihe des Instituts für Demographie der Österreichischen Akademie der Wissenschaften* 4 (with H. Vogelsang), 1978.

67. Cohort trends in the timing of the family life cycle in Austria and West-Germany. In: J. Dupâquier et al. (Eds.), *'Historische Demographie'*. Paris, 1979.

68. Optimale intertemporale Allokationen mittels des Maximumprinzips. In: J. Schwarze et al. (Eds.), *'Proceedings in Operations Research 9'*, 484-489, Physica, Würzburg, 1979.

69. Analyse des Familienzyklus auf demographischer Basis. In: *'Bericht über die Situation der Familie in Österreich'*. Familienbericht 1, Bundeskanzleramt Wien, 1979.

70. Von sinkender zu steigender Fertilität: Bevölkerungsdynamik bei einer Trendumkehr der Fruchtbarkeit. *Zeitschrift für Bevölkerungswissenschaft* 3, 327-340, 1979.

71. Zum Begriff "Schwung des Bevölkerungswachstums". *Jahrbücher für Nationalökonomie und Statistik* 194, 399-400, 1979.

72. Exponentielle Niveau- und Musteränderung der Fertilität. Eine Analyse mittels stabiler Vergleichsbevölkerungen. *Zeitschrift für Bevölkerungswissenschaft* 1, 31-63 (with H. Vogelsang), 1979.

73. Optimale Allokation von Ausbildung und Berufsausübung in einem nichtlinearen Kontrollmodell. *Zeitschrift für Operations Research* **25**, 25-34, 1981.

74. Optimierung von Instandhaltungsinvestitionen, Produktionsintensität und Nutzungsdauer maschineller Produktionsanlagen: Anwendungen der Kontrolltheorie in der Instandhaltungs- und Produktionsplanung. In: K. Brockhoff and W. Krelle (Eds.), *'Unternehmensplanung'*, 213-234, Springer, Berlin, 1981.

75. Ein nichtlineares Kontrollproblem der Instandhaltung. *OR-Spektrum* **3**, 49-58 (with R.F. Hartl), 1981.

76. Saddle-point analysis in a price-advertising model. *Journal of Economic Dynamics and Control* **4**, 319-340, 1982.

77. Optimal repair policy for a machine service problem. *Optimal Control Applications and Methods* **3**, 15-22, 1982.

78. Optimal pricing in a diffusion model with nonlinear price-dependent market potential. *Operations Research Letters* **1**, 236-240, 1982.

79. Anwendungen des Maximumprinzips im Operations Research. Teil 1 und 2. *OR-Spektrum* **4**, 171-190, 195-212, 1982.

80. The Nash solution of a maintenance-production differential game. *European Journal of Operational Research* **10**, 165-172, 1982.

81. Optimal research policies in a noncooperative project undertaken by two firms. In: G. Feichtinger (Eds.), *'Optimal Control Theory and Economic Analysis'*, 373-397, North-Holland, Amsterdam, 1982.

82. Optimal bimodal harvest policies in age-specific bioeconomic models. In: G. Feichtinger and P. Kall (Eds.), *'Operations Research in Progress'*, 285-299, Reidel, Dordrecht, 1982.

83. Ein Differentialspiel für den Markteintritt einer Firma. In: B. Fleischmann et al. (Eds.), *'Operations Research Proceedings 1981'*, 636-644, Springer, Berlin, 1982.

84. Optimale dynamische Preispolitik bei drohender Konkurrenz. *Zeitschrift für Betriebswirtschaft* **53**, 155-177, 1983.

85. The Nash solution of an advertising differential game: Generalization of a model by Leitmann and Schmitendorf. *IEEE Transactions on Automatic Control* **AC-28**, 1044-1048, 1983.

86. A differential games solution to a model of competition between a thief and the police. *Management Science* **29**, 686-699, 1983.

87. Pseudostabile Bevölkerungen. In: S. Rupp and K. Schwarz (Eds.), *Beiträge aus der bevölkerungswissenschaftlichen Forschung*. Festschrift für Hermann Schubnell, Schriftenreihe des Bundesinstituts für Bevölkerungsforschung **11**, 535-544, 1983.

88. Differential game models in management science. *European Journal of Operational Research* **14**, 137-155 (with S. Jorgensen), 1983.

89. Eine Fruchtbarkeitstafel auf Paritätsbasis. *Zeitschrift für Bevölkerungswissenschaft* **3**, 363-376 (with W. Lutz), 1983.

90. Optimal employment strategies of profit-maximizing and labour-managed firms. *Optimal Control Applications and Methods* **5**, 235-253, 1984.

91. On the synergistic influence of two control variables on the state of nonlinear optimal control models. *Journal of Operational Research Society* **35**, 907-914, 1984.

92. On optimal control models for the dynamics of the firm. In: R. Henn (Eds.), *'Methods of Operations Research* **51**'. Hain, Athenäum, 1984.

93. A note to Jorgensen's logarithmic advertising differential game. *Zeitschrift für Operations Research* **28**, B 133-B 153 (with E.J. Dockner), 1984.

94. Alterstruktureffekt bei der Schätzung schließlicher Paritätsverteilungen. In: C. Höhn et al. (Eds.), *'Demographie in der Bundesrepublik Deutschland'*. Festschrift für Karl Schwarz. Boppard, H. Boldt (with W. Lutz), 1984.

95. Lagerhaltungs- und Beschäftigungsstrategien einer Firma. *OR-Spektrum* **6**, 93-107 (with A. Steindl), 1984.

96. Optimal modification of machine reliability by maintenance and production. *OR-Spektrum* **7**, 43-50, 1985.

97. Optimal pricing in a duopoly: a noncooperative differential games solution. *Journal of Optimization Theory and Applications* **45**, 199-218 (with E.J. Dockner), 1985.

98. Optimal pricing and production in an inventory model. *European Journal of Operational Research* **19**, 45-56 (with R.F. Hartl), 1985.

99. On the use of Hamiltonian and maximized Hamiltonian in non-differentiable control theory. *Journal of Optimization Theory and Applications* **46**, 493-504 (with R.F. Hartl), 1985.

100. Tractable classes of nonzero-sum open-loop Nash differential games: theory and examples. *Journal of Optimization Theory and Applications* **45**, 179-197 (with E.J. Dockner and S. Jorgensen), 1985.

101. Periodic optimal control: can oscillations be optimal in autonomous economic control models? In: W. Domschke et al. (Eds.), *'Methods of Operations Research* **57**', 349-366, Athenäum, Königstein, 1986.

102. Dynamic advertising and pricing in an oligopoly: a Nash equilibrium approach. *Journal of Economic Dynamics and Control* **10**, 37-39 (with E.J. Dockner), 1986.

103. Planning the unusual: applications of control theory to non-standard problems. *Acta Applicandae Mathematicae* **7**, 79-102 (with A. Mehlmann), 1986.

104. On the exploitation of curious resources. In: M.J. Beckmann et al. (Eds.), *'Methods in Operations Research* **54**', 127-145, Hain, Athenäum (with A. Mehlmann), 1986.

105. Optimal oscillations in control models: how can constant demand lead to cyclical production? *Operations Research Letters* **5**, 277-281 (with G. Sorger), 1986.

106. On the optimality of cyclical employment policies: a numerical investigation. *Journal of Economic Dynamics and Control* **10**, 457-466 (with A. Steindl, R.F. Hartl and G. Sorger), 1986.

107. Limit cycles in economic control models. In: R. Bulirsch et al. (Eds.), *'Optimal Control'*. Proceedings of the Conference on Optimal Control and Variational Calculus Oberwolfach, West-Germany, June 15-21, 1986, 46-55, Springer, Berlin, 1987.

108. Intertemporal optimization of wine consumption at a party: an unusual optimal control model. In: G. Gandolfo and F. Marzano (Eds.), *'Keynesian Theory Planning Models and Quantitive Economics'*. Essays in Memory of Vittorio Marrama, Vol. **II**, 777-797, Giuffre, Milano, 1987.

109. Zur Anwendung der Kontrolltheorie auf die Optimierung ökonomischer Prozesse. In: Th. Fischer (Ed.), *'Betriebswirtschaftliche Systemforschung und ökonomische Kybernetik'*, 271-289, Duncker & Humblot, Berlin, 1987.

110. The statistical measurement of the family life cycle: table methods for analyzing the tempo and structure of the family life cycle. In: J. Bongaarts (Hrsg.), *'Family Demography: Methods and Their Applications'*, 81-101, Clarendon Press, Oxford, 1987.

111. A new sufficient condition for most rapid approach paths. *Journal of Optimization Theory and Applications* **54**, 403-411 (with R.F. Hartl), 1987.

112. Optimal production and abatement policies of a firm. *European Journal of Operational Research* **29**, 274-285 (with M. Luptacik), 1987.

113. Optimal employment and wage policies of a monopolistic firm. *Journal of Optimization Theory and Applications* **53**, 59-83 (with M. Luptacik), 1987.

114. Intertemporal sharecropping: a differential game approach. In: G. Bamberg and K. Spremann (Eds.), *'Agency Theory, Information and Incentives'*, 415-438, Springer, Berlin (with G. Sorger), 1987.

115. Optimal persistent oscillations and adjustment costs. In: P. Flaschel and M. Krüger (Eds.), *'Recent Approaches to Economic Dynamics'*, 119-131, P. Lang, Frankfurt, 1988.

116. Ein dynamisches Modell des Intensitätssplittings. *Zeitschrift für Betriebswirtschaft* **11**, 1242-1258 (with K.-P. Kistner and A. Luhmer), 1988.

117. Optimal price and advertising policy in convenience goods retailing. *Marketing Science* **7**, 187-201 (with A. Luhmer and G. Sorger), 1988.

118. ADPULS in continuous time. *European Journal of Operational Research* **34**, 171-177 (with A. Luhmer, A. Steindl, R.F. Hartl and G. Sorger), 1988.

119. Periodic research and development. In: G. Feichtinger (Ed.), *'Optimal Control Theory and Economic Analysis* 3'. Third Viennese Workshop on Optimal Control Theory and Economic Analysis, Vienna, May 20 - 22, 1987, 121-141, North-Holland, Amsterdam (with G. Sorger), 1988.

120. Demographische Prognosen und populationsdynamische Modelle. In: B. Felderer (Hrsg.), *'Bevölkerung und Wirtschaft'*, 71-92, Duncker & Humblot, Berlin, 1989.

121.On cyclical inventory and marketing decisions. In: U. Rieder et al, *'Methods of Operations Research'* **62**, 387-388, Hain (with E.J. Dockner), 1989.

122.Noncooperative solutions for a differential game model of fishery. *Journal of Economic Dynamics and Control* **13**, 1-20 (with E.J. Dockner and A. Mehlmann), 1989.

123.Self-generated fertility waves in a non-linear continuous overlapping generations model. *Journal of Population Economics* **2**, 267-280 (with G. Sorger), 1989.

124.Optimal periodic policies in dynamic economic systems. *National Contribution of Austria*. IFORS 1990, Athen. In: H.E. Bradley (Ed.), *'Operational Research '90'*, 771-782, Pergamon Press, Oxford, 1990.

125.Capital accumulation, endogenous population growth, and Easterlin cycles. *Journal of Population Economics* **3**, 73-87 (with E.J. Dockner), 1990.

126.Oligopolistisches Preismanagement bei Lerneffekten und Überkapazitäten. *Zeitschrift für Betriebswirtschaft* **1**, 7-19 (with E.J. Dockner), 1990.

127.Interaction of price and advertising under dynamic conditions. In: K.-P. Kistner et al. (Eds.), *'Operations Research Proceedings 1989'*, 547-554, Springer, Berlin, (with E.J. Dockner and G. Sorger), 1990.

128.Capital accumulation, aspiration adjustment, and population growth: limit cycles in an Easterlin-type model. *Mathematical Population Studies* **2**, 93-103 (with G. Sorger), 1990.

129.On the optimality of limit cycles in dynamic economic systems. *Journal of Economics* **53**, 31-50 (with E.J. Dockner), 1991.

130.Cyclical production and marketing decisions: application of Hopf bifurcation theory. *International Journal of Systems Science* **22**, 1035-1046 (with E.J. Dockner and A. Novak), 1991.

131.A note on the optimal use of environmental resources by an indebted country. *Journal of Institutional and Theoretical Economics* **147**, 547-555 (with A. Novak), 1991.

132.Politico-economic cycles of regulation and deregulation. *European Journal of Political Economy* **7**, 469-485 (with F. Wirl), 1991.

133.Hopf bifurcation in an advertising diffusion model. *Journal of Economic Behavior and Organisation* **17**, 401-411, 1992.

134.Rational addictive cycles ('binges') under a budget constraint. *Optimal Control Applications and Methods* **13**, 95-104, 1992.

135.Optimal control of economic systems. In: S. Tzafestas (Ed.), *'Automatic Control Handbook'*. M. Dekker, New York, 1023-1044, 1992.

136.Limit cycles in dynamic economic systems. In: G. Feichtinger and R.F. Hartl (Eds.), *'Nonlinear Methods in Economic Dynamics and Optimal Control'*. Annals of Operations Research **37**, Baltzer, Basel, 313-344, 1992.

137.Nonlinear threshold dynamics: Further examples for chaos in social sciences. In: G. Haag, U. Mueller and K.G. Troitzsch (Eds.), *'Economic Evolution and Demographic Change, Formal Models in Social Sciences: Formal Models in*

Social Sciences'. Lecture Notes in Economics and Mathematical Systems **395**, Springer, Berlin, 141-154, 1992.

138.Optimal recycling of tailings for the production of building materials. *Czechoslovak Journal for Operations Research* **1**, 181-192 (with R.F. Hartl and G.T. Kirakossian), 1992.

139.Stable resource-employment limit cycles in an optimally regulated fishery. In: G. Feichtinger (Ed.): *'Dynamic Economic Models and Optimal Control'.* North-Holland, Amsterdam, 163-184 (with V. Kaitala and A. Novak), 1992.

140.A note on the optimal exploitation of migratory fish stocks. *Dynamics and Control* **2**, 255-263 (with A. Novak), 1992.

141.Optimal consumption, training, working time and leisure over the life cycle. *Journal of Optimization Theory and Applications* **75**, 369-388 (with A. Novak), 1992.

142.Seltsames Verhalten nichtlinearer demographischer Prozesse. *Acta Demographica*, 131-156 (with A. Prskawetz), 1992.

143.Immigration into a population with fertility below replacement level - the case of Germany. *Population Studies* **46**, 275-284 (with G. Steinmann), 1992.

144.Strange addictive behaviour: Periodic and chaotic binges. In: W.E. Diewert, K. Spremann and F. Stehling (Eds.): *'Mathematical Modelling in Economics'.* Essays in Honor of Wolfgang Eichhorn. Springer, Berlin, 149-162, 1993.

145.Cyclical consumption patterns and rational addiction. *American Economic Review* **83**, 256-263 (with E.J. Dockner), 1993.

146.Dynamic R & D competition with memory. *Journal of Evolutionary Economics* **3**, 145-152 (with E.J. Dockner and A. Mehlmann), 1993.

147.Chaos in nonlinear dynamical systems exemplified by an R & D model. *European Journal of Operational Research* **68**, 145-159 (with M. Kopel), 1993.

148.Optimal treatment of cancer diseases. *International Journal of Systems Science* **24**, 1253-1263 (with A. Novak), 1993.

149.A dynamic variant of the battle of the sexes. *International Journal of Game Theory* **22**, 359-380 (with F. Wirl), 1993.

150.Optimal complex dynamics in environmental economics. *Proceedings of International Conference of Dynamical Systems and Chaos*, Tokyo. World Scientific, 1994.

151.Environmentalists versus resources exploiters: a dynamic game analysis. In: G.L. Paredes V. (Ed.), *International Symposium on Systems Analysis and Management Decisions in Forestry.* Universidad Austral of Chile Press, Valdivia, 392-398 (with E.J. Dockner), 1994.

152.Nonconcavity and proper optimal periodic control. *Journal of Economic Dynamics and Control* **18**, 975-990 (with M. Han and R.F. Hartl), 1994.

153.Dynamic optimal control models in advertising: recent developments. *Management Science* **40**, 195-226 (with R.F. Hartl and S.P. Sethi), 1994.

154.Chaos in a simple deterministic queueing system. *Zeitschrift für Operations Research* **40**, 109-119 (with C.H. Hommes and W. Herold), 1994.

155. Complex dynamics in a threshold advertising model. *OR-Spektrum* **16**, 101-111 (with C.H. Hommes and A. Milik), 1994.

156. Nichtlineare dynamische Systeme und Chaos: Neue Impulse für die Betriebswirtschaftslehre? *Zeitschrift für Betriebswirtschaft* **64**, 7-34 (with M. Kopel), 1994.

157. Differential game model of the dynastic cycle: 3-D canonical system with a stable limit cycle. *Journal of Optimization Theory and Applications* **80**, 407-423 (with A. Novak), 1994.

158. Optimal pulsing in an advertising diffusion model. *Optimal Control Applications and Methods* **15**, 267-276 (with A. Novak), 1994.

159. How stock-dependent flow rates imply chaos in educational planning. *Mathematical Population Studies* **5**, 75-85 (with A. Novak), 1994.

160. Limit cycles in intertemporal adjustment models - theory and applications. *Journal of Economic Dynamics and Control* **18**, 353-380 (with A. Novak and F. Wirl), 1994.

161. Endogenous population growth and the exploitation of renewable resources. *Mathematical Population Studies* **5**, 87-106 (with A. Prskawetz and F. Wirl), 1994.

162. A nonlinear model of population growth and renewable resources. *Chaotische Nachrichten* **31**, 16-21 (with A. Prskawetz and F. Wirl), 1994.

163. On the stability and potential cyclicity of corruption in governments subject to popularity constraints. *Mathematical Social Sciences* **28**, 113-131 (with F. Wirl), 1994.

164. On the optimal trade-off between consumption and pollution. *Izvestiya of the Academy of Sciences. Control Theory and Systems* **6**, 235-239, 1995.

165. Complex optimal policies in an advertising diffusion model. *Chaos, Solitons and Fractals* **5**, 45-53 (with H. Dawid), 1995.

166. Optimal resource exploitation may be chaotic. *Central European Journal for Operations Research and Economics* **3**, 111-122 (with H. Dawid and G. Tragler), 1995.

167. Chaotic behavior in an advertising diffusion model. *International Journal of Bifurcation and Chaos* **5**, 255-263 (with L.L. Ghezzi and C. Piccardi), 1995.

168. Slow-fast limit cycles in controlled drug markets. *Proceedings of 3rd European Control Conference* **4**, 3031-3034 (with A. Gragnani and S. Rinaldi), 1995.

169. Habit formation with threshold adjustment: addiction may imply complex dynamics. *Journal of Evolutionary Economics* **5**, 157-172 (with W. Herold, A. Prskawetz and P. Zinner), 1995.

170. Rationalität und komplexes Verhalten. Homo oeconomicus und Chaos - Ein Widerspruch? *Zeitschrift für Betriebswirtschaftliche Forschung* **6**, 545-557 (with M. Kopel), 1995.

171. Resource leasing and optimal periodic capital investments. *Zeitschrift für Operations Research* **42**, 47-67 (with A. Novak and V. Kaitala), 1995.

172.Endogenous population growth may imply chaos. *Journal of Population Economics* **8**, 59-80 (with A. Prskawetz), 1995.

173.Persistent cyclical consumption: variations on the Becker-Murphy model of addiction. *Rationality and Society* **7**, 156-166 (with F. Wirl), 1995.

174.Chaos theory in Operations Research. *International Transactions of Operations Research* **3**, 23-36, 1996.

175.Optimal allocation of drug control efforts: a differential game analysis. *Journal of Optimization Theory and Applications* **91**, 279-297 (with H. Dawid), 1996.

176.On the persistence of corruption. *Journal of Economics* **64**, 177-193 (with H. Dawid), 1996.

177.Despotism and anarchy in ancient China: Visualizing the dynastic cycle. *Jahrbuch für Wirtschaftswissenschaften* **47**, 1-13 (with G. Fischel, E. Gröller and A. Prskawetz), 1996.

178.A nonlinear dynamical model for the dynastic cycle. *Chaos, Solitons and Fractals* **7**, 257-271 (with C. Forst and C. Piccardi), 1996.

179.The Geometry of Wonderland. *Chaos, Solitons and Fractals* **7**, 1989-2006 (with E. Gröller, R. Wegenkittl, A. Milik, A. Prskawetz and W.C. Sanderson), 1996.

180.Slow-fast dynamics in Wonderland. *Environmental Modeling and Assessment* **1**, 3-17 (with A. Milik, A. Prskawetz and W.C. Sanderson), 1996.

181.Persistent oscillations in a threshold adjustment model. *Mathematical Modelling of Systems* **2**, 197-211 (with A. Novak), 1996.

182.Komplexe Dynamik in den Sozialwissenschaften. In: F. Stadler (hrsg.): *Wissenschaft als Kultur. Österreichs Beitrag zur Moderne.* Springer, Wien, 197-214, 1997.

183.Optimal control of law enforcement. In: F. Udwadia, A. Kryazhimskii and G. Leitmann: *Advances in Dynamics and Control.* Gordon & Breach Science Publ, London, 1997.

184.Complex dynamics and control of arms race. *European Journal of Operational Research* **100**, 192-215 (with D. Behrens and A. Prskawetz), 1997.

185.Complex solutions of nonconcave dynamic optimization models. *Economic Theory* **9**, 427-439 (with H. Dawid and M. Kopel), 1997.

186.Indeterminacy of open-loop Nash equilibria: The ruling class versus the tabloid press. In: H.G. Natke and Y. Ben-Haim: *Uncertainty: Models and Measures.* Akademie-Verlag, Berlin, 124-136 (with H. Dawid, A. Novak and F. Wirl), 1997.

187.Non-linear dynamics and predictability in the Austrian stock market. In: Ch. Heij et al. ed.: *System Dynamics in Economic and Financial Models.* Wiley, Chichester, 43-70 (with E.J. Dockner and A. Prskawetz), 1997.

188.Complex dynamics in romantic relationships. *International Journal of Bifurcations and Chaos* **11**, 2611-2619 (with A. Gragnani and S. Rinaldi), 1997.

189.Dynamics of drug consumption: a theoretical model. *Socio-Economic Planning Science* **31**, 127-137 (with A. Gragnani and S. Rinaldi), 1997.

190.Chaotic consumption patterns in a simple 2-D addiction model. *Economic Theory* **10**, 147-173 (with C.H. Hommes and A. Milik), 1997.

191.Einfache dynamische Werbestrategien im Kampf um Marktanteile. *Marketing Zeitschrift für Forschung und Praxis* **4**, 221-231 (with M. Kopel and F. Wirl), 1997.

192.Optimale Kontrolle des Drogenkonsums. *OR News* **3**, 6-12 (with D. Behrens and G. Tragler), 1998.

193.Solution techniques for periodic control problems: a case study in production planning. *Optimal Control Applications and Methods* **19**, 185-203 (with C. Büskens and H. Maurer) , 1998.

194.Periodic and chaotic programs of intertemporal optimization models with non-concave net benefit function. *Journal of Economic Behavior and Organization* **33**, 435-447 (with H. Dawid and M. Kopel), 1998.

195.Optimal enforcement policies (crackdowns) on a illicit drug market. *Optimal Control Applications and Methods* **19**, 169-184 (with R.F. Hartl, J.L. Haunschmied and P. Kort), 1998.

196.The accomplishment of the Maastricht criteria with respect to initial debt. *Journal of Economics* **68**, 93-110 (with M. Luptacik and A. Prskawetz), 1998.

197.A model on the escape from the Malthusian trap. *Journal of Population Economics* **11**, 535-550 (with A. Prskawetz and G. Steinmann), 1998.

198.Corruption dynamics in democratic societies. *Complexity* **3**, 53-64 (with S. Rinaldi and F. Wirl), 1998.

199.Dynamic economic models of optimal law enforcement. In: U. Leopold-Wildburger, G. Feichtinger, K.-P. Kistner: *Modelling and Decisions in Economis*. Essays in Honor of Franz Ferschl. Springer-Verlag, Heidelberg, 1999.

200.Variances of population projections: Comparison of two approaches. IIASA Report (with D. Bauer, W. Lutz, W. Sanderson), 1999.

201.A resource-constrained optimal control model for crackdown on illicit drug markets. Submitted to *JMAA* (with A. Baveja, R.F. Hartl, J.L. Haunschmied and P.M. Kort), 1999.

202.A dynamic model of drug initiation: Implications for treatment and drug control. *Mathematical Biosciences* **159**, 1-20 (with D. Behrens, J.P. Caulkins, J. Haunschmied and G. Tragler), 1999.

203.Controlling the US cocaine epidemic: Finding the optimal mix of drug prevention and treatment. Forthcoming in *Management Science* (with D. Behrens, J.P. Caulkins and G. Tragler), 1999.

204.Present oriented societies favor the occurrence of cycles of drug epidemics. Submitted for publication (with D. Behrens, J.P. Caulkins and G. Tragler), 1999.

205. Optimal enforcement on a pure seller's market of illicit drugs. Forthcoming in *Journal of Optimization Theory and Applications* (with V. Borisov and A. Kryazhimskii), 1999.

206. Price-raising drug enforcement and property crime: A dynamic model. Forthcoming in *Journal of Economics* (with J.P. Caulkins, M. Dworak and G. Tragler), 1999.

207. The impact of enforcement and treatment on illicit drug consumption. Forthcoming in *Operations Research* (with J.P. Caulkins and G. Tragler), 1999.

208. Crime and law enforcement: A multistage game. Forthcoming in *Annals of Dynamic Games* (with H. Dawid and S. Jorgensen), 1999.

209. Optimal offending in view of the offender's criminal record. *Central European Journal of Operations Research* 7, 111-127 (with T. Fent and M. Zalesak), 1999.

210. Environmental effects of tourism industry investments: an intertemporal trade-off. Submitted for publication (with A. Greiner, R.F. Hartl, J.L. Haunschmied, P.M. Kort and A. Novak), 1999.

211. Optimal periodic development of a pollution generating tourism industry. Submitted for publication (with A. Greiner, R.F. Hartl, J.L. Haunschmied and P.M. Kort), 1999.

212. Optimal firm investment in security. *Annals of Operations Research* **88**, 81-98 (with J.L. Haunschmied and P.M. Kort), 1999.

213. Endogenous growth of population and income depending on resource and knowledge. *European Journal of Population* **14**, 305-331 (with F.X. Hof, M. Luptacik, W. Lutz, A. Milik, A. Prskawetz and F. Wirl), 1999.

214. Petrarch's "Canzoniere": Rational addiction and amorous cycles. *Journal of Mathematical Sociology* **23**, 225-240 (with S. Jorgensen and A. Novak), 1999.

215. Threshold advertising rules in a duopoly framework. *CEJOR* 7, 39-52 (with M. Kopel and F. Wirl), 1999.

216. Human capital, technological progress and the demographic transition. Forthcoming in *Mathematical Population Studies* (A. Prskawetz and G. Steinmann), 1999.

217. Interfamiliar consumption and saving under altruism and wealth considerations. Submitted for publication (with F. Wirl), 1999.

218. History Dependence due to Unstable Steady States in Concave Intertemporal Optimizations. Forschungsbericht **238** des Instituts für Ökonometrie, OR und Systemtheorie, TU Wien. (with F. Wirl), 1999.

219. Instabilities in concave, dynamic, economic optimizations. Submitted for publication (with F. Wirl), 1999.

220. Indeterminacy due to relative investment costs. Submitted for publication (with R. F. Hartl, P. M. Kort and F. Wirl), 2000